Chromatin and Disease

Subcellular Biochemistry
Volume 41

SUBCELLULAR BIOCHEMISTRY

Recent Volumes in this Series

Chromatin and Disease

Subcellular Biochemistry Volume 41

Edited by

Tapas K. Kundu
Jawaharlal Centre for Advanced Scientific Research,
Jakkur,
Bangalore,
India

and

Dipak Dasgupta
Saha Institute of Nuclear Physics,
Kolkata,
India

Springer

This series is a continuation of the journal Sub-Cellular Biochemistry. Volume 1 to 4 of which were published quarterly from 1972 to 1975

ISBN-10 1-4020-5465-3 (HB)
ISBN-13 978-1-4020-5465-5 (HB)
ISBN-10 1-4020-5466-1 (e-book)
ISBN-13 978-1-4020-5466-2 (e-book)

(BS/DH)

9 8 7 6 5 4 3 2 1

springer.com

CONTENTS

PREFACE

It is more evident now than ever before that dynamic organization of human genome into nucleoprotein structure, chromatin confers the unique regulatory mechanisms for most of the cellular phenomena, which include replication, transcription, DNA repair, recombination and also apoptosis. The dynamic nature of the chromatin is regulated by chromatin modifications (epigenetic alterations), remodeling, histone chaperones and functional interactions of different chromatin interacting non-histone proteins. Dysfunction of this highly inter connected machineries disturb the cellular homoeostasis, and thereby causes several diseases. As we advance in our knowledge of chromatin function and also disease mechanisms in more details, their causal relationship is becoming more evident. This has lead to the identification of chromatin function as target for new generation therapeutics. In the light of these advances, it happens to be the right time to explore current insights into various aspect of chromatin and disease connection under one cover.

Authors who are actively involved in chromatin research and have made several original contributions to develop latest paradigms in the field have written the chapters of this book. Significantly, the authors' repertoire is truly international. They come from eight different countries of Asia, Europe and America. The book has been divided into three different parts. Part I introduces the reader to the dynamic nature of chromatin structure and its link to diseases. First two chapters in this part deal with the chromatin architecture, chromatin dynamics in the cell cycle and molecular mechanism of chromatin remodeling. The next chapter describes the role of Poly (ADP-Ribose) Polymerase-1(PARP-1) in the regulation of chromatin structure and transcription in response to specific cellular signals. This chapter also highlights the potential therapeutic use of drug that target PARP-1's enzymatic activity for the treatment of diseases. The incorporation of histone variants (non-allelic form of conventional histones) within a nucleosome could affect the overall nucleosome structure and generate a non-canonical nucleosome particle with novel properties. Following two chapters discuss the mechanism of histone variant exchange, the unusual structure of the chromatin domains containing histone variants, their functional significance as epigenetic markers and link to congenital anomalies (cancer and other diseases). Recent research has indicated that histone chaperones play a major role in the chromatin dynamics, ranging from replication dependent histone deposition to the replication independent histone exchange or removal. Therefore, we have included two, chapters expanding the present understanding of histone chaperones with an emphasis upon their involvement in cancer and other diseases. In the last chapter of this part, the importance of the chromatin structure and dynamics has been discussed in connection with the mode of action of DNA binding chemotherapeutic drugs.

A broader definition of chromatin is not limited to being a complex of histones and DNA, rather it is a dynamic organization of histone, DNA, RNA and rapidly interacting non-histone chromatin associated proteins (CAPs). Post translational modifications of non histone chromatin proteins, therefore dramatically alter the chromatin structure-function. The first chapter of the Part II, describes the role of reversible acetylation of non histone proteins (several of which are also component of chromatin) in cellular function and disease. Chromatin is compartmentalized into various domains by a series of loops tethered onto the base of nuclear matrix. Scaffold or Matrix Attachment Regions (S/MAR) punctuate these attachment sites and govern the nuclear architecture by establishing chromatin boundaries. The second chapter of Part II, enumerates the role of MARs and MAR binding proteins in the alteration of local chromatin structure during transcription regulation, viral integration and also disease manifestation.

Part III focuses on the epigenetic modifications of chromatin that are linked to disease and also the development of putative therapeutic approaches which are not far from reality. Reversible acetylation is one of the most widely studied chromatin modifications involved in the regulation of several cellular phenomena. Dysfunction of histone acetyltransferases and deacetylases lead to several diseases (Chapter 11) ranging from neurodegenerative diseases to cancer (Chapters 11, 12 and 13). Therefore small molecule modulators (activator and inhibitors) of HATs and HDACs are now being recognized as potential therapeutic tools (Chapter 12). Apart from acetylation, rapidly increasing number of literature suggest that phosphorylation of histones plays a pivotal role in regulation of acetylation and thereby transcription and DNA repair. Aberrant histone phosphorylation and histone kinases activity are often associated with diseases (Chapter 14). Histone methylation is one of the most versatile and stable epigenetic markers. In Chapter 15 the repressive chromatin marker, H3K9 methylation has been discussed. The role of chromatin modifications including methylation in the regulation of tumor suppressor p53 function has been discussed in Chapter 16. Chromatin remodeling and modifications (acetylation, methylation and phosphorylation) are essential for the gene expression of human genome integrated HIV. Therefore, the modifying enzymes could be the target for combinatorial therapy (Chapter 17). Apart from discussing each of the posttranslational modification in connection to disease manifestation, Chapter 18 also high lights the possible therapeutic approaches targeting enzymes involved in the process of modifications.

In writing and editing the chapters we have put our best effort to make the materials accessible to the scientist not familiar with the chromatin field and to students those who are beginning their career. We hope that this book will provide a stimulating overview for investigators who are working in the field and as well as scientists in related fields (e.g. Virology, Neurobiology, Chemical Biology,

Pharmaceutical Chemistry and Nanobiology). We also hope that this book attracts the pharmaceutical industry, who can utilize the knowledge of chromatin dynamics and its link to disease for the betterment of mankind. We thank all the authors for their outstanding contributions in preparing this book and also Ashok Reddy and Marie Johnson for helping in the processing of the manuscript and all the correspondence.

J N Centre For Advanced Scientific Research T.K.K.
Jakkur PO
Bangalore-64

Biophysics Division DD
Saha Institute of Nuclear Physics
Bidhannagar, Kolkata – 700 064
June, 2006

PART I

CHROMATIN STRUCTURE, DYNAMICS AND DISEASE

CHAPTER 1

STRUCTURAL ORGANIZATION OF DYNAMIC CHROMATIN

KOHJI HIZUME, SHIGE H. YOSHIMURA, MASAHIRO KUMETA AND KUNIO TAKEYASU*

Laboratory of Plasma Membrane and Nuclear Signaling, Graduate School of Biostudies, Kyoto University, Kitashirakawa-oiwake-cho, Sakyo-ku, Kyoto, 606-8502, Japan

Eukaryotic genomic DNA is packed as chromatin in a nucleus in a well-organized manner. Although the folding mechanism of genomic DNA is not yet fully understood, it has been accepted that there are several folding steps starting from naked DNA followed by the formation of "beads-on-a-string" and thicker fibers, with a help of a set of proteins (Fig. 1). The genome structures play critical roles in the regulation of various genomic events: the nucleosome structure restricts the access of regulatory proteins and prevents the progress of enzymatic reactions along DNA (Horn and Peterson, 2002; Jackson and Cook, 1995; Owen-Hughes and Workman, 1994). Structural changes of the chromatin could directly lead to the different states of the genomic reactions. The higher-order architecture of the chromatin fiber also contributes to the successful chromosome passage to daughter cells; the chromatin-fiber undergoes a series of dynamic structural changes into condensed chromosome during the cell division (Hernandez-Verdun and Gautier, 1994). Here, we review the chromatin structure and dynamics from the viewpoints of its architectural components, regulatory proteins and post-transcriptional modifications.

*E-mail: takeyasu@lif.kyoto-u.ac.jp

T.K. Kundu and D. Dasgupta (eds.), Chromatin and Disease, 3–28.

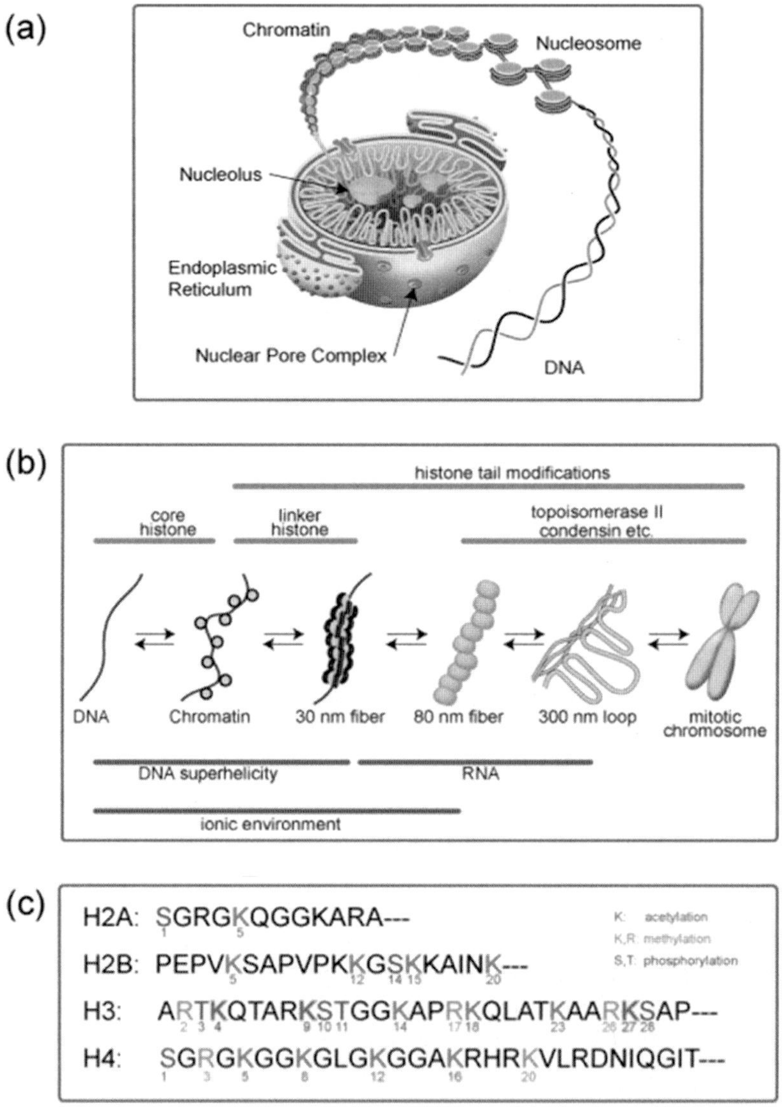

Figure 1. Hierarchical model of chromosome structure. (a) In interphase cells, DNA is packed in a nucleus as forming nucleosome and chromatin. (b) DNA forms nucleosome structure together with core histone octamer, which is then folded up into 30 nm fiber with a help of linker histone H1. This 30 nm fiber is further folded into 80 nm fiber and 300 nm loop structures in a nucleus. In mitosis, chromosome is highly condensed. Proteins which are involved in each folding step are indicated above and non-protein factors are indicated below. (c) The amino acid sequences of histone tails (H2A, H2B, H3 and H4) are shown to indicate acetylation, methylation and phosphorylation sites. (See Colour Plate 1.)

1. BUILDING BLOCKS OF CHROMOSOME

1.1. Microscopic Observations of Chromatin and Chromosome

During the cell cycle, chromosome structures shuttle between de-condensed interphase and condensed mitosis states. Dynamic changes also occur at the lower levels of architectures, i.e., at the chromatin and nucleosome levels. Upon gene activation and inactivation, folding and unfolding of the nucleosome structure and the chromatin fibers occur at limited loci of the genome. Namely, the structures of the chromosome are dynamic and mobile. Nevertheless, there are basic structural units that remain stable and constitute the fundamental chromosome architecture.

The sub-structures of the chromosome have been investigated by means of different types of microscopy. The electron microscopy (EM) visualized a helical coiling of 200∼300 nm fibers that compose the metaphase chromosome arms (Rattner and Lin, 1985). Taniguchi and Takayama reported that the metaphase chromosome possesses 200 nm fibers, which consist of spiralized 30 nm fibers (Taniguchi and Takayama, 1986). To understand the packing mechanisms from 30 nm fibers to 200 nm fibers, some models have been proposed, in which 30 nm fibers form helical coils or axis-attached loops (Ushiki *et al.*, 2002). In the metaphase chromosome treated with dextran sulfate and heparin, naked DNA loops (10∼30 μm) protruds out of the chromosome axis (Paulson and Laemmli, 1977). The EM observation of a thin-section of metaphase chromosomes revealed that a 20–30 nm fiber forms a loop of 3∼4 μm contour length (Marsden and Laemmli, 1979). Length of these loops corresponds to the DNA of ∼100 kb (Marsden and Laemmli, 1979; Paulson and Laemmli, 1977).

A 80–100 nm fiber have also been identified in the chromosome by EM (Belmont and Bruce, 1994) and AFM (Yoshimura *et al.*, 2003) (Fig. 2a). The 80 nm fiber exists also in yeast and chicken erythrocyte nuclei (Kobori *et al.*, 2006) (see section 2.4).

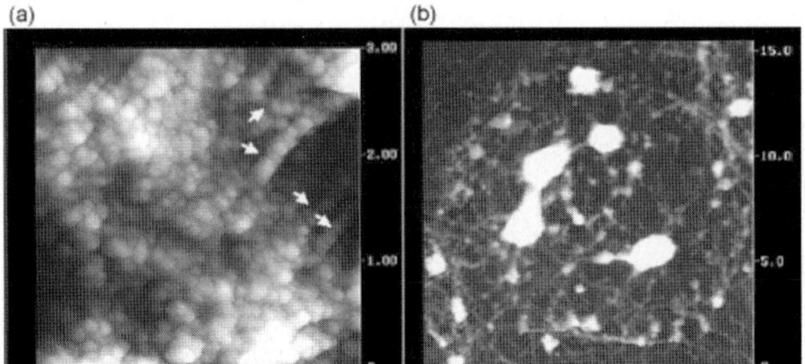

Figure 2. (a) 80 nm granular fibers observed in the interphase chromosome. HeLa cells grown on the cover slide was successively treated with detergent, high-salt solution and low concentration of DNaseI. The sample was fixed and observed by AFM. Arrows indicate granular fibers of 80 nm width. (b) Nuclear scaffold. HeLa cells on a cover slide was successively treated with detergent, high-salt solution, and high concentration of DNaseI. The specimen was fixed and observed by AFM

The partially digested chromatin fiber isolated from interphase nuclei contains ~30 nm fiber observed by EM (Rattner and Hamkalo, 1979; Rattner *et al.*, 1982; Thoma and Koller, 1977; Thoma *et al.*, 1979; Widom and Klug, 1985) and AFM (Leuba *et al.*, 1994). These reports suggest that the ~30 nm fiber is folded into a one-step higher-order unit (80–100 nm fiber) in the interphase nuclei, although its molecular mechanism is not yet well-characterized. On the other hand, the ~30 nm fibers are known to be unfolded into the "beads-on-a-string" structure of nucleosomes (Thoma and Koller, 1977).

1.2. A Number of Proteins are Involved in Establishing the Chromosome Architecture

Proteomic analyses of the metaphase chromosome have identified 209 proteins (Uchiyama *et al.*, 2005). More than half of the proteins (65% in molar ratio) are core histones (histone H2A, H2B, H3, and H4) (Table 1). Core histones are lysine- and arginine-rich proteins that are involved in the formation of nucleosomes

Table 1. Component of the metaphase chromosome

	Molar ratio	Mw (kDa)
Histones		
histone H4	100	11.4
histone H2A.5	41.23	14.0
H2A histone family, member Q	41.43	14.0
histone H2A.X	5.66	15.1
MacroH2A	2.98	39.2
histone H2B.1	49.81	11.3
H2B histone family, member E	40.35	14.0
H3 histone, family 3A	73.73	15.4
linker histones	43.5	10.8
Non-histone proteins		
HMGN2	23.14	9.39
HMGA1	5.68	10.7
HMGB2	0.57	22.3
HP1γ	0.51	20.8
HP1α	0.36	22.1
Ribosomal proteins		
Ribosomal protein L23a	1.77	17.7
Ribosomal protein L11	0.96	20.1
Ribosomal protein L5	0.81	34.3
Ribosomal protein L35	0.66	14.6
Ribosomal protein L38	0.63	8.21
Ribosomal protein L17	0.61	21.3
Ribosomal protein S6	0.58	28.6
Ribosomal protein L7	0.51	30.0
Ribosomal protein S3	0.39	29.9
Ribosomal protein L13	0.28	24.2

Ribosomal protein L8	0.26	22.9
Ribosomal protein S4	0.23	29.6
Ribosomal protein L3	0.11	46.1
Ribosomal protein L4	0.11	47.7
Heat shock protein		
heat shock 70 kDa protein 8 isoform 1	0.87	70.9
BiP (heat shock 70 kDa protein 5)	0.78	72.3
heat shock 70 kDa protein 9B precursor	0.33	73.6
Skeletal proteins		
Keratin 17	7.34	48.1
β-Actin	3.28	41.0
Cytokeratin 18	2.73	47.3
Cytokeratin 8	2.40	53.7
β-tublin	1.32	48.8
Lamin A/C	0.78	65.1
Vimentin	0.75	53.6
α-actinin 4	0.09	10.2
Transcriptional factors		
transcription factor 6 like 1	1.20	29.1
transcription factor ELYS	0.07	25.2
Nucleolus proteins		
Fibrillarin	1.01	33.4
Ku70	0.87	69.8
Nucleophosmin/B23.2	0.78	28.4
nucleolin	0.26	76.3
Scaffold/matrix components		
topoisomerase II α	0.72	174.3
hCAP-C (SMC subunit of condensin complex)	0.40	147.1
hCAP-E (SMC subunit of condensin complex)	0.60	135.7
hCAP-D2 (non-SMC subunit of condensin I complex)	0.40	157.1
hCAP-G (non-SMC subunit of condensin I complex)	0.25	114.3
hCAP-H (non-SMC subunit of condensin I complex)	0.15	82.5
hCAP-D3 (non-SMC subunit of condensin II complex)	0.32	169.6
hCAP-G2 (non-SMC subunit of condensin II complex)	0.27	130.9
hCAP-H2 (non-SMC subunit of condensin II complex)	0.20	62.4
hnRNP G (RNA binding motif protein, X chromosome)	0.82	42.3
hnRNP H (RNA binding motif protein)	0.22	49.2
Others		
DNA topoisomerase I	1.31	90.2
INCENP	0.60	105.5
CENP-C	0.40	106.9
hSNF2H (chromatin remodeling)	0.40	121.9

The molar ratio of the components of metaphase chromosome was determined by Uchiyama and his colleagues (Uchiyama *et al.*, 2005). Metaphase chromosomes has been isolated from synchronized human cell line and then purified by sucrose density gradient centrifugation and Percoll density gradient centrifugation. The major components described in this chapter are summarized here. Molar ratio is provided per 100 histone H4 molecules.

(described in section 2.1) (Kornberg, 1974). The core histones are well conserved in eukaryotes. (Prokaryotes have "histone-like" proteins, such as HU and IHF, but these proteins have no homologous amino acid sequence to histones.) Each core histone has a molecular weight from 11 to 15 kDa and 17~20 positive charges at pH 7.0. All of the four core histones carry the N-terminal tail regions which are rich in lysine residues. These lysine-rich tail regions bind to negatively charged DNA, and play a role in nucleosome-nucleosome interactions. As described in section 2.2, the histone tails undergo post-transcriptional modifications such as acetylation, phosphorylation, and methylation. These modifications are involved in the regulation of chromatin structures and gene activities. The middle and the C-terminus regions constitute a "core" domain. The four core domains of two H3 and two H4 molecules assemble into a tetramer. H2A and H2B form a dimer, and two H2A/H2B dimers assemble with a H3/H4 tetramer, to form an octamer. DNA wraps around this octamer for 1.65 turns to form a nucleosome (see section 2.1).

Linker histones (H1, H5 and others) are also major components of metaphase chromosome, and occupy 5.8% of the total protein amount (Uchiyama *et al.*, 2005). They play an important role in the formation of the 30 nm fiber (see also section 2.3). These linker histones carry more lysine residues (~30% of the total amino acids) than the core histones and have a core domain in the middle part that binds to a nucleosome. The linker histones could be easily extracted from the chromatin with 0.5 M NaCl, whereas the core histone octamers need more than 0.8 M NaCl to dissociate from nucleosomes.

There are more than 100 different proteins in mitotic chromosome other than histones. The HMG (High Mobility Group) proteins are second abundant proteins. Like histones, they also have small molecular weights less than 30 kDa, and exhibit *high mobility* in a gel electrophoresis (Goodwin *et al.*, 1973). The HMG proteins have been classified into three families (HMGA, HMGB, and HMGN) based on their domain structures, and thought to participate in the maintenance of chromatin structures as well as in the transcriptional regulation (Agresti and Bianchi, 2003; Bustin, 2001; Ellwood *et al.*, 2000). The HMGA proteins have three AT-hook regions, which bind to a run of 4 or 5 A/Ts in a minor groove of DNA double strand, and work at enhanceosome. According to the studies on the interferon β (IFN-β) promoter, HMGA1 binds to the A/T sequence and distorts DNA, which recruits transcription factors to the promoter region. The expression levels of HMGAs are high in the early embryonic stage and low in adult. Since HMGA1 is highly expressed in several tumor cells, it is used as a tumor marker. HMGB proteins have two HMG box domains, each of which contains three α-helices and binds to a minor groove of DNA. It has been proposed that the binding of HMGBs to the nucleosome bends the nucleosomal DNA and, thus, loosens the nucleosomal structure (Bonaldi *et al.*, 2002). HMGBs also interact with TBP (Lu *et al.*, 2000; Sutrias-Grau *et al.*, 1999) and other general transcription factors (Agresti and Bianchi, 2003). These properties of HMGBs are thought to be important for the up-regulation of the gene activity. HMGNs have a nucleosome-binding domain in the N-terminus and a chromatin-unfolding domain in the C-terminus. Interestingly, HMGNs associate with the nucleosome by binding to the space between DNA and the histone octamer (Bianchi and Agresti, 2005). When HMGNs are phosphorylated in the M phase, they dissociate from the chromatin (Prymakowska-Bosak *et al.*, 2001). Histone H1 and dephosphorylated HMGNs bind to chromatin competitively (Catez *et al.*, 2002).

Another group of non-histone proteins have been identified as essential components for the formation of the condensed chromosome (Table 1). Topoisomerase II (topo II) localizes in the scaffold/matrix fraction of the interphase nuclear (Berrios *et al.*, 1985) and the mitotic chromosome (Maeshima and Laemmli, 2003) (see section 3.1). Topo II forms a ring-shaped homodimer (Berger *et al.*, 1996; Nettikadan *et al.*, 1998) and catalyzes the decatenation and relaxation of DNA double strand (Wang, 2002). In fission yeast, chromosomes cannot be condensed without functional topo II (Uemura *et al.*, 1987). In addition, in *in vitro* experiment, mitotic extracts containing topo II induce chromatin condensation in the isolated nuclei from HeLa and chicken erythrocyte cells (Adachi *et al.*, 1991).

The condensin complex has been identified in the same scaffold fraction (Maeshima and Laemmli, 2003) and shown to be essential for the mitotic chromosome condensation (Hirano *et al.*, 1997). The frog condensin complex exhibits ATP-dependent DNA-supercoiling activity (Kimura and Hirano, 1997). It consists of a heterodimer of SMC and a trimer of non-SMC proteins (Hirano *et al.*, 1997). The SMC complex has a globular head domain and a coiled-coil tail region (Anderson *et al.*, 2002; Melby *et al.*, 1998; Yoshimura *et al.*, 2002). In vertebrates, two types of condensin complex, condensin I and condensin II, exist; they are composed of the same SMC subunits but with different non-SMC subunits (Ono *et al.*, 2003).

1.3. Physical Properties of DNA Affect the Nucleosome Stability and Dynamics

The higher-order architectures of DNA/protein complexes are achieved on the balance of the physical properties of DNA and the interactive forces of DNA-binding proteins (Hizume *et al.*, 2002; Yoshimura *et al.*, 2000). The superhelical strain of the DNA has been suggested to play a critical role in a variety of genome events such as transcriptional regulation and DNA replication (Opel *et al.*, 2001).

The effect of the superhelical strain of the DNA template on the nucleosome structure can be investigated from the *in vitro* chromatin reconstitution system (for the detail of *in vitro* chromatin reconstitution, see sections 2.1 and 2.3). Interestingly, the efficiency of the reconstitution becomes higher as the lengths of the DNA used are longer (Hizume *et al.*, 2004) (Fig. 3a–c). In the 3 kb reconstituted chromatin, one nucleosome could be formed in every 826 bp DNA on average, while in the 106 kb chromatin fibers, one nucleosome can be formed in every 260 bp of DNA. The chromatin reconstituted on the any length of linearized plasmid, the efficiency of the reconstitution becomes one nucleosome per ~800 bp DNA. The treatment of the

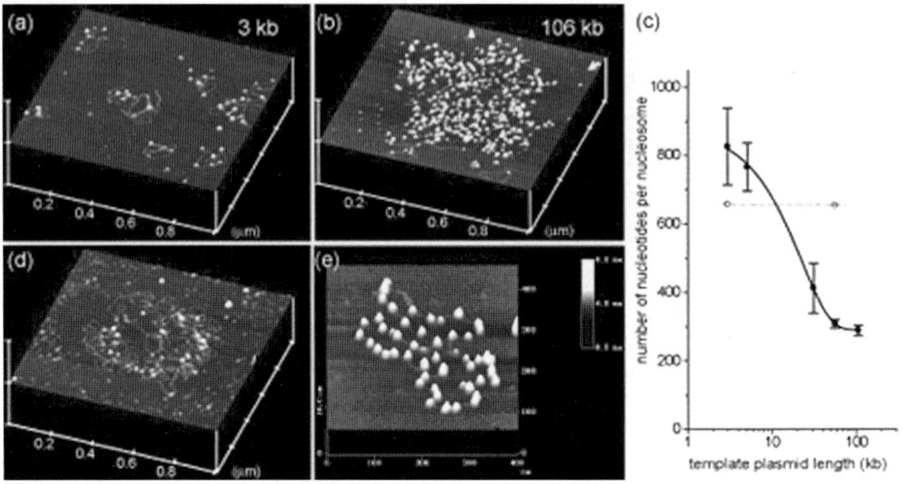

Figure 3. The stability of the nucleosome is affected by the length and the superhelicity of DNA. (a–b) The chromatin fibers were reconstituted from the purified plasmids and the histone octamers by a salt-dialysis method and observed under AFM. The 3 kb (a) or 106 kb (e) supercoiled circular plasmid was used as a template. (c) Relationship between the plasmid length and the frequency of nucleosome formation in the reconstitution process. The nucleosome frequency *is* represented as the number of base pairs per nucleosome and plotted against the length of the template DNA in supercoiled (filled circle) and linear (open circle) forms. (d) AFM image of the chromatin fiber reconstituted on the topoisomerase I-treated plasmid. (e) Chromatin fiber reconstituted with *Drosophila* embryo extract. The chromatin fiber was reconstituted from plasmid DNA of ~10 kband the embryo extract of Drosophila, and was observed by AFM

plasmids with topoisomerase I completely removed the superhelical strain. When the topoisomerase I-relaxed 106 kb plasmid was used for the chromatin reconstitution, only 30 nucleosomes could be formed (Fig. 3d), the efficiency was found to be much lower than the linearized plasmid. It is interesting that the chromatin fiber was highly tangled on the mica surface with a number of DNA loops (Fig. 3d), probably due to the accumulation of the positive supercoiling in the closed circular DNA upon the nucleosome formation.

DNA sequence has significant effect for the nucleosome formation. The DNA fragment of 5S ribosomal RNA gene has been identified as good template on which nucleosomes easily form. (Simpson and Stafford, 1983; Simpson *et al.*, 1985) This is called "nucleosome positioning signal" (Hayes and Lee, 1997). AFM observation revealed that two nucleosomes are placed on the tandem fragments (Sato *et al.*, 1999). DNA methylation has been implicated in transcriptional silencing (Buschhausen *et al.*, 1987) and chromatin structures (Karymov *et al.*, 2001). When the nucleosomes formed on the DNA which has been methylated by the treatment with SssI DNA-methylase, the distance of nucleosomes is shorter than that of nucleosomes formed on unmethylated DNA, suggesting that DNA methylation induces chromatin compaction.

2. CHROMATIN STRUCTURES AND DYNAMICS

2.1. Nucleosome – The Most Fundamental Structural Unit of Chromatin

The most fundamental and well-characterized structural unit of the chromosome is the nucleosome (Kornberg, 1974; McGhee and Felsenfeld, 1980). The nucleosome/chromatin structures play critical roles in the regulation of various genomic events; the nucleosome restricts the access of regulatory proteins and prevents the progress of enzymatic reactions along the DNA (Felts *et al.*, 1990; O'Neill *et al.*, 1992). Not only the nucleosome structures but also the higher-order arrangement of the nucleosome array plays important roles in a number of genome functions; transcriptional regulation is known to be tightly coupled with the dynamic structural changes of the nucleosomal arrays (Ding *et al.*, 1997),

The nucleosome is composed of 146 bp DNA wrapping around the core histone octamer (two of each H2A, H2B, H3 and H4) for ~1.65 turns (Luger *et al.*, 1997). X-ray crystallography has revealed that histone octamer has a disk-like shape with the diameter of 6.5 nm and the thickness of 6.0 nm (Arents *et al.*, 1991). The tails of H2A and H4 stick out of the disk-shaped nucleosome and the tails of H3 and H2B extend between DNA gyres (Luger *et al.*, 1997). DNA wrapping around the octamer (nucleosomal DNA) has a helical pitch of 10.2 bp per turn (Satchwell *et al.*, 1986), which is shorter than that of free B-type DNA (10.5 bp per turn) (Rhodes and Klug, 1981). Therefore, about 1 positive supercoiling is introduced by one nucleosome assemble (Germond *et al.*, 1975; Keller, 1975), because 146 bp of DNA wraps around the octamer. The length of nucleosomal DNA can be examined by using micrococcal nuclease (MNase) that digests only linker DNA. In the case of a correctly assembled nucleosome, a ~146 bp band should be detected by the gel-electrophoresis after the MNase digestion, because the nucleosomal DNA is protected from MNase digestion. A complete dissociation of the histone octamer and DNA needs a high-salt treatment (2 M NaCl) in biochemical experiments (Tatchell and Van Holde, 1977). The order of dissociation reflects the structural stability of the histone octamer. H2A and H2B dissociate first, and then H3 and H4 dissociate from DNA when treated with high-salt or hydroxylapatite (Simon and Felsenfeld, 1979).

The structure of the nucleosome is affected by ionic environment. The low ion concentration makes a nucleosomal array well spread whereas 100 mM NaCl induces nucleosome-nucleosome interaction (Hizume *et al.*, 2005; Nakai *et al.*, 2005; Olins and Olins, 1972; Thoma *et al.*, 1979). *In vitro* transcription system has been applied to understand the relationship between the nucleosome assembly and transcription activity. The results demonstrated that a positive supercoiling introduced during the progress of RNA polymerase makes the nucleosomes ahead of the polymerase unstable (Pfaffle *et al.*, 1990).

In order to investigate the structural and functional characteristics of the chromatin fiber, several methods for *in vitro* nucleosome reconstitution have been developed (Lusser and Kadonga, 2004). Among them, the salt-dialysis method is the simplest because it uses only the histone octamer and DNA (Tatchell and Van Holde, 1977). The efficiency of the reconstitution is not very high unless highly supercoiled long (~100 kb) DNA is used (section 1.3) and a constant "spacing" between nucleosomes is hardly obtained unless the template DNA contains nucleosome positioning signals. A highly efficient reconstitution can be achieved when the *Drosophila* embryo extract is used; nucleosomes are formed efficiently and evenly spaced (one nucleosome per ~200 bp) (Becker and Wu, 1992) (Fig. 3e). The most recently-established method utilizes chromatin assembly factors (NAP-1 protein and ACF complex), with which a correct spacing of the nucleosome is also obtained (Ito *et al.*, 1997). Extensive studies employing these *in vitro* reconstitution methods have brought new insights into how the transcriptional activation is achieved by altering the nucleosome structure in the promoter region (Croston and Kadonaga, 1993; Narlikar *et al.*, 2002).

The nucleosome structures have been analyzed not only by biochemical approaches but also by biophysical techniques. Using optical tweezers, the chromatin fibers can be mechanically stretched and the mechanical force to disrupt the nucleosome characterized. Bennink and his colleague have reported that the individual nucleosome can be disrupted under a force in the range of 20–40 pN (Bennink *et al.*, 2001). Small angle x-ray scattering has demonstrated that an increase in the monovalent salt concentration up to 200 mM enlarges the extension of histone tails (Mangenot *et al.*, 2002). This salt-dependent histone tails extension leads to an interaction between the histone tail and DNA or between the histone tails (Zheng and Hayes, 2003). Divalent cations also induce such interactions of nucleosomes as revealed by the phase diagram (de Frutos *et al.*, 2001).

2.2. Post-transcriptional Modifications of Histone Tails

In addition to the role for the nucleosome–nucleosome interactions, the histone tails are known as the region that undergoes post-transcriptional modifications, such as acetylation, phosphorylation, and methylation (Fig. 1c) (Peterson and Laniel, 2004). These modifications trigger the formation of euchromatin (acetylation), heterochromatin (methylation), or metaphase chromosome (phosphorylation). The details of these modifications will be described in chapters 8–11.

Acetylation of the histone tails correlates with the activities of genes (Kimura *et al.*, 2005). However, the detailed analyses of the acetylation on the individual lysine residue have revealed that the relationship between the acetylation and the chromatin-compaction is not simple. There are 1–6 lysine residues in each histone subunit, that could be acetylated; the Lys^5 of H2A, Lys^5, Lys^{12}, Lys^{15}, and Lys^{20} of H2B, Lys^4, Lys^9, Lys^{14}, Lys^{18}, Lys^{23}, and Lys^{27} of H3, and Lys^5, Lys^8, Lys^{12}, and Lys^{16} of H4. In mammal, more than ten HATs (*H*istone *A*cetyl *T*ransferases) have been identified, each of which acetylates a specific lysine residue. Acetylation frequently occurs in euchromatin regions, and some in heterochromatin regions. For example, the acetylation of Lys^{12} of H4 leads to a telomeric silencing (Kelly *et al.*, 2000). In *Drosophila*, Lys^{16} of H4 is acetylated specifically in the transcriptionally hyperactive X chromosome in the male larvae (Turner *et al.*, 1992). These reports suggest that the acetylation of the histone tails play a role in the regulation of higher-order chromatin structures and the "signaling" in the specific genomic regions as well.

The methylation of the histone tails occurs at the arginine and lysine residues. The amino group of the side chain is methylated. There are mono-, di and tri-methylated states depending on the number of methyl groups attached to the nitrogen atom. Arginine is methylated by the protein R (arginine) methyltransferases (PRMTases) (Chen *et al.*, 1999), which stimulates the transcription process. Arg^{3} of H4 is methylated by PRMT1 (Wang *et al.*, 2001) and Arg^{17} of H3 by CARM1 (Coactivator Associated R methyltransferase) (Bauer *et al.*, 2002; Schurter *et al.*, 2001). The methylation of the lysine residues (Lys^{4}, Lys^{9}, Lys^{27}, Lys^{36} and Lys^{79} of H3, and Lys^{20} and Lys^{59} of H4) occurs more frequently than the arginine methylation. Since some lysine residues are subjected to both acetylation and methylation, these two modifications sometimes compete with each other. For example, the methylation of Lys^{9} of H3 competes with the acetylation.

Belmont and the colleagues have developed a system to label a specific chromosome region in a living cell by integrating *lac* operator into the chromosome (Li *et al.*, 1998). When GFP-lac repressor is expressed in the cell which carries the *lac* operator array in the heterochromatic HSR (*h*omogeneously *s*taining *r*egion), the inner nuclear localization of the HSR can be monitored by GFP signal in the living cell. Verschure and colleagues have applied this GFP-lac repressor/operator system for understanding how the HP1 (*h*eterochromatin *p*rotein 1) is involved in the histone modifications (Verschure *et al.*, 2005). HP1 associates specifically with heterochromatic regions (James and Elgin, 1986). When GFP-lac repressor-HP1 fusion protein is expressed in the cell which carries the *lac* operator array in an euchromatic region, this region becomes more condensed than in the absence of HP1 fusion protein. Targeting of HP1 is sufficient to induce heterochromatin formation. Histone methyltransferase is recruited to this region and tri-methylates Lys^{9}. Namely SUV39H1 is a methyltransferase for histone H3 Lys^{9} (Rea *et al.*, 2000), and binds to HP1 (Aagaard *et al.*, 1999). Methylated Lys^{9} –modified H3 also binds directly to HP1 (Bannister *et al.*, 2001; Lachner *et al.*, 2001). Heterochromatic regions apparently maintain the "heterochromatic" status by a cooperative regulation; HP1 targeting induces histone H3 Lys^{9} methylation and vice versa.

Phosphorylation has been thought to be correlated to the mitotic chromatin condensation and the transcriptional regulation in interphase (Nowak and Corces, 2004). The mitotic phosphorylation, which was first identified in 1978 (Gurley *et al.*, 1978), occurs at Ser^{10} (Wei *et al.*, 1998), Ser^{28} (Goto *et al.*, 1999), and Thr^{11} (Preuss *et al.*, 2003) in histone H3. The Ser^{10} phosphorylation is catalyzed by the aurora kinase family (de la Barre *et al.*, 2000), and is required for the initiation of chromosome condensation but not for its maintenance (dephosphorylation of mitotic chromosomes does not induce chromosome decondensation) (Van Hooser *et al.*,1998). In meiosis, Ser^{10} phosphorylation is also required for the cohesion of sister chromatids rather than the condensation (Kaszas and Cande, 2000).

2.3. Structures and Dynamics of 30 nm Chromatin Fiber

Although the existence of the ~30 nm fiber is widely accepted, several models have been proposed for its structure (Felsenfeld and McGhee, 1986). Thoma and colleagues proposed a solenoid model, in which the nucleosomes are ordered in a spiral manner (Thoma *et al.*, 1979). Woodcock and colleagues postulated a helical ribbon model, in which the nucleosomes are arranged in a zig-zag manner and the sheet of the zig-zag nucleosomes winds up helically to form a ribbon-like structure (Woodcock *et al.*, 1984).

A number of biochemical and microscopic studies have demonstrated that the linker histone plays an important role in the higher-order folding of the "beads-on-a-string" fiber. Histone H1 binds to linker DNA, especially to the region where the DNA enters and exits the nucleosome core particle (McGhee and Felsenfeld, 1980; Zlatanova *et al.*, 1999). A complex of the nucleosome and histone H1 (named a "chromatosome") contains 168 bp of DNA (Simpson, 1978). Removal of histone H1 results in an unfolding of the 30-nm fiber into the "beads-on-a-string" fiber (Thoma and Koller, 1977; Thoma *et al.*, 1979). A trypsin-digestion of histone H1 and the N-terminus of histone H3 leads to a loss of the "zig-zag" arrangement of the nucleosomes and deforms the 30-nm fiber (Leuba *et al.*, 1998). Therefore, the role of histone H1 seems to be to tighten the nucleosome-nucleosome interaction. Huynh and colleagues have reconstituted a nucleosome array on the 12 and 19 tandem repeats of a nucleosome positioning signal, and have studied the effect of linker histone H5 by EM imaging and sedimentation velocity assays (Huynh *et al.*, 2005). The nucleosomal array undergoes a compaction into the 30 nm fiber in the high NaCl concentrations or 1 mM $MgCl_2$.

Biochemical reconstitution of the 30 nm fiber has recently been succeeded by using a salt-dialysis procedure with a long DNA template (>100 kb) (Hizume *et al.*, 2005). AFM imaging of the reconstituted chromatin has shown that the beads-on-a-string structure of the nucleosomes (~400 nucleosomes on 100 kb DNA) are converted to a thicker fiber in the presence of histone H1. The thickness of the fiber changes reversibly between 20 nm and 30 nm, depending on the salt environment (in 50 mM and 100 mM NaCl, respectively) (Fig. 4); namely, the linker histone directly promotes a thicker fiber formation in a salt-dependent manner.

It is interesting and important to note that several species, such as *Schizosaccharomyces pombe*, lack histone H1 (Wood *et al.* 2002) (see also section 2.4). The nucleosome-repeat length is slightly shorter in *S. pombe* than in human (Godde and Widom, 1992). The contribution of histone H1 to the mitotic chromosome condensation has been examined with the use of a cell-free system of *Xenopus* eggs, in which the condensed sperm nuclei can be transformed into metaphase chromosomes. Even when histone H1 is removed from the extract, the metaphase chromosomes can still be formed (Ohsumi *et al.*, 1993). In addition, an elimination of all H1 genes in *Tetrahymena* exerts no phenotypic effect (Shen *et al.*, 1995).

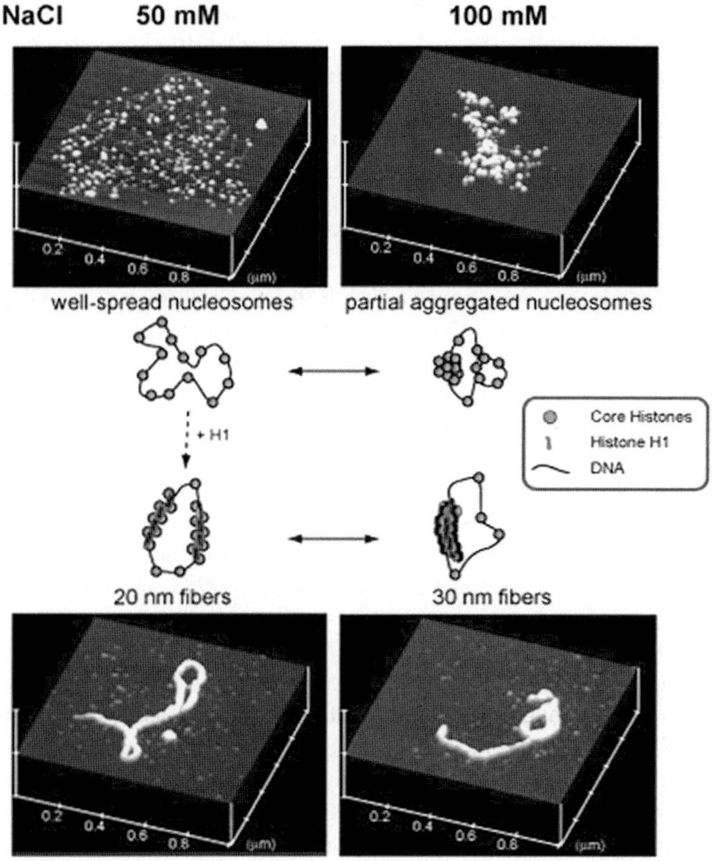

Figure 4. In vitro reconstituted 30 nm chromatin fiber. Dynamic structural changes in the chromatin fiber in the absence (top) or presence (bottom) of linker histone H1 with different NaCl concentration were observed by AFM. Nucleosomes were reconstituted on the 106 kb plasmid and then fixed in the buffer containing 50 mM (top left) or 100 mM NaCl (top right). Nucleosomes were well-spread in 50 mM NaCl but attached each other and partially aggregated in 100 mM NaCl. After the addition of histone H1, the thicker fibers were formed. The width of the fibers *is* 20 nm in 50 mM NaCl (bottom left) or 30 nm in 100 mM NaCl (bottom right)

2.4. How is the 30 nm Fiber Folded Up?

When the surface of the purified HeLa cell nucleus is observed by AFM (Fig. 5a), relatively smooth surface is seen, although many small projections and cavities can be identified (Fig. 5b). When the isolated nuclei are successively subjected to the detergent treatment and high-salt treatment on the glass substrate to remove the nuclear membrane and nucleoplasm, fibrous structures appear (Fig. 5c, d). A close examination by AFM classified the existence of granular structures (~80 nm width)

(Fig. 5e) and ~80 nm fibers inside and outside the nucleus, respectively and the 30 nm fibers are occasionally detected (Fig. 5f). Similar to the HeLa cell nucleus, the surface of the isolated chicken erythrocyte nucleus is smooth and no internal structure can be observed without the detergent treatment (Fig. 5g). A successive treatment with a detergent and high-salt reveals many chromatin fibers released from the nucleus (Fig. 5h, i). A statistical analyses identifies ~80 nm beads (Fig. 5j), and ~30 nm and ~80 nm fibers (Fig. 5k), similar to those in the HeLa cell nucleus.

Since *S. pombe* lacks linker histone H1, the structural comparison of its genome and other eukaryotic genomes would provide significant information. AFM observation of isolated *S. pombe* nucleus has revealed that the isolated nucleus has a smooth nuclear envelope (Fig. 5l), and that the removal of the nuclear membranes by the detergent treatment greatly changes the surface morphology (Fig. 5m) (Kobori *et al.*, 2006). A high-salt treatment also causes the chromatin fiber spreading out of the nucleus (Fig. 5n), as observed in the HeLa cells (Fig. 5d) and chicken erythrocytes (Fig. 5i). The majority of the fibers exhibit the width of ~30 nm (Fig. 5p), and the granular structures inside the nucleus possess the diameter of ~100 nm (Fig. 5o). Since the yeast nucleus requires longer incubation with high-salt solution to unfold the chromatin fiber, the yeast chromosome may be more tightly folded than the chromosomes from other two species.

The high-salt treatment releases the ~30 nm fibers from the nuclei of other eukaryotic cells including plant cells (Sugiyama *et al.*, 2003, 2004). Therefore, the 30 nm fiber seems to be the thinnest chromatin structure relatively stable in the eukaryotic genome, although the intrinsic characteristics of chromatin, such as genome size, nucleosome spacing, and histone composition, vary among them. It is intriguing that *Saccharomyces cerevisiae*, in which the nucleosome-repeat length (165 bp) is shorter than those of higher eukaryotes (~200 bp), forms a similar 30–40 nm chromatin fiber (Bystricky *et al.*, 2004). The ~80 nm granular structures was previously identified in the interphase nucleus (Yoshimura *et al.*, 2003) and the mitotic chromosome (Adolph *et al.*, 1986; Hoshi and Ushiki, 2001; Tamayo and miles, 2000). No structural population other than the 30 nm fiber and the 80 nm beads was clearly found, suggesting that the 40 nm fiber would be folded up directly to the 80–100 nm granular fiber (Fig. 1).

RNA has been known to contribute to the higher-order architectures (Worcel and Burgi, 1972; Zamore and Haley, 2005). Such higher-order architectures and their hierarchical transition seem to be critical to achieve the genomic activities in cells. To investigate the role of RNA in the formation of the 30 nm and/or 80 nm units, the nucleus isolated from HeLa cells has been treated with RNase (5 μg/ml) (Ohniwa *et al.*, In Press). The RNase treatment disrupts the edge of the nucleus, and exposes fibrous structures (Fig. 5q). Section analysis of the AFM image showed that the majority of the fiber widths was ~30 nm, indicating that this corresponds to a thick chromatin fiber induced by linker histone H1 (Hizume *et al.*, 2005). It is reasonable to suggest that RNA is critical in maintaining the architectures higher than the 30 nm chromatin fiber (Fig. 1).

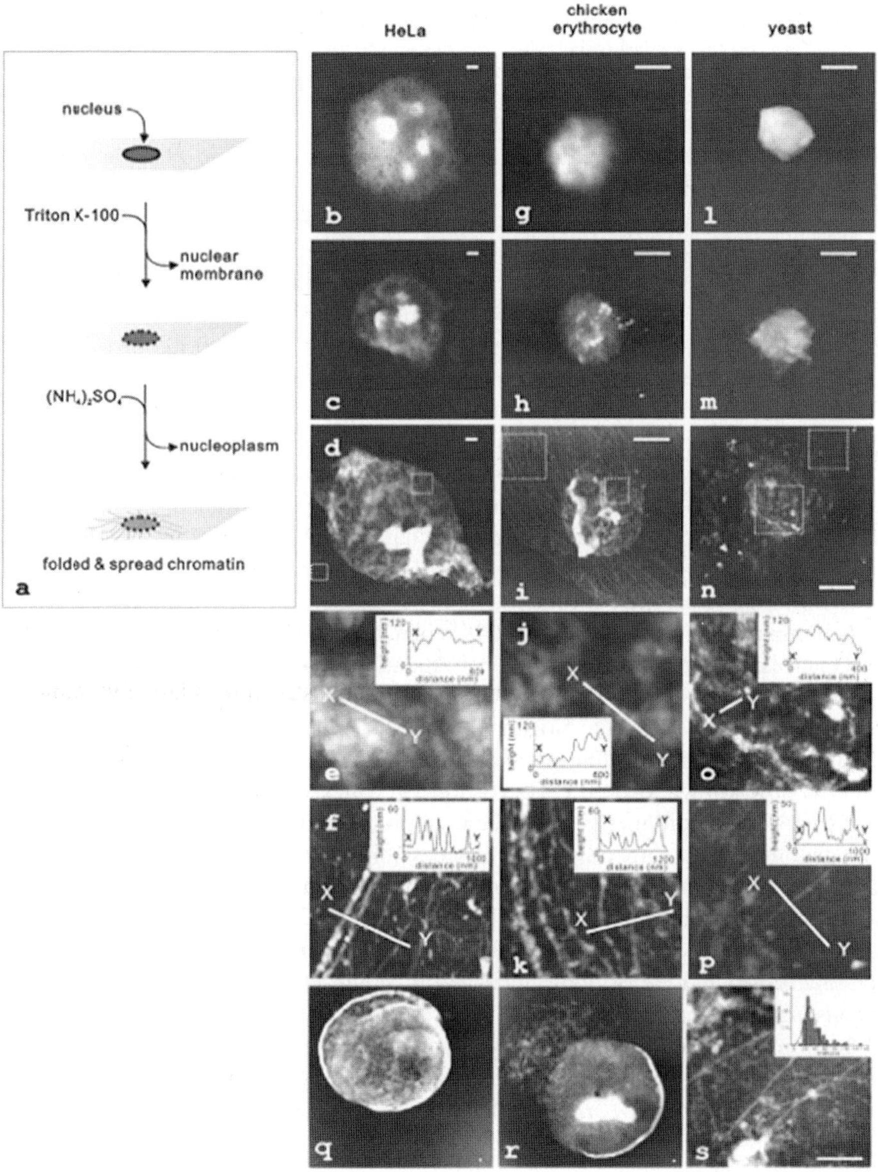

Figure 5. 80 nm fibers and granules. The 'on-substrate' lysis makes it possible to observe a chromatin structure by removing nuclear membrane and nucleoplasm from nuclei on a cover slide (a). HeLa, chicken erythrocyte, and yeast nuclei were subject to the 'on-substrate' lysis (panels b–f for HeLa nucleus, g–k for chicken erythrocyte, l–p for yeast). AFM visualized the isolated nucleus without detergent treatment (b, g, l), after the detergent treatment (c, h, m) and after the high-salt treatment (d, i, n). Scale bars indicate 2 μm. The high-salt treated nuclei (d, i, n) were rescanned for magnification, and shown in

3. NON-CHROMOSOMAL ARCHITECTURES SUPPORTING CHROMOSOME ARRANGEMENT IN THE NUCLEUS

3.1. Scaffolds in Nucleus

Individual chromosomes appear to be organized into discrete "territories" in the nucleus (Mahy *et al.*, 2002; Williams, 2003). The microscopic observation of a yeast cell carrying the GFP-*lac* repressor/operator system (lac operator is integrated in *LEU2* locus) has demonstrated that the chromatin keeps moving within a limited subdomain of the nucleus (Marshall *et al.*, 1997). When the lac operator is integrated in condensed chromatin regions such as nucleoli and the nuclear periphery, a relatively slow movement is observed (Chubb *et al.*, 2002). The mobility of the chromosome region near the nucleolus is increased as sequel to the disruption of nucleoli by a transcriptional inhibitor, 5,6-dichloro-D-ribohuranosylbenzimidazole (DRB) (Chubb *et al.*, 2002). These facts suggest that the positioning of the chromatin in the nucleus in related to the physical attachment to the nuclear compartment. The chromatin mobility decreases as the cell cycle progresses from early G2 to late G2, indicating that the chromatin attachment on a fixed subdomain in the nucleus is regulated by the cell cycle-dependent mechanisms (Vazquez *et al.*, 2001).

It has been assumed that the nucleus contains an "immobile structure", in which chromatin fibers are partially attached and fixed. These structures are called "nuclear matrix" or "nuclear scaffold". When cells are successively treated with detergent, high-salt solution and DNase I, the nuclear scaffold can be observed as a fibrous network in the cell nucleus (Fey *et al.*, 1986; Nickerson, 2001; Yoshimura *et al.*, 2003) (Fig. 2b). The biochemical analyses of the nuclear scaffold have identified a ~174 kDa protein as a major component (Fisher *et al.*, 1982). This protein is now known as topo II (Berrios *et al.*, 1985).

SAR/MAR (scaffold/matrix associated region) has been experimentally identified as the genome DNA segment which is inaccessible to the nuclease digestion (Gasser and Laemmli, 1986) or the DNA segment which remains attached to the insoluble fraction of the nucleus after the restriction endonuclease digestion (Jarman and Higgs, 1988). The well-known SAR is located in the histone gene cluster (Gasser and Laemmli, 1986) and the β-globin locus (Jarman and Higgs, 1988), which contains the consensus sequence for the interaction with topo II (Adachi *et al.*, 1989; Gasser and Laemmli, 1986). SAR/MAR also functions as an "insulator" (Nabirochkin *et al.*, 1998; Namciu *et al.*, 1998). The insulator is thought to be the border between

Figure 5. (Continued) different panels (e and f for HeLa, j and k for chicken erythrocyte, and o and p for yeast). A section profile obtained along X-Y line shows a typical granular structure in the nucleus (e, j, o), and the peak-to-peak distance between the granular structure was distributed from 60 nm to 120 nm (e). The thickness of the chromatin fibers released out of the nucleus varied possibly due to the assembly of thinner fibers (f, k, p). A section profile for the spread fibers was obtained along X-Y line (f, k, p). Isolated HeLa cell nucleus was treated with (r, s) or without (q) RNase. The treatment releases ~30 nmfiber from the nucleus. The histogram of the fiber width is shown in an inset of (s). Bars, 250 nm. (See Colour Plate 2.)

heterochromatin and euchromatin (Bell and Felsenfeld, 1999), and can prevent the enhancer actions toward promoters (Udvardy *et al.*, 1985). Thus, in a probable model, insulators are placed in the border between the condensed chromatin and decondensed chromatin.

3.2. Cytoskeletal Proteins in the Nucleus

The proteomic analyses of nuclear proteins and immunofluorescence microscopic analyses have revealed that a nucleus contains a substantial number of cytoskeletal proteins and cytoskeleton-related proteins (Table 1). Keratin forms intermediate filaments in the cytoplasm. A detailed analysis of keratin subtypes has shown that, in HeLa cells, keratin 8 and keratin 18 form a dimer and compose a filament in the cytoplasm (Moll *et al.*, 1982). Interestingly, only keratin 18 can be detected in the nucleus (Fig. 6), and the immunoreactive signal of keratin 18 is stronger in the nucleus than in the cytoplasm. When the soluble proteins are removed by a mild detergent treatment before the fixation, the nuclear signals of keratin 18 are completely disappeared (Fig. 6), whereas the signal as the cytoskeletal filament still remained. This result suggests that there are two types of keratin 18 molecules: some molecules are involved in the construction of the intermediate filament together with keratin 8, and others localize in the nucleus and have not-yet-identified functions.

Actin was found in the nucleus in 1969 (Lane, 1969). However, this result was not immediately accepted because of the idea that it was just a contamination during the sample preparation. Nevertheless, many studies have demonstrated the presence

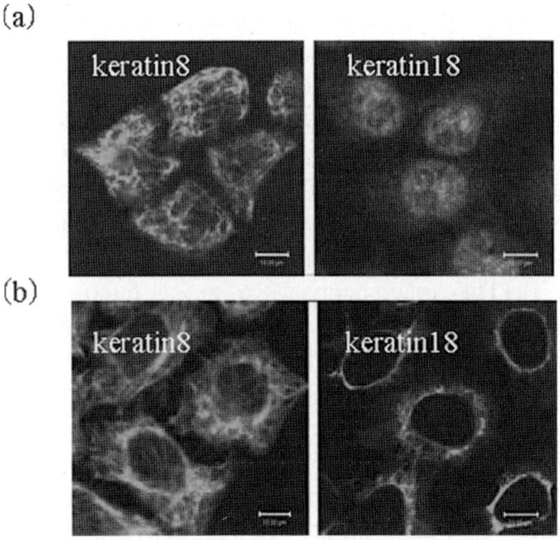

Figure 6. Cytoskeletal proteins in nucleus. HeLa cells were immunostained with anti-keratin 8 or anti-keratin 18 antibodies. (a) Paraformaldehyde-fixed cells. (b) Cells treated with detergent before the fixation. (Scale bars: 10 μm)

of actin in the nucleus and shown that it has various functions in the nucleus; it is involved in the transcription by RNA polymerases, chromatin remodeling, nuclear transport, and nuclear envelope reassembly (de Lanerolle *et al.*, 2005). A study using *f*luorescence *r*ecovery *a*fter *p*hotobleaching (FRAP) and mutant cell lines has demonstrated that both cytoplasmic and nuclear actin molecules are in the polymerized form (Hofmann and de Lanerolle, 2006). Experiments using monoclonal antibodies have suggested that the polymerization mechanisms are different in these two types of actin filaments (Gonsior *et al.*, 1999).

Actin-related proteins (Arps) have been shown to contribute to the organization and regulation of chromatin structures. In budding yeast, Arp 1–10 subfamilies have been identified according to their similarity to actin; Arp1is the most similar and Arp10 is the least (Poch and Winsor, 1997). Among them, Arp4–9 predominantly localize in the nucleus. Vertebrates also possess these Arps. However, only Arp4 subfamily proteins have been known to localize in the nucleus (Harata *et al.*, 1999). Recently vertebrate Arp6 has also been shown to localize in the nucleus and interact with isoforms of HP1 (Ohfuchi *et al.*, 2006). These nuclear Arps are found in various chromatin remodeling and histone acetyltransferese (HAT) complexes (Olave *et al.*, 2002; Schafer and Schroer, 1999; Zhao *et al.*, 1998).

3.3. Nucleolar Organization and Heterochromatin

The proteomic analysis of "insoluble fraction" in the cell nucleus has been hampered by the difficulties in its sample preparation. A mass production of monoclonal antibodies is one approach to identify and characterize such nuclear proteins. Proteins extracted from the mitotic chromosomes, nucleoplasmic proteins and the nuclear "insoluble fraction" were injected into mice and a series of monoclonal antibodies were obtained. Immunostaining of the HeLa cells with these antibodies revealed a variety of intracellular localization of the antigen inside and outside the nucleus (Fig. 7). They recognize several dots inside the nucleolus (Fig. 7a), whole nucleolus (Fig. 7b), nuclear foci (Fig. 7c), nucleoplasm (Fig. 7d), and the edge of the nucleus (Fig. 7e). There are also many antibodies that recognize cytoplasmic architectures, for example, cytoplasm (Fig. 7f), cytoskeleton (Fig. 7g), plasma membrane (Fig. 7h), and mitochondria (Fig. 7i). This result suggests a strong relationship between the nuclear and cytoplasm proteins. Some antibodies recognize multiple regions inside the cells: the nucleus and cytoplasm (Fig. 7j), the paranuclear structure and nucleus (Fig. 7k), and the paranuclear structure and nucleoplasm (Fig. 7l).

Close observations of immunofluorescence signals showed that there are three different staining patterns, which correspond to three different nucleolar compartments; FC (Fibrillar center), DFC (dense fibrillar component), and GC (granular component). Nucleolus is surrounded by heterochromatin. When the cells are in very active state of its proliferation, the nucleolar compartments and heterochromatin are integrated into a highly intricate structure called "nucleolonema". A recent study has suggested that the chromatin associated with the nucleolus is less mobile than

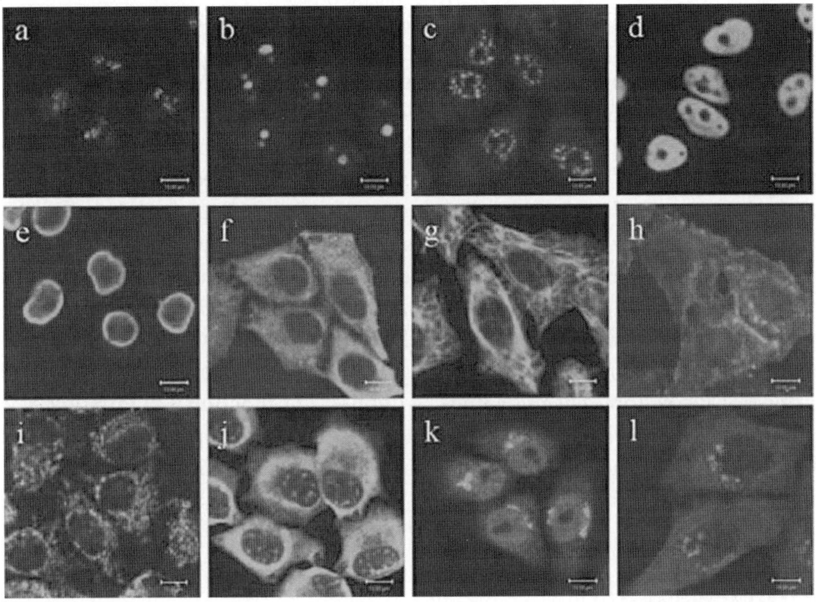

Figure 7. A series of monoclonal antibodies raised against nuclear proteins. HeLa cells fixed with 4% paraformaldehyde were immunostained with monoclonal antibodies raised against nuclear proteins. The intracellular localization of these antigens are (a) dot inside the nucleolus, (b) whole nucleolus, (c) nuclear foci, (d) nucleoplasm, (e) the edge of the nucleus, (f) cytoplasm, (g) cytoskeleton, (h) plasma membrane, (i) mitochondria, (j) nucleus and cytoplasm, (k) nucleus and the paranuclear structure, and (l) paranuclear structure and nucleoplasm

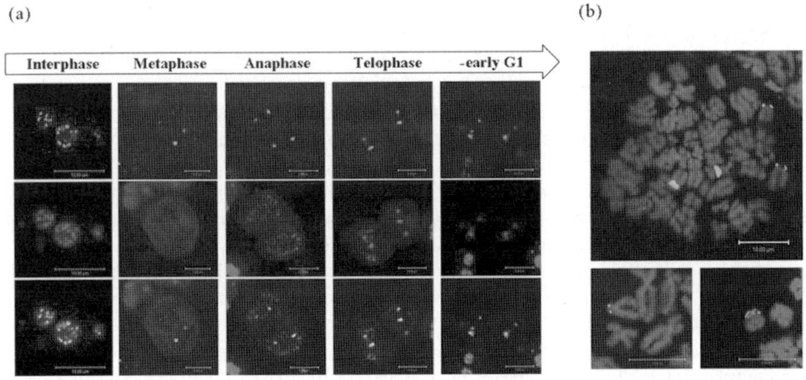

Figure 8. Localization of a nucleolar protein in FC regin. (a) Localization in mitotic cells. Green: antigen, Red: fibrillarin (DFC marker). (b) Localization on the metaphase chromosomes. Green: antigen, Red: DNA (PI). (Scale bars: 10 μm). (See Colour Plate 3.)

the chromatin in other regions, indicating its role for the regulation of chromatin dynamics (Chubb *et al.*, 2002).

Nucleolar proteins exhibit a variety of dynamic movement during mitosis. Several proteins in FC region (UBF, RNA polymerase I and 2–30C antigen) localize in several foci on the mitotic chromosomes (Fig. 8a). When the HeLa metaphase chromosomes are spread on a cover glass and immunostained with anti-2-30C antibody, the immunoreactive foci are detected on two pairs of acrocentric chromosomes (Fig. 8b). Since the rDNA genes are known to be distributed among 5 pairs of acrocentric chromosomes and 3 pairs of them are active during mitosis (Roussel *et al.*, 1996), this FC antigen is involved in the rDNA inactivation. Most of the proteins in the GC region are dispersed into the cytoplasm in mitosis. On the other hand, several proteins in the DFC region (ObgH2 and Ki-67) are known to localize on the surface of the mitotic chromosomes. Although the functional significance of these unique localizations is not fully understood, the understanding of the dynamics of these antigens might lead to the elucidation of the reorganization mechanisms of nucleolus in the telophase.

ACKNOWLEDGEMENT

Fig. 5q–s are data of R. L. Ohniwa at Kyoto University.

REFERENCES

Aagaard L, Laible G, Selenko P, Schmid M, Dorn R, Schotta G, Kuhfittig S, Wolf A, Lebersorger A, Singh PB et al (1999) Functional mammalian homologues of the Drosophila PEV-modifier Su(var)3–9 encode centromere-associated proteins which complex with the heterochromatin component M31. Embo J 18:1923–1938

Adachi Y, Kas E, Laemmli UK (1989) Preferential, cooperative binding of DNA topoisomerase II to scaffold-associated regions. Embo J 8:3997–4006

Adachi Y, Luke M, Laemmli UK (1991) Chromosome assembly *in vitro*: topoisomerase II is required for condensation. Cell 64:137–148

Adolph KW, Kreisman LR, Kuehn RL (1986) Assembly of chromatin fibers into metaphase chromosomes analyzed by transmission electron microscopy and scanning electron microscopy. Biophys J 49:221–231

Agresti A, Bianchi ME (2003) HMGB proteins and gene expression. Curr Opin Genet Dev 13:170–178

Anderson DE, Losada A, Erickson HP, Hirano T (2002) Condensin and cohesin display different arm conformations with characteristic hinge angles. J Cell Biol 156:419–424

Arents G, Burlingame RW, Wang BC, Love WE, Moudrianakis EN (1991) The nucleosomal core histone octamer at 3.1 A resolution: a tripartite protein assembly and a left-handed superhelix. Proc Natl Acad Sci U S A 88:10148–10152

Bannister AJ, Zegerman P, Partridge JF, Miska EA, Thomas JO, Allshire RC, Kouzarides T (2001) Selective recognition of methylated lysine 9 on histone H3 by the HP1 chromo domain. Nature 410:120–124

Bauer UM, Daujat S, Nielsen SJ, Nightingale K, Kouzarides T (2002) Methylation at arginine 17 of histone H3 is linked to gene activation. EMBO Rep 3:39–44

Becker PB, Wu C (1992) Cell-free system for assembly of transcriptionally repressed chromatin from Drosophila embryos. Mol Cell Biol 12:2241–2249

Bell AC, Felsenfeld G (1999) Stopped at the border: boundaries and insulators. Curr Opin Genet Dev 9:191–198

Belmont AS, Bruce K (1994) Visualization of G1 chromosomes: a folded, twisted, supercoiled chromonema model of interphase chromatid structure. J Cell Biol 127:287–302

Bennink ML, Leuba SH, Leno GH, Zlatanova J, de Grooth BG, Greve J (2001) Unfolding individual nucleosomes by stretching single chromatin fibers with optical tweezers. Nat Struct Biol 8:606–610

Berger JM, Gamblin SJ, Harrison SC, Wang JC (1996) Structure and mechanism of DNA topoisomerase II. Nature 379:225–232

Berrios M, Osheroff N, Fisher PA (1985) In situ localization of DNA topoisomerase II, a major polypeptide component of the Drosophila nuclear matrix fraction. Proc Natl Acad Sci U S A 82:4142–4146

Bianchi ME, Agresti A (2005) HMG proteins: dynamic players in gene regulation and differentiation. Curr Opin Genet Dev 15:496–506

Bonaldi T, Langst G, Strohner R, Becker PB, Bianchi ME (2002) The DNA chaperone HMGB1 facilitates ACF/CHRAC-dependent nucleosome sliding. Embo J 21:6865–6873

Buschhausen G, Wittig B, Graessmann M, Graessmann A (1987) Chromatin structure is required to block transcription of the methylated herpes simplex virus thymidine kinase gene. Proc Natl Acad Sci U S A 84:1177–1181

Bustin M (2001) Chromatin unfolding and activation by HMGN(*) chromosomal proteins. Trends Biochem Sci 26:431–437

Bystricky K, Heun P, Gehlen L, Langowski J, Gasser SM (2004) Long-range compaction and flexibility of interphase chromatin in budding yeast analyzed by high-resolution imaging techniques. Proc Natl Acad Sci U S A 101:16495–16500

Catez F, Brown DT, Misteli T, Bustin M (2002) Competition between histone H1 and HMGN proteins for chromatin binding sites. EMBO Rep 3:760–766

Chen D, Ma H, Hong H, Koh SS, Huang SM, Schurter BT, Aswad DW, Stallcup MR (1999) Regulation of transcription by a protein methyltransferase. Science 284:2174–2177

Chubb JR, Boyle S, Perry P, Bickmore WA (2002) Chromatin motion is constrained by association with nuclear compartments in human cells. Curr Biol 12:439–445

Croston GE, Kadonaga JT (1993) Role of chromatin structure in the regulation of transcription by RNA polymerase II. Curr Opin Cell Biol 5:417–423

de Frutos M, Raspaud E, Leforestier A, Livolant F (2001) Aggregation of nucleosomes by divalent cations. Biophys J 81:1127–1132

de la Barre AE, Gerson V, Gout S, Creaven M, Allis CD, Dimitrov S (2000) Core histone N-termini play an essential role in mitotic chromosome condensation. Embo J 19:379–391

de Lanerolle P, Johnson T, Hofmann WA (2005) Actin and myosin I in the nucleus: what next? Nat Struct Mol Biol 12:742–746

Ding HF, Bustin M, Hansen U (1997) Alleviation of histone H1-mediated transcriptional repression and chromatin compaction by the acidic activation region in chromosomal protein HMG-14. Mol Cell Biol 17:5843–5855

Ellwood KB, Yen YM, Johnson RC, Carey M (2000) Mechanism for specificity by HMG-1 in enhanceosome assembly. Mol Cell Biol 20:4359–4370

Felsenfeld G, McGhee JD (1986) Structure of the 30 nm chromatin fiber. Cell 44:375–377

Felts SJ, Weil PA, Chalkley R (1990) Transcription factor requirements for *in vitro* formation of transcriptionally competent 5S rRNA gene chromatin. Mol Cell Biol 10:2390–2401

Fey EG, Krochmalnic G, Penman S (1986) The nonchromatin substructures of the nucleus: the ribonucleoprotein (RNP)-containing and RNP-depleted matrices analyzed by sequential fractionation and resinless section electron microscopy. J Cell Biol 102:1654–1665

Fisher PA, Berrios M, Blobel G (1982) Isolation and characterization of a proteinaceous subnuclear fraction composed of nuclear matrix, peripheral lamina, and nuclear pore complexes from embryos of Drosophila melanogaster. J Cell Biol 92:674–686

Gasser SM, Laemmli UK (1986) The organisation of chromatin loops: characterization of a scaffold attachment site. Embo J 5:511–518

Germond JE, Hirt B, Oudet P, Gross-Bellark M, Chambon P (1975) Folding of the DNA double helix in chromatin-like structures from simian virus 40. Proc Natl Acad Sci U S A 72: 1843–1847
Godde JS, Widom J (1992) Chromatin structure of Schizosaccharomyces pombe. A nucleosome repeat length that is shorter than the chromatosomal DNA length. J Mol Biol 226:1009–1025
Gonsior SM, Platz S, Buchmeier S, Scheer U, Jockusch BM, Hinssen H (1999) Conformational difference between nuclear and cytoplasmic actin as detected by a monoclonal antibody. J Cell Sci 112(Pt 6):797–809
Goodwin GH, Sanders C, Johns EW (1973) A new group of chromatin-associated proteins with a high content of acidic and basic amino acids. Eur J Biochem 38:14–19
Goto H, Tomono Y, Ajiro K, Kosako H, Fujita M, Sakurai M, Okawa K, Iwamatsu A, Okigaki T, Takahashi T, Inagaki M (1999) Identification of a novel phosphorylation site on histone H3 coupled with mitotic chromosome condensation. J Biol Chem 274:25543–25549
Gurley LR, D'Anna, JA, Barham SS, Deaven LL, Tobey RA (1978) Histone phosphorylation and chromatin structure during mitosis in Chinese hamster cells. Eur J Biochem 84:1–15
Harata M, Oma Y, Mizuno S, Jiang YW, Stillman DJ, Wintersberger U (1999) The nuclear actin-related protein of Saccharomyces cerevisiae, Act3p/Arp4, interacts with core histones. Mol Biol Cell 10:2595–2605
Hayes JJ, Lee KM (1997) *In vitro* reconstitution and analysis of mononucleosomes containing defined DNAs and proteins. Methods 12:2–9
Hernandez-Verdun D, Gautier T (1994) The chromosome periphery during mitosis. Bioessays 16:179–185
Hirano T, Kobayashi R, Hirano M (1997) Condensins, chromosome condensation protein complexes containing XCAP-C, XCAP-E and a Xenopus homolog of the Drosophila Barren protein. Cell 89:511–521
Hizume K, Yoshimura SH, Maruyama H, Kim J, Wada H, Takeyasu K (2002) Chromatin reconstitution: development of a salt-dialysis method monitored by nano-technology. Arch Histol Cytol 65:405–413
Hizume K, Yoshimura SH, Takeyasu K (2004) Atomic force microscopy demonstrates a critical role of DNA superhelicity in nucleosome dynamics. Cell Biochem Biophys 40:249–262
Hizume K, Yoshimura SH, Takeyasu K (2005) Linker histone H1 per se can induce three-dimensional folding of chromatin fiber. Biochemistry 44:12978–12989
Hofmann WA, de Lanerolle P (2006) Nuclear actin: to polymerize or not to polymerize. J Cell Biol 172:495–496
Horn PJ, Peterson CL (2002) Molecular biology: Chromatin higher order folding – Wrapping up transcription. Science 297:1824–1827
Hoshi O, Ushiki T (2001) Three-dimensional structure of G-banded human metaphase chromosomes observed by atomic force microscopy. Arch Histol Cytol 64:475–482
Huynh VA, Robinson PJ, Rhodes D (2005) A method for the *in vitro* reconstitution of a defined "30 nm" chromatin fibre containing stoichiometric amounts of the linker histone. J Mol Biol 345:957–968
Ito T, Bulger M, Pazin MJ, Kobayashi R, Kadonaga JT (1997) ACF, an ISWI-containing and ATP-utilizing chromatin assembly and remodeling factor. Cell 90:145–155
Jackson DA, Cook PR (1995) The structural basis of nuclear function. Int Rev Cytol 162A:125–149
James TC, Elgin SC (1986) Identification of a nonhistone chromosomal protein associated with heterochromatin in Drosophila melanogaster and its gene. Mol Cell Biol 6:3862–3872
Jarman AP, Higgs DR (1988) Nuclear scaffold attachment sites in the human globin gene complexes. Embo J 7:3337–3344
Karymov MA, Tomschik M, Leuba SH, Caiafa P, Zlatanova J (2001) DNA methylation-dependent chromatin fiber compaction *in vivo* and *in vitro*: requirement for linker histone. Faseb J 15:2631–2641
Kaszas E, Cande WZ (2000) Phosphorylation of histone H3 is correlated with changes in the maintenance of sister chromatid cohesion during meiosis in maize, rather than the condensation of the chromatin. J Cell Sci 113(Pt 18):3217–3226
Keller W (1975) Determination of the number of superhelical turns in simian virus 40 DNA by gel electrophoresis. Proc Natl Acad Sci U S A 72:4876–4880

Kelly TJ, Qin S, Gottschling DE, Parthun MR (2000) Type B histone acetyltransferase Hat1p participates in telomeric silencing. Mol Cell Biol 20:7051–7058

Kimura A, Matsubara K, Horikoshi M (2005) A decade of histone acetylation: marking eukaryotic chromosomes with specific codes. J Biochem (Tokyo) 138:647–662

Kimura K, Hirano T (1997) ATP-dependent positive supercoiling of DNA by 13S condensin: a biochemical implication for chromosome condensation. Cell 90:625–634

Kobori T, Kodama M, Hizume K, Yoshimura SH, Ohtani T, Takeyasu K (2006) Comparative structural biology of the genome: nano-scale imaging of single nucleus from different kingdoms reveals the common physicochemical property of chromatin with a 40 nm structural unit. J Electron Microsc (Tokyo) 55:31–40

Kornberg RD (1974) Chromatin structure: a repeating unit of histones and DNA. Science 184:868–871

Lachner M, O'Carroll D, Rea S, Mechtler K, Jenuwein T (2001) Methylation of histone H3 lysine 9 creates a binding site for HP1 proteins. Nature 410:116–120

Lane NJ (1969) Intranuclear fibrillar bodies in actinomycin D-treated oocytes. J Cell Biol 40:286–291

Leuba SH, Yang G, Robert C, Samori B, van Holde K, Zlatanova J, Bustamante C (1994) Three-dimensional structure of extended chromatin fibers as revealed by tapping-mode scanning force microscopy. Proc Natl Acad Sci U S A 91:11621–11625

Leuba SH, Bustamante C, van Holde K, Zlatanova J (1998) Linker histone tails and N-tails of histone H3 are redundant: scanning force microscopy studies of reconstituted fibers. Biophys J 74:2830–2839

Li G, Sudlow G, Belmont AS (1998) Interphase cell cycle dynamics of a late-replicating, heterochromatic homogeneously staining region: precise choreography of condensation/decondensation and nuclear positioning. J Cell Biol 140:975–989

Lu W, Peterson R, Dasgupta A, Scovell WM (2000) Influence of HMG-1 and adenovirus oncoprotein E1A on early stages of transcriptional preinitiation complex assembly. J Biol Chem 275:35006–35012

Luger K, Mader AW, Richmond RK, Sargent DF, Richmond TJ (1997) Crystal structure of the nucleosome core particle at 2.8 A resolution. Nature 389:251–260

Lusser A, Kadonaga JT (2004) Strategies for the reconstitution of chromatin. Nat Methods 1:19–26

Maeshima K, Laemmli UK (2003) A two-step scaffolding model for mitotic chromosome assembly. Dev Cell 4:467–480

Mahy NL, Perry PE, Gilchrist S, Baldock RA, Bickmore WA (2002) Spatial organization of active and inactive genes and noncoding DNA within chromosome territories. J Cell Biol 157:579–589

Mangenot S, Leforestier A, Vachette P, Durand D, Livolant F (2002) Salt-induced conformation and interaction changes of nucleosome core particles. Biophys J 82:345–356

Marsden MP, Laemmli UK (1979) Metaphase chromosome structure: evidence for a radial loop model. Cell 17:849–858

Marshall WF, Straight A, Marko JF, Swedlow J, Dernburg A, Belmont A, Murray AW, Agard DA, Sedat JW (1997) Interphase chromosomes undergo constrained diffusional motion in living cells. Curr Biol 7:930–939

McGhee JD, Felsenfeld G (1980) Nucleosome structure. Annu Rev Biochem 49:1115–1156

Melby TE, Ciampaglio CN, Briscoe G, Erickson HP (1998) The symmetrical structure of structural maintenance of chromosomes (SMC) and MukB proteins: long, antiparallel coiled coils, folded at a flexible hinge. J Cell Biol 142:1595–1604

Moll R, Franke WW, Schiller DL, Geiger B, Krepler R (1982) The catalog of human cytokeratins: patterns of expression in normal epithelia, tumors and cultured cells. Cell 31:11–24

Nabirochkin S, Ossokina M, Heidmann T (1998) A nuclear matrix/scaffold attachment region co-localizes with the gypsy retrotransposon insulator sequence. J Biol Chem 273:2473–2479

Nakai T, Hizume K, Yoshimura SH, Takeyasu K, Yoshikawa K (2005) Phase transition in reconstituted chromatin. Europhysics Letters 69:1024–1030

Namciu SJ, Blochlinger KB, Fournier RE (1998) Human matrix attachment regions insulate transgene expression from chromosomal position effects in Drosophila melanogaster. Mol Cell Biol 18:2382–2391

Narlikar GJ, Fan HY, Kingston RE (2002) Cooperation between complexes that regulate chromatin structure and transcription. Cell 108:475–487

Nettikadan SR, Furbee CS, Muller MT, Takeyasu K (1998) Molecular structure of human topoisomerase II alpha revealed by atomic force microscopy. J Electron Microsc (Tokyo) 47:671–674

Nickerson J (2001) Experimental observations of a nuclear matrix. J Cell Sci 114:463–474

Nowak SJ, Corces VG (2004) Phosphorylation of histone H3: a balancing act between chromosome condensation and transcriptional activation. Trends Genet 20:214–220

O'Neill TE, Roberge M, Bradbury EM (1992) Nucleosome arrays inhibit both initiation and elongation of transcripts by bacteriophage T7 RNA polymerase. J Mol Biol 223:67–78

Ohfuchi E, Kato M, Sasaki M, Sugimoto K, Oma Y, Harata M (2006) Vertebrate Arp6, a novel nuclear actin-related protein, interacts with heterochromatin protein 1. Eur J Cell Biol 85:411–421

Ohniwa RL, Morikawa K, Kim J, Kobori T, Hizume K, Matsumi R, Atomi H, Imanaka T, Ohta T, Yoshimura SH, Takeyasu K (2006) Atomic Force Microscopy Dissects the Hierarchy of Genome Architectures in Eukaryote, Prokaryote and Chloroplast. Microscopy & Microanalysis 13:1–10

Ohsumi K, Katagiri C, Kishimoto T (1993) Chromosome condensation in Xenopus mitotic extracts without histone H1. Science 262:2033–2035

Olave IA, Reck-Peterson SL, Crabtree GR (2002) Nuclear actin and actin-related proteins in chromatin remodeling. Annu Rev Biochem 71:755–781

Olins DE, Olins AL (1972) Physical studies of isolated eucaryotic nuclei. J Cell Biol 53:715–736

Ono T, Losada A, Hirano M, Myers MP, Neuwald AF, Hirano T (2003) Differential contributions of condensin I and condensin II to mitotic chromosome architecture in vertebrate cells. Cell 115:109–121

Opel ML, Arfin SM, Hatfield GW (2001) The effects of DNA supercoiling on the expression of operons of the ilv regulon of Escherichia coli suggest a physiological rationale for divergently transcribed operons. Mol Microbiol 39:1109–1115

Owen-Hughes T, Workman JL (1994) Experimental analysis of chromatin function in transcription control. Crit Rev Eukaryot Gene Expr 4:403–441

Paulson JR, Laemmli UK (1977) The structure of histone-depleted metaphase chromosomes. Cell 12:817–828

Peterson CL, Laniel MA (2004) Histones and histone modifications. Curr Biol 14, R546–551

Pfaffle P, Gerlach V, Bunzel L, Jackson V (1990) *In vitro* evidence that transcription-induced stress causes nucleosome dissolution and regeneration. J Biol Chem 265:16830–16840

Poch O, Winsor B (1997) Who's who among the Saccharomyces cerevisiae actin-related proteins? A classification and nomenclature proposal for a large family. Yeast 13:1053–1058

Preuss U, Landsberg G, Scheidtmann KH (2003) Novel mitosis-specific phosphorylation of histone H3 at Thr11 mediated by Dlk/ZIP kinase. Nucleic Acids Res 31:878–885

Prymakowska-Bosak M, Misteli T, Herrera JE, Shirakawa H, Birger Y, Garfield S, Bustin M (2001) Mitotic phosphorylation prevents the binding of HMGN proteins to chromatin. Mol Cell Biol 21:5169–5178

Rattner JB, Hamkalo BA (1979) Nucleosome packing in interphase chromatin. J Cell Biol 81: 453–457

Rattner JB, Lin CC (1985) Radial loops and helical coils coexist in metaphase chromosomes. Cell 42:291–296

Rattner JB, Saunders C, Davie JR, Hamkalo BA (1982) Ultrastructural organization of yeast chromatin. J Cell Biol 93:217–222

Rea S, Eisenhaber F, O'Carroll D, Strahl BD, Sun ZW, Schmid M, Opravil S, Mechtler K, Ponting CP, Allis CD, Jenuwein T (2000) Regulation of chromatin structure by site-specific histone H3 methyltransferases. Nature 406:593–599

Rhodes D, Klug A (1981) Sequence-dependent helical periodicity of DNA. Nature 292:378–380

Roussel P, Andre C, Comai L, Hernandez-Verdun D (1996) The rDNA transcription machinery is assembled during mitosis in active NORs and absent in inactive NORs. J Cell Biol 133: 235–246

Satchwell SC, Drew HR, Travers AA (1986) Sequence periodicities in chicken nucleosome core DNA. J Mol Biol 191:659–675

Sato MH, Ura K, Hohmura KI, Tokumasu F, Yoshimura SH, Hanaoka F, Takeyasu K (1999) Atomic force microscopy sees nucleosome positioning and histone H1-induced compaction in reconstituted chromatin. FEBS Lett 452:267–271

Schafer DA, Schroer TA (1999) Actin-related proteins. Annu Rev Cell Dev Biol 15:341–363

Schurter BT, Koh SS, Chen D, Bunick GJ, Harp JM, Hanson BL, Henschen-Edman A, Mackay DR, Stallcup MR, Aswad DW (2001) Methylation of histone H3 by coactivator-associated arginine methyltransferase 1. Biochemistry 40:5747–5756

Shen X, Yu L, Weir JW, Gorovsky MA (1995) Linker histones are not essential and affect chromatin condensation *in vivo*. Cell 82:47–56

Simon RH, Felsenfeld G (1979) A new procedure for purifying histone pairs H2A + H2B and H3 + H4 from chromatin using hydroxylapatite. Nucleic Acids Res 6:689–696

Simpson RT (1978) Structure of the chromatosome, a chromatin particle containing 160 base pairs of DNA and all the histones. Biochemistry 17:5524–5531

Simpson RT, Stafford DW (1983) Structural features of a phased nucleosome core particle. Proc Natl Acad Sci U S A 80:51–55

Simpson RT, Thoma F, Brubaker JM (1985) Chromatin reconstituted from tandemly repeated cloned DNA fragments and core histones: a model system for study of higher order structure. Cell 42:799–808

Sugiyama S, Yoshino T, Kanahara H, Kobori T, Ohtani T (2003) Atomic force microscopic imaging of 30 nm chromatin fiber from partially relaxed plant chromosomes. Scanning 25:132–136

Sugiyama S, Yoshino T, Kanahara H, Shichiri M, Fukushi D, Ohtani T (2004) Effects of acetic acid treatment on plant chromosome structures analyzed by atomic force microscopy. Anal Biochem 324:39–44

Sutrias-Grau M, Bianchi ME, Bernues J (1999) High mobility group protein 1 interacts specifically with the core domain of human TATA box-binding protein and interferes with transcription factor IIB within the pre-initiation complex. J Biol Chem 274:1628–1634

Tamayo J, Miles M (2000) Human chromosome structure studied by scanning force microscopy after an enzymatic digestion of the covering cell material. Ultramicroscopy 82:245–251

Taniguchi T, Takayama S (1986) High-order structure of metaphase chromosomes: evidence for a multiple coiling model. Chromosoma 93:511–514

Tatchell K, Van Holde KE (1977) Reconstitution of chromatin core particles. Biochemistry 16:5295–5303

Thoma F, Koller T (1977) Influence of histone H1 on chromatin structure. Cell 12:101–107

Thoma F, Koller T, Klug A (1979) Involvement of histone H1 in the organization of the nucleosome and of the salt-dependent superstructures of chromatin. J Cell Biol 83:403–427

Turner BM, Birley AJ, Lavender J (1992) Histone H4 isoforms acetylated at specific lysine residues define individual chromosomes and chromatin domains in Drosophila polytene nuclei. Cell 69:375–384

Uchiyama S, Kobayashi S, Takata H, Ishihara T, Hori N, Higashi T, Hayashihara K, Sone T, Higo D, Nirasawa T, *et al* (2005) Proteome analysis of human metaphase chromosomes. J Biol Chem 280:16994–17004

Udvardy A, Maine E, Schedl P (1985) The 87A7 chromomere. Identification of novel chromatin structures flanking the heat shock locus that may define the boundaries of higher order domains. J Mol Biol 185:341–358

Uemura T, Ohkura H, Adachi Y, Morino K, Shiozaki K, Yanagida M (1987) DNA topoisomerase II is required for condensation and separation of mitotic chromosomes in S. pombe. Cell 50:917–925

Ushiki T, Hoshi O, Iwai K, Kimura E, Shigeno M (2002) The structure of human metaphase chromosomes: its histological perspective and new horizons by atomic force microscopy. Arch Histol Cytol 65:377–390

Van Hooser A, Goodrich DW, Allis CD, Brinkley BR, Mancini MA (1998) Histone H3 phosphorylation is required for the initiation, but not maintenance, of mammalian chromosome condensation. J Cell Sci 111(Pt 23):3497–3506

Vazquez J, Belmont AS, Sedat JW (2001) Multiple regimes of constrained chromosome motion are regulated in the interphase Drosophila nucleus. Curr Biol 11:1227–1239

Verschure PJ, van der Kraan I, de Leeuw W, van der Vlag J, Carpenter AE, Belmont AS, van Driel R (2005) *In vivo* HP1 targeting causes large-scale chromatin condensation and enhanced histone lysine methylation. Mol Cell Biol 25:4552–4564

Wang H, Huang ZQ, Xia L, Feng Q, Erdjument-Bromage H, Strahl BD, Briggs SD, Allis CD, Wong J, Tempst P, Zhang Y (2001) Methylation of histone H4 at arginine 3 facilitating transcriptional activation by nuclear hormone receptor. Science 293:853–857

Wang JC (2002) Cellular roles of DNA topoisomerases: a molecular perspective. Nat Rev Mol Cell Biol 3:430–440

Wei Y, Mizzen CA, Cook RG, Gorovsky MA, Allis CD (1998) Phosphorylation of histone H3 at serine 10 is correlated with chromosome condensation during mitosis and meiosis in Tetrahymena. Proc Natl Acad Sci U S A 95:7480–7484

Widom J, Klug A (1985) Structure of the 300A chromatin filament: X-ray diffraction from oriented samples. Cell 43:207–213

Williams RR (2003) Transcription and the territory: the ins and outs of gene positioning. Trends Genet 19:298–302

Wood V, Gwilliam R, Rajandream MA, Lyne M, Lyne R, Stewart A, Sgouros J, Peat N, Hayles J, Baker S *et al* (2002) The genome sequence of Schizosaccharomyces pombe. Nature 415:871–880

Woodcock CL, Frado LL, Rattner JB (1984) The higher-order structure of chromatin: evidence for a helical ribbon arrangement. J Cell Biol 99:42–52

Worcel A, Burgi E (1972) On the structure of the folded chromosome of Escherichia coli. J Mol Biol 71:127–147

Yoshimura SH, Ohniwa RL, Sato MH, Matsunaga F, Kobayashi G, Uga H, Wada C, Takeyasu K (2000) DNA phase transition promoted by replication initiator. Biochemistry 39:9139–9145

Yoshimura SH, Hizume K, Murakami A, Sutani T, Takeyasu K, Yanagida M (2002) Condensin architecture and interaction with DNA. Regulatory non-SMC subunits bind to the head of SMC heterodimer. Curr Biol 12:508–513

Yoshimura SH, Kim J, Takeyasu K (2003) On-substrate lysis treatment combined with scanning probe microscopy revealed chromosome structures in eukaryotes and prokaryotes. J Electron Microsc (Tokyo) 52:415–423

Zamore PD, Haley B (2005) Ribo-gnome: the big world of small RNAs. Science 309:1519–1524

Zhao K, Wang W, Rando OJ, Xue Y, Swiderek K, Kuo A, Crabtree GR (1998) Rapid and phosphoinositol-dependent binding of the SWI/SNF-like BAF complex to chromatin after T lymphocyte receptor signaling. Cell 95:625–636

Zheng C, Hayes JJ (2003) Structures and interactions of the core histone tail domains. Biopolymers 68:539–546

Zlatanova J, Leuba SH, van Holde K (1999) Chromatin structure revisited. Crit Rev Eukaryot Gene Expr 9:245–255

CHAPTER 2

ATP-DEPENDENT CHROMATIN REMODELLING

Action and Reaction

PARUL CHOUDHARY AND PATRICK VARGA-WEISZ*

Chromatin & Gene Expression, Babraham Institute, Cambridge CB2 4AT, UK

Abstract: Alterations of chromatin structure play an important role in gene regulation. One way of doing so involves ATP-dependent chromatin remodelling enzymes that act as molecular machines coupling ATP-hydrolysis to structural changes of the nucleosome. Several recent studies shed important insights into the mechanism of these factors and indicate that they couple DNA translocation within the nucleosome to DNA loop propagation through the nucleosome. This reaction causes the movement of a nucleosome with respect to a given DNA sequence and also drives its disassembly. It is becoming clear that the biology of these factors is very complex considering the plethora of known ATP-dependent nucleosome remodelling factors and their many, in part overlapping functions and varied ways of regulation and targeting. Finally, nucleosome remodelling may only be one aspect of the function of these enzymes, because they may impart or regulate higher order levels of chromatin organization. The importance of these enzymes for normal growth and development is illustrated by disorders and neoplasias linked to mutations of those factors or their misregulation. Given that these enzymes have such profound roles in gene expression and cell proliferation, they may constitute important drug targets for clinical applications in the future

Keywords: ATP, Nucleosome, Chromatin, Helicase, RAD54, Imitation Switch, Tumour suppressor

Abbreviations: ATP: Adenosine triphosphate; ATR-X: Alpha-thalassemia/mental retardation syndrome, X-Linked; BRG1: *B*rahma-*R*elated *G*ene 1; CHD: Chromodomain, Helicase, DNA-binding; ISWI: Imitation Switch; HDAC: Histone deacetylase; PCNA: *P*roliferating *C*ell *N*uclear *A*ntigen;RSC: *R*emodels the *S*tructure of *C*hromatin; PHD: Plant Homeodomain; SWI: Switch; SNF: *S*ucrose *N*on-*F*ermenting; SNF2H: *S*ucrose *N*on-*f*ermenting *h*omologue

*E-mail: Patrick.varga-weisz@bbsrc.ac.uk

T.K. Kundu and D. Dasgupta (eds.), Chromatin and Disease, 29–43.

1. INTRODUCTION

A high degree of organization is required to confine the eukaryotic genome into the nucleus. In order to achieve this, DNA is associated with proteins that extensively fold and condense it into a structure called chromatin. The fundamental unit of chromatin, termed the nucleosome, is composed of 147 base pairs DNA wrapped in about 2 superhelical turns around a histone octamer (reviewed in Luger and Hansen, 2005). Nucleosomes are arrayed along the genome to form a nucleofilament which can adopt higher levels of compaction ultimately resulting in the metaphase chromosome. Despite this enormous degree of compaction, chromatin is not an unwavering and static complex. In contrast, it is a dynamic assembly that can adopt many different conformations and substructures and allows DNA to be accessed as needed while simultaneously being packaged. Non-canonical histone variants (such as H2A.Z and H3.3) and non-histone chromosomal proteins (such as High Mobility Group proteins and Heterochromatin Protein 1, HP1) are incorporated to modify the structure and properties of chromatin (reviewed in Henikoff and Ahmad, 2005; Hiragami and Festenstein, 2005). Nucleosomes containing histone variants have an altered propensity for subsequent remodelling or polymerase passage events and may affect the higher order chromatin structure. Histone chaperones also play a key role in chromatin dynamics by facilitating nucleosome assembly and disassembly, transiently shielding the positively charged histones to regulate their interaction with negatively charged DNA (reviewed in Adkins and Tyler, 2004; Akey and Luger, 2003).

Key factors that render chromatin dynamic and regulate access to packaged genes are enzymes which affect the function, stability, and positioning of nucleosomes. These are broadly divided into two major classes: histone modification enzymes and ATP-dependent chromatin remodelling factors. This review highlights recent progress in our understanding of the mode of action of ATP-dependent nucleosome remodelling.

2. HISTONE MODIFICATIONS

The so-called histone tails, spanning about 25 amino acids, protrude out of the nucleosome body and are accessible to a range of post translational modifications such as acetylation, phosphorylation, methylation, ADP-ribosylation and ubiquitination (Peterson and Laniel, 2004). These modifications may impart epigenetic information, defining gene expression patterns through multiple cell generations. It is thought that specific combinations of modifications in one or more histones form a ‘code’ which is recognized by particular regulatory proteins and trans-acting factors including ATP-dependent remodelling enzymes (see below), leading to additional downstream events (Jenuwein and Allis, 2001). Whereas acetylation, phosphorylation and ubiquitination are readily reversible, lysine and arginine methylation are comparatively more stable marks. Reversal of some forms of methylation have been demonstrated only recently by Peptidyl Arginine Deiminase (PADI4/PAD4)

(Cuthbert *et al.*, 2004), nuclear amine oxidase LSD1 (Shi *et al.*, 2004) and JmjC-domain containing proteins – JHDM1 and JMJD2 (Tsukada *et al.*, 2006; Whetstine *et al.*, 2006;). These enzymes differ in their substrate specifity and demethylate mono-, di- or trimethylated histones specifically. Besides the histone tails, many modifications have been mapped to lie in the structured globular histone core which may interfere with the wrapping of DNA around the histones lateral surface. Such modifications could play a direct role in modulating histone-DNA interactions within the nucleosome, thereby determining the ease with which a nucleosome can be translocated along DNA (Cosgrove and Wolberger, 2005). Histone modifications and ATP-dependent remodelling enzymes could therefore regulate nucleosome mobility in a concerted action, providing an energetically efficient way to generate accessible DNA.

3. ATP-DEPENDENT CHROMATIN REMODELLING FACTORS

The DNA in the core nucleosome contacts the histone surface at 14 sites with clusters of hydrogen bonds and salt links (Luger and Richmond, 1998). Collectively, these weak interactions render the nucleosome a rather stable structure. Movement of the nucleosome would require these interactions to be broken and reformed which would lead to a free energy penalty. Nucleosome remodelling factors are able to lower the energy barrier by coupling ATP hydrolysis to the disruption of histone-DNA contacts. They mediate a range of alterations to chromatin structure that in effect regulate the access to the underlying DNA. These include the (1) sliding and spacing of nucleosomes, (2) altering the path of the DNA wrapped around nucleosomes, (3) transferring the histone octamer in trans to DNA molecules, (4) exchanging the canonical histones of the core with variants, (5) creating dinucleosome like structures, and (6) generating superhelical torsion in DNA (reviewed in Becker and Hörz, 2002; Langst and Becker, 2004). These enzymes are usually complexes of diverse proteins but have a common subunit of a Snf2p-related ATP hydrolysing enzyme sharing sequence homology with the DEXX-box helicase superfamily 2 (SF2) (Eisen *et al.*, 1995). Based on characteristic domain features and functional properties, chromatin remodellers are subdivided into at least four major subfamilies: SWI2/SNF2, ISWI, INO80 and CHD (reviewed in Eberharter and Becker, 2004). The SWI2/SNF2 subfamily consists of proteins that share the most extensive homology with Snf2p (also named Swi2p) and include a bromodomain. Homologues of SWI2/SNF2 are Brahma (in *Drosophila*), Brm and Brg1 (in mammals). The ISWI proteins contain a C-terminal SANT and SLIDE domain which interacts with both DNA and proteins (Grune *et al.*, 2003). The CHD family proteins contain PHD fingers and a chromodomain while the Ino80 family contains a split ATPase domain. Differences between enzymes in each class are reflected by their different ATPase activities. For example, the Swi2 ATPase is maximally stimulated by free DNA whereas the ATPase Mi-2 (belonging to the CHD family) is induced in the presence of nucleosomal but not free DNA (Brehm *et al.*, 2000).

4. THE MANY FUNCTIONS OF ATP-DEPENDENT CHROMATIN REMODELLING FACTORS

Given that around 80% of nuclear DNA is packaged by nucleosomes, it is not surprising that ATP-dependent chromatin remodelling factors function in all processes involving DNA such as transcription, replication, recombination and repair (reviewed in Corona and Tamkun, 2004). They affect transcription by changing the accessibility of transcription factors to genes leading either to activation or repression. A compelling link with transcriptional repression has emerged with the discovery of the NURD complexes (containing the CHD-type ATPase Mi-2) which have both *nu*cleosome *r*emodelling and histone *d*eacetylation activities (reviewed in Bowen *et al.*, 2004). NoRC (for *N*ucle*o*lar chromatin *R*emodeling *C*omplex) containing the ISWI-homologue SNF2H is involved in the repression of PolI transcription through the recruitment of the SIN3/HDAC co-repressor to the ribosomal DNA promoter (Santoro and Grummt, 2005 and references therein). The role of remodellers in gene regulation impacts development. An illustration of this are the roles of two distinct ATP-dependent chromatin remodelling factors, ISWI and DOM, in controlling specific stem cell self renewal (Xi and Xie, 2005). They do this by regulating responses to peptide factor signalling in the stem cell microenvironment ('niche'). In Drosophila, ISWI was found to control Germline Stem Cell self-renewal and DOM was shown to be essential for Somatic Stem Cell self renewal. These findings suggest that different stem cell types depend on different chromatin remodelling factors to control their self-renewal, at least in part by regulating their gene expression response to the niche signals. Because these chromatin remodelling complexes are highly conserved, it is possible they may play a role in stem cell self-renewal in other organisms, including humans.

Remodelling factors play important regulatory roles to facilitate the many steps of eukaryotic DNA replication. (reviewed in Falbo and Shen, 2006). For example, the mammalian ISWI isoform SNF2H has been shown to be required for efficient DNA replication from a viral origin of replication and through heterochromatin (Collins *et al.*, 2002; Zhou *et al.*, 2005). SNF2H may have also a role in chromatin maturation and the maintenance of epigenetic patterns through replication by being targeted to replication sites in a complex with the Williams Syndrome Transcription Factor (WSTF) that in turn binds directly to replication factor PCNA (Poot *et al.*, 2004). The remodelling event might serve to keep an open state of chromatin after the replication fork passes, thereby creating an opportunity for the epigenetic marks to be copied and transmitted to the next generation (Poot *et al.*, 2005). Roles of various ATP-dependent nucleosome remodelling factors in DNA repair and recombination have also been identified (reviewed in Huang *et al.*, 2005; Shaked *et al.*, 2006). Finally, remodelling factors may also play an important regulatory and architectural role in the maintenance of higher order structure of chromatin (reviewed in Varga-Weisz and Becker, 2006; see also MacCallum *et al.*, 2002).

Several links have emerged between remodelling factors and oncogenesis (reviewed in Cairns, 2001). Subunits of the mammalian SWI/SNF complex possess intrinsic tumour suppressor function or are required for the activity of other tumour suppressor genes. SNF5 (INI1), a core subunit of SWI/SNF is inactivated in malignant rhabdoid tumours, a highly aggressive cancer of early childhood (reviewed in Roberts and Orkin, 2004). Specific mutations in the mammalian SWI2/SNF2 homologue BRG1 have been identified in pancreatic, breast, lung and prostrate cancer cell lines (Wong *et al.*, 2000). SWI/SNF also directly interacts with tumour suppressors and proto-oncogenes such as RB, BRCA1, c-Myc and MLL (Bochar *et al.*, 2000; Cheng *et al.*, 1999; Dunaief *et al.*, 1994; Nie *et al.*, 2003). However, the mechanisms by which the remodelling complex contributes to tumour suppression are yet to be fully understood. Mutations in other ATP-dependent chromatin remodelling factors have been linked to disease, such as in ATR-X, causing X-linked mental retardation syndromes. The phenotypes include facial dysmorphism, urogenital defects, and alpha-thalassaemia (resulting from reduced alpha-globin expression (Xue *et al.*, 2003 and references therein). The SWI/SNF-related SMARCAL1 is mutated in Schimke immuno-osseous dysplasia, a pleiotropic disorder with the diagnostic features of T-cell immunodeficiency, spondyloepiphyseal dysplasia, renal failure, hypothyroidism, episodic cerebral ischemia, and bone-marrow failure (Boerkoel *et al.*, 2002).

5. INSIGHTS FROM STRUCTURES OF SWI2/SNF2-FAMILY MEMBERS

The ATPase domains of SWI/SNF-family members exhibit a conserved architecture with an N-terminal domain containing conserved motifs I, Ia, II and III, required for ATP-hydrolysis, and a C-terminal domain (motifs IV to VI) (Eisen *et al.*, 1995). Functional analysis of these conserved sequence motifs have shown that whereas many of them play key roles in ATP binding and hydrolysis, certain residues within the conserved motif V are specifically required to couple ATP hydrolysis to chromatin remodelling activity: Deletion of eight amino acids in Motif V cripples the chromatin remodelling activity of SWI/SNF without altering its DNA-stimulated ATPase activity (Smith and Peterson, 2005). Therefore, this particular motif seems to be required specifically for coupling the energy from ATP-hydrolysis to the actual biomechanical force required for nucleosome remodelling. Remarkably, Motif V of the human SWI2/SNF2 homolog Brg1 has been shown to be a possible hot spot for mutational alterations associated with cancers (Medina *et al.*, 2004).

Helicases catalyze the processive separation of duplex DNA into single strands. Despite sharing similarity to helicases, none of the chromatin remodelling factors, with the exception of the INO80 complex, have been shown to catalyze the separation of DNA strands (Shen *et al.*, 2000). Instead, they can translocate on double-stranded (ds) DNA in an ATP-hydrolysis dependent manner and are characterized by their ability to generate superhelical torsional strain in DNA (Havas *et al.*, 2000; Saha *et al.*, 2002; Whitehouse *et al.*, 2003). The crystal structure of Rad54, a member of the SWI/SNF family has been solved for both *S. solfataricus* and zebrafish which helps to understand the mechanism of the SWI/SNF ATPase domain in remodelling processes (Durr *et al.*, 2005; Thoma *et al.*, 2005). It reveals a striking similarity to the SF2 helicases suggesting that SWI2/SNF2 proteins use a mechanism analogous to helicases to translocate on DNA. Moreover, the motifs for ATP and Mg^{2+} binding and interlobe-interaction are related to those found in other helicases. However, they contain two SWI2/SNF2-family specific insertions where other helicases have accessory domains and these are likely to play a central role in the remodelling mechanism. Studies on RecG and PcrA helicases have shown that translocation on DNA and strand separation are separable activities (reviewed in Caruthers and McKay, 2002; Singleton and Wigley, 2002). The translocation activity resides within the bi-lobal ATPase core whereas strand separation requires the wedge-like DNA binding accessory domains specific for the particular helicases. The lack of an equivalent wedge-like structure may explain the absence of helicase activity in SWI/SNF ATPases.

The similarities of DNA recognition by the Rad54 ATPase domain and helicases explain how the biochemical activities of SWI2/SNF2 enzymes are generated by a mechanism that is related to that of Dexx-box helicases (Durr *et al.*, 2005). The Rad54 catalytic core consists of two domains, Domain 1 and 2, separated by a deep cleft. Prior to translocation, DNA is bound at Domain 1 at the high affinity DNA binding site. In the presence of ATP, Domain 2 would undergo a conformational change that would push on the upstream minor groove of DNA advancing the ds DNA in the active site of SWI2/SNF2 enzymes. Following advancement of DNA, ATP hydrolysis might relax the structure, allowing rebinding of Domain 1 to DNA, ADP to ATP exchange and rebinding of Domain 2 at a new translocated upstream DNA binding site. It is important to note that even a relatively moderate translocation of DNA by sliding along the minor groove would include a substantial rotation of DNA along the helical axis. Both translocation and rotation could be used by remodelling factors for force generation. Remodelling factors might bind to substrate DNA-protein complexes both by the catalytic domain and by additional substrate binding domains. The screw motion of the DNA at the catalytic domain could not only transport DNA but also generate torque that would lead to disruption of DNA-protein interfaces. The precise mechanism of remodelling is yet not understood completely. However, some models have been proposed recently which will be discussed below.

6. MECHANICS OF NUCLEOSOME REMODELLING BY ATP-DEPENDENT CHROMATIN REMODELLING FACTORS

The image of two DNA windings around the highly charged histone octamer implies a stable, if not static structure for the nucleosome. However, the nucleosome is actually surprisingly dynamic on its own account (reviewed in Luger, 2006). Fluorescence Resonance Energy Transfer (FRET) experiments show that mononucleosomes exist in a dynamic equilibrium conformational transition in which the ends of the DNA unpeel from the surface of the histone octamer (Li *et al.*, 2005). Measurement of the equilibrium constants for this transition at millisecond timescales showed that under physiological conditions, a nucleosome is in its altered conformation 2–10% of the time. A single molecule approach to study nucleosome structure in solution also revealed that the nucleosome dynamically switches between an 'open' and 'closed' state (Tomschik *et al.*, 2005). Under the conditions tested, nucleosomes preferred the closed state and made brief, ~100 ms excursions into the open state. This timescale is sufficient to permit binding of regulatory proteins to nucleosomal DNA and allows access for the transcriptional machinery. Binding of an ATP-dependent remodeller or a transcription factor can capture and stabilize the partially dissociated DNA. Knowledge of the rate of site exposure will help to distinguish whether remodelling factors act simply as Brownian ratchets, trapping and harnessing spontaneous nucleosome unwrapping events or whether they actively drive DNA off the nucleosome.

Nucleosomes can also slide along DNA spontaneously and such movements were first explained by the 'twist diffusion model' that suggests that thermal energy fluctuations would be sufficient to twist the DNA helix at the edge of the nucleosomes, replacing histone-DNA interactions by neighbouring DNA base pairs (reviewed in van Holde and Yager, 2003; see Fig. 1). This distortion within the nucleosome is propagated over the histone octamer surface changing the translational position of the nucleosome. According to this model, the DNA rotates around its axis as it 'screws' over the surface of the nucleosome in a base-by-base reaction. Remodelling complexes have been shown to introduce superhelical torsion into nucleosomal DNA supporting a twisting mechanism (Havas *et al.*, 2000). However, single stranded DNA nicks, abasic sites and bulky adducts that should inhibit DNA twisting did not inhibit remodelling by several enzymes, indicating that the twisting mechanism alone does not explain nucleosome remodelling by ATPases (Aoyagi and Hayes, 2002; Langst and Becker, 2001; Strohner *et al.*, 2005; Zofall *et al.*, 2006).

Several studies of the action of ISWI-, SWI/SNF- and the SWI/SNF-related RSC-complexes support an alternative mechanism – the 'loop recapture' (Kassabov *et al.*, 2003; Lorch *et al.*, 2005; Saha *et al.*, 2005; Schwanbeck *et al.*, 2004; Strohner *et al.*, 2005; Zofall *et al.*, 2006), see Fig. 1. According to this proposal, a detachment of a segment of DNA occurs at the entry site of the nucleosome. Dissociation of a DNA segment and interaction of freed protein surfaces with neighbouring DNA segments would lead to the formation of a DNA loop on the histone octamer. The propagation of the DNA loop over the histone octamer (like in a caterpillar motion) would change the translational position of the nucleosome corresponding to the size of the DNA loop. Loop size is expected to be a function of how far the enzyme reaches into the nucleosomal linker in order to pull in DNA. Mapping of the nucleosome remodelling intermediates shows that nucleosomes move in about 10 bp increments (a full turn of the DNA superhelix), from 10 to 50 bp, without detectable intermediates (Zofall *et al.*, 2006). This predicts that a loop of 10–50 bp is propagated over the nucleosome surface by the action of the remodeller.

How is the property of DNA translocation applied to nucleosomes to break histone-DNA contacts, enabling nucleosome mobility? Insights into this come from

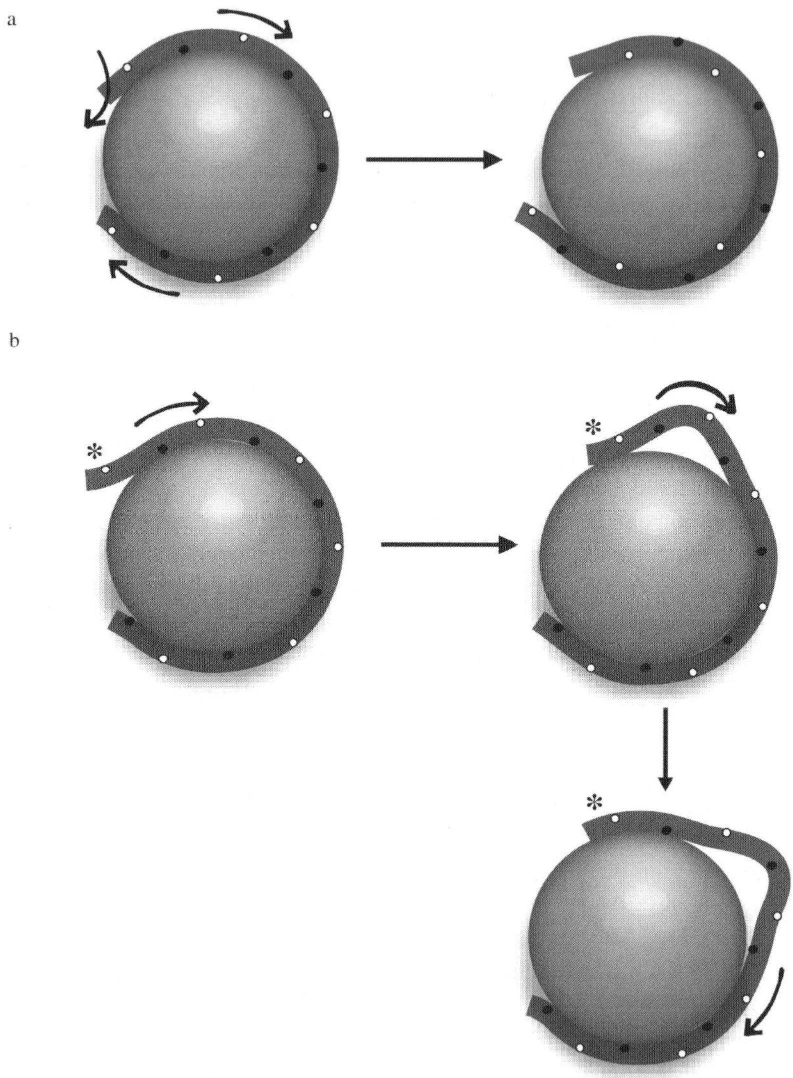

Figure 1. Schematic representation of remodelling mechanisms. (Adapted form Langst and Becker, 2004.) The schemes show nucleosomes from the top. (a) The twist diffusion model – Twisting of DNA moves it over the histone surface in one base pair increments. This changes the position of the DNA with respect to the histone, as shown by the open and closed circles. (b) The Loop recapture model – Extranucleosomal DNA is pulled into the nucleosomes to replace a DNA segment which consequently loops out. This loop is then propragated over the histone surface like ripples of a wave. The star, *, indicates how this leads to a change in the position of DNA relative to the nucleosome. (See Colour Plate 4.)

experiments on Sth1, the catalytic core of RSC (Saha *et al.*, 2005). These studies indicate that Sth1 binds the nucleosome core in the absence of ATP and then uses ATP hydrolysis to track along a DNA strand in the 3′–5′ direction, leading to translational movement of DNA along the octamer surface. It is proposed that Sth1 binds to DNA at a fixed site inside the nucleosome, about two DNA helical turns off the dyad axis. The suggestion is that directional DNA translocation by Sth1 breaks histone-DNA contacts creating a DNA wave intermediate that then propagates to the distal entry/exit site by diffusion. However, in contrast to the related movements of DNA helicases, Sth1 remains anchored at the histone octamer at a position where the Histone H4 N-terminus passes between two gyres of nucleosomal DNA. This site was also found to be essential for remodelling by the ISWI containing NURF complex as DNA nicks in this region compromise nucleosome sliding (Schwanbeck *et al.*, 2004). ISW2 (a yeast ISWI complex) and the SWI/SNF complex also show similar DNA translocation from an internal site (Zofall *et al.*, 2006). The latter study was done by placing DNA gaps at the nucleosome dyad to determine whether it interfered with nucleosome mobilization. Changes in histone-DNA interactions were mapped by attaching a photoreactive group to a unique amino acid residue on the surface of the histone octamer and cross-linking it to a proximal nucleotide. The movement of the DNA on the histone surface can therefore be tracked throughout the remodelling reaction with base pair resolution. These experiments showed that gaps interfered with nucleosome movement only when positioned adjacent to the internal translocation site. Therefore, the earlier view that remodellers translocate along extranucleosomal DNA to displace DNA from nucleosomes has to be amended in light of these new findings (Becker and Hörz, 2002). Further studies on the ISW2-complex indicate that ATP-hydrolysis mediated conformational changes of the remodelling enzymes and DNA-binding and -release cycles are likely to be important aspects of the nucleosome remodelling mechanism (Fitzgerald *et al.*, 2004). A future challenge will be to understand the coordination and temporal sequence of the various aspects of ATP-dependent nucleosome remodelling, such as conformation transitions, DNA loop formation and translocation. Finally, it should be emphasized that nucleosome remodelling is likely only one of several functional aspects of many chromatin remodelling complexes, because these are often quite abundant chromatin constituents and large multiprotein assemblies, harbouring several activities (MacCallum *et al.*, 2002; Varga-Weisz and Becker, 2006). They may also affect the chromatin fiber by folding DNA into quite large (400 bp or more) loops as recently shown by single molecule analysis (Lia *et al.*, 2006).

7. TARGETING AND REGULATION OF CHROMATIN REMODELLING FACTORS

Remodelling complexes appear to have no intrinsic DNA sequence specificity. Targeting may therefore occur primarily via interactions with other regulatory proteins or to epitopes on the histones marked by specific modifications. Both these means of recruitment are important and may not be mutually exclusive. They are illustrated by several studies of ISWI-complexes. These complexes show two modes of binding to chromatin: a basal level of binding globally throughout the genome, and a more target specific interaction (Fazzio *et al.*, 2005). The target specific binding *in vivo* requires the presence of sequence specific DNA binding proteins like Ume6p or Bdp1p (Bachman *et al.*, 2005). Another mode of targeting is seen with the Williams Syndrome Transcription Factor (WSTF) which interacts with PCNA directly to target chromatin remodelling by SNF2H to replication foci (Poot *et al.*, 2004).

ISWI complexes also 'sense' histone modification. They have been shown to require the H4 'basic' patch of amino acids R17H18R19 to specifically associate with the target sites on chromatin (Clapier *et al.*, 2002). In *Drosophila* ISWI is essential and ISWI mutants show chromosome defects in the male X-chromosome but not in the autosomes or female X-chromosomes (Deuring *et al.*, 2000). Ordinarily, the transcription of genes on the male X-chromosome is increased two-fold relative to females due to dosage compensation, a process involving the acetylation of histone H4 at Lysine 16 (H4K16), next to the basic patch that is so critical for ISWI function. Blocking this H4K16 acetylation suppresses the X-chromosome defects resulting from loss of ISWI function in males (Corona *et al.*, 2002). Therefore, it appears that H4K16 acetylation reduces the ability of ISWI to interact productively with its substrate and directly counteracts chromatin compaction mediated by the ISWI ATPase. It is possible that the acetylation of the adjacent Lysine 16 modulates the structure of this patch and hence interferes with ISWI function. ISWI is also targeted to nucleosomes that are decorated by specific methylation marks, however, the mechanism of interaction with methylated histones is not understood (Mellor and Morillon, 2004; Santos-Rosa *et al.*, 2003). Another example of specific recognition of histone modifications is the interaction of CHD1 with methylated Lysine 4 of histone H3 (Flanagan *et al.*, 2005; Pray-Grant *et al.*, 2005; Sims *et al.*, 2005). In summary, site-specific histone modifications are a mechanism of targeting and regulating remodelling factor action.

Besides the localization of remodelling complexes, the activities of the complexes themselves must be tightly regulated, because aberrant activity could have deleterious effects on the organization and expression of eukaryotic genomes. For example, Mi-2 is shown to be regulated by phosphorylation by Casein Kinase 2 in Drosophila cell extracts (Bouazoune and Brehm, 2005). Dephosphorylation increases its affinity for the nucleosomal substrate, nucleosome stimulated ATPase and ATP-dependent nucleosome mobilization activities. These findings suggest that constitutive phosphorylation serves to restrain enzymatic activity and that Mi2 is fully activated by an unidentified phosphatase following recruitment to chromatin. This presents a possible mechanism to rapidly and reversibly control Mi2 activity, subsequent to chromatin association. Phosphorylation also inactivates human forms of the SWI/SNF complex as cells traverse mitosis and dephosphorylation restores the nucleosome remodelling activity. This transitional inactivation and reactivation of human SWI/SNF may be required for formation of mitotic chromatin and reformation of an active and open chromatin structure as cells exit mitosis (Muchardt *et al.*, 1996; Sif *et al.*, 1998). Secondary messengers like Inositol polyphosphates have been implicated in regulating the activity of several yeast nucleosome remodelling complexes (Rando *et al.*, 2002; Shen *et al.*, 2003; Steger *et al.*, 2003). Protein-protein interactions and protein quality control systems may also regulate the activity of multimeric enzyme complexes by maintaining stoichiometric levels of the subunit proteins as seen in the human SWI/SNF complexes (Chen and Archer, 2005). Many more means of regulation of the remodelling factors exist, other than those discussed above. Understanding these regulatory mechanisms is essential given the varied functions performed by the remodellers.

8. CONCLUSION

Chromatin remodelling factors are molecular motors that play a crucial role in maintaining the dynamic eukaryotic genome. However, there are still many avenues that are unexplored and questions to be answered such as: Why do we need so many remodelling complexes? Why has nature evolved different classes of remodelling factors which act through different mechanisms? Do these complexes play distinct or overlapping roles? How is the cross talk between histone modifying enzymes and ATP-dependent remodelling factors mediated? Is the final outcome an effect of the cooperativity between these two factors? Complete understanding of the functional diversification of these factors, their targeting, regulation and mechanism of action is an important goal for the future. However, what has become clear is the importance of these factors in gene regulation and cell proliferation. Given that other pleiotropically acting ATPases such as topoisomerase II are prime drug targets for cancer therapy, it is likely that at least some of the ATP-dependent nucleosome remodelling factors could be rewarding targets for future development as well.

ACKNOWLEDGEMENTS

Parul Choudhary is supported by a fellowship from the Commonwealth Scholarship Commission in the United Kingdom. Research in the VW laboratory is supported by the Biotechnology and Biosciences Research Council, the Association for International Cancer Research, St Andrews, and the European Epigenome Network of Excellence. We thank Leila Barki-Celli and Ludmila Bozhenok for comments that improved this manuscript.

REFERENCES

Adkins MW, Tyler JK (2004) The histone chaperone Asf1p mediates global chromatin disassembly in vivo. J Biol Chem 279:52069–52074

Akey CW, Luger K (2003) Histone chaperones and nucleosome assembly. Curr Opin Struct Biol 13:6–14

Aoyagi S, Hayes JJ (2002) hSWI/SNF-catalyzed nucleosome sliding does not occur solely via a twist-diffusion mechanism. Mol Cell Biol 22:7484–7490

Bachman N, Gelbart ME, Tsukiyama T, Boeke JD (2005) TFIIIB subunit Bdp1p is required for periodic integration of the Ty1 retrotransposon and targeting of Isw2p to S. cerevisiae tDNAs. Genes Dev 19:955–964

Becker PB, Hörz W (2002) ATP-dependent nucleosome remodeling. Annu Rev Biochem 71:247–273

Bochar DA, Wang L, Beniya H, Kinev A, Xue Y, Lane WS, Wang W, Kashanchi F, Shiekhattar R (2000) BRCA1 is associated with a human SWI/SNF-related complex: linking chromatin remodeling to breast cancer. Cell 102:257–265

Boerkoel CF, Takashima H, John J, Yan J, Stankiewicz P, Rosenbarker L, Andre JL, Bogdanovic R, Burguet A, Cockfield S *et al* (2002) Mutant chromatin remodeling protein SMARCAL1 causes Schimke immuno-osseous dysplasia. Nat Genet 30:215–220

Bouazoune K, Brehm A (2005) dMi-2 chromatin binding and remodeling activities are regulated by dCK2 phosphorylation. J Biol Chem 280:41912–41920

Bowen NJ, Fujita N, Kajita M, Wade PA (2004) Mi-2/NuRD: multiple complexes for many purposes. Biochim Biophys Acta 1677:52–57

Brehm A, Langst G, Kehle J, Clapier CR, Imhof A, Eberharter A, Muller J, Becker PB (2000) dMi-2 and ISWI chromatin remodelling factors have distinct nucleosome binding and mobilization properties. Embo J 19:4332–4341

Cairns BR (2001) Emerging roles for chromatin remodeling in cancer biology. Trends Cell Biol 11:S15–21

Caruthers JM, McKay DB (2002) Helicase structure and mechanism. Curr Opin Struct Biol 12:123–133

Chen J, Archer TK (2005) Regulating SWI/SNF subunit levels via protein-protein interactions and proteasomal degradation: BAF155 and BAF170 limit expression of BAF57. Mol Cell Biol 25:9016–9027

Cheng SW, Davies KP, Yung E, Beltran RJ, Yu J, Kalpana GV (1999) c-MYC interacts with INI1/hSNF5 and requires the SWI/SNF complex for transactivation function. Nat Genet 22:102–105

Clapier CR, Nightingale KP, Becker PB (2002) A critical epitope for substrate recognition by the nucleosome remodeling ATPase ISWI. Nucleic Acids Res 30:649–655

Collins N, Poot RA, Kukimoto I, Garcia-Jimenez C, Dellaire G, Varga-Weisz PD (2002) An ACF1-ISWI chromatin-remodeling complex is required for DNA replication through heterochromatin. Nat Genet 32:627–632

Corona DF, Clapier CR, Becker PB, Tamkun JW (2002) Modulation of ISWI function by site-specific histone acetylation. EMBO Rep 3:242–247

Corona DF, Tamkun JW (2004) Multiple roles for ISWI in transcription, chromosome organization and DNA replication. Biochim Biophys Acta 1677:113–119

Cosgrove MS, Wolberger C (2005) How does the histone code work? Biochem Cell Biol 83: 468–476

Cuthbert GL, Daujat S, Snowden AW, Erdjument-Bromage H, Hagiwara T, Yamada M, Schneider R, Gregory PD, Tempst P, Bannister AJ, *et al* (2004) Histone deimination antagonizes arginine methylation. Cell 118:545–553

Deuring R, Fanti L, Armstrong JA, Sarte M, Papoulas O, Prestel M, Daubresse G, Verardo M, Moseley SL, Berloco M *et al* (2000) The ISWI chromatin-remodeling protein is required for gene expression and the maintenance of higher order chromatin structure in vivo. Mol Cell 5:355–365

Dunaief JL, Strober BE, Guha S, Khavari PA, Alin K, Luban J, Begemann M, Crabtree GR, Goff SP (1994) The retinoblastoma protein and BRG1 form a complex and cooperate to induce cell cycle arrest. Cell 79:119–130

Durr H, Korner C, Muller M, Hickmann V, Hopfner KP (2005) X-ray structures of the Sulfolobus solfataricus SWI2/SNF2 ATPase core and its complex with DNA. Cell 121:363–373

Eberharter A, Becker PB (2004) ATP-dependent nucleosome remodelling: factors and functions. J Cell Sci 117:3707–3711

Eisen JA, Sweder KS, Hanawalt PC (1995) Evolution of the SNF2 family of proteins: subfamilies with distinct sequences and functions. Nucleic Acids Res 23:2715–2723

Falbo KB, Shen X (2006) Chromatin remodeling in DNA replication. J Cell Biochem 97:684–689

Fazzio TG, Gelbart ME, Tsukiyama T (2005) Two distinct mechanisms of chromatin interaction by the Isw2 chromatin remodeling complex in vivo. Mol Cell Biol 25:9165–9174

Fitzgerald DJ, DeLuca C, Berger I, Gaillard H, Sigrist R, Schimmele K, Richmond TJ (2004) Reaction cycle of the yeast Isw2 chromatin remodeling complex. Embo J 23:3836–3843

Flanagan JF, Mi LZ, Chruszcz M, Cymborowski M, Clines KL, Kim Y, Minor W, Rastinejad F, Khorasanizadeh S (2005) Double chromodomains cooperate to recognize the methylated histone H3 tail. Nature 438:1181–1185

Grune T, Brzeski J, Eberharter A, Clapier CR, Corona DF, Becker PB, Muller CW (2003) Crystal structure and functional analysis of a nucleosome recognition module of the remodeling factor ISWI. Mol Cell 12:449–460

Havas K, Flaus A, Phelan M, Kingston R, Wade PA, Lilley DM, Owen-Hughes T (2000) Generation of superhelical torsion by ATP-dependent chromatin remodeling activities. Cell 103:1133–1142

Henikoff S, Ahmad K (2005) Assembly of variant histones into chromatin. Annu Rev Cell Dev Biol 21:133–153

Hiragami K, Festenstein R (2005) Heterochromatin protein 1: a pervasive controlling influence. Cell Mol Life Sci 62:2711–2726

Huang J, Liang B, Qiu J, Laurent BC (2005) ATP-dependent chromatin-remodeling complexes in DNA double-strand break repair: remodeling, pairing and (re)pairing. Cell Cycle 4:1713–1715

Jenuwein T, Allis CD (2001) Translating the histone code. Science 293:1074–1080

Kassabov SR, Zhang B, Persinger J, Bartholomew B (2003) SWI/SNF unwraps, slides, and rewraps the nucleosome. Mol Cell 11:391–403

Langst G, Becker PB (2001) ISWI induces nucleosome sliding on nicked DNA. Mol Cell 8:1085–1092

Langst G, Becker PB (2004) Nucleosome remodeling: one mechanism, many phenomena? Biochim Biophys Acta 1677:58–63

Li G, Levitus M, Bustamante C, Widom J (2005) Rapid spontaneous accessibility of nucleosomal DNA. Nat Struct Mol Biol 12:46–53

Lia G, Praly E, Ferreira H, Stockdale C, Tse-Dinh YC, Dunlap D, Croquette V, Bensimon D, Owen-Hughes T (2006) Direct observation of DNA distortion by the RSC complex. Mol Cell 21:417–425

Lorch Y, Davis B, Kornberg RD (2005) Chromatin remodeling by DNA bending, not twisting. Proc Natl Acad Sci U S A 102:1329–1332

Luger K (2006) Dynamic nucleosomes. Chromosome Res 14:5–16

Luger K, Hansen JC (2005) Nucleosome and chromatin fiber dynamics. Curr Opin Struct Biol 15:188–196

Luger K, Richmond TJ (1998) DNA binding within the nucleosome core. Curr Opin Struct Biol 8:33–40

MacCallum DE, Losada A, Kobayashi R, Hirano T (2002) ISWI remodeling complexes in Xenopus egg extracts: identification as major chromosomal components that are regulated by INCENP-aurora B. Mol Biol Cell 13:25–39

Medina PP, Carretero J, Fraga MF, Esteller M, Sidransky D, Sanchez-Cespedes M (2004) Genetic and epigenetic screening for gene alterations of the chromatin-remodeling factor, SMARCA4/BRG1, in lung tumors. Genes Chromosomes Cancer 41:170–177

Mellor J, Morillon A (2004) ISWI complexes in Saccharomyces cerevisiae. Biochim Biophys Acta 1677:100–112

Muchardt C, Reyes JC, Bourachot B, Leguoy E, Yaniv M (1996) The hbrm and BRG-1 proteins, components of the human SNF/SWI complex, are phosphorylated and excluded from the condensed chromosomes during mitosis. Embo J 15:3394–3402

Nie Z, Yan Z, Chen EH, Sechi S, Ling C, Zhou S, Xue Y, Yang D, Murray D, Kanakubo E *et al* (2003) Novel SWI/SNF chromatin-remodeling complexes contain a mixed-lineage leukemia chromosomal translocation partner. Mol Cell Biol 23:2942–2952

Peterson CL, Laniel MA (2004) Histones and histone modifications. Curr Biol 14:R546–551

Poot RA, Bozhenok L, van den Berg DL, Hawkes N, Varga-Weisz PD (2005) Chromatin remodeling by WSTF-ISWI at the replication site: opening a window of opportunity for epigenetic inheritance? Cell Cycle 4:543–546

Poot RA, Bozhenok L, van den Berg DL, Steffensen S, Ferreira F, Grimaldi M, Gilbert N, Ferreira J, Varga-Weisz PD (2004) The Williams syndrome transcription factor interacts with PCNA to target chromatin remodelling by ISWI to replication foci. Nat Cell Biol 6:1236–1244

Pray-Grant MG, Daniel JA, Schieltz D, Yates JR, 3rd, Grant PA (2005) Chd1 chromodomain links histone H3 methylation with SAGA- and SLIK-dependent acetylation. Nature 433:434–438

Rando OJ, Zhao K, Janmey P, Crabtree GR (2002) Phosphatidylinositol-dependent actin filament binding by the SWI/SNF-like BAF chromatin remodeling complex. Proc Natl Acad Sci U S A 99:2824–2829

Roberts CW, Orkin SH (2004) The SWI/SNF complex – chromatin and cancer. Nat Rev Cancer 4:133–142

Saha A, Wittmeyer J, Cairns BR (2002) Chromatin remodeling by RSC involves ATP-dependent DNA translocation. Genes Dev 16:2120–2134

Saha A, Wittmeyer J, Cairns BR (2005) Chromatin remodeling through directional DNA translocation from an internal nucleosomal site. Nat Struct Mol Biol 12:747–755

Santoro R, Grummt I (2005) Epigenetic mechanism of rRNA gene silencing: temporal order of NoRC-mediated histone modification, chromatin remodeling, and DNA methylation. Mol Cell Biol 25:2539–2546

Santos-Rosa H, Schneider R, Bernstein BE, Karabetsou N, Morillon A, Weise C, Schreiber SL, Mellor J, Kouzarides T (2003) Methylation of histone H3 K4 mediates association of the Isw1p ATPase with chromatin. Mol Cell 12:1325–1332

Schwanbeck R, Xiao H, Wu C (2004) Spatial contacts and nucleosome step movements induced by the NURF chromatin remodeling complex. J Biol Chem 279:39933–39941

Shaked H, Avivi-Ragolsky N, Levy AA (2006) Involvement of the Arabidopsis SWI2/SNF2 chromatin remodeling gene family in DNA damage response and recombination. Genetics

Shen X, Mizuguchi G, Hamiche A, Wu C (2000) A chromatin remodelling complex involved in transcription and DNA processing. Nature 406:541–544

Shen X, Xiao H, Ranallo R, Wu WH, Wu C (2003) Modulation of ATP-dependent chromatin-remodeling complexes by inositol polyphosphates. Science 299:112–114

Shi Y, Lan F, Matson C, Mulligan P, Whetstine JR, Cole PA, Casero RA (2004) Histone demethylation mediated by the nuclear amine oxidase homolog LSD1. Cell 119:941–953

Sif S, Stukenberg PT, Kirschner MW, Kingston RE (1998) Mitotic inactivation of a human SWI/SNF chromatin remodeling complex. Genes Dev 12:2842–2851

Sims RJ, 3rd, Chen CF, Santos-Rosa H, Kouzarides T, Patel SS, Reinberg D (2005) Human but not yeast CHD1 binds directly and selectively to histone H3 methylated at lysine 4 via its tandem chromodomains. J Biol Chem 280:41789–41792

Singleton MR, Wigley DB (2002) Modularity and specialization in superfamily 1 and 2 helicases. J Bacteriol 184:1819–1826

Smith CL, Peterson CL (2005) A conserved Swi2/Snf2 ATPase motif couples ATP hydrolysis to chromatin remodeling. Mol Cell Biol 25:5880–5892

Steger DJ, Haswell ES, Miller AL, Wente SR, O'Shea EK (2003) Regulation of chromatin remodeling by inositol polyphosphates. Science 299:114–116

Strohner R, Wachsmuth M, Dachauer K, Mazurkiewicz J, Hochstatter J, Rippe K, Langst G (2005) A 'loop recapture' mechanism for ACF-dependent nucleosome remodeling. Nat Struct Mol Biol 12:683–690

Thoma NH, Czyzewski BK, Alexeev AA, Mazin AV, Kowalczykowski SC, Pavletich NP (2005) Structure of the SWI2/SNF2 chromatin-remodeling domain of eukaryotic Rad54. Nat Struct Mol Biol 12:350–356

Tomschik M, Zheng H, van Holde K, Zlatanova J, Leuba SH (2005) Fast, long-range, reversible conformational fluctuations in nucleosomes revealed by single-pair fluorescence resonance energy transfer. Proc Natl Acad Sci USA 102:3278–3283

Tsukada Y, Fang J, Erdjument-Bromage H, Warren ME, Borchers CH, Tempst P, Zhang Y (2006) Histone demethylation by a family of JmjC domain-containing proteins. Nature 439:811–816

van Holde K, Yager T (2003) Models for chromatin remodeling: a critical comparison. Biochem Cell Biol 81:169–172

Varga-Weisz PD, Becker PB (2006) Regulation of higher-order chromatin structures by nucleosome-remodelling factors. Curr Opin Genet Dev 16:151–156

Whetstine JR, Nottke A, Lan F, Huarte M, Smolikov S, Chen Z, Spooner E, Li E, Zhang G, Colaiacovo M *et al* (2006) Reversal of histone lysine trimethylation by the JMJD2 family of histone demethylases. Cell 125:467–481

Whitehouse I, Stockdale C, Flaus A, Szczelkun MD, Owen-Hughes T (2003) Evidence for DNA translocation by the ISWI chromatin-remodeling enzyme. Mol Cell Biol 23:1935–1945

Wong AK, Shanahan F, Chen Y, Lian L, Ha P, Hendricks K, Ghaffari S, Iliev D, Penn B, Woodland AM *et al* (2000) BRG1, a component of the SWI-SNF complex, is mutated in multiple human tumor cell lines. Cancer Res 60:6171–6177

Xi R, Xie T (2005) Stem cell self-renewal controlled by chromatin remodeling factors. Science 310:1487–1489

Xue Y, Gibbons R, Yan Z, Yang D, McDowell TL, Sechi S, Qin J, Zhou S, Higgs D, Wang W (2003) The ATRX syndrome protein forms a chromatin-remodeling complex with Daxx and localizes in promyelocytic leukemia nuclear bodies. Proc Natl Acad Sci U S A 100:10635–10640

Zhou J, Chau CM, Deng Z, Shiekhattar R, Spindler MP, Schepers A, Lieberman PM (2005) Cell cycle regulation of chromatin at an origin of DNA replication. Embo J 24:1406–1417

Zofall M, Persinger J, Kassabov SR, Bartholomew B (2006) Chromatin remodeling by ISW2 and SWI/SNF requires DNA translocation inside the nucleosome. Nat Struct Mol Biol 13:339–346

CHAPTER 3

REGULATION OF CHROMATIN STRUCTURE AND CHROMATIN-DEPENDENT TRANSCRIPTION BY POLY(ADP-RIBOSE) POLYMERASE-1

Possible targets for drug-based therapies

DAVID A. WACKER[1], KRISTINE M. FRIZZELL[1], TONG ZHANG[1] AND W. LEE KRAUS[1,2,*]

[1] *Department of Molecular Biology and Genetics, Cornell University, 465 Biotechnology Building, Ithaca, NY 14853*

[2] *Department of Pharmacology, Weill Medical College of Cornell University, New York, NY 10021*

Abstract: Poly(ADP-Ribose) Polymerase-1 (PARP-1) is the prototypical and most abundantly expressed member of a family of PARPs that catalyze the polymerization of ADP-ribose (ADPR) units from donor NAD^+ molecules on target proteins. PARP-1 plays roles in a variety of genomic processes, including the regulation of chromatin structure and transcription in response to specific cellular signals. PARP-1 also plays important roles in many stress-induced disease states. In this chapter, we review the molecular and cellular aspects of PARP-1's chromatin-modulating activities, as well as the impact that these chromatin-modulating activities have on the regulation of gene expression. In addition, we highlight the potential therapeutic use of drugs that target PARP-1's enzymatic activity for the treatment of human diseases

Abbreviations: 3AB: 3-aminobenzamide; ADPR: ADP-ribose; AMD: Automodification domain; BRCT: BRCA1 C-terminus-like; DBD: DNA-binding domain; dPARP: *Drosophila* PARP; MPTP: 1-methyl-4-phenyl-1,2,3,6-tetrahydropyridine; NAD^+: Nicotinamide adenine dinucleotide; NAm: Nicotinamide; NLS: Nuclear localization signal; OAADPR: O-acetyl-ADP-ribose; PAR: Poly(ADP-ribose); PARG: Poly(ADP-ribose) glycohydrolase; PARP: Poly(ADP-ribose) polymerase; PARylation: poly(ADP-ribosyl)ation

[*]E-mail: wlk5@cornell.edu

T.K. Kundu and D. Dasgupta (eds.), Chromatin and Disease, 45–69.

1. OVERVIEW OF THE BIOLOGY OF PARP-1 AND PAR

1.1. PARP-1: A Multifunctional Nuclear Protein Involved in the Regulation of Chromatin Structure and Transcription

Poly(ADP-ribose) polymerase-1 (PARP-1) is a ubiquitous and abundant (~1 to 2 million molecules per cell) nuclear protein involved in a variety of nuclear processes. Although PARP-1 has historically been studied as a DNA damage detection and repair protein, more recent studies have demonstrated a clear role for PARP-1 under non-pathological cell states. For example, PARP-1 has been shown to function as a sequence-specific DNA-binding transcriptional regulator, a DNA-binding-independent transcriptional coregulator, and a modulator of chromatin structure (Table 1) (Kraus and Lis, 2003; Kim *et al.*, 2005). The role of PARP-1 as a modulator of chromatin structure, as well as its chromatin-dependent effects on gene expression, will be the focus of this review. The reader is directed to a number of available reviews for more information on other transcription-related, as well as non-transcription-related, functions of PARP-1 (D'Amours *et al.*, 1999; Burkle, 2001, 2005; Smith, 2001; Kraus and Lis, 2003; Amé *et al.*, 2004; Diefenbach and Burkle, 2005; Kim *et al.*, 2005).

1.2. PARP-1 Structure and Function

PARP-1 is the founding and most abundantly expressed member of the PARP family, which contains as many as 18 distinct proteins in humans (Amé *et al.*, 2004). PARP-1 and its enzymatically active paralogs (*e.g.*, PARP-2, tankyrases) catalyze the polymerization of ADP-ribose (ADPR) units from donor NAD^+ molecules on target proteins through a process called poly(ADP-ribosyl)ation (or PARylation) (Fig. 1a, b) (D'Amours *et al.*, 1999). PARP family members share a conserved catalytic domain, including a highly conserved ~50 amino acid sequence called the "PARP signature" motif, which defines the PARP family. In addition, PARP family members typically contain one or more additional motifs or domains [*e.g.*, DNA-binding domains, "BRCA1 C-terminus-like" (BRCT) motifs, ankyrin repeats, macro domains, WWE domains] that "functionalize" the PARP proteins and dictate their unique activities. For example, PARP-1 contains (1) an amino-terminal DNA-binding domain (DBD) with two zinc finger motifs, (2) a nuclear localization signal

Table 1. Molecular and cellular functions of PARP-1

- DNA damage detection and repair
- Cell death pathways: apoptosis and necrosis
- Chromatin modulation
- Transcriptional regulation
- Insulator function
- Mitotic apparatus function

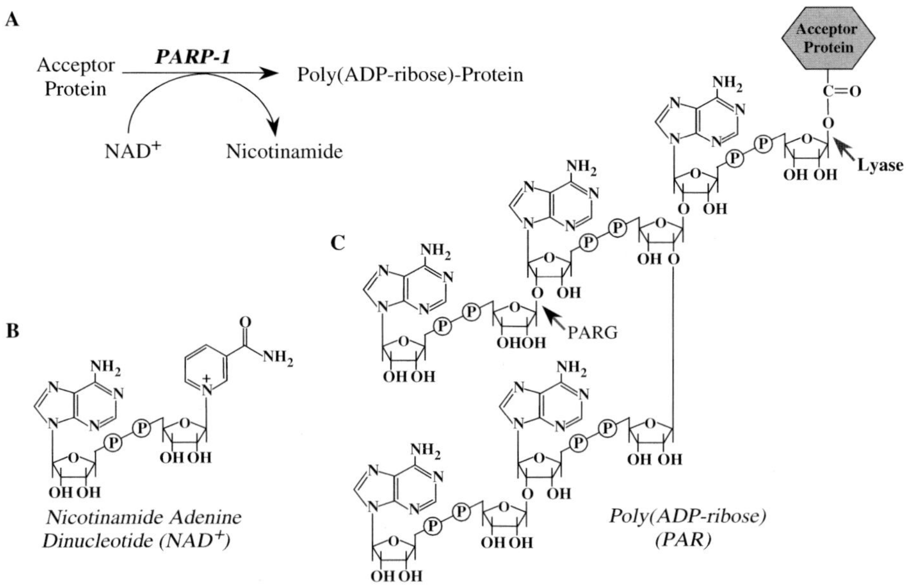

Figure 1. Synthesis of PAR from NAD^+ on target proteins by PARP-1. (a) PARP-1 catalyzes the polymerization of ADP-ribose units from donor NAD^+ molecules on target proteins, resulting in the attachment of PAR and the release of nicotinamide. (b) The chemical structure of NAD^+. (c) The chemical structure of PAR. PAR is a branched polymer synthesized by PARPs on acceptor proteins. The ADPR units in the ADPR chain are linked by $1'' \Rightarrow 2'$ ribose–ribose glycosidic bonds, whereas the ADPR units at the branch points are linked by $1''' \Rightarrow 2''$ ribose–ribose glycosidic bonds. The degradation of PAR is catalyzed by PARG, an enzyme with both exoglycosidase and endoglycosidase activities, that hydrolyzes the glycosidic linkages between the ADPR units of PAR (see red arrow for one example of a cleavage site). The remaining protein-proximal ADPR monomers are removed by ADP-ribosyl protein lyase (blue arrow)

(NLS), (3) a central automodification domain (AMD) with a BRCT motif, and (4) a carboxyl-terminal PARP catalytic domain (Fig. 2) (D'Amours *et al.*, 1999; Amé *et al.*, 2004).

PARP-1 binds through its DBD to a variety of DNA structures, including single and double strand breaks, crossovers, cruciforms, supercoils, and specific double-stranded DNA sequences (Rolli *et al.*, 2000). In the absence of an allosteric activator, PARP-1's basal enzymatic activity is very low. In contrast, in the presence of an allosteric activator, such as damaged DNA, some undamaged DNA structures, nucleosomes, and a variety of protein binding partners, PARP-1's enzymatic activity can be stimulated as much as 500-fold (D'Amours *et al.*, 1999; Oei and Shi, 2001; Kun *et al.*, 2002, 2004; Kim *et al.*, 2004). The targets of PARP-1's enzymatic activity include PARP-1 itself, which is the primary target *in vivo*, as well as a variety of secondary targets such as core histones, the linker histone H1, and a variety of transcription-related factors that interact with PARP-1 (Ogata *et al.*, 1981; Huletsky *et al.*, 1989; D'Amours *et al.*, 1999; Kraus and Lis, 2003).

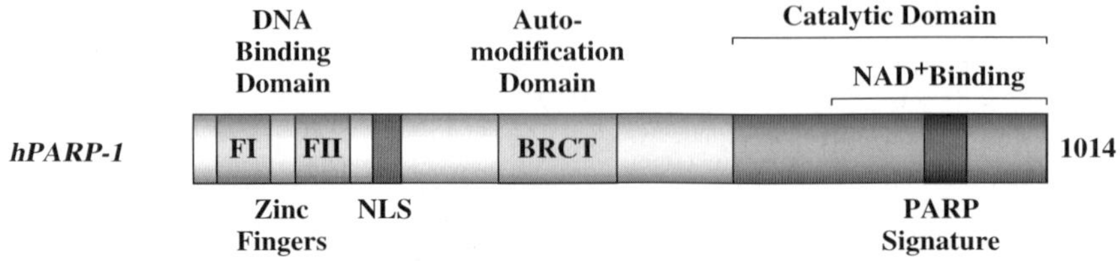

Figure 2. Structural and functional domains of PARP-1. PARP-1 has a highly conserved structural and functional organization including: (1) an N-terminal DNA binding domain with two Cys-Cys-His-Cys zinc finger motifs (FI and FII), (2) a nuclear localization signal (NLS), (3) a central automodification domain containing a BRCT ("BRCA1 C-terminus-like") protein-protein interaction motif, and (4) a C-terminal catalytic domain with a contiguous 50 amino acid sequence, the "PARP signature" motif, that forms the active site

Automodification of PARP-1 occurs in the AMD, which contains a number of glutamate residues that are the likely targets for PAR attachment, as well as a BRCT motif that is thought to function in protein-protein interactions (D'Amours *et al.*, 1999; Rolli *et al.*, 2000). AutoPARylated PARP-1 exhibits dramatically reduced DNA-binding activity and, hence, altered activities in a number of cellular endpoints. Together, the various biochemical activities of PARP-1 make it ideally suited as a regulator of a variety of signal-dependent nuclear processes.

1.3. Effects of PARylation on Target Proteins

Poly(ADP-ribose) (PAR) is a heterogenous branched polymer of repeating ADPR units linked via glycosidic ribose-ribose $1'' \Rightarrow 2'$ bonds in the linear chain and glycosidic ribose-ribose $1''' \Rightarrow 2''$ bonds at the branchpoints (Fig. 1c) (D'Amours *et al.*, 1999; Kim *et al.*, 2005). A single PAR molecule may contain as many as 200 ADPR units and has approximately one branch per 20 to 50 ADPR units. Based on its chemical similarity to DNA and RNA, PAR has been referred to as the "third type of nucleic acid" (D'Amours *et al.*, 1999). Each ADPR residue in PAR contains an adenine moiety capable of base stacking and hydrogen bonding, as well as two negatively charged phosphate groups (Amé *et al.*, 2000). Thus, PAR may form definitive structures through intramolecular interactions (Minaga and Kun, 1983a, b), and these structures have the potential for non-covalent attractive or repulsive interactions with other molecules (Mathis and Althaus, 1987; Wesierska-Gadek and Sauermann, 1988; Panzeter *et al.*, 1992).

PAR may alter the activity of PARylated proteins by functioning as a site-specific covalent modification or a steric blocker. For example, inhibition of PARP-1's DNA binding activity by autoPARylation may be the result of charge repulsion between PAR and DNA, or steric effects of PAR that mask PARP-1's DBD (D'Amours *et al.*, 1999). Furthermore, PAR may act as a protein binding matrix for a variety of nuclear proteins. In this regard, proteomic approaches have been used to identify a 20 amino acid PAR binding motif in a heterogeneous group of PAR-binding proteins, including core histones, p53, and XRCC-1 (Pleschke *et al.*, 2000;

Gagne *et al.*, 2003). Some non-sequence-specific DNA-binding proteins, such as the linker histone H1, may even have a greater affinity for PAR than DNA (Malanga *et al.*, 1998). Thus, an understanding of PAR is an important aspect of understanding the biology of PARP-1 and other PARPs.

1.4. PAR Catabolism by PARG

The steady-state levels of PAR *in vivo* are regulated by the opposing actions of PARPs and poly(ADP-ribose) glycohydrolase (PARG). PARG is an enzyme with both exo- and endoglycosidase activities that hydrolyzes the glycosidic linkages between the ADP-ribose units of PAR producing free ADP-ribose (Amé *et al.*, 2000; Davidovic *et al.*, 2001) (Fig. 1c, see red arrow). Although PARP-1 may be present at a 5- to 20-fold molar excess over PARG in some cell types, PARG has a higher specific activity than PARP-1 and its enzymatic activity increases with increased PAR length (D'Amours *et al.*, 1999). Furthermore, nucleo-cytoplasmic shuttling of the PARG protein may modulate the level of nuclear PARG activity (Bonicalzi *et al.*, 2003; Ohashi *et al.*, 2003). Ultimately, the activity of PARP-1 is intimately tied to the regulatory actions of PARG, yet this is an area of research that has not been explored in sufficient detail. Where appropriate below, we have described results that tie the actions of PARG to the biological function of PARP-1. For more details about the catabolism of PAR by PARG, as well as the interplay between PARP-1 and PARG, the reader is directed to the additional references listed in the bibliography (*e.g.*, Davidovic *et al.*, 2001).

2. REGULATION OF CHROMATIN STRUCTURE AND CHROMATIN-DEPENDENT TRANSCRIPTION BY PARP-1 AND PAR

The role of PARP-1 as a modulator of chromatin structure, as well as its chromatin-dependent effects on transcription, has been well established in the literature (D'Amours *et al.*, 1999; Kraus and Lis, 2003; Rouleau *et al.*, 2004; Kim *et al.*, 2005). In this section, we will highlight some of the historical studies, as well as more recent studies, that have characterized the role of PARP-1 and PAR in these processes. Note that, when describing this work, the term "PARP" in the absence of any additional identifiers refers to PAR-generating proteins of ambiguous identity (*e.g.*, as in early studies, which were conducted without the knowledge that multiple PARP family members exist, or when multiple splice variants of a PARP-1-like protein exist in a single organism, as in *Drosophila*).

2.1. Early Studies Examining the Effects of PARylation on Chromatin Structure

In the past three decades, numerous biochemical studies have shown that PARP enzymatic activity can dramatically alter chromatin structure and nucleosome stability (D'Amours *et al.*, 1999; Kraus and Lis, 2003; Rouleau *et al.*, 2004; Kim *et al.*, 2005). Poirier *et al.* (1982) demonstrated by using electron microscopy that native chromatin incubated in the presence of PARP-1 and NAD^+ has a more open conformation than chromatin incubated with PARP-1 in the absence of NAD^+. This work was later confirmed and extended by de Murcia *et al.* (1986), who demonstrated rapid kinetics for NAD^+-dependent decondensation of chromatin by PARP-1 (*i.e.*, beginning immediately and peaking between 30 and 60 minutes). In addition, Huletsky *et al.* (1989) demonstrated that PARylation of linker histone H1-depleted chromatin prevents recondensation by exogenously added H1, despite binding of the added H1. The specific role of PARylation in these PARP-induced chromatin changes is evident from studies showing that PARG activity can reverse the PARP-dependent effects (de Murcia *et al.*, 1986; Realini and Althaus, 1992). Collectively, the results from these studies indicate that PARP enzymatic activity promotes chromatin decondensation. This conclusion has been well supported by additional biochemical studies using gradient sedimentation assays (de Murcia *et al.*, 1986; Huletsky *et al.*, 1989), epitope accessibility assays (Huletsky *et al.*, 1989; Thibeault *et al.*, 1992), and *in vitro* reconstitutions (Kim *et al.*, 2004; see below), as well as a variety of *in vivo* assays (Tulin *et al.*, 2002, 2006; Tulin and Spradling, 2003; see below).

2.2. Histones and Chromatin-Associated Proteins as Targets for PARylation

While the precise mechanisms for the aforementioned PARP-mediated effects on chromatin structure are unclear, several clues have been provided from the literature. First, multiple chromatin proteins are targets for PARylation, including histone proteins (Krupitza and Cerutti, 1989), especially histones H1 and H2B (Adamietz and Rudolph, 1984), some high-mobility group proteins (Tanuma and Johnson, 1983), and other DNA-binding proteins such as DNA ligases and topoisomerases (D'Amours *et al.*, 1999). PARylation may alter the activity of these proteins, thereby changing the way they interact with chromatin and alter chromatin structure. For example, PARylation of H1, a key target of PARP-1 activity *in vivo*, may block H1 binding to nucleosomes and inhibit chromatin compaction by H1 (de Murcia *et al.*, 1986) (Fig. 3a). This may be accomplished either through steric hindrance caused by the PAR chain, or electrostatic repulsion between the negatively charged DNA and PAR. Likewise, PARylation of core histones may disrupt histone-DNA contacts and destabilize nucleosomes (Mathis and Althaus, 1987; Realini and Althaus, 1992). In addition to these direct effects on target proteins, PAR chains may also form an anionic scaffold that can bind displaced histones and other chromatin proteins (Mathis and Althaus, 1987; Realini and Althaus, 1992) (Fig. 3b). Removing these proteins from the DNA to a PAR scaffold would allow access for transcription, replication, and repair enzymes, as well as increase accessibility of the chromatin proteins to enzymes that could covalently modify them (Tulin and Spradling, 2003). While these mechanisms are likely to play a role in PARP-mediated alteration of

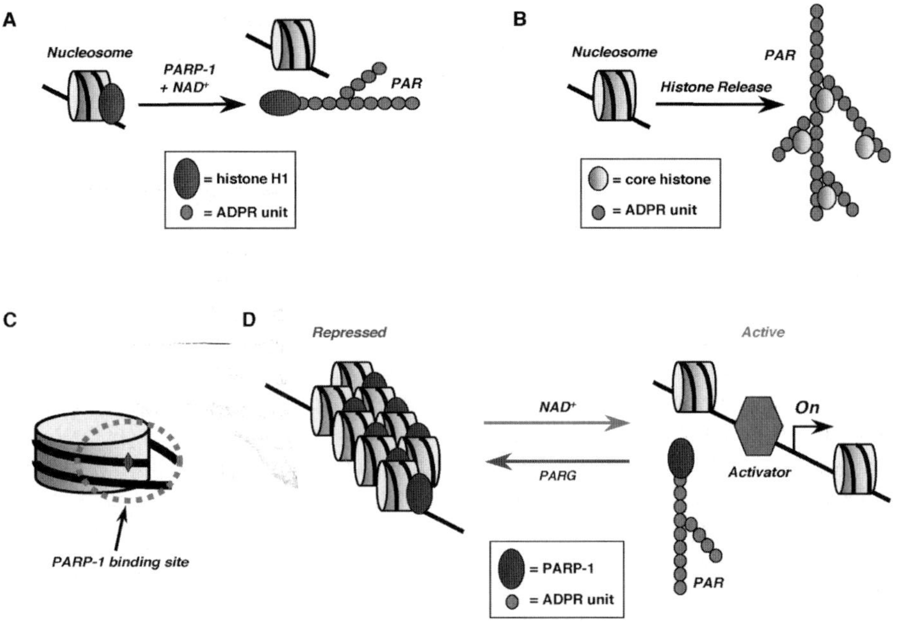

Figure 3. Possible mechanisms for the modulation of chromatin structure by PARP-1 and PAR. Multiple models for the modulation of chromatin structure by PARP-1 and PAR are shown. (a) A number of chromatin proteins are targets for PARylation by PARP-1, including the linker histone H1. PARylation may alter the interaction of these proteins with chromatin (as shown for H1 in this panel) and, as a result, alter chromatin structure. (b) PAR chains may form an attractive anionic scaffold that can bind displaced histones and other chromatin proteins. Removing these proteins from the DNA to a PAR scaffold would allow access for transcription, replication, and repair enzymes. (c) PARP-1 is a nucleosome-binding protein that binds at or near the dyad axis of the nucleosome (blue diamond) and contacts the linker DNA where it exits the nucleosome. (d) PARP-1, through its nucleosome-binding activity, functions as a structural component of chromatin and promotes the compaction of chromatin into transcriptionally repressed structures. Upon autoPARylation in the presence of NAD^+, PARP-1 is released from chromatin and the compaction is reversed, promoting the formation of transcriptionally active chromatin structures and allowing activator binding. PARG cleaves the PAR chains from PARP-1, allowing PARP-1 to re-bind to the nucleosomes. (See Colour Plate 5.)

chromatin structure, more recent discoveries highlighted below have elucidated new functions for PARP-1 and suggest new mechanisms for PARP-1's influence on chromatin structure.

2.3. The Role of PARP-1 as a Specific Nucleosome-Binding Factor: Effects on Chromatin Compaction and Chromatin-dependent Transcription

In addition to its role in the covalent modification of chromatin proteins, PARP-1 also functions as a nucleosome-binding protein and, hence, a structural component of chromatin. A recent study from our lab using an *in vitro* chromatin reconstitution system demonstrated a specific nucleosome-binding activity of PARP-1, which resembles the nucleosome-binding activity of the linker histone H1 (Kim *et al.*, 2004). Like H1, PARP-1 binds to nucleosomes at or near the dyad axis and contacts the linker DNA where it exits the nucleosome (Fig. 3c). Through this binding, PARP-1 acts as a structural component of chromatin and promotes the compaction of nucleosomal arrays into higher-order structures (Fig. 3d). In accordance with this role, PARP-1 binding protects *in vitro* assembled chromatin arrays from nuclease digestion and increases their sedimentation velocity in glycerol gradients, both of which are indicative of a change in the chromatin to a more condensed or compacted form. Imaging by atomic force microscopy has confirmed these results (Fig. 4). Furthermore, the length of nucleosomal DNA protected in limiting micrococcal nuclease digests (*i.e.*, the nucleosomal "repeat length") increases in the presence of PARP-1, much like it does in the presence of H1, indicating that PARP-1 binding protects the nucleosomal DNA where it exits the nucleosome (Kim *et al.*, 2004). Interestingly, the addition of NAD^+ to PARP-1-associated chromatin arrays can reverse these effects by promoting the autoPARylation of PARP-1 and the dissociation of PARP-1 from chromatin.

The enzymatic activity of PARP-1 is not required for its binding to nucleosomes, but it is required for NAD^+-dependent release and, hence, dynamic NAD^+-modulated association/dissociation (Fig. 3d) (Kim *et al.*, 2004). This suggests an exciting difference between PARP-1 and other nucleosome-binding proteins, such as the linker histone H1. While the binding of either induces chromatin compaction, PARP-1 is unique in that it can facilitate its own release via its enzymatic activity. PARP-1's primary enzymatic substrate *in vivo* is itself (Ogata *et al.*, 1981), and autoPARylation of PARP-1 blocks its DNA-binding activity (Ferro and Olivera, 1982). In this regard, nucleosomes themselves are potent stimulators of PARP-1 enzymatic activity (Kim *et al.*, 2004). AutoPARylation causes PARP-1 to dissociate from nucleosomes and the chromatin to relax or decompact to a more open conformation. These effects are reversed by PARG, which cleaves the PAR chains from PARP-1 and allows it to bind again to nucleosomes (Kim *et al.*, 2004) (Fig. 3d). This mechanism for the NAD^+-dependent modulation of chromatin structure is consistent with previous biochemical assays showing that the formation and maintenance of PAR chains is critical for maintaining an open chromatin conformation (de Murcia *et al.*, 1986; Realini and Althaus, 1992).

A system such as this can provide a powerful mechanism for transcriptional control. While bound to nucleosomes, PARP-1 promotes a transcriptionally repressed state, but one that is simultaneously poised for activation because of the allosteric activating effect of nucleosomes on PARP-1 enzymatic activity. In the presence of NAD^+, PARP-1 can autoPARylate and release from nucleosomes, shifting the chromatin to a more transcriptionally active conformation. PARG can reset the system by cleaving the PAR chains from PARP-1, allowing PARP-1 to re-bind the nucleosomes and re-establish a transcriptionally repressed state.

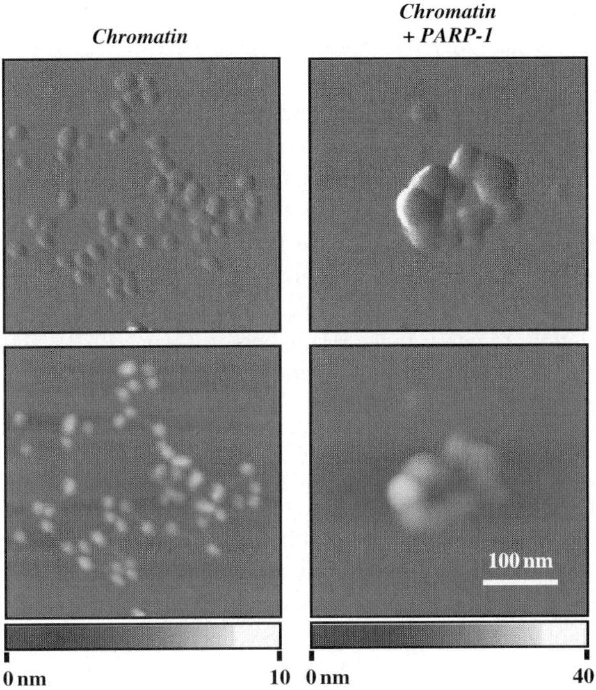

Figure 4. Visualization of PARP-1-mediated chromatin compaction by atomic force microscopy. Chromatin assembled *in vitro* on a circular ~10.5 kb plasmid DNA was purified, incubated with or without recombinant human PARP-1, and imaged by atomic force microscopy. Two types of images are shown: scan probe oscillation amplitude *(top)* and topography (*bottom*; height scale is indicated). The length scale is indicated. (See Colour Plate 6.)

As expected, *in vitro* transcription assays involving PARP-1, NAD^+, and PARG illustrate these predicted outcomes (Kim *et al.*, 2004). Even when driven by a transcriptional activator, such as estradiol-bound estrogen receptor, transcription is repressed when PARP-1 is added to chromatin templates. The repression is reversed by NAD^+, and the NAD^+-dependent effects are reversed by PARG (Kim *et al.*, 2004). This system for transcriptional control shifts new importance onto the enzymes responsible for synthesis of NAD^+ in the nucleus, such as nicotinamide mononucleotide adenylyltransferase-1 (Magni *et al.*, 2004). Because NAD^+ facilitates the decompaction of chromatin and the derepression of transcription, nuclear NAD^+ biosynthetic enzymes may play critical roles as cofactors.

2.4. The Role of PARP-1 as a Modulator of Chromatin Structure and Chromatin-dependent Transcription: *in vivo* Studies

In a recent series of studies, Tulin *et al.* (Tulin *et al.*, 2002, 2006; Tulin and Spradling, 2003) used *Drosophila* as a model system to study the role of PARP-1 in the modulation of chromatin structure *in vivo*. *Drosophila* is a tractable model for studying PARP-1 activity since it has only two PARP genes, PARP-1-like, which expresses three isoforms, and tankyrase-like (Hanai *et al.*, 1998; Miwa *et al.*, 1999). Using a *Drosophila* line devoid of PARP-1-like molecules (referred to as $Parp^{CH1}$), Tulin *et al.* demonstrate that loss of dPARP causes alterations in chromatin structure and transcription at some loci (Tulin *et al.*, 2002). Notably, transcription of *copia*, a multicopy retrotransposon, is increased by more than 50-fold in $Parp^{CH1}$ animals compared to those that express dPARP. This increase in transcription is accompanied by an increase in nuclease accessibility in the chromatin of the *copia* repeats and at other heterochromatic loci, but not in euchromatic regions. While it is not trivial to draw a direct connection, together these results suggest that dPARP acts to repress transcription at certain loci by promoting the formation or helping to maintain heterochromatin structures.

Tulin and Spradling also investigated the role of dPARP in chromatin structure and transcription by probing *Drosophila* polytene chromosomes, particularly the transcriptionally active "puffs" (or decondensed regions), for dPARP and PAR (Tulin and Spradling, 2003). They showed that although dPARP is broadly distributed on polytene chromosomes, PAR appears only at a limited number of loci. At signal-regulated loci known to puff in response to external stimuli, such as 75A and 75B (ecdysone-inducible) and 87A and 87C (heat shock-inducible), dPARP is present before and during puffing. In contrast, PAR accumulates only after puff formation and begins to decline just before regression of the puff (*i.e.*, recondensation). As might be expected, knockout (*i.e.*, $Parp^{CH1}$) or chemical inhibition (*i.e.*, 3-aminobenzidine) of dPARP prevents puffing at heat shock-induced loci and may block expression of the genes located in the puffs. Collectively, these results demonstrate a role for dPARP in the regulation of chromatin structure *in vivo*. Furthermore, they demonstrate a role for dPARP enzymatic activity in the NAD^+-dependent decondensation of dPARP-containing chromatin *in vivo*, consistent with the biochemical studies of Kim *et al.* described in the preceding section.

Our lab has also used *Drosophila* polytene chromosomes to examine the localization of dPARP, as well as the linker histone H1, on genomic chromatin. Although both PARP-1 and H1 show broad distribution on the chromosomes, they localize in patterns that are distinct from each other and from sites of active transcription (Kim *et al.*, 2004). These results are supported by mononucleosome-chromatin immunoprecipitation-Western blotting assays using human (*i.e.*, HeLa) cells, which demonstrate that PARP-1 and H1 occupy distinct nucleosomal fractions (Kim *et al.*, 2004). Together, these results suggest that although PARP-1 and H1 both localize to chromatin, they are directed to distinct chromatin domains or nucleosomal populations by distinct signals. Specific post-translational modifications of histone tails (*i.e.*, a "histone code"), which have been shown to direct factor binding to specifically modified nucleosomes (Bannister *et al.*, 2001; Jenuwein and Allis, 2001; Lachner *et al.*, 2001), may be one of these signals.

2.5. Connections between PARP-1 and SIRT1, Two NAD^+-Utilizing, Chromatin-Modulating Enzymes

Several recent studies have suggested interesting connections between PARP-1 and the NAD^+-dependent histone deacetylase SIRT1 (Zhang, 2003; Kim *et al.*, 2005). For example, PARP-1 and SIRT1 are both: (1) NAD^+-utilizing nuclear enzymes, (2) chromatin-modulating proteins, (3) potential substrates for each other's enzymatic activities, and (4) involved in NAD^+ metabolic pathways that generate potential ligands for the histone variant macroH2A (D'Amours *et al.*, 1999; Moazed, 2001; Parsons *et al.*, 2003; Karras *et al.*, 2005; Kim *et al.*, 2005; Kustatscher *et al.*, 2005). These similarities underlie a possible interplay between the two proteins that has the potential to control certain aspects of nuclear function, perhaps at the level of chromatin structure (Fig. 5).

Activity and function of SIRT1. SIRT1 is the mammalian homolog of yeast Sir2, the founding member of a family of NAD^+-dependent protein deacetylases (the sirtuins) conserved from prokaryotes to mammals (Blander and Guarente, 2004). Yeast Sir2 regulates chromatin-dependent silencing at the mating type loci, telomeres, and the rDNA locus (Sinclair and Guarente, 1997; Kaeberlein *et al.*, 1999), and deacetylation of histone substrates is critical for these

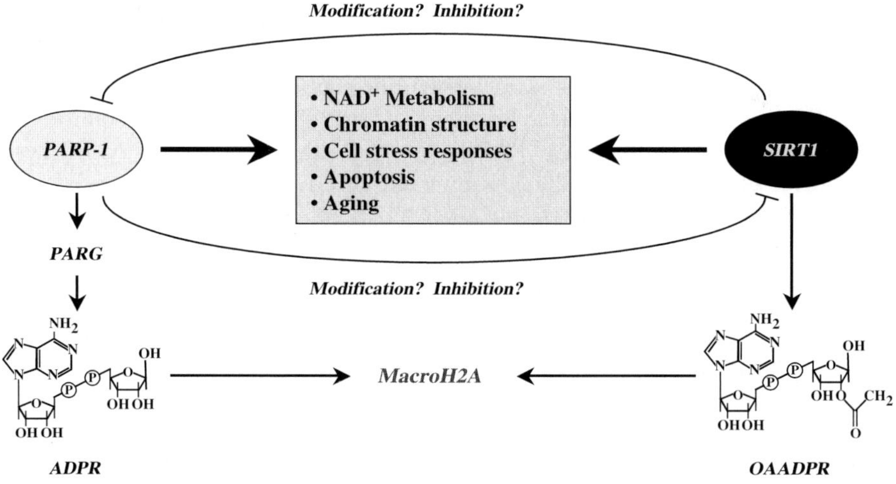

Figure 5. Interplay between PARP-1 and SIRT1, two NAD^+-utilizing, chromatin-modulating enzymes. PARP-1 and SIRT1 are both: (1) NAD^+-utilizing nuclear enzymes, (2) chromatin-modulating proteins, (3) potential substrates for each other's enzymatic activities, and (4) involved in NAD^+ metabolic pathways that generate potential ligands for the histone variant macroH2A. These similarities underlie a possible interplay between the two proteins that has the potential to control certain aspects of nuclear function, perhaps at the level of chromatin structure. In this regard, recent studies have suggested that increased PARP-1 activity in stressed cardiac myocytes leads to a depletion of NAD^+ levels and a reduction in SIRT1 deacetylase activity. Conversely, activation of SIRT1 by the small-molecule activator resveratrol leads to a decrease in PARP-1 activity in response to DNA damage

biological functions (Imai *et al.*, 2000; Blander and Guarente, 2004). Mammalian SIRT1 has been implicated in metabolic processes, apoptosis, and cellular stress responses (Blander and Guarente, 2004). The role of global histone deacetylation in the biological functions of mammalian SIRT1 is unclear, although localized histone deacetylation may be important in SIRT1 function. In addition, deacetylation of non-histone targets by mammalian SIRT1 plays important roles as well. For example, SIRT1 deacetylates and regulates the activity of a wide range of transcription factors and nuclear regulatory proteins (Blander and Guarente, 2004), including the p53 tumor suppressor (Luo *et al.*, 2001; Vaziri *et al.*, 2001), the FOXO transcription factors (Brunet *et al.*, 2004; Motta *et al.*, 2004) and the DNA repair and anti-apoptotic factor Ku70 (Cohen *et al.*, 2004), all of which are involved in cellular stress responses and organismal aging. Interestingly, p53 and Ku70 are also targets for modification and regulation by PARP-1 (Wesierska-Gadek *et al.*, 1996; Ruscetti *et al.*, 1998; Smith and Grosovsky, 1999; Li *et al.*, 2004). SIRT1-dependent deacetylation also represses PPAR-γ regulated fat storage (Picard *et al.*, 2004) and estrogen receptor α DNA binding (Kim *et al.*, 2006).

One area of intense research is the regulation of lifespan by Sir2/SIRT1 in a wide range of organisms. Sir2/SIRT1 overexpression extends lifespan in yeast (Kaeberlein *et al.*, 1999), *C. elegans* (Tissenbaum and Guarente, 2001) and *Drosophila* (Rogina and Helfand, 2004). Furthermore, Sir2 is required for caloric restriction (CR)-induced lifespan extension in yeast (Lin *et al.*, 2000) and *Drosophila* (Rogina and Helfand, 2004), and SIRT1 mediates CR-induced anti-apoptotic effects in cultured human cells (Cohen *et al.*, 2004). Direct activation of Sir2/SIRT1 by small molecules also extends lifespan (Howitz *et al.*, 2003; Wood *et al.*, 2004). Overall, the nuclear functions of Sir2/SIRT1 appear to control cell survival and aging. Interestingly, PARP-1 plays a role in the regulation of these same processes (Burkle *et al.*, 2005; Kim *et al.*, 2005).

Connections between PARP-1 and SIRT1. Recent studies have suggested an intriguing connection between the activities of PARP-1 and SIRT1 at the molecular level (Zhang, 2003). For example, Pillai *et al.* have examined the competition between SIRT1 and PARP-1 for NAD^+ stores, focusing on its consequences for myocyte cell survival (Pillai *et al.*, 2005). Their results indicate that increased PARP-1 activity in stressed cardiac myocytes leads to a depletion of NAD^+ levels and a reduction in SIRT1 deacetylase activity, conditions that promote death of the myocytes. Conversely, Kolthur-Seetharam *et al.* (2006) have shown that activation of SIRT1 by the small-molecule activator resveratrol leads to a decrease in PARP-1 activity in response to DNA damage. Furthermore, the loss of SIRT1 in SIRT1 knockout cells leads to a dramatic increase in PAR synthesis and ultimately results in AIF (apoptosis inducing factor)-mediated cell death. Collectively, these results suggest an important functional interplay between PARP-1 and SIRT1 in DNA damage and cell survival responses (Fig. 5).

The extent to which the chromatin-dependent activities of PARP-1 and SIRT1 are involved in the functional interplay between these two proteins has not yet been determined. However, the catalytic activity of both enzymes in the presence of

NAD^+ is likely to play a role. As shown by Kim *et al.*, autoPARylation blocks the nucleosome-binding activity of PARP-1, promoting the formation of decondensed, transcriptionally active chromatin (Kim *et al.*, 2004). In contrast, deacetylation of nucleosomal histones by SIRT1 might promote the formation of more condensed, transcriptionally inactive chromatin (Parsons *et al.*, 2003). Furthermore, ADPR and O-acetyl-ADPR (OAADPR), enzymatic products of PARP-1/PARG and SIRT1, respectively, bind to the macro domain of the histone variant macroH2A (Karras *et al.*, 2005; Kustatscher *et al.*, 2005). Although the functional consequences of ADPR and OAADPR binding to macroH2A have not been determined, it could provide a signal for the control of chromatin structure by PARP-1 and SIRT1 through macroH2A (Fig. 5). PARP-1 and SIRT1 may also directly regulate each other's activities through post-translational modification. PARP-1, which is acetylated by the p300 acetyltransferase, may be a target for deacetylation by SIRT1 (Hassa *et al.*, 2005). Likewise, SIRT1 may be a target for PARylation by PARP-1, at least in PARG-deficient *Drosophil*a (Tulin *et al.*, 2006). Collectively, these studies suggest multiple ways in which PARP-1 and SIRT1 may interact to modulate the structure and activity of chromatin.

3. ADDITIONAL ROLES FOR PAR AND PARPs IN THE NUCLEUS: TRANSCRIPTIONAL INSULATOR FUNCTION, NUCLEAR ARCHITECTURE, AND CHROMOSOME DYNAMICS

In addition to PAR and PARP-1's ability to modulate local chromatin structure, recent studies suggest that PAR and a variety of PARPs may play additional roles in the regulation of genomic function, including the regulation of insulator function, nuclear architecture, and chromosome dynamics. For example, Tulin *et al.* showed that a loss of dPARP function in *Drosophila* causes chromatin to appear more diffuse upon visualization by DAPI staining and inhibits the formation of nucleoli (Tulin *et al.*, 2002). These gross changes in chromatin and nuclear organization are suggestive of more global effects on chromatin-dependent functions.

PARP-1 and Insulators. A recent study from Yu *et al.* (2004) showed that PARylation regulates the activity of CTCF, a ubiquitous DNA binding protein, at transcriptional insulators. Insulators are elements that organize the genomic chromatin into discrete regulatory domains by limiting the actions of enhancers and silencers through a "positional blocking" mechanism (Bell *et al.*, 2001). Yu *et al.* showed that the function of the insulator contained within the imprinting control region (ICR) of the *Igf2-H19* locus, an element that binds CTCF and limits the extent of expression in a parent-of-origin-specific manner, is dependent on PARP activity. The presence of PAR at the ICR was only detected on the maternally inherited allele and was dependent on functional CTCF binding sites. PAR and CTCF were found at nearly 80 percent of the ~180 CTCF target sites examined by chromatin immunoprecipitation assays. Although CTCF was not directly shown to be the PARylated protein at these sites, CTCF was shown to be PARylated by

PARP-1 *in vitro*. The insulator function of most of the same CTCF target sites was sensitive to the general PARP inhibitor 3-aminobenzamide. Collectively, these results support a role for PAR and PARP(s) in the activity of insulators, with an insulator-bound protein, possibly CTCF, as a target for PARylation.

PARP-1 and the nuclear matrix. Several studies have found PARP-1 at regions of genomic DNA responsible for connection to the nuclear matrix. For example, Yanagisawa *et al.* observed a ~114 kDal protein (later identified as PARP-1; [Galande and Kohwi-Shigematsu, 1999]) associated with matrix attachment regions in breast carcinoma samples (Yanagisawa *et al.*, 1996). This association was specific to the breast carcinoma samples, however, and was not observed in healthy breast tissue, or breast tissue with benign disease. PARP-1 has also been detected in association with the nuclear matrix (Vidakovic *et al.*, 2004), and components of the nuclear matrix are targets for PARylation (Cardenas-Corona *et al.*, 1987). Further studies will be needed to understand the functional consequences of these associations.

PARP-1 and the mitotic apparatus. A number of studies have suggested a role for PARP-1 and other PARP family members in mitotic apparatus function. PARP-1 and at least five other PARPs (*i.e.*, PARP-2, PARP-3, VPARP, tankyrase 1 and tankyrase 2) are associated with components of the mitotic apparatus, the cellular machinery required for accurate chromosome segregation during cell division (Smith, 2001; Amé *et al.*, 2004). For example, both PARP-1 and PARP-2 localize to mitotic centromeres, the chromosomal regions where kinetochores form to capture microtubules from the mitotic spindle (Earle *et al.*, 2000; Saxena *et al.*, 2002a, b). Several PARPs, including PARP-1, PARP-3 and tankyrases, are associated with centrosomes, the cellular microtubule organizing center that functions as the spindle pole during mitosis (Smith and de Lange, 1999; Kaminker *et al.*, 2001; Augustin *et al.*, 2003; Kanai *et al.*, 2003). Furthermore, VPARP and PARG localize to the mitotic spindle (Kickhoefer *et al.*, 1999; Chang *et al.*, 2004). Recent studies by Chang *et al.* suggest that increased production of PAR during mitosis plays an essential role in the assembly and structure of bipolar spindles (Chang *et al.*, 2004). These studies implicate PAR and PAR-metabolizing enzymes in the function of the mitotic apparatus, although their specific roles have not yet been determined.

4. PARP-1 AS A THERAPEUTIC TARGET IN DISEASE

4.1. Role of PARP-1 in Physiological and Pathophysiological Processes: Genome Maintenance, Cell Death, and Inflammation

PARP-1 has been implicated in a variety of physiological and pathophysiological processes, including genome maintenance, carcinogenesis, aging, immunity, inflammation, and neurological function (Kim *et al.*, 2005) (Table 2). PARP-1's roles in DNA repair, cell death pathways, and pro-inflammatory gene expression underlie many of the contributions PARP-1 makes to the aforementioned processes and, as such, have been the focus for therapeutic targeting. Less clear, however, is

Table 2. Physiological and pathophysiological processes involving PARP-1

• Genome maintenance
• Carcinogenesis
• Cell death
• Aging
• Inflammatory responses
• Neuronal function

the connection between PARP-1's role in regulating chromatin structure and its contribution to these physiological and pathophysiological processes.

DNA damage detection and repair. PARP-1 has a well-characterized role in DNA damage detection and repair. The binding of PARP-1 to damaged DNA (*e.g.*, single strand and double strand breaks) potently activates PARP-1 enzymatic activity (D'Amours *et al.*, 1999), allowing PARP-1 to function as a DNA damage sensor. Protein PARylation by PARP-1 is an immediate response to DNA damage induced by oxidation, alkylation, and ionizing radiation. With low levels of DNA damage, PARP-1 acts as a survival factor that can promote the repair of damaged DNA (Fig. 6a). PARP-1 has been implicated in multiple DNA repair pathways, including the single strand break, double strand break, and base excision repair pathways (Burkle, 2001; Masutani *et al.*, 2003), making PARP-1 a key regulator of genomic integrity. In this regard, Parp-1$^{-/-}$ mice or Parp-1$^{-/-}$ embryonic fibroblasts exhibit defective single strand break repair, as well as increased homologous recombination, sister chromatid exchange, and micronuclei formation (de Murcia *et al.*, 1997; Wang *et al.*, 1997; D'Amours *et al.*, 1999; Burkle, 2001). In addition, PARP-1 deficiency causes increased deletion mutations and insertions/rearrangements after treatment with alkylating agents (Shibata *et al.*, 2005). Thus, as expected, Parp-1$^{-/-}$ mice show increased tumor formation in a number of chemically-induced and transgenic cancer models (Masutani *et al.*, 2005), suggesting a role for PARP-1 in preventing tumor formation in response to DNA damage.

Cell death pathways. In contrast to its role as a survival factor in response to low levels of DNA damage, PARP-1 acts to promote cell death under conditions of extensive DNA damage (Burkle *et al.*, 2005) (Fig. 6a). Thus, chemical inhibition or genetic ablation of PARP-1 can provide protection from a number of DNA damage-dependent pathophysiological conditions leading to cell death, including (1) ischemia-reperfusion injury, (2) glutamate excitotoxicity in the central nervous system, (3) 1-methyl-4-phenyl-1,2,3,6-tetrahydropyridine (MPTP)-induced Parkinsonism, (4) cardiac infarction, (5) inflammatory injury, and (6) streptozotocin-induced diabetes (Szabo, 1998; Mandir *et al.*, 1999; Shall and de Murcia, 2000; Beneke *et al.*, 2004). PARP-1 plays a role in both necrotic and apoptotic cell death pathways. In the necrotic pathway, hypersynthesis of PAR by PARP-1 in response to extensive DNA damage causes the depletion of cellular NAD^+ and ATP, subsequent cellular energy failure, and ultimately cell death (Decker and Muller, 2002; Bouchard *et al.*, 2003). In contrast, PARP-1 can promote a

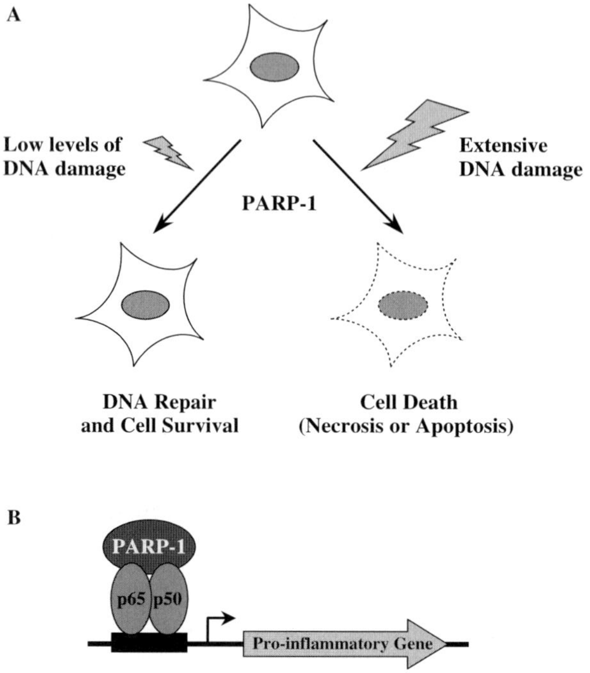

Figure 6. PARP-1 plays important roles in genome maintenance, cell death, and inflammatory responses. PARP-1's roles in DNA repair, cell death pathways, and pro-inflammatory gene expression underlie many of the contributions of PARP-1 to various disease states. (a) PARP-1 has well-characterized roles in DNA repair and cell death. With low levels of DNA damage, PARP-1 acts as a survival factor that can promote the repair of damaged DNA. Under conditions of extensive DNA damage, PARP-1 acts to promote cell death through necrotic and apoptotic pathways. (b) PARP-1 has been implicated in pathophysiological inflammatory responses through its role as a coactivator of transcription factors that regulate immune and inflammatory response genes (*e.g.*, NF-κ B and AP-1). (See Colour Plate 7.)

caspase-independent apoptotic pathway through apoptosis inducing factor (AIF), a proapoptotic flavoprotein that is among the most powerful triggers of apoptosis (Chiarugi and Moskowitz, 2002; Yu *et al.*, 2002). The mechanisms underlying the choice between necrosis and apoptosis in response to genotoxic stimuli are unclear, but may be influenced by the type, strength, and duration of the stimuli, as well as the cell type (Virag, 2005).

Inflammatory responses. In addition to DNA repair and cell death pathways, PARP-1 has also been implicated in pathophysiological inflammatory responses, in which it plays two roles. First, as noted above, PARP-1 induces necrotic cell death in response to genotoxins. This leads to the release of intracellular components into the surrounding tissue, causing tissue damage and promoting inflammation. Second, PARP-1 functions as a coactivator of transcription factors that regulate immune and inflammatory response genes (*e.g.*, NF-κ B and AP-1) (Oliver *et al.*, 1999; Hassa and Hottiger, 2002; Andreone *et al.*, 2003) (Fig. 6b). In accordance with these observations, *Parp-1*$^{-/-}$ mice show resistance to inflammation, including streptozotocin-induced diabetes and lipopolysaccharide-induced septic shock (Oliver *et al.*, 1999; Mabley *et al.*, 2001; Hassa and Hottiger, 2002). Furthermore, NF-κB- and AP-1-regulated pro-inflammatory genes are down regulated in *Parp-1*$^{-/-}$ mice or *Parp-1*$^{-/-}$ immune cells (Zingarelli *et al.*, 1998; Shall and de Murcia, 2000; Hassa and Hottiger, 2002) and up-regulated in immune cells treated with a PARG inhibitor (Rapizzi *et al.*, 2004).

How does PARP-1's role as a nucleosome-binding protein and modulator of chromatin structure, which is evident under normal physiological conditions, impact PARP-1-dependent DNA repair, cell death, and inflammatory response pathways, which occur under pathophysiological conditions? A number of different scenarios are possible. For example, PARP-1's chromatin-dependent activities may be critical for its function as a DNA repair protein, since the repair of genomic DNA must occur in the context of chromatin. In addition, nucleosome-stimulated autoPARylation may play a role in depleting cellular NAD^+ pools in response to cellular stresses. Furthermore, PARP-1's chromatin-dependent activities may help to regulate the expression of immune and inflammatory response genes. These possibilities will need to be examined in the future.

4.2. Therapeutic Applications of PARP-1 Inhibitors

Chemical inhibition of PARP-1 activity has potential therapeutic applications in both acute and chronic diseases. For example, PARP-1 inhibitors can potentially enhance the cytotoxicity of DNA-damaging anticancer drugs, reduce parenchymal cell necrosis in conditions such as stroke or myocardial infarction, and inhibit inflammation and tissue injury pathways in conditions such as circulatory shock and diabetes mellitus (Jagtap and Szabo, 2005). As noted above, PARP-1 deficiency or inhibition in mouse and cell models exerts protective effects against pathological cell death and inflammatory responses. Thus, PARP-1 has been an attractive therapeutic target for chemical inhibition. One concern, however, is that chronic use of PARP-1 inhibitors could pose a long-term risk for genomic instability and secondary cancers (Jagtap and Szabo, 2005). Clinical trials currently underway will help to resolve if this concern is justified. Studies examining the interplay between PARP-1 and SIRT1 (see above) suggest that chemical modulators of SIRT1 activity may also be useful as therapeutic agents for treating PARP-1-dependent diseases.

Chemistry. Nicotinamide (NAm), a product of PARP-1 enzymatic action on NAD^+ and a weak inhibitor of PARP-1 activity (Curtin, 2006), was used as a model for the first PARP-1 inhibitors, including benzamide and 3-aminobenzamide (3AB) (Shall, 1975) (Fig. 7a). Benzamide and 3AB inhibit PARP-1 activity in the mM range and may also inhibit other PARP family members (*e.g.*, PARP-2 and tankyrase; Smith *et al.*, 1998; Amé *et al.*, 1999), thus lacking the potency and specificity required for therapeutic purposes. Banasik *et al.* developed 1,5-dihydroisoquinoline (Fig. 7c) and a variety of other inhibitors (Banasik *et al.*, 1992), which were used in turn to develop even more inhibitors with greater specificity,

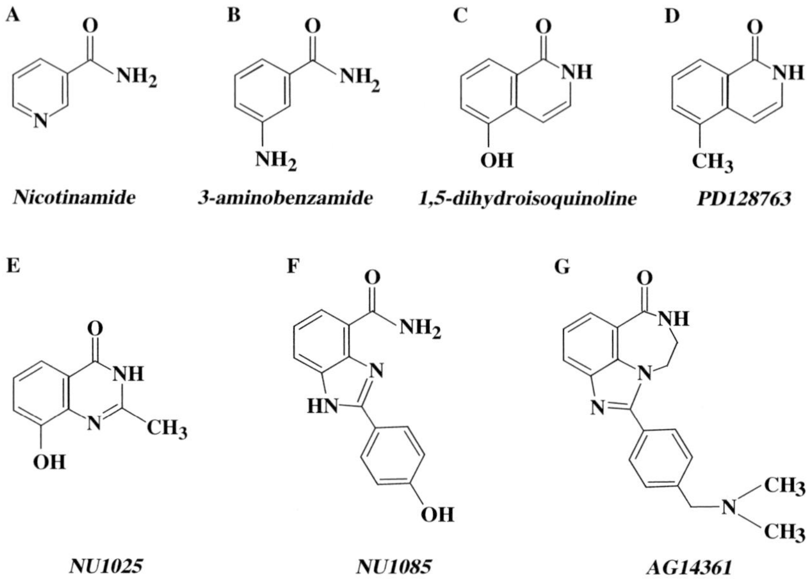

Figure 7. Structures of PARP-1 inhibitors. The structures of PARP-1 inhibitors discussed in the text are shown. Nicotinamide (a), a product of PARP-1 enzymatic action on NAD^+ and a weak inhibitor of PARP-1 activity, served as a model for the first synthetic PARP-1 inhibitors, including benzamide and 3-aminobenzamide (b). Later synthetic compounds, such as 1,5-dihydroisoquinoline (c), were used in turn to develop even more inhibitors with greater specificity, potency, and solubility, including PD128763 (d), NU1025 (e), and NU1085 (f). The incorporation of bi-, tri-, and tetra-cyclic frameworks, and cyclic lactam rings, which interact more strongly with the catalytic residue Glu988, as in AG14361 (g), has generated even more potent and specific PARP-1 inhibitors

potency, and solubility, including PD128763 (Fig. 7d) (Arundel-Suto *et al.*, 1991; Suto *et al.*, 1991), NU1025 (Fig. 7e) (Griffin *et al.*, 1995), and NU1085 (Fig. 7f) (Bryant and Helleday, 2004; Curtin, 2006). These inhibitors are 50- to 100-fold more potent than benzamides, inhibiting PARP-1 activity in the μM to nM concentration range. Although the improvements are striking, PD128763, NU1025, and NU1085 may still not be potent enough for clinical use.

The "benzamide" style of PARP-1 inhibitor has recently been improved based on X-ray crystal structure data. The incorporation of bi-, tri-, and tetra-cyclic frameworks, and cyclic lactam rings, which interact more strongly with the catalytic residue Glu988 (*e.g.*, as in AG14361; Fig. 7g), has generated even more potent and specific PARP-1 inhibitors (Jagtap and Szabo, 2005; Curtin, 2006; Tao *et al.*, 2006; Wells *et al.*, 2006). X-ray crystal structures have revealed that most PARP-1 inhibitors bind to the active site and mimic the nicotinamide moiety of NAD^+ (Ruf *et al.*, 1996, 1998; Oliver *et al.*, 2004; Jagtap and Szabo, 2005). The key structural features for high potency inhibition include an electron-rich aromatic ring system, a non-cleavable bond at the 3-position of the benzamide moiety, and a carboxamide group placed in a favorable position so as to restrict free rotation of the molecule

(Curtin, 2006). In summary, PARP-1 inhibitors have been improved considerably since the development of the benzamide and 3AB prototypes, possessing greater specificity, potency, solubility, and pharmokinetics.

Therapeutic applications. PARP-1 inhibitors may ultimately have utility as therapeutic agents for the treatment of cancers, cardiovascular diseases, diabetes mellitus and other diseases. For example, PARP-1 inhibitors promote cell death in chemically-treated or radiation therapy-treated cancers by impairing DNA damage response and repair pathways, and ultimately promoting apoptosis (Delaney *et al.*, 2000; Tentori *et al.*, 2002b; Bryant and Helleday, 2004; Curtin *et al.*, 2004). The efficacy of some anti-cancer chemotherapies or radiotherapies is improved when they are administered with PARP-1 inhibitors. For example, in mouse models and human cancer cell lines, the efficacy of temozolomide, an anti-cancer alkylating agent that damages DNA and inhibits DNA replication, is improved when co-administered with the PARP-1 inhibitor NU1025, which blocks PARP-1's DNA repair activities (Delaney *et al.*, 2000; Tentori *et al.*, 2002a, b). Some PARP-1 inhibitors, such as AG14361, are being tested in clinical trials as therapeutic agents for cancers (Curtin *et al.*, 2004).

PARP-1 inhibitors can provide protection from cardiovascular diseases. Acute and chronic cardiovascular disease can lead to the release of oxidants and free radicals that promote DNA damage within cells, as well as the release of toxic mediators to nearby tissues. Inhibition of PARP-1 can attenuate these effects, promoting cell survival and improved cellular function within the affected area. The PARP-1 inhibitor INO1001 has passed Phase I clinical trials and is currently being tested in Phase II trials as a therapeutic agent for cardiovascular diseases (Graziani and Szabo, 2005). Some PARP-1 inhibitors, such as PJ-34, may also have utility as therapeutic agents in diabetes mellitus, acting to protect against islet necrosis by reducing the levels of nitric oxide and reactive oxygen species through reduced cytokine production (Tentori *et al.*, 2002b; Beneke *et al.*, 2004). Additionally, PJ-34 can reduce hyperactivation of PARP-1 in response to massive DNA damage, thereby blocking the PARP-1-dependent necrotic cell death pathway. The ability of PARP-1 inhibitors to block pathological inflammatory responses may have therapeutic utility. The examples presented here are only a few of the many examples of the potential therapeutic applications of PARP-1 inhibitors for the treatment of diseases. Current clinical trials will provide an indication of whether PARP-1 inhibitors will ultimately become widely used therapeutic agents.

5. CONCLUSIONS

PARP-1 is an abundant ($\sim$1 to 2 million molecules per cell) and ubiquitous nuclear protein that plays important roles in a variety of cellular functions. One aspect of PARP-1 biology is the modulation of chromatin structure through direct nucleosome binding, covalent modification of chromatin proteins, or the production of PAR which can serve as a polyanionic matrix for the binding of chromatin proteins. Given its role in a variety of physiological and pathophysiological processes, PARP-1 has become an attractive target for drug-based therapies. Although the role of PARP-1's nucleosome-binding and chromatin-modulating activities in PARP-1-dependent DNA repair, cell death, and inflammatory response pathways have not yet been defined, they are likely to play a critical role. Continued biochemical and cell-based studies will help to clarify PARP-1's role as a chromatin-modulating protein, as well as elucidate new ways to target PARP-1 for the treatment of disease.

REFERENCES

Adamietz P, Rudolph A (1984) ADP-ribosylation of nuclear proteins *in vivo*. Identification of histone H2B as a major acceptor for mono- and poly(ADP-ribose) in dimethyl sulfate-treated hepatoma AH 7974 cells. J Biol Chem 259:6841–6846

Amé J-C, Jacobson EL, Jacobson MK (2000) ADP-ribose polymer metabolism. In: de Murcia G, Shall S (eds) From DNA damage and stress signalling to cell death: poly ADP-ribosylation reactions. Oxford University Press, New York, pp1–34

Amé JC, Rolli V, Schreiber V, Niedergang C, Apiou F, Decker P, Muller S, Hoger T, Menissier-de Murcia J, de Murcia G (1999) PARP-2, A novel mammalian DNA damage-dependent poly(ADP-ribose) polymerase. J Biol Chem 274:17860–17868

Amé, JC, Spenlehauer C, de Murcia G (2004) The PARP superfamily. Bioessays 26:882–893

Andreone TL, O'Connor, M, Denenberg A, Hake PW, Zingarelli B (2003) Poly(ADP-ribose) polymerase-1 regulates activation of activator protein-1 in murine fibroblasts. J Immunol 170:2113–2120

Arundel-Suto CM, Scavone SV, Turner WR, Suto MJ, Sebolt-Leopold JS (1991) Effect of PD 128763, a new potent inhibitor of poly(ADP-ribose) polymerase, on X-ray-induced cellular recovery processes in Chinese hamster V79 cells. Radiat Res 126:367–371

Augustin A, Spenlehauer C, Dumond H, Menissier-De Murcia J, Piel M, Schmit AC, Apiou F, Vonesch JL, Kock M, Bornens M, De Murcia G (2003) PARP-3 localizes preferentially to the daughter centriole and interferes with the G1/S cell cycle progression. J Cell Sci 116:1551–1562

Banasik M, Komura H, Shimoyama M, Ueda K (1992) Specific inhibitors of poly(ADP-ribose) synthetase and mono(ADP-ribosyl)transferase. J Biol Chem 267:1569–1575

Bannister A, Zegerman P, Partridge J, Miska E, Thomas J, Allshire R, Kouzarides T (2001) Selective recognition of methylated lysine 9 on histone H3 by the HP1 chromo domain. Nature 410: 120–124

Bell AC, West AG, Felsenfeld G (2001) Insulators and boundaries: versatile regulatory elements in the eukaryotic. Science 291:447–450

Beneke S, Diefenbach J, Burkle A (2004) Poly(ADP-ribosyl)ation inhibitors: promising drug candidates for a wide variety of pathophysiologic conditions. Int J Cancer 111:813–818

Blander G and Guarente L (2004) The Sir2 family of protein deacetylases. Annu Rev Biochem 73:417–435

Bonicalzi ME, Vodenicharov M, Coulombe M, Gagne JP, Poirier GG (2003) Alteration of poly(ADP-ribose) glycohydrolase nucleocytoplasmic shuttling characteristics upon cleavage by apoptotic proteases. Biol Cell 95:635–644

Bouchard VJ, Rouleau M, Poirier GG (2003) PARP-1, a determinant of cell survival in response to DNA damage. Exp Hematol 31:446–454

Brunet A, Sweeney LB, Sturgill JF, Chua KF, Greer PL, Lin Y, Tran H, Ross SE, Mostoslavsky R, Cohen HY, Hu LS, Cheng HL, Jedrychowski MP, Gygi SP, Sinclair DA, Alt FW, Greenberg ME (2004) Stress-dependent regulation of FOXO transcription factors by the SIRT1 deacetylase. Science 303:2011–2015

Bryant HE and Helleday T (2004) Poly(ADP-ribose) polymerase inhibitors as potential chemotherapeutic agents. Biochem Soc Trans 32:959–961

Burkle A (2001) Physiology and pathophysiology of poly(ADP-ribosyl)ation. Bioessays 23:795–806

Burkle A (2005) Poly(ADP-ribose). The most elaborate metabolite of NAD^+. Febs J 272:4576–4589
Burkle A, Brabeck C, Diefenbach J, Beneke S (2005) The emerging role of poly(ADP-ribose) polymerase-1 in longevity. Int J Biochem Cell Biol 37:1043–1053
Cardenas-Corona M, Jacobson E, Jacobson M (1987) Endogenous polymers of ADP-ribose are associated with the nuclear matrix. J Biol Chem 262:14863–14866
Chang P, Jacobson MK, Mitchison TJ (2004) Poly(ADP-ribose) is required for spindle assembly and structure. Nature 432:645–649
Chiarugi A and Moskowitz MA (2002) Cell biology. PARP-1 – a perpetrator of apoptotic cell death? Science 297:200–201
Cohen HY, Miller C, Bitterman KJ, Wall NR, Hekking B, Kessler B, Howitz KT, Gorospe M, de Cabo R, Sinclair DA (2004) Calorie restriction promotes mammalian cell survival by inducing the SIRT1 deacetylase. Science 305:390–392
Curtin NJ (2006) PARP inhibitors and cancer therapy. In: Bürkle A (ed) Poly(ADP-Ribosyl)ation. Landes Bioscience, Georgetown, pp218–233
Curtin NJ, Wang LZ, Yiakouvaki A, Kyle S, Arris CA, Canan-Koch S, Webber SE, Durkacz BW, Calvert HA, Hostomsky Z, Newell DR (2004) Novel poly(ADP-ribose) polymerase-1 inhibitor, AG14361, restores sensitivity to temozolomide in mismatch repair-deficient cells. Clin Cancer Res 10:881–889
D'Amours, D, Desnoyers S, D'Silva, I, Poirier G (1999) Poly(ADP-ribosyl)ation reactions in the regulation of nuclear functions. Biochem J 342:249–268
Davidovic L, Vodenicharov M, Affar EB, Poirier GG (2001) Importance of poly(ADP-ribose) glycohydrolase in the control of poly(ADP-ribose) metabolism. Exp Cell Res 268:7–13
de Murcia G, Huletsky A, Lamarre D, Gaudreau A, Pouyett J, Daune M, Poirier G (1986) Modulation of chromatin superstructure induced by poly(ADP-ribose) synthesis and degradation. J Biol Chem 261:7011–7017
de Murcia JM, Niedergang C, Trucco C, Ricoul M, Dutrillaux B, Mark M, Oliver FJ, Masson M, Dierich A, LeMeur M, Walztinger C, Chambon P, de Murcia G (1997) Requirement of poly(ADP-ribose) polymerase in recovery from DNA damage in mice and in cells. Proc Natl Acad Sci U S A 94:7303–7307
Decker P, Muller S (2002) Modulating poly (ADP-ribose) polymerase activity: potential for the prevention and therapy of pathogenic situations involving DNA damage and oxidative stress. Curr Pharm Biotechnol 3:275–283
Delaney CA, Wang LZ, Kyle S, White AW, Calvert AH, Curtin NJ, Durkacz BW, Hostomsky Z, Newell DR (2000) Potentiation of temozolomide and topotecan growth inhibition and cytotoxicity by novel poly(adenosine diphosphoribose) polymerase inhibitors in a panel of human tumor cell lines. Clin Cancer Res 6:2860–2867
Diefenbach J, Burkle A (2005) Introduction to poly(ADP-ribose) metabolism. Cell Mol Life Sci 62:721–730
Earle E, Saxena A, MacDonald A, Hudson DF, Shaffer LG, Saffery R, Cancilla MR, Cutts SM, Howman E, Choo KH (2000) Poly(ADP-ribose) polymerase at active centromeres and neocentromeres at metaphase. Hum Mol Genet 9:187–194
Ferro A, Olivera B (1982) Poly(ADP-ribosylation) *in vitro*. J Biol Chem 257:7808–7813
Gagne JP, Hunter JM, Labrecque B, Chabot B, Poirier GG (2003) A proteomic approach to the identification of heterogeneous nuclear ribonucleoproteins as a new family of poly(ADP-ribose)-binding proteins. Biochem J 371:331–340
Galande S, Kohwi-Shigematsu T (1999) Poly(ADP-ribose) Polymerase and Ku autoantigen form a complex and synergistically bind to matrix attachment sequences. J Biol Chem 274:20521–20528
Graziani G, Szabo C (2005) Clinical perspectives of PARP inhibitors. Pharmacol Res 52:109–118
Griffin RJ, Pemberton LC, Rhodes D, Bleasdale C, Bowman K, Calvert AH, Curtin NJ, Durkacz BW, Newell DR, Porteous JK *et al* (1995) Novel potent inhibitors of the DNA repair enzyme poly(ADP-ribose)polymerase (PARP). Anticancer Drug Des 10:507–514
Hanai S, Uchida M, Kobayashi S, Miwa M, Uchida K (1998) Genomic organization of Drosophila poly(ADP-ribose) polymerase and distribution of its mRNA during development. J Biol Chem 273:11881–11886

Hassa P, Haenni S, Buerki C, Meier N, Lane W, Owen H, Gersbach M, Imhof R, Hottiger M (2005) Acetylation of poly(ADP-ribose) polymerase-1 by p300/CREB-binding protein regulates coactivation of NF-κB-dependent transcription. J Biol Chem 280:40450–40464

Hassa PO, Hottiger MO (2002) The functional role of poly(ADP-ribose)polymerase 1 as novel coactivator of NF-kappaB in inflammatory disorders. Cell Mol Life Sci 59:1534–1553

Howitz KT, Bitterman KJ, Cohen HY, Lamming DW, Lavu S, Wood JG, Zipkin RE, Chung P, Kisielewski A, Zhang LL, Scherer B, Sinclair DA (2003) Small molecule activators of sirtuins extend Saccharomyces cerevisiae lifespan. Nature 425:191–196

Huletsky A, de Murcia G, Muller S, Hengartner M, Menard L, Lamarre D, Poirier G (1989) The effect of Poly(ADP-ribosyl)ation on native and H1-depleted chromatin. J Biol Chem 264: 8878–8886

Imai S, Johnson FB, Marciniak RA, McVey M, Park PU, Guarente L (2000) Sir2: an NAD-dependent histone deacetylase that connects chromatin silencing, metabolism, and aging. Cold Spring Harb Symp Quant Biol 65:297–302

Jagtap P, Szabo C (2005) Poly(ADP-ribose) polymerase and the therapeutic effects of its inhibitors. Nat Rev Drug Discov 4:421–440

Jenuwein T, Allis CD (2001) Translating the histone code. Science 293:1074–1080

Kaeberlein M, McVey M, Guarente L (1999) The SIR2/3/4 complex and SIR2 alone promote longevity in Saccharomyces cerevisiae by two different mechanisms. Genes Dev 13:2570–2580

Kaminker PG, Kim SH, Taylor RD, Zebarjadian Y, Funk WD, Morin GB, Yaswen P, Campisi J (2001) TANK2, a new TRF1-associated poly(ADP-ribose) polymerase, causes rapid induction of cell death upon overexpression. J Biol Chem 276:35891–35899

Kanai M, Tong WM, Sugihara E, Wang ZQ, Fukasawa K, Miwa M (2003) Involvement of poly(ADP-Ribose) polymerase 1 and poly(ADP-Ribosyl)ation in regulation of centrosome function. Mol Cell Biol 23:2451–2462

Karras G, Kustatscher G, Buhecha H, Allen M, Pugieux C, Sait F, Bycroft M, Ladurner A (2005) The macro domain is an ADP-ribose binding module. EMBO J 24:1911–1920

Kickhoefer VA, Siva AC, Kedersha NL, Inman EM, Ruland C, Streuli M, Rome LH (1999) The 193-kD vault protein, VPARP, is a novel poly(ADP-ribose) polymerase. J Cell Biol 146:917–928

Kim MY, Mauro S, Gevry N, Lis J, Kraus WL (2004) NAD^+-dependent modulation of chromatin strucuture and tanscription by nucleosome binding properties of PARP-1. Cell 119:803–814

Kim MY, Zhang T, Kraus WL (2005) Poly(ADP-ribosyl)ation by PARP-1: "PAR-laying" NAD^+ into a nuclear signal. Genes Dev 19:1951–1967

Kim MY, Woo EM, Chong YT, Homenko DR, Kraus WL (2006) Acetylation of estrogen receptor alpha by p300 at lysines 266 and 268 enhances the DNA binding and transactivation activities of the receptor. Mol Endocrinol

Kolthur-Seetharam U, Dantzer F, McBurney MW, de Murcia G, Sassone-Corsi P (2006) Control of AIF-mediated Cell Death by the Functional Interplay of SIRT1 and PARP-1 in Response to DNA Damage. Cell Cycle 5

Kraus WL, Lis J (2003) PARP goes transcription. Cell 113:677–683

Krupitza G, Cerutti P (1989) Poly(ADP-ribosylation) of histones in intact human keratinocytes. Biochemistry 28:4054–4060

Kun E, Kirsten E, Ordahl CP (2002) Coenzymatic activity of randomly broken or intact double-stranded DNAs in auto and histone H1 trans-poly(ADP-ribosylation), catalyzed by poly(ADP-ribose) polymerase (PARP I). J Biol Chem 277:39066–39069

Kun E, Kirsten E, Mendeleyev J, Ordahl CP (2004) Regulation of the enzymatic catalysis of poly(ADP-ribose) polymerase by dsDNA, polyamines, Mg2+, Ca2+, histones H1 and H3, and ATP. Biochemistry 43:210–216

Kustatscher G, Hothorn M, Pugieux C, Scheffzek K, Ladurner A (2005) Splicing regulates NAD metabolite binding to histone macroH2A. Nat Struct Mol Biol 12:390–392

Lachner M, O'Carroll, D, Rea S, Mechtler K, Jenuwein T (2001) Methylation of histone H3 lysine 9 creates a binding site for HP1 proteins. Nature 410:116–120

Li B, Navarro S, Kasahara N, Comai L (2004) Identification and biochemical characterization of a Werner's syndrome protein complex with Ku70/80 and poly(ADP-ribose) polymerase-1. J Biol Chem 279:13659–13667

Lin SJ, Defossez PA, Guarente L (2000) Requirement of NAD and SIR2 for life-span extension by calorie restriction in Saccharomyces cerevisiae. Science 289:2126–2128

Luo J, Nikolaev AY, Imai S, Chen D, Su F, Shiloh A, Guarente L, Gu W (2001) Negative control of p53 by Sir2alpha promotes cell survival under stress. Cell 107:137–148

Mabley JG, Suarez-Pinzon WL, Hasko G, Salzman AL, Rabinovitch A, Kun E, Szabo C (2001) Inhibition of poly (ADP-ribose) synthetase by gene disruption or inhibition with 5-iodo-6-amino-1,2-benzopyrone protects mice from multiple-low-dose-streptozotocin-induced diabetes. Br J Pharmacol 133:909–919

Magni G, Amici A, Emanuelli M, Orsomando G, Raffaelli N, Ruggieri S (2004) Structure and function of nicotinamide mononucleotide adenylyltransferase. Curr Med Chem 11:873–885

Malanga M, Atorino L, Tramontano F, Farina B, Quesada P (1998) Poly(ADP-ribose) binding properties of histone H1 variants. Biochim Biophys Acta 1399:154–160

Mandir AS, Przedborski S, Jackson-Lewis V, Wang ZQ, Simbulan-Rosenthal CM, Smulson ME, Hoffman BE, Guastella DB, Dawson VL, Dawson TM (1999) Poly(ADP-ribose) polymerase activation mediates 1-methyl-4-phenyl-1, 2,3,6-tetrahydropyridine (MPTP)-induced parkinsonism. Proc Natl Acad Sci U S A 96:5774–5779

Masutani M, Nakagama H, Sugimura T (2003) Poly(ADP-ribose) and carcinogenesis. Genes Chromosomes Cancer 38:339–348

Masutani M, Gunji A, Tsutsumi M, Ogawa K, Kamada N, Shirai T, Jishage K, Nakagama H, Sugimura T (2005) Role of poly-ADP-ribosylation in cancer development. In: Burkle A (ed) Poly(ADP-Ribosyl)ation? Landes Bioscience, Georgetown, TX

Mathis G, Althaus F (1987) Release of core DNA from nucleosomal core particles following (ADP-ribose)n-modification *in vitro*. Biochem Biophys Res Commun 143:1049–1054

Minaga T, Kun E (1983a) Probable helical conformation of poly(ADP-ribose). The effect of cations on spectral properties. J Biol Chem 258:5726–5730

Minaga T, Kun E (1983b) Spectral analysis of the conformation of polyadenosine diphosphoribose. Evidence indicating secondary structure. J Biol Chem 258:725–730

Miwa M, Hanai S, Poltronieri P, Uchida M, Uchida K (1999) Functional analysis of poly(ADP-ribose) polymerase in Drosophila melanogaster. Mol Cell Biochem 193:103–107

Moazed D (2001) Common themes in mechanisms of gene silencing. Mol Cell 8:489–498

Motta MC, Divecha N, Lemieux M, Kamel C, Chen D, Gu W, Bultsma Y, McBurney M, Guarente L (2004) Mammalian SIRT1 represses forkhead transcription factors. Cell 116:551–563

Oei SL, Shi Y (2001) Transcription factor Yin Yang 1 stimulates poly(ADP-ribosyl)ation and DNA repair. Biochem Biophys Res Commun 284:450–454

Ogata N, Ueda K, Kawaichi M, Hayaishi O (1981) Poly(ADP-ribose) synthetase, a main acceptor of poly(ADP-ribose) in isolated nulcei. J Biol Chem 256:4135–4137

Ohashi S, Kanai M, Hanai S, Uchiumi F, Maruta H, Tanuma S, Miwa M (2003) Subcellular localization of poly(ADP-ribose) glycohydrolase in mammalian cells. Biochem Biophys Res Commun 307:915–921

Oliver AW, Amé JC, Roe SM, Good V, de Murcia G, Pearl LH (2004) Crystal structure of the catalytic fragment of murine poly(ADP-ribose) polymerase-2. Nucleic Acids Res 32:456–464

Oliver FJ, Menissier-de Murcia J, Nacci C, Decker P, Andriantsitohaina R, Muller S, de la Rubia G, Stoclet JC, de Murcia G (1999) Resistance to endotoxic shock as a consequence of defective NF-kappaB activation in poly (ADP-ribose) polymerase-1 deficient mice. EMBO J 18:4446–4454

Panzeter PL, Realini CA, Althaus FR (1992) Noncovalent interactions of poly(adenosine diphosphate ribose) with histones. Biochemistry 31:1379–1385

Parsons XH, Garcia SN, Pillus L, Kadonaga JT (2003) Histone deacetylation by Sir2 generates a transcriptionally repressed nucleoprotein complex. Proc Natl Acad Sci U S A 100:1609–1614

Picard F, Kurtev M, Chung N, Topark-Ngarm A, Senawong T, Machado De Oliveira R, Leid M, McBurney MW, Guarente L (2004) Sirt1 promotes fat mobilization in white adipocytes by repressing PPAR-gamma. Nature 429:771–776

Pillai J, Isbatan A, Imai S-I, Gupta M (2005) Poly(ADP-ribose) Polymerase-1-dependent cardiac myocyte cell death during heart failure is mediated by NAD^+ depletion and reduced Sir2α deacetylase activity. J Biol Chem 280:43121–43130

Pleschke JM, Kleczkowska HE, Strohm M, Althaus FR (2000) Poly(ADP-ribose) binds to specific domains in DNA damage checkpoint proteins. J Biol Chem 275:40974–40980

Poirier G, de Murcia G, Jongstra-Bilen J, Niedergang C, Mandel P (1982) Poly(ADP-ribosyl)ation of polynucleosomes causes relaxation of chromatin structure. Proc Natl Acad Sci U S A 79:3423–3427

Rapizzi E, Fossati S, Moroni F, Chiarugi A (2004) Inhibition of poly(ADP-ribose) glycohydrolase by gallotannin selectively up-regulates expression of proinflammatory genes. Mol Pharmacol 66:890–898

Realini C, Althaus F (1992) Histone shuttling by poly(ADP-ribosylation). J Biol Chem 267:18858–18865

Rogina B, Helfand SL (2004) Sir2 mediates longevity in the fly through a pathway related to calorie restriction. Proc Natl Acad Sci U S A 101:15998–16003

Rolli V, Ruf A, Augustin A, Schulz GE, Ménissier-de Murcia J, de Murcia G (2000) Poly(ADP-ribose) polymerase: structure and function. In: de Murcia G, Shall S (eds) From DNA damage and stress signalling to cell death: poly ADP-ribosylation reactions. Oxford University Press, New York, pp35–79

Rouleau M, Aubin R, Poirier G (2004) Poly(ADP-ribosyl)ated chromatin domains: access granted. J Cell Science 117:815–825

Ruf A, Mennissier de Murcia J, de Murcia G, Schulz GE (1996) Structure of the catalytic fragment of poly(AD-ribose) polymerase from chicken. Proc Natl Acad Sci U S A 93:7481–7485

Ruf A, de Murcia G, Schulz GE (1998) Inhibitor and NAD^+ binding to poly(ADP-ribose) polymerase as derived from crystal structures and homology modeling. Biochemistry 37:3893–3900

Ruscetti T, Lehnert BE, Halbrook J, Le Trong H, Hoekstra MF, Chen DJ, Peterson SR (1998) Stimulation of the DNA-dependent protein kinase by poly(ADP-ribose) polymerase. J Biol Chem 273:14461–14467

Saxena A, Saffery R, Wong L, Kalitsis P, Choo KHA (2002a) Centromere proteins Cenpa, Cenpb, and Bub3 interact with Poly(ADP-ribose) Polymerase-1 protein and are poly(ADP-ribosyl)ated. J Biol Chem 277:26921–26926

Saxena A, Wong L, Kalitsis P, Earle E, Schaffer L, Choo KHA (2002b) Poly(ADP-ribose) polymerase 2 localizes to mammalian active centromeres and interacts with PARP-1, Cenpa, Cenpb, Bub3, but not Cenpc. Hum Mol Genet 11:2319–2329

Shall S (1975) Proceedings: Experimental manipulation of the specific activity of poly(ADP-ribose) polymerase. J Biochem (Tokyo) 77:2p.

Shall S, de Murcia G (2000) Poly(ADP-ribose) polymerase-1: what have we learned from the deficient mouse model? Mutat Res 460:1–15

Shibata A, Kamada N, Masumura K, Nohmi T, Kobayashi S, Teraoka H, Nakagama H, Sugimura T, Suzuki H, Masutani M (2005) Parp-1 deficiency causes an increase of deletion mutations and insertions/rearrangements *in vivo* after treatment with an alkylating agent. Oncogene 24:1328–1337

Sinclair DA, Guarente L (1997) Extrachromosomal rDNA circles – a cause of aging in yeast. Cell 91:1033–1042

Smith HM, Grosovsky AJ (1999) PolyADP-ribose-mediated regulation of p53 complexed with topoisomerase I following ionizing radiation. Carcinogenesis 20:1439–1443

Smith S (2001) The world according to PARP. Trends Biochem Sci 26:174–179

Smith S, de Lange T (1999) Cell cycle dependent localization of the telomeric PARP, tankyrase, to nuclear pore complexes and centrosomes. J Cell Sci 112(Pt 21):3649-3656

Smith S, Giriat I, Schmitt A, de Lange T (1998) Tankyrase, a poly(ADP-ribose) polymerase at human telomeres. Science 282:1484–1487

Suto MJ, Turner WR, Arundel-Suto CM, Werbel LM, Sebolt-Leopold JS (1991) Dihydroisoquinolinones: the design and synthesis of a new series of potent inhibitors of poly(ADP-ribose) polymerase. Anticancer Drug Des 6:107–117

Szabo C (1998) Role of poly(ADP-ribose)synthetase in inflammation. Eur J Pharmacol 350:1–19

Tanuma S, Johnson GS (1983) ADP-ribosylation of nonhistone high mobility group proteins in intact cells. J Biol Chem 258:4067–4070

Tao M, Park CH, Bihovsky R, Wells GJ, Husten J, Ator MA, Hudkins RL (2006) Synthesis and structure-activity relationships of novel poly(ADP-ribose) polymerase-1 inhibitors. Bioorg Med Chem Lett 16:938–942

Tentori L, Leonetti C, Scarsella M, d'Amati, G, Portarena I, Zupi G, Bonmassar E, Graziani G (2002a) Combined treatment with temozolomide and poly(ADP-ribose) polymerase inhibitor enhances survival of mice bearing hematologic malignancy at the central nervous system site. Blood 99:2241–2244

Tentori L, Portarena I, Graziani G (2002b) Potential clinical applications of poly(ADP-ribose) polymerase (PARP) inhibitors. Pharmacol Res 45:73–85

Thibeault L, Hengartner M, Lagueux J, Poirier G, Muller S (1992) Rearrangements of the nucleosome structure in chromatin by poly (ADP-ribose). Biochim Biophys Acta 1121:317–324

Tissenbaum HA, Guarente L (2001) Increased dosage of a sir-2 gene extends lifespan in Caenorhabditis elegans. Nature 410:227–230

Tulin A, Spradling A (2003) Chromatin loosening by poly(ADP)-ribose polymerase (PARP) at Drosophila puff loci. Science 299:560–562

Tulin A, Stewart D, Spradling A (2002) The Drosophila heterochromatic gene encoding poly(ADP-ribose) polymerase (PARP) is required to modulate chromatin structure during development. Genes Dev 16:2108–2119

Tulin A, Naumova N, Menon A, Spradling A (2006) Drosophila poly(ADP-ribose) glycohydrolase mediates chromatin structure and Sir2-dependent silencing. Genetics 172:363–371

Vaziri H, Dessain SK, Ng Eaton E, Imai SI, Frye RA, Pandita TK, Guarente L, Weinberg RA (2001) hSIR2(SIRT1) functions as an NAD-dependent p53 deacetylase. Cell 107:149–159

Vidakovic M, Grdovic N, Quesada P, Bode J, Poznanovic G (2004) Poly(ADP-Ribose) Polymerase-1: association with nuclear lamins in rodent liver cells. J Cell Biochem 93:1155–1168

Virag L (2005) PARP-1 and the shape of cell death. In: Burkle A (ed) Poly(ADP-Ribosyl)ation. Landes Bioscience, Georgetown, TX

Wang ZQ, Stingl L, Morrison C, Jantsch M, Los M, Schulze-Osthoff K, Wagner EF (1997) PARP is important for genomic stability but dispensable in apoptosis. Genes Dev 11:2347–2358

Wells GJ, Bihovsky R, Hudkins RL, Ator MA, Husten J (2006) Synthesis and structure-activity relationships of novel pyrrolocarbazole lactam analogs as potent and cell-permeable inhibitors of poly(ADP-ribose)polymerase-1 (PARP-1). Bioorg Med Chem Lett 16:1151–1155

Wesierska-Gadek J, Sauermann G (1988) The effect of poly(ADP-ribose) on interactions of DNA with histones H1, H3 and H4. Eur J Biochem 173:675–679

Wesierska-Gadek J, Schmid G, Cerni C (1996) ADP-ribosylation of wild-type p53 *in vitro*: binding of p53 protein to specific p53 consensus sequence prevents its modification. Biochem Biophys Res Commun 224:96–102

Wood JG, Rogina B, Lavu S, Howitz K, Helfand SL, Tatar M, Sinclair D (2004) Sirtuin activators mimic caloric restriction and delay ageing in metazoans. Nature 430:686–689

Yanagisawa J, Ando J, Nakayama J, Kohwi Y, Kohwi-Shigematsu T (1996) A matrix attachment region (MAR)-binding activity due to a p114 kilodalton protein is found only in human breast carcinomas and not in normal and benign breast disease tissues. Cancer Res 56:457–462

Yu SW, Wang H, Poitras MF, Coombs C, Bowers WJ, Federoff HJ, Poirier GG, Dawson TM, Dawson VL (2002) Mediation of poly(ADP-ribose) polymerase-1-dependent cell death by apoptosis-inducing factor. Science 297:259–263

Yu W, Ginjala V, Pant V, Chernukhin I, Whitehead J, Docquier F, Farrar D, Tavoosidana G, Mukhopadhyay R, Kanduri C, Oshimura M, Feinberg A, Lobanenkov V, Klenova E, Ohlsson R (2004) Poly(ADP-ribosyl)ation regulates CTCF-dependent chromatin insulation. Nat Genet 36:1105–1110

Zhang J (2003) Are poly(ADP-ribosyl)ation by PARP-1 and deacetylation by Sir2 linked? Bioessays 25:808–814

Zingarelli B, Salzman AL, Szabo C (1998) Genetic disruption of poly (ADP-ribose) synthetase inhibits the expression of P-selectin and intercellular adhesion molecule-1 in myocardial ischemia/reperfusion injury. Circ Res 83:85–94

CHAPTER 4

HISTONE VARIANT NUCLEOSOMES

Structure, function and implication in disease

MATHIEU BOULARD[1,3], PHILIPPE BOUVET[1,2], TAPAS K. KUNDU[4] AND STEFAN DIMITROV[1,3,*]

[1] *Laboratoire Joliot-Curie, CNRS-UMR 5161/INRA 1237/IFR128 Biosciences Lyon-Gerland Ecole Normale Supérieure de Lyon, 46 Allée d'Italie, 69007 Lyon, France*

[2] *Laboratoire de Biologie Moléculaire de la Cellule, CNRS-UMR 5161/INRA 1237/IFR128 Biosciences Lyon-Gerland Ecole Normale Supérieure de Lyon, 46 Allée d'Italie, 69007 Lyon, France*

[3] *Institut Albert Bonniot, INSERM U309, 38706 La Tronche cedex, France*

[4] *Transcription and Disease Laboratory, Molecular Biology and Genetics Unit, Jawaharlal Nehru Centre for Advanced Scientific Research, Jakkur P. O. Bangalore 560064, India*

Abstract: Histone variant are non-allelic forms of the conventional histones. They are expressed at very low levels compared to their conventional counterparts. All the conventional histones, except H4, have histone variants. Together with histone modifications and chromatin remodeling machines, the incorporation of histone variants into the nucleosome is one of the main strategies that the cell uses to regulate transcription, repair, chromosome assembly and segregation. The exact role of the histone variants in these processes is far from clear, but the emerging picture is that the presence of histone variants confers novel structural and functional properties of the nucleosome which affect the chromatin dynamics. In this article we will discuss the functional significance of histone variants on chromatin function and its link to disease manifestation

1. INTRODUCTION

DNA is packaged in the nucleus into the form of chromatin. Chromatin is a nucleoprotein complex composed of histone and non-histone proteins, DNA and RNA and it exhibits a repeating structure (van Holde, 1988). The basal unit of chromatin, the nucleosome, is composed of a histone octamer (two each of H2A, H2B, H3 and H4) around which two superhelical turns of DNA are wrapped (van Holde, 1988). The structure of both the histone octamer (Arents *et al*, 1991)

*E-mail: stefan.dimitrov@ujf-grenoble.fr

T.K. Kundu and D. Dasgupta (eds.), Chromatin and Disease, 71–89.

and the nucleosome core particle (Luger *et al*, 1997) was solved by X-ray crystallography. The histone octamer exhibits a complex outer surface (Arents *et al*, 1991). The $(H3\text{-}H4)_2$ tetramer formed a central kernel, which is flanked by two H2A-H2B dimers. All four histones within the octamer consist of an elongated structured domain, termed histone fold, and unstructured N-terminal tails (Arents and Moudrianakis, 1993; Arents *et al*, 1991). The repeating structural motifs at the histone octamer surface allowed the suggestion of a specific organization of the nucleosomal DNA (Arents and Moudrianakis, 1993). The crystal structure of the nucleosome core particle showed that DNA is wrapped around the histone octamer in a flat, left-handed non-uniform superhelix, which formed multiple contacts with the histone octamer (Luger *et al*, 1997). The histone N-terminal tails protruded from the nucleosome core particle and appeared to be highly flexible (Luger *et al*, 1997).

The nucleosome is a barrier for numerous cellular processes, which required accessibility of distinct protein factors to their cognate DNA sequences (Beato and Eisfeld, 1997). To overcome the nucleosome barrier the cell uses different strategies including histone modifications, chromatin remodelers, histone chaperones and histone variants (Becker, 2002; Henikoff and Ahmad, 2005; Henikoff *et al*, 2004; Strahl and Allis, 2000). The roles of histone modifications and chromatin remodelers are widely studied and were the topics of numerous reviews in the literature, while how histone variants function remains elusive. In this review we will summarize the reported data on the structure of histone variant nucleosomes. We will further discuss the emerging picture of the role of histone variants in DNA repair, chromosome segregation, transcription and spermatogenesis and their possible associations with pathologies such as tumorigenesis and infertility.

2. HISTONE VARIANTS: GENERAL DESCRIPTION

Histone variants are non-allelic isoforms of conventional histones, which are able to substitute for the canonical histones within the nucleosome. The mRNA of histone variants, in contrast to that of conventional histones, is polyadenylated (Tsanev *et al*, 1993; van Holde, 1988). The histone variant mRNA lacks a stem-loop in the 3′end of the transcript, a feature required for cell-cycle regulated degradation (Pandey and Marzluff, 1987). Instead, transcripts encoding histone variants are characterized by a longer poly-A tail, which results in higher transcript stability (Moss *et al*, 1989). Histone variants coding genes are not organized in cluster as the conventional histone genes. The variants exhibit three major characteristics, which made them distinct from conventional histones: (i) the timing of expression, (ii) the association with specific genomic DNA sequences and, (iii) few cases, the tissue specific expression. Interestingly, the expression of histone variant is not coupled with replication as in the case of conventional histones and they can be deposited in the nucleosomes at different phases of the cell cycle (Tsanev *et al*, 1993; van Holde, 1988). Some histone variants can be synthesized even in differentiated, non-dividing cells (Bosch and Suau, 1995). Distinct histone variants are specifically associated with distinct DNA sequences and remained associated with chromatin in mature spermatozoa (Palmer *et al*, 1990). Several testis-specific histone variants are described to date (for a recent review see, Govin *et al*, 2005). Though it is not clearly understood yet, presumably different histone variants might use distinct chaperone complexes for deposition at the specific sites (reviewed in Loyola and Almouzni, 2004).

3. H2A HISTONE VARIANT NUCLEOSOMES

The histone variants of H2A form the largest family of identified histone variants (Redon *et al*, 2002; Sarma and Reinberg, 2005). This could be associated with both the strategic position that has the histone H2A within the histone octamer and the less stable interaction of the H2A-H2B dimmer with both DNA and the $(H3\text{-}H4)_2$ tetramer within the nucleosome (Luger *et al*, 1997). Most of the histone H2A variants exhibit a unique property: in addition to the N-terminal tail domain, they also posses an unstructured C-terminal tail. To date four variants of histone H2A have been discovered. These include, H2AZ, H2A.X, macroH2A and H2A.Bbd. The highest differences in the primary structure of these H2A variants are observed in their C-terminal portion. Each of these variants could be efficiently incorporated in the nucleosome *in vitro* and *in vivo*. The presence of these variants alter the structural and functional properties of the nucleosome distinctly.

3.1. H2A.Z Histone Variant Nucleosome

H2A.Z is one of the best studied histone variants. H2A.Z is a highly conserved histone variant, which is found in all eukaryotes. In budding yeast H2A.Z is a non-essential gene, while in both *Drososphila* and mouse it is required for survival (Clarkson *et al*, 1999; Dhillon and Kamakaka, 2000; Faast *et al*, 2001; Santisteban *et al*, 2000). To date only H2A.Z nucleosome structure has been solved by X-ray crystallography (Suto *et al*, 2000). The overall H2A.Z nucleosome structure was quite similar to the structure of the conventional nucleosome and the protein-DNA interactions were found to be essentially unaffected upon the replacement (Luger *et al*, 1997). In agreement with the crystallographic data, the DNase I footprinting analysis also suggested that the structure of the H2A.Z nucleosome in solution closely resembles the conventional core particle (Abbott *et al*, 2001). Some local changes between the variant H2A.Z-H2B dimmer and the $(H3\text{-}H4)_2$ tetramer were, however, observed in the crystal structure. It was suggested that these changes could alter the stability of the H2A.Z nucleosome. The measurements of the sedimentation coefficient at different ionic strength suggested that the variant particle is relatively less stable as compared to the conventional one (Abbott *et al*, 2001). The salt dependent folding of the nucleosome arrays was also altered upon replacement of conventional H2A by H2A.Z (Abbott *et al*, 2001). These observations indicate that the presence of the H2A.Z histone variant may play a chromatin destabilizing role (Abbott *et al*, 2001).

In contrast, according to Fan and colleagues (Fan *et al*, 2002), the intramolecular folding ability of nucleosome arrays appeared to be facilitated by H2A.Z.

The most intriguing feature of the H2A.Z nucleosome is the increased acidic nature of the nucleosome surface (the acidic patch) (Luger *et al*, 1997; Suto *et al*, 2000). HP1α, a mediator of chromatin compaction and a marker of heterochromatin, was able to induce the folding of nucleosomal array, which was enhanced by H2A.Z in an acidic patch dependent manner (Fan *et al*, 2004). In agreement with this, the suppression of the expression of H2A.Z resulted in perturbations in both the structure of constitutive heterochromatin and HP1α–chromatin interactions leading to strong mitotic defects (Rangasamy *et al*, 2004). In *Drosophila* H2A.Z was associated with the establishment of heterochromatin and HP1 recruitment (Swaminathan *et al*, 2005). In addition, H2A.Z may directly interact with the passenger protein INCENP (Rangasamy *et al*, 2003), a member of the Aurora B kinase complex, which plays a crucial role in chromosome segregation and cytokinesis (Adams *et al*, 2001). In yeast, Htz1 the orthologue of H2A.Z, was found to be associated with the actively transcribed regions that flanked Sir silencing complex containing heterochromatin. In absence of Htz1, the heterochromatin spreads *in vivo*, suggesting that it may be involved in the maintenance or establishment of the boundary between heterochromatic or euchromatic regions. Though at this stage it is very difficult to propose a unified model for the role of H2A.Z in chromatin organization, all these data support the view that H2A.Z is significantly important for both heterochromatin assembly and maintenance and chromosome segregation through a specific higher order chromatin folding.

H2A.Z appeared to be involved not only in mitosis, but also in transcriptional regulation. Genome-wide high-resolution localization of H2A.Z in yeast was studied by several groups, but the reported data exhibited some controversial character (Guillemette *et al*, 2005; Li *et al*, 2005; Raisner *et al*, 2005 ; Zhang *et al*, 2005a). All authors agreed that H2A.Z occupied promoters genome-wide. One group mapped H2A.Z to the 5′ends of nearly all promoters of both inactive and active genes (Raisner *et al*, 2005). Intriguingly, the H2A.Z nucleosomes exhibited a very specific promoter localization pattern with two precisely positioned nucleosomes, which flanked a nucleosome free transcription start site region (Raisner *et al*, 2005). A 22 bp DNA sequence was involved in the formation of this pattern (Raisner *et al*, 2005). Another study, using also chromatin immunoprecipitation (ChIP) combined with hybridization with DNA microarrays, pointed to a preferential deposition of H2A.Z to repressed/basal Pol II promoter. In this case preferential deposition of H2A.Z was observed at the TATA-less promoters (Zhang *et al*, 2005a). The analysis of six selected promoters with quantitative PCR has revealed that the amount of H2A.Z associated with the promoters peaked from ~100 to ~400 bp upstream of the transcription start site. Other two groups reported a H2A.Z preferential occupancy on the promoter of inactive genes all across the genome and these groups have also used a ChIP-DNA microarray based approach (Guillemette *et al*, 2005; Li *et al*, 2005). In addition, it was found that in general H2A.Z was not a determinant for the nucleosome positioning (Li *et al*, 2005). It is difficult to point out the reasons for these disagreements at present. The reagents, especially antibodies used in the chromatin immunoprecipitation assays, might have made the differences.

In *Tetrahymena thermophila* H2A.Z was found exclusively in the transcriptionally active macronucleus, which indicated that the H2A.Z might be involved in the activation of gene expression. Curiously, H2A.Z was shown to be important for both activation and repression of transcription in yeast and was found to be partially redundant with chromatin remodeling complexes (Dhillon and Kamakaka, 2000; Meneghini *et al*, 2003; Santisteban *et al*, 2000). The C-terminal region of H2A.Z was crucial for its role in transcription and some results indicated that H2A.Z acted as an activator, which was required for preferential binding to intergenic DNA (Adam *et al*, 2001; Larochelle and Gaudreau, 2003). The H2A.Z location correlated with both particular transcription factors and particular histone acetylation pattern (Raisner *et al*, 2005; Zhang *et al*, 2005a). The presence of H2A.Z was inversely correlated with the transcription rates (Li *et al*, 2005; Zhang *et al*, 2005a). Recent report suggested that H2A.Z was not associated with a direct regulation of transcriptional repression, but instead it might play a crucial role for optimal activation of a subset of genes (Li *et al*, 2005). Interestingly in yeast chromatin, the histone H2A.Z was more susceptible to release than the conventional histones H2A and H3 and transcriptional activation was associated with H2A.Z loss, while repression was associated with H2A.Z acquisition (Zhang *et al*, 2005a). The deposition of H2A.Z requires the Swr1 histone exchange complex, which is able to exchange the H2A-H2B conventional dimmer for the H2A.Z-H2B variant dimmer (Mizuguchi *et al*, 2003). The Swr1 complex contains several polypeptides, some of which are common to INO80 complex.

3.2. H2AX Containing Nucleosomes

H2AX is highly conserved histone variant, which represents up to 25% (depending of cell type and tissue studied) of the mammalian histone H2A. H2AX exhibited an even distribution across the genome (Siino *et al*, 2002). The primary H2AX sequence differs from the conventional H2A in its C-terminus. The C-terminus of H2AX contained, however, a sequence termed the SQ motif, which is highly homologous in all eukaryotes from yeast to human, pointing to a conserved function (Redon *et al*, 2002). A seminal manuscript from William Bonner's lab was first to demonstrate that the serine of this motif (serine four from the carboxyl terminus of H2AX) was phosphorylated when double strand breaks (DBS) were introduced into DNA by ionizing irradiation of different cells in culture and mouse (Rogakou *et al*, 1998). The phosphorylated specie of H2AX was called γ-H2AX (Rogakou *et al*, 1998). The yeast did not have a H2AX variant, but the yeast H2A is also phosphorylated as a result of DSB induction (for simplicity further in the text we will used also the term γ-H2AX for the phosphorylated form of yeast H2A). The phosphorylation of H2AX was shown to be a general phenomenon correlating with the induction with DSB in all eukaryotes suggesting that γ-H2AX is associated with the repair of DSB (Redon *et al*, 2002). γ-H2AX formed foci in cells containing DSB, which could be easily visualized by using anti-γ-H2AX antibodies (Rogakou *et al*, 1999).

The invariant SQ motif in the C-terminus of H2AX is a consensus sequence for the 3 kinases belonging to the PIKK family, namely ATM, DNA-PK and ATR (Stiff *et al*, 2004). These kinases are involved in DNA repair. ATM [ataxia telangiectasia (A-T) mutated protein] is a crucial kinase for the signal transduction DSB pathway (Savitsky *et al*, 1995) and it is widely accepted that ATM is the major kinase involved in the *in vivo* phosphorylation of H2AX (Burma *et al*, 2001; Fernandez-Capetillo *et al*, 2002; Redon *et al*, 2002). The two other kinases were also associated with the generation γ-H2AX, but they appeared not to be dominant (Redon *et al*, 2002; Stiff *et al*, 2004).

The appearance of γ-H2AX is one of the earliest cellular responses to the induction of DSB (Celeste *et al*, 2003b; Rogakou *et al*, 1999). The phosphorylated serine in γ-H2AX is localized at the end of the unstructured C-terminus, which made it accessible to protein factors. The induction of γ-H2AX foci resulted in the accumulation of several protein factors (including ATM, Mdc1 and 53BP1) involved in DNA repair and their colocalization with the γ-H2AX foci (Bekker-Jensen *et al*, 2006; Fernandez-Capetillo *et al*, 2002). Some of these factors are shown to directly interact with the γ-H2AX (reviewed in (Bassing and Alt, 2004; Fernandez-Capetillo *et al*, 2004)).

Chromatin within the nucleus is in a highly condensed state and in order the DSB to be repaired it would be necessary it to decondense. This suggests a potential involvement of chromatin remodelers to alter the chromatin structure, to decrease the chromatin compaction and to facilitate repair *in vivo*. And indeed, a series of recent papers using yeast as a model system demonstrated that this is really the case (Downs *et al*, 2004; Morrison *et al*, 2004; van Attikum *et al*, 2004). These papers used a common approach for the generation of a HO-nuclease induced DSB in the yeast genome. This has allowed the use of ChIP for localization of the remodeling chromatin machineries in vicinity of the single DSB (Downs *et al*, 2004; Morrison *et al*, 2004; van Attikum *et al*, 2004). The data showed that the chromatin remodeler INO80 was recruited to the DSB via a specific direct interaction with γ-H2AX. The Nhp10 subunit of INO80 was required for this interaction (Morrison *et al*, 2004). In addition, it also found that the two actin related subunits of INO80, Arp5 and Arp8, bind near sites of DNA damages (van Attikum *et al*, 2004). The presence of a non-phosphorylable form of H2A.X compromised the recruitment of INO80 (van Attikum *et al*, 2004). The NuA4 HAT complex, involved in the acetylation of histone H4, was shown to associate specifically with γ-H2AX, suggesting that the chromatin acetylation by NuA4 was important for DSB repair (Downs *et al*, 2004). The interaction of NuA4 with γ-H2AX required Arp4, a NuA4 subunit common also for both INO80 and Swr1 chromatin remodeling complexes, which were found to interact with γ-H2AX (Downs *et al*, 2004). This indicated that γ-H2AX could be viewed as a mark of the DSB, which are recognized by different remodelers (Downs *et al*, 2004).

3.3. The Unusual MacroH2A Nucleosomes

Histone macroH2A (mH2A) is an unusual histone variant. Its size is about three times the size of the conventional H2A. It consists of H2A-like domain, which showed ~65% identity with the conventional histone H2A, fused to a large Non-Histone Region (NHR). The NHR is also termed macrodomain (Pehrson and Fried, 1992). In humans, two mH2A genes, mH2A1 and mH2A2, were identified, which coded for two distinct, but closely related proteins (Chadwick *et al*, 2001; Costanzi and Pehrson, 2001; Pehrson and Fuji, 1998). mH2A1 has two spliced variants, mH2A1.1 and mH2A1.2, which differed only by a short aminoacid sequence in the macrodomain (Pehrson and Fuji, 1998).

The macro domain is characteristic for numerous bacterial, archaebacterial and eukaryotic proteins, indicating that it might play important roles in several evolutionary conserved phenomena (Allen *et al*, 2003). AF1521, a Archaeoglobus fulgidus protein, consisting of a stand-alone macrodomain, was successfully crystallized and its crystal structure was solved with 1.7 Å resolution (Allen *et al*, 2003). The structure is characterized by a mixed alpha/beta fold and exhibited similarity to the N-terminal binding domain of the *E. Coli* leucine aminopeptidase PepA and to members of the P-loop family of nucleotide hydrolysases (Allen *et al*, 2003). Both AF1521 and macrodomain yeast protein (the product of the YBR022w gene) had an ADP-ribose-1″-phosphate (Appr-1″-p)-processing activity (Allen *et al*, 2003; Martzen *et al*, 1999). Based on this and on some other structural considerations, it was proposed that the macrodomain of mH2A could act as an ADP-ribose phosphoesterase (Allen *et al*, 2003; Ladurner, 2003). No experimental data confirming this interesting hypothesis were reported to date.

Recently, evidence was provided that macrodomains are ADP ribose binding modules (Karras *et al*, 2005; Kustatscher *et al*, 2005). It was demonstrated that the macro domain fold possessed a conserved pocket, which binds to ADP-ribose with high affinity (Karras *et al*, 2005). Intriguingly, the macro domain of mH2A1.1, but not this of mH2A.2, binds the SirT1-metabolite O-acetyl-ADP-ribose (Kustatscher *et al*, 2005). Since the two proteins are spliced variants and they differ only by a short amino-acid stretch embedded within the macro domain, this points to a regulation of the of the binding of NAD (nicotinamide adenine dinucleotide) metabolites through alternative splicing (Kustatscher *et al*, 2005). Experimental data supporting this suggestion are still missing.

Several reports in the literature suggest a function of mH2A in the assembly and maintenance of heterochromatin (Chadwick and Willard, 2002; Choo *et al*, 2006; Costanzi and Pehrson, 1998; Grigoryev *et al*, 2004; Ma *et al*, 2005; Perche *et al*, 2000; Rasmussen *et al*, 2000). The location of mH2A was modified during the induced reorganization of heterochromation upon DNA CpG hypomethylation

(Ma *et al*, 2005). The localization of both mH2A and HP1α was dramatically changed during lymphocyte reactivation. Reciprocal changes in the localization of both proteins were observed in activated fibroblasts induced into quiescence (Grigoryev *et al*, 2004). In senescent cells the silent senescence-associated heterochromatin foci (SAHF) were found to be enriched in mH2A (Zhang *et al*, 2005b).

For many years it was believed that the inactive X chromosome (Xi) is enriched of mH2A (Chadwick and Willard, 2001a; Costanzi and Pehrson, 1998, 2001; Mermoud *et al*, 1999). Indeed, immunofluorescence data showed much stronger staining of the Xi compared to other regions of the nucleus, suggesting that mH2A is selectively accumulated at the Xi (Chadwick and Willard, 2001a; Costanzi and Pehrson, 1998, 2001; Mermoud *et al*, 1999). This preferential staining could reflect, however, the higher density of the nucleosome within the Xi and not an enrichment of mH2A (Perche *et al*, 2000).

The *in vitro* studies with reconstituted mH2A chromatin templates was highly useful to understand the function of mH2A (Angelov *et al*, 2003; Doyen *et al*, 2006). The reconstituted nucleosomes containing mH2A were refractive to both SWI/SNF and ACF chromatin remodeling and binding of transcription factor to their cognate sequence incorporated into the nucleosomal DNA (Angelov *et al*, 2003). In addition, the presence of mH2A interferes with the initiation of Pol II transcription and histone acetylation (Doyen *et al*, 2006). The macrodomain of mH2A was found to be essential for this interference (Doyen *et al*, 2006). These data suggest that mH2A could be viewed *in vivo* as a major stopper of transcription by negatively regulating two different chromatin associated pathways, namely nucleosome remodeling and histone acetylation (Doyen *et al*, 2006). In agreement with this, transient transfection experiments with the macrodomain of mH2A showed a macro domain-mediated inhibition of Pol II transcription (Perche *et al*, 2000). A general model for mH2A mediated transcriptional repression is shown in Fig. 1.

3.4. The H2A.Bbd Nucleosomes

The recently identified histone variant H2A.Bbd (*B*arr *b*ody *d*eficient) is the least studied histone variant of the H2A family and, as its name indicates, it was found excluded from the inactive X chromosome (Chadwick and Willard, 2001b). This variant is very divergent compared to the conventional H2A (it exhibits only 48% identity with H2A). In contrast to mH2A, H2A.Bbd is relatively short and it lacks both the typical for the H2A-family flexible C-terminus and the very last fragment of the docking domain, which is responsible for the interaction of histone H3 within the histone octamer (Chadwick and Willard, 2001b; Luger *et al*, 1997). In addition, its N-terminal tail contains a row of six arginines, indicating that it could interact stronger than the N-tail of H2A with nucleosomal DNA (Chadwick and Willard, 2001b). Nucleosomes, reconstituted with H2A.Bbd (H2A.Bbd nucleosomes) instead of conventional H2A, exhibited altered structure and these structural alterations appeared to determine the inability

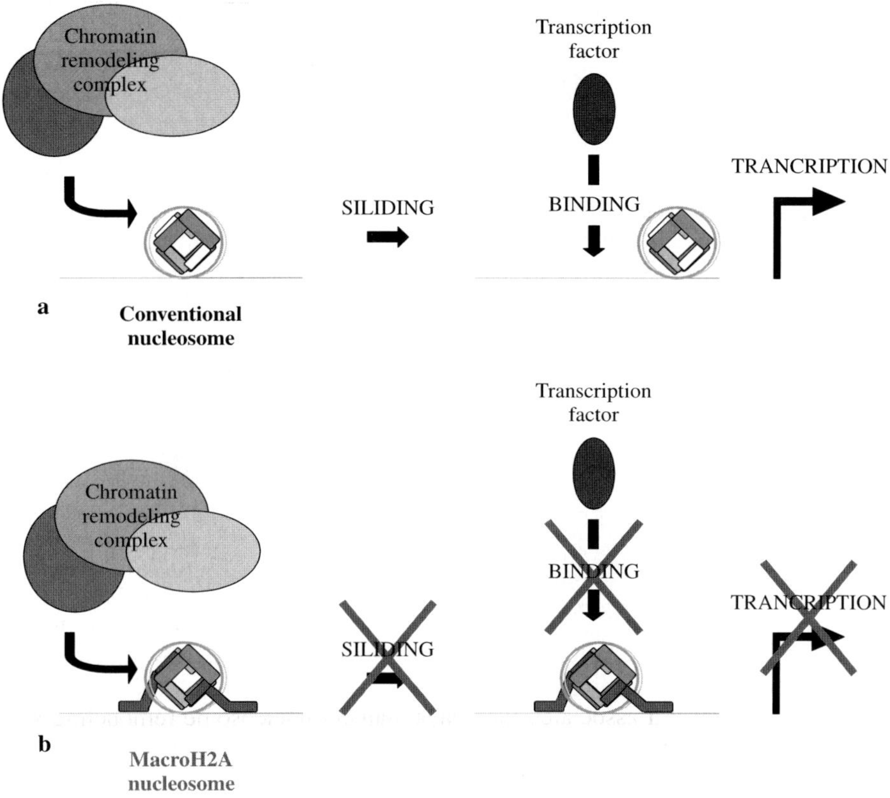

Figure 1. Schematics of the effect of mH2A on transcription. (a) The mobilization of conventional promoter nucleosome by the chromatin remodeling complex generated a nucleosome-free promoter, which allowed the binding of the transcription factors and transcription to proceed. (b) MacroH2A nucleosome cannot be mobilized by the chromatin remodeling complex and the transcriptions factors are unable to bind macroH2A nucleosome containing promoter, which results in inhibition of the initiation of transcription. (See Colour Plate 8.)

of the chromatin remodeling machines SWI/SNF and ACF to remodel them (Angelov *et al*, 2004). Microccocal nuclease digestion data indicated that the variant H2A.Bbd octamer organized only 118 base pairs of DNA and the unusual H2A.Bbd docking domain determined the looser nucleosomal DNA organization (Bao *et al*, 2004). FRET, sedimentation analysis and biochemical studies suggested that the incorporation of H2A.Bbd in the nucleosome resulted in its destabilization (Angelov *et al*, 2004; Bao *et al*, 2004; Gautier *et al*, 2004). The destabilization effect of H2A.Bbd was further confirmed *in vivo* in a series of FRAP studies (Gautier *et al*, 2004). The *in vitro* transcription studies showed that H2A.Bbd nucleosomal arrays, contrary to mH2A arrays, are easily transcribed and acetylated (Angelov *et al*, 2004). These data suggest a role of H2A.Bbd in gene

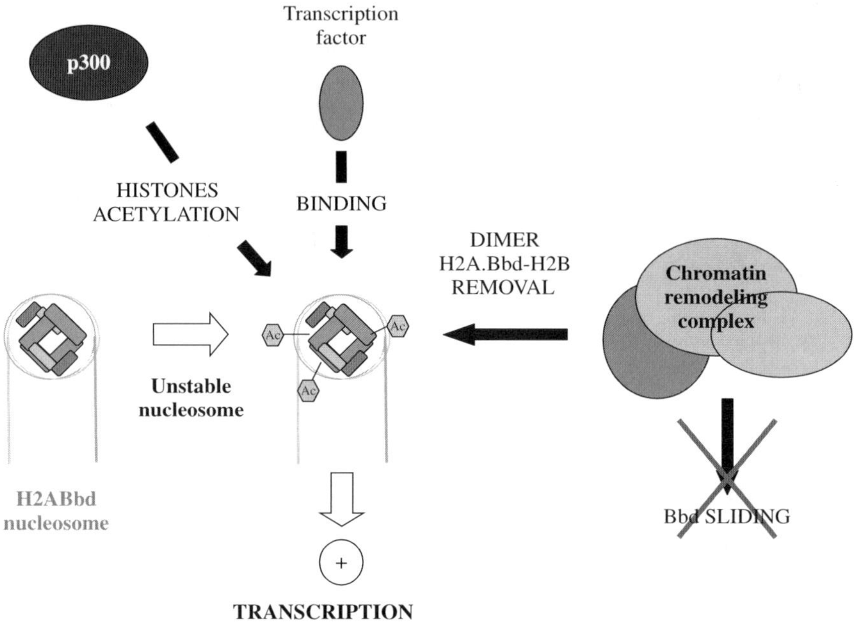

Figure 2. Effect of H2A.Bbd on transcription. The presence of H2A.Bbd confers lower stability and more loose structure to the nucleosomes, which allows the transcription factors binding to this variant nucleosome and thereby recruitment of p300 and acetylation of the promoter proximal histones. The remodeling complex can not mobilize the variant nucleosome, but instead helps in the removal of H2A.Bbd-H2B dimer. All these events facilitate transcription. (See Colour Plate 9.)

activation by a specific, but still unknown mechanism (Fig. 2). The *in vivo* colocalization of H2A.Bbd with regions of the nucleus containing chromatin with hyperacetylated histones further supports this hypothesis (Chadwick and Willard, 2001b).

4. THE HISTONE H3 VARIANT NUCLEOSOMES

The two major histone variants of H3, CENP-A (centromeric protein A) and H3.3, are the object of several studies. CENP-A is present in all eukaryotes and it has different names (Kamakaka and Biggins, 2005), but here for simplicity we will use the name CenH3 (centromeric H3) for this protein. CenH3 is specifically localized to the centromeres and it replaces the canonical H3 in the centromeric chromatin (Palmer *et al*, 1989, 1990). The CenH3 gene is essential and its disruption resulted in very early embryonic lethality (Howman *et al*, 2000). The structured region of CenH3 is well evolutionary conserved, while its N-terminal tail exhibits high divergence between the different organisms (Malik and Henikoff, 2003). *In vitro* studies using reconstituted nucleosome, where conventional H3 was substituted with CenH3 (CenH3 nucleosome), showed that the overall structure of this particle was very similar to the conventional nucleosome (Yoda *et al*, 2000).

Within the centromere the conventional and CenH3 nucleosomes are interspersed (Blower *et al*, 2002). It is widely accepted that CenH3 is absolutely required for the assembly of the kinetochores (Blower and Karpen, 2001; Howman *et al*, 2000). The depletion of CenH3 in worm, fly and mouse resulted in massive mislocalization of the kinetochore proteins (Blower and Karpen, 2001; Howman *et al*, 2000; Oegema *et al*, 2001). The reported data indicated that CenH3 interacts directly or indirectly with several kinetochore proteins (Van Hooser *et al*, 2001). When CenH3 was mistargeted to non-centromeric regions of chromatin, kinetochore proteins were also mistargeted to these regions by a mechanism, which required the N-terminal tail of CenH3 (Van Hooser *et al*, 2001).

CenH3 is selectively and quantitatively retained in the mature spermatozoa and it forms discrete foci in the nucleus (Palmer *et al*, 1990). This suggests that information on the centromeric organization of chromatin is transmitted through generations and thus, CenH3 could act as a specific epigenetic marker.

Three recent reports addressed the mechanism of deposition of CenH3 and the assembly of CenH3 nucleosomes (Foltz *et al*, 2006; Furuyama *et al*, 2006; Okada *et al*, 2006). In *Drosophila* cells, a CenH3 soluble assembly complex was isolated. This complex consists of CenH3, histone H4 and the protein RbAp48 (Furuyama *et al*, 2006). The single protein RbAp48 exhibited histone chaperone properties and it has the ability to promote the assembly of CenH3 nucleosomes *in vitro*. This observation made the chaperone (which is active on its own) distinct from the other chromatin assembly complexes, which consists of several subunits (Furuyama *et al*, 2006). By using tandem affinity purification, a human multiprotein complex (CenH3 NAC) directly recruited by CenH3 nucleosome, was identified (Foltz *et al*, 2006). CenH3 NAC comprised three new centromere proteins (CENP-M, CENP-N and CENP-T) and the assembly of CenH3 NAC depended on these proteins (Foltz *et al*, 2006). In addition, the transcription implicated protein FACT and the histone chaperone nucleophosmin were found associated with CenH3 independent of CenH3 NAC (Foltz *et al*, 2006). Another study claimed that the CenH3 NAC consists of three subcomplexes and each one of them was found to be essential for faithfull chromosome segregation (Fukagawa *et al*, 2004).

H3.3, the other major histone variant of H3, differs from H3 by only four aminoacids, three of which are in the histone fold (globular) domain of the protein. These three residues are crucial for the distinctive deposition of H3.3 during the cell cycle. The epigenetic markers of H3.3, such as di- and tri-methylation of lysine4 (K4), acetylation of lysine9, 18, and −23 and methylation at K79 suggest its important role in the transcriptional activation. H3.3 could be deposited at all stages of cell cycle, unlike histone H3 which assembles only in the S-phase. In *Drosophila*, a good correlation between the transcriptional gene activity and genome-wide localization of H3.3 was demonstrated (Mito *et al*, 2005). In contrast to H3, it is deposited exclusively by the histone chaperone HIRA in a replication-independent manner (Bosch and Suau, 1995; Tagami *et al*, 2004). However, involvement of histone chaperone(s) in the deposition as well as removal of H3.3 during transcriptional activation and repair could not be ruled out.

5. THE H2B VARIANT NUCLEOSOMES

Very few variants of H2B, have been identified so far. The identified H2B variants play some yet not well understood role during gametogenesis (Doenecke *et al*, 1997; Poccia and Green, 1992; Tanphaichitr *et al*, 1978). Recently, H2BFWT, a novel H2B variant, was cloned (Churikov *et al*, 2004). This variant, has very low homology (45% identity) with the conventional H2B but could replace H2B efficiently in the nucleosome (Boulard *et al*, 2006). The H2BFWT nucleosome showed identical DNase I footprinting pattern of the conventional nucleosome, indicating that the variant nucleosome also may have very similar solution structure (Boulard *et al*, 2006). Intriguingly, the highly divergent N-terminal tail of H2AFWT, was unable to recruit chromosome assembly factors and to participate in the formation of condensed mitotic chromosomes (Boulard *et al*, 2006).

The involvement of a testis specific H2B variant in specialized higher order chromosomal domain such as telomeric heterochromatin has been documented (Gineitis *et al*, 2000). Indeed, an uncharacterized H2B variant was identified as a component of the telomere binding complex, which is implicated in telomere membrane attachment in human spermatozoa (Gineitis *et al*, 2000). Presumably, H2BFWT is a sperm specific histone variant associated with telomeric sequences (Churikov *et al*, 2004). Nevertheless, it has been also suggested that H2BFWT could be a component of the large telomeric chromatin of some somatic cells (Churikov *et al*, 2004). Taken together these data suggest that H2BFWT, as CenH3, could be an epigenetic marker required for the transmission of distinct chromatin structure through generations.

6. HISTONE VARIANTS AND DISEASE

6.1. Histone Variants and Cancer

The involvement of histone variants in disease is just beginning to emerge. Some causal relationships between H2AX and tumorigenesis are already well established in a series of experiments with H2AX knockout mice (Celeste *et al*, 2002). H2AX$^{-/-}$ mice are immune deficient, radiosensitive, exhibited repair defects, chromosomal instability and enhanced susceptibility to cancer in the absence of p53 (Bassing *et al*, 2003; Celeste *et al*, 2003a). In particular, mice deficient for both H2AX and p53, developed rapidly immature T and B lymphomas, sarcomas, leukemia and solid tumors (Bassing *et al*, 2003; Celeste *et al*, 2003a). The lymphomas in these mice exhibited an increased frequency of clonal nonreciprocal translocations and amplifications (Celeste *et al*, 2003a). Rescue with H2AX restores both genomic stability and radiation resistance (Celeste *et al*, 2003a). Importantly, if replacement of the H2AX null allele was carried out with a non-phosphorylable H2AX (a single

H2AX aminoacid mutant bearing a serine within the SQ motif substituted to either alanine or glutamic acid residues), no rescue of genome stability and radiation resistance was achieved (Celeste *et al*, 2003a). These data support the hypothesis that H2AX functions as a genome caretaker and tumor suppressor. This is consistent with the cytogenetic location of the H2AX gene in the region 11q23, which is altered in numerous cancers (Monni and Knuutila, 2001).

A recent study claimed that H2AX could be used as a potential target for radiotherapy (Taneja *et al*, 2004). When a peptide of the H2AX C-terminal tail, containing the SQ motif (containing the phosphorylable serine), was used to compete for phophorylation with the endogenous H2AX and to affect the efficiency of DNA repair, the radiotherapy effect on resistant tumor cell lines was clearly increased (Taneja *et al*, 2004). Of note is that in contrast to the cytotoxic agents commonly used in chemotherapy, the targeting of H2AX would not affect the non-irradiated tissue (Taneja *et al*, 2004).

A well characterized feature of human cancers is aneuploidy. CenH3 is, as described above, a key player in kinetochore assembly, maintenance and function. CenH3 was overexpressed in primary human colorectal cancer tissues and this occurred at the transcriptional level (Tomonaga *et al*, 2003). Immunofluorescence studies showed that the tumor cells in these tissues contained the overexpressed CenH3, which in addition was mistargetted to noncentromeric chromatin (Tomonaga *et al*, 2003). In *Drosophila*, mistargetting of CenH3 resulted in the promotion of the formation of ectopic centromeres and multicentric chromosomes, which in turn led to aneuploidy and growth defects (Heun *et al*, 2006). These results strongly suggest that CenH3 could be involved in genomic instability during cancer progression and, in particular, it could play an important role in aneuploidy in colorectal cancers (Tomonaga *et al*, 2003).

6.2. Histone Variants and Infertility

Since, H2AX deficient males, but not female mice, are infertile, H2AX could be involved in male infertility (Celeste *et al*, 2002). The size of the testes of male $H2AX^{-/-}$ mice was twofold reduced, the diameter of the seminuferous tubules was smaller and at early pachytene stage the $H2AX^{-/-}$ cells showed apoptotic features, pointing that the $H2AX^{-/-}$ spermatocytes were arrested in the pachytene stage of meiosis I (Celeste *et al*, 2002). An accumulation of γ-H2AX in the sex body (a condensed structure observed in the nuclei of the mammalian spermatocytes at meotic prophase I, which contains the highly condensed X-Y chromosome pair) was also observed (Fernandez-Capetillo *et al*, 2003). This accumulation did not depend on the meiotic recombination-associated DSB (Fernandez-Capetillo *et al*, 2003). Importantly, the loss of H2AX resulted in the absence of sex-body and no initiation of meiotic sex chromosome inactivation was detected (Fernandez-Capetillo *et al*, 2003). In addition, the X and Y chromosomes in $H2AX^{-/-}$ mice exhibited severe defects in chromosome pairing. All these data evidence for participation of H2AX in remodeling of chromatin and silencing in male meiosis (Fernandez-Capetillo *et al*, 2003).

To date the identified and characterized H2B variants are specifically express in the testes at the early (for instance in spermatogonia) or later stages of spermatogenesis (Moss *et al*, 1989; Unni *et al*, 1995). During spermatogenesis somatic histones are initially replaced by the testis specific histone variants, then by transitional proteins and finally by protamines. It was found that spermatozoa from infertile males contained more histones than these of fertile males, indicating that insufficient removal of histone or the lacking of one variant might be involved in infertility (Chevaillier *et al*, 1987; Foresta *et al*, 1992; Hofmann and Hilscher, 1991). Thus, the human testis H2B variant TH2B has been observed in variable amount among several subfertile males witch are characterized by abnormality of spermatozoa morphology (van Roijen *et al*, 1998). This example points an association between an altered regulation in a testis specific histone variant and infertility.

ACKNOWLEDGEMENTS

This work was supported supported by CNRS, INSERM, ATIP Plus to PB, Région Rhône-Alpes and grants from the Ministère de la Recherche: BCM0070; DRAB 04 2 136; ANR Project n° NT05-1_41978 and the ARCUS-INDIA program.

REFERENCES

Abbott DW, Ivanova VS, Wang X, Bonner WM, Ausio J (2001) Characterization of the stability and folding of H2A.Z chromatin particles: implications for transcriptional activation. J Biol Chem 276:41945–41949

Adam M, Robert F, Larochelle M, Gaudreau L (2001) H2A.Z is required for global chromatin integrity and for recruitment of RNA polymerase II under specific conditions. Mol Cell Biol 21:6270–6279

Adams RR, Carmena M, Earnshaw WC (2001) Chromosomal passengers and the (aurora) ABCs of mitosis. Trends Cell Biol 11:49–54

Allen MD, Buckle AM, Cordell SC, Lowe J, Bycroft M (2003) The crystal structure of AF1521 a protein from Archaeoglobus fulgidus with homology to the non-histone domain of macroH2A. J Mol Biol 330:503–511

Angelov D, Molla A, Perche PY, Hans F, Cote J, Khochbin S, Bouvet P, Dimitrov S (2003) The histone variant macroH2A interferes with transcription factor binding and SWI/SNF nucleosome remodeling. Mol Cell 11:1033–1041

Angelov D, Verdel A, An W, Bondarenko V, Hans F, Doyen CM, Studitsky VM, Hamiche A, Roeder RG, Bouvet P, Dimitrov S (2004) SWI/SNF remodeling and p300-dependent transcription of histone variant H2ABbd nucleosomal arrays. Embo J 23:3815–3824

Arents G, Moudrianakis EN (1993) Topography of the histone octamer surface: repeating structural motifs utilized in the docking of nucleosomal DNA. Proc Natl Acad Sci U S A 90:10489–10493

Arents G, Burlingame RW, Wang B-C, Love WE, Moudrianakis EN (1991) The nucleosomal core histone octamer at 3.1 A resolution: A tripartite protein assembly and a left-handed superhelix. Proc Natl Acad Sci U S A 88:10148–10152

Bao Y, Konesky K, Park YJ, Rosu S, Dyer PN, Rangasamy D, Tremethick DJ, Laybourn PJ, Luger K (2004) Nucleosomes containing the histone variant H2A.Bbd organize only 118 base pairs of DNA. Embo J 23:3314–3324

Bassing CH, Alt FW (2004) H2AX may function as an anchor to hold broken chromosomal DNA ends in close proximity. Cell Cycle 3:149–153

Bassing CH, Suh H, Ferguson DO, Chua KF, Manis J, Eckersdorff M, Gleason M, Bronson R, Lee C, Alt FW (2003) Histone H2AX: a dosage-dependent suppressor of oncogenic translocations and tumors. Cell 114:359–370

Beato M, Eisfeld K (1997) Transcription factor access to chromatin. Nucleic Acids Res 25:3559–3563

Becker PB (2002) Nucleosome sliding: facts and fiction. Embo J 21:4749–4753

Bekker-Jensen S, Lukas C, Kitagawa R, Melander F, Kastan MB, Bartek J, Lukas J (2006) Spatial organization of the mammalian genome surveillance machinery in response to DNA strand breaks. J Cell Biol 173:195–206

Blower MD, Karpen GH (2001) The role of Drosophila CID in kinetochore formation, cell-cycle progression and heterochromatin interactions. Nat Cell Biol 3:730–739

Blower MD, Sullivan BA, Karpen GH (2002) Conserved organization of centromeric chromatin in flies and humans. Dev Cell 2:319–330

Bosch A, Suau P (1995) Changes in core histone variant composition in differentiating neurons: the roles of differential turnover and synthesis rates. Eur J Cell Biol 68:220–225

Boulard M, Gautier T, Mbele GO, Gerson V, Hamiche A, Angelov D, Bouvet P, Dimitrov S (2006) The NH2 tail of the novel histone variant H2BFWT exhibits properties distinct from conventional H2B with respect to the assembly of mitotic chromosomes. Mol Cell Biol 26:1518–1526

Burma S, Chen BP, Murphy M, Kurimasa A, Chen DJ (2001) ATM phosphorylates histone H2AX in response to DNA double-strand breaks. J Biol Chem 276:42462–42467

Celeste A, Petersen S, Romanienko PJ, Fernandez-Capetillo O, Chen HT, Sedelnikova OA, Reina-San-Martin B, Coppola V, Meffre E, Difilippantonio MJ, Redon C, Pilch DR, Olaru A, Eckhaus M, Camerini-Otero RD, Tessarollo L, Livak F, Manova K, Bonner WM, Nussenzweig MC, Nussenzweig A (2002) Genomic instability in mice lacking histone H2AX. Science 296:922–927

Celeste A, Difilippantonio S, Difilippantonio MJ, Fernandez-Capetillo O, Pilch DR, Sedelnikova OA, Eckhaus M, Ried T, Bonner WM, Nussenzweig A (2003a) H2AX haploinsufficiency modifies genomic stability and tumor susceptibility. Cell 114:371–383

Celeste A, Fernandez-Capetillo O, Kruhlak MJ, Pilch DR, Staudt DW, Lee A, Bonner RF, Bonner WM, Nussenzweig A (2003b) Histone H2AX phosphorylation is dispensable for the initial recognition of DNA breaks. Nat Cell Biol 5:675–679

Chadwick BP, Willard HF (2001a) Histone H2A variants and the inactive X chromosome: identification of a second macroH2A variant. Hum Mol Genet 10:1101–1113

Chadwick BP, Willard HF (2001b) A novel chromatin protein, distantly related to histone H2A, is largely excluded from the inactive X chromosome. J Cell Biol 152:375–384

Chadwick BP, Willard HF (2002) Cell cycle-dependent localization of macroH2A in chromatin of the inactive X chromosome. J Cell Biol 157:1113–1123

Chadwick BP, Valley CM, Willard HF (2001) Histone variant macroH2A contains two distinct macrochromatin domains capable of directing macroH2A to the inactive X chromosome. Nucleic Acids Res. 29:2699–2705

Chevaillier P, Mauro N, Feneux D, Jouannet P, David G (1987) Anomalous protein complement of sperm nuclei in some infertile men. Lancet 2:806–807

Choo JH, Kim JD, Chung JH, Stubbs L, Kim J (2006) Allele-specific deposition of macroH2A1 in imprinting control regions. Hum Mol Genet 15:717–724

Churikov D, Siino J, Svetlova M, Zhang K, Gineitis A, Morton Bradbury E, Zalensky A (2004) Novel human testis-specific histone H2B encoded by the interrupted gene on the X chromosome. Genomics 84:745–756

Clarkson MJ, Wells JR, Gibson F, Saint R, Tremethick DJ (1999) Regions of variant histone His2AvD required for Drosophila development. Nature 399:694–697

Costanzi C, Pehrson JR (1998) Histone macroH2A1 is concentrated in the inactive X chromosome of female mammals. Nature 393:599–601

Costanzi C, Pehrson JR (2001) MacroH2A2, a new member of the macroH2A core histone family. J Biol Chem 276:21776–21784

Dhillon N, Kamakaka RT (2000) A histone variant, Htz1p, and a Sir1p-like protein, Esc2p, mediate silencing at HMR. Mol Cell 6:769–780

Doenecke D, Albig W, Bode C, Drabent B, Franke K, Gavenis K, Witt O (1997) Histones: genetic diversity and tissue-specific gene expression. Histochem Cell Biol 107:1–10

Downs JA, Allard S, Jobin-Robitaille O, Javaheri A, Auger A, Bouchard N, Kron SJ, Jackson SP, Cote J (2004) Binding of chromatin-modifying activities to phosphorylated histone H2A at DNA damage sites. Mol Cell 16:979–990

Doyen CM, An W, Angelov D, Bondarenko V, Mietton F, Studitsky VM, Hamiche A, Roeder RG, Bouvet P, Dimitrov S (2006) Mechanism of polymerase II transcription repression by the histone variant macroH2A. Mol Cell Biol 26:1156–1164

Faast R, Thonglairoam V, Schulz TC, Beall J, Wells JR, Taylor H, Matthaei K, Rathjen PD, Tremethick DJ, Lyons I (2001) Histone variant H2A.Z is required for early mammalian development. Curr Biol 11:1183–1187

Fan JY, Gordon F, Luger K, Hansen JC, Tremethick DJ (2002) The essential histone variant H2A.Z regulates the equilibrium between different chromatin conformational states. Nat Struct Biol 9:172–176

Fan JY, Rangasamy D, Luger K, Tremethick DJ (2004) H2A.Z alters the nucleosome surface to promote HP1alpha-mediated chromatin fiber folding. Mol Cell 16:655–661

Fernandez-Capetillo O, Chen HT, Celeste A, Ward I, Romanienko PJ, Morales JC, Naka K, Xia Z, Camerini-Otero RD, Motoyama N, Carpenter PB, Bonner WM, Chen J, Nussenzweig A (2002) DNA damage-induced G2-M checkpoint activation by histone H2AX and 53BP1. Nat Cell Biol 4:993–997

Fernandez-Capetillo O, Mahadevaiah SK, Celeste A, Romanienko PJ, Camerini-Otero RD, Bonner WM, Manova K, Burgoyne P, Nussenzweig A (2003) H2AX is required for chromatin remodeling and inactivation of sex chromosomes in male mouse meiosis. Dev Cell 4:497–508

Fernandez-Capetillo O, Lee A, Nussenzweig M, Nussenzweig A (2004) H2AX: the histone guardian of the genome. DNA Repair (Amst) 3:959–967

Foltz DR, Jansen LE, Black BE, Bailey AO, Yates JR, Cleveland DW (2006) The human CENP-A centromeric nucleosome-associated complex. Nat Cell Biol 8:458–469

Foresta C, Zorzi M, Rossato M, Varotto A (1992) Sperm nuclear instability and staining with aniline blue: abnormal persistence of histones in spermatozoa in infertile men. Int J Androl 15:330–337

Fukagawa T, Nogami M, Yoshikawa M, Ikeno M, Okazaki T, Takami Y, Nakayama T, Oshimura M (2004) Dicer is essential for formation of the heterochromatin structure in vertebrate cells. Nat Cell Biol 6:784–791

Furuyama T, Dalal Y, Henikoff S (2006) Chaperone-mediated assembly of centromeric chromatin in vitro. Proc Natl Acad Sci U S A 103:6172–6177

Gautier T, Abbott DW, Molla A, Verdel A, Ausio J, Dimitrov S (2004) Histone variant H2ABbd confers lower stability to the nucleosome. EMBO Rep 5:715–720

Gineitis AA, Zalenskaya IA, Yau PM, Bradbury EM, Zalensky AO (2000) Human sperm telomere-binding complex involves histone H2B and secures telomere membrane attachment. J Cell Biol 151:1591–1598

Govin J, Caron C, Rousseaux S, Khochbin S (2005) Testis-specific histone H3 expression in somatic cells. Trends Biochem Sci 30:357–359

Grigoryev SA, Nikitina T, Pehrson JR, Singh PB, Woodcock CL (2004) Dynamic relocation of epigenetic chromatin markers reveals an active role of constitutive heterochromatin in the transition from proliferation to quiescence. J Cell Sci 117:6153–6162

Guillemette B, Bataille AR, Gevry N, Adam M, Blanchette M, Robert F, Gaudreau L (2005) Variant histone H2A.Z is globally localized to the promoters of inactive yeast genes and regulates nucleosome positioning. PLoS Biol, 3, e384

Henikoff S, Ahmad K (2005) Assembly of variant histones into chromatin. Annu Rev Cell Dev Biol 21, 133–153

Henikoff S, Furuyama T, Ahmad K (2004) Histone variants, nucleosome assembly and epigenetic inheritance. Trends Genet 20:320–326

Heun P, Erhardt S, Blower MD, Weiss S, Skora AD, Karpen GH (2006) Mislocalization of the Drosophila centromere-specific histone CID promotes formation of functional ectopic kinetochores. Dev Cell 10:303–315

Hofmann N, Hilscher B (1991) Use of aniline blue to assess chromatin condensation in morphologically normal spermatozoa in normal and infertile men. Hum Reprod 6:979–982

Howman EV, Fowler KJ, Newson AJ, Redward S, MacDonald AC, Kalitsis P, Choo KH (2000) Early disruption of centromeric chromatin organization in centromere protein A (Cenpa) null mice. Proc Natl Acad Sci U S A 97:1148–1153

Kamakaka RT, Biggins S (2005) Histone variants: deviants? Genes Dev 19:295–310

Karras GI, Kustatscher G, Buhecha HR, Allen MD, Pugieux C, Sait F, Bycroft M, Ladurner AG (2005) The macro domain is an ADP-ribose binding module. Embo J 24:1911–1920

Kustatscher G, Hothorn M, Pugieux C, Scheffzek K, Ladurner AG (2005) Splicing regulates NAD metabolite binding to histone macroH2A. Nat Struct Mol Biol 12:624–625

Ladurner AG (2003) Inactivating chromosomes: a macro domain that minimizes transcription. Mol Cell 12:1–3

Larochelle M, Gaudreau L (2003) H2A.Z has a function reminiscent of an activator required for preferential binding to intergenic DNA. Embo J 22:4512–4522

Li B, Pattenden SG, Lee D, Gutierrez J, Chen J, Seidel C, Gerton J, Workman JL (2005) Preferential occupancy of histone variant H2AZ at inactive promoters influences local histone modifications and chromatin remodeling. Proc Natl Acad Sci U S A 102:18385–18390

Loyola A, Almouzni G (2004) Histone chaperones, a supporting role in the limelight. Biochim Biophys Acta 1677:3–11

Luger K, Mäder, AW., Richmond RK, Sargent DF, Richmond TJ (1997) Crystal structure of the nucleosome core particle at 2.8 A resolution. Nature 389:251–260

Ma Y, Jacobs SB, Jackson-Grusby L, Mastrangelo MA, Torres-Betancourt JA, Jaenisch R, Rasmussen TP (2005) DNA CpG hypomethylation induces heterochromatin reorganization involving the histone variant macroH2A. J Cell Sci 118:1607–1616

Malik HS, Henikoff S (2003) Phylogenomics of the nucleosome. Nat Struct Biol 10:882–891

Martzen MR, McCraith SM, Spinelli SL, Torres FM, Fields S, Grayhack EJ, Phizicky EM (1999) A biochemical genomics approach for identifying genes by the activity of their products. Science 286:1153–1155

Meneghini MD, Wu M, Madhani HD (2003) Conserved histone variant H2A.Z protects euchromatin from the ectopic spread of silent heterochromatin. Cell 112:725–736

Mermoud JE, Costanzi C, Pehrson JR, Brockdorff N (1999) Histone macroH2A1.2 relocates to the inactive X chromosome after initiation and propagation of X-inactivation. J Cell Biol 147:1399–1408

Mito Y, Henikoff JG, Henikoff S (2005) Genome-scale profiling of histone H3.3 replacement patterns. Nat Genet 37:1090–1097

Mizuguchi G, Shen X, Landry J, Wu WH, Sen S, Wu C (2003) ATP-Driven Exchange of Histone H2AZ Variant Catalyzed by SWR1 Chromatin Remodeling Complex. Science

Monni O, Knuutila S (2001) 11q deletions in hematological malignancies. Leuk Lymphoma 40:259–266

Morrison AJ, Highland J, Krogan NJ, Arbel-Eden A, Greenblatt JF, Haber JE, Shen X (2004) INO80 and gamma-H2AX interaction links ATP-dependent chromatin remodeling to DNA damage repair. Cell 119:767–775

Moss SB, Challoner PB, Groudine M (1989) Expression of a novel histone 2B during mouse spermiogenesis. Dev Biol 133:83–92

Oegema K, Desai A, Rybina S, Kirkham M, Hyman AA (2001) Functional analysis of kinetochore assembly in Caenorhabditis elegans. J Cell Biol 153:1209–1226

Okada M, Cheeseman IM, Hori T, Okawa K, McLeod IX, Yates JR, Desai A, Fukagawa T (2006) The CENP-H-I complex is required for the efficient incorporation of newly synthesized CENP-A into centromeres. Nat Cell Biol

Palmer DK, O'Day K, Margolis RL (1989) Biochemical analysis of CENP-A, a centromeric protein with histone-like properties. Prog Clin Biol Res 318:61–72

Palmer DK, O'Day K, Margolis RL (1990) The centromere specific histone CENP-A is selectively retained in discrete foci in mammalian sperm nuclei. Chromosoma 100:32–36

Pandey NB, Marzluff WF (1987) The stem-loop structure at the 3' end of histone mRNA is necessary and sufficient for regulation of histone mRNA stability. Mol Cell Biol 7:4557–4559

Pehrson J, Fuji RN (1998) Evolutionary conservation of histone macroH2A subtypes and domains. Nucl Acids Res 26:2837–2842

Pehrson JR, Fried VA (1992) MacroH2A, a core histone containing a large nonhistone region. Science 257:1398–1400

Perche P, Vourch C, Souchier C, Robert-Nicoud M, Dimitrov S, Khochbin C (2000) Higher concentrations of histone macroH2A in the Barr body are correlated with higher nucleosome density. Curr Biol 10:1531–1534

Poccia DL, Green GR (1992) Packaging and unpackaging the sea urchin sperm genome. Trends Biochem Sci 17:223–227

Raisner RM, Hartley PD, Meneghini MD, Bao MZ, Liu CL, Schreiber SL, Rando OJ, Madhani HD (2005) Histone Variant H2A.Z Marks the 5' Ends of Both Active and Inactive Genes in Euchromatin. Cell 123:233–248

Rangasamy D, Berven L, Ridgway P, Tremethick DJ (2003) Pericentric heterochromatin becomes enriched with H2A.Z during early mammalian development. Embo J 22:1599–1607

Rangasamy D, Greaves I, Tremethick DJ (2004) RNA interference demonstrates a novel role for H2A.Z in chromosome segregation. Nat Struct Mol Biol 11:650–655

Rasmussen TP, Mastrangelo MA, Eden A, Pehrson JR, Jaenisch R (2000) Dynamic relocalization of histone MacroH2A1 from centrosomes to inactive X chromosomes during X inactivation. J Cell Biol 150:1189–1198

Redon C, Pilch D, Rogakou E, Sedelnikova O, Newrock K, Bonner W (2002) Histone H2A variants H2AX and H2AZ. Curr Opin Genet Dev 12:162–169

Rogakou EP, Pilch DR, Orr AH, Ivanova VS, Bonner WM (1998) DNA double-stranded breaks induce histone H2AX phosphorylation on serine 139. J Biol Chem 273:5858–5868

Rogakou EP, Boon C, Redon C, Bonner WM (1999) Megabase chromatin domains involved in DNA double-strand breaks in vivo. J Cell Biol 146:905–916

Santisteban MS, Kalashnikova T, Smith MM (2000) Histone H2A.Z regulates transcription and is partially redundant with nucleosome remodeling complexes. Cell 103:411–422

Sarma K, Reinberg D (2005) Histone variants meet their match. Nat Rev Mol Cell Biol 6:139–149

Savitsky K, Bar-Shira A, Gilad S, Rotman G, Ziv Y, Vanagaite L, Tagle DA, Smith S, Uziel T, Sfez S *et al* (1995) A single ataxia telangiectasia gene with a product similar to PI-3 kinase. Science 268:1749–1753

Siino JS, Nazarov IB, Svetlova MP, Solovjeva LV, Adamson RH, Zalenskaya IA, Yau PM, Bradbury EM, Tomilin NV (2002) Photobleaching of GFP-labeled H2AX in chromatin: H2AX has low diffusional mobility in the nucleus. Biochem Biophys Res Commun 297:1318–1323

Stiff T, Shtivelman E, Jeggo P, Kysela B (2004) AHNAK interacts with the DNA ligase IV-XRCC4 complex and stimulates DNA ligase IV-mediated double-stranded ligation. DNA Repair (Amst) 3:245–256

Strahl BD, Allis CD (2000) The language of covalent histone modifications. Nature 403:41–45

Suto RK, Clarkson MJ, Tremethick DJ, Luger K (2000) Crystal structure of a nucleosome core particle containing the variant histone H2A.Z. Nature Struct. Biol. 7:1121–1124

Swaminathan J, Baxter EM, Corces VG (2005) The role of histone H2Av variant replacement and histone H4 acetylation in the establishment of Drosophila heterochromatin. Genes Dev 19: 65–76

Tagami H, Ray-Gallet D, Almouzni G, Nakatani Y (2004) Histone H3.1 and H3.3 complexes mediate nucleosome assembly pathways dependent or independent of DNA synthesis. Cell 116:51–61

Taneja N, Davis M, Choy JS, Beckett MA, Singh R, Kron SJ, Weichselbaum RR (2004) Histone H2AX phosphorylation as a predictor of radiosensitivity and target for radiotherapy. J Biol Chem 279:2273–2280

Tanphaichitr N, Sobhon P, Taluppeth N, Chalermisarachai P (1978) Basic nuclear proteins in testicular cells and ejaculated spermatozoa in man. Exp Cell Res 117:347–356

Tomonaga T, Matsushita K, Yamaguchi S, Oohashi T, Shimada H, Ochiai T, Yoda K, Nomura F (2003) Overexpression and mistargeting of centromere protein-A in human primary colorectal cancer. Cancer Res 63:3511–3516

Tsanev R, Russev G, Pashev I, Zlatanova J (1993) Replication and transcription of chromatin. CRC Press, Boca Raton, Fl

Unni E, Zhang Y, Kangasniemi M, Saperstein W, Moss SB, Meistrich ML (1995) Stage-specific distribution of the spermatid-specific histone 2B in the rat testis. Biol Reprod 53:820–826

van Attikum H, Fritsch O, Hohn B, Gasser SM (2004) Recruitment of the INO80 complex by H2A phosphorylation links ATP-dependent chromatin remodeling with DNA double-strand break repair. Cell 119:777–788

van Holde K (1988) Chromatin. Springer-Verlag KG, Berlin, Germany

Van Hooser AA, Ouspenski II, Gregson HC, Starr DA, Yen TJ, Goldberg ML, Yokomori K, Earnshaw WC, Sullivan KF, Brinkley BR (2001) Specification of kinetochore-forming chromatin by the histone H3 variant CENP-A. J Cell Sci 114:3529–3542

van Roijen HJ, Ooms MP, Spaargaren MC, Baarends WM, Weber RF, Grootegoed JA, Vreeburg JT (1998) Immunoexpression of testis-specific histone 2B in human spermatozoa and testis tissue. Hum Reprod 13:1559–1566

Yoda K, Ando S, Morishita S, Houmura K, Hashimoto K, Takeyasu K, Okazaki T (2000) Human centromere protein A (CENP-A) can replace histone H3 in nucleosome reconstitution in vitro. Proc Natl Acad Sci U S A 97:7266–7271

Zhang H, Roberts DN, Cairns BR (2005a) Genome-wide dynamics of Htz1, a histone H2A variant that poises repressed/basal promoters for activation through histone loss. Cell 123:219–231

Zhang R, Poustovoitov MV, Ye X, Santos HA, Chen W, Daganzo SM, Erzberger JP, Serebriiskii IG, Canutescu AA, Dunbrack RL, Pehrson JR, Berger JM, Kaufman PD, Adams PD (2005b) Formation of MacroH2A-containing senescence-associated heterochromatin foci and senescence driven by ASF1a and HIRA. Dev Cell 8:19–30

CHAPTER 5

HISTONE VARIANTS AND COMPLEXES INVOLVED IN THEIR EXCHANGE

THOMAS KUSCH[1,*] AND JERRY L. WORKMAN[2]

[1] *Department of Biochemistry and Molecular Biology, Room #123A, Rutgers University, 604 Allison Road, Piscataway, NJ 08854, USA*

[2] *Stowers Institute for Medical Research, 1000 E. 50th Street, Kansas City, MO 64110, USA*

Abstract: In contrast to canonical histones, which are assembled into nucleosomes during DNA replication, histone variants can be incorporated into specific regions of the genome throughout the cell cycle. Recent findings suggest that histone variants associate with factors mediating their deposition into specialized chromatin domains. The mechanisms of their targeted deposition, their turnover, and their posttranslational modification are not yet fully understood. Emerging evidence indicates that histone variants and associated factors are essential for the epigenetic control of gene expression and other cellular responses. Thus, histone variants and complexes involved in their exchange are likely to play major roles in controlling chromosomal architecture, and their deregulation is expected to be linked to cancers, infertility, mental disorders, ageing, and degenerative diseases

1. INTRODUCTION

The genetic information of eukaryotic cells is propagated in the form of chromosomal DNA. Besides the nucleic acid component, chromosomes contain architectural proteins as stoichiometric components, which are involved in the protective compaction of the fragile DNA double strands. Together, the DNA and proteins form a nucleoprotein structure called chromatin. The fundamental repeating unit of chromatin is the nucleosome core particle. It consists of about 147 base pairs of DNA wrapped around a histone octamer of a $(H3/H4)_2$ tetramer and two (H2A-H2B) heterodimers. One molecule of the linker histone H1 (or H5) binds the linker DNA region between two nucleosome core particles (Bates and Thomas 1981).

*E-mail: TNK@stowers-institute.org

T.K. Kundu and D. Dasgupta (eds.), Chromatin and Disease, 91–109.

H1 is believed to regulate the further compaction of the nucleosomes and to have influence on gene regulation (Parseghian and Hamkalo 2001). All four core histones are synthesized during DNA replication, and they are deposited by histone chaperone complexes to fill the gaps forming behind the progressing replication fork. The $(H3/H4)_2$ tetramers are deposited prior to the incorporation of the heterodimers into the nucleosomes.

Nucleosomes are considered to be some of the most stable nucleoprotein complexes in a cell (Luger and Richmond 1998); however, photobleaching and other studies revealed that chromatin structure is highly dynamic and changes locally in response to gene expression, cell cycle progression, DNA damage, replication, and recombination (reviewed in Jin *et al.* 2005 and references therein). The removal of histones from an octamer as well as the destabilization of the histone-DNA interactions within a nucleosome can increase the efficiency of the above mentioned processes and other events requiring increased access to the DNA (reviewed in Ehrenhofer-Murray 2004). A number of multiprotein complexes introducing structural changes within nucleosomes have been identified. They roughly can be subdivided into (1) enzyme complexes that posttranslationally modify histones or (2) factors that introduce structural changes to the histone DNA interface in an ATP-dependent manner. A third, less understood class of complexes incorporates variants of histones into certain areas of chromosomes.

While some histone variants can become deposited during DNA replication, certain variants also are assembled into chromatin in a replication-independent manner (reviewed in Jin *et al.* 2005). This allows the incorporation of histone variant into chromosomal regions with high levels of histone turnover. Histone variants can distinguish the affected nucleosomes from their canonical counterparts and are likely to have important function in the specialization of chromatin domains and their epigenetic maintenance.

Variants for all four histones as well as the linker histones have been identified, which are distinguished from their canonical counterparts by specific deviations in their amino acid sequences. These specific amino acid exchanges can lead to the generation of structural differences between nucleosomes, but also allow the differential interaction of variants with distinct histone depositing factors, chromatin remodeling and modifying enzyme complexes. We are only beginning to understand the complex roles of histone variant incorporation, their posttranslational modifications, and their turnover in the epigenetic control of chromatin organization. In the following, we attempt to provide an overview of the histone variants thus far identified, function, and the complexes involved in their dynamic turnover.

2. VARIANTS OF HISTONE H3

Initial studies in mice lead to a classification of histone variants into three main subgroups on the basis of their expression and integration into chromosomes (Isenberg 1979). These are DNA replication-dependent, replication-independent, and tissue-specific histones. Variants of histone H3 are represented in all three classes as outlined below. H3 variants have important roles in gene regulation, heterochromatin formation, and faithful transmission of the genome during replication.

2.1. Histones H3.1 and H3.2

The histones H3.1 and H3.2 are expressed and deposited strictly during DNA replication (Table 1). H3.2 is found in all eukaryotes except for *S. cerevisiae*, which only possesses histone H3.3 (Malik and Henikoff 2003). H3.1 and a testis-specific variant H3.1t are solely present in mammals (Witt *et al.* 1996). H3.1 and H3.2 are almost identical and differ in only one amino acid. Recent studies demonstrated that these histone variants might have different biological functions (Hake *et al.* 2006): H3.2 is mostly enriched in posttranslational modifications associated with gene silencing such di- and trimethylation of lysine 27, while H3.1 was shown to be enriched for posttranslational modifications linked to gene activation (lysine 14 acetylation) as well as silencing (dimethylation of lysine 9). These findings support distinct function and chromosomal distribution for each of the two H3 variants, and demonstrate that a single amino acid exchange can have fundamental consequences for subsequent patterns of posttranslational modifications of the affected nucleosomes. It is possible that this single amino acid exchange mediates the interaction of each of the variants with distinct loading factors. The amino acid difference between H3.1 and H3.2 lies within a region that appears to be required for the interaction of H3-relatives and their deposition chaperones. While it has been shown that H3.1 is deposited by a complex containing the histone chaperone CAF-1 (chromatin assembly factor 1), a H3.2-specific chaperone remains yet to be identified (Tagami *et al.* 2004). It also needs to be determined whether these two replication-dependent histone variants indeed are incorporated into different chromosomal regions as suggested by their differential posttranslational modifications.

2.2. Histone H3.3

H3.3 is a replication-independent histone and therefore is available for the chromosomal deposition during interphase (Ahmad and Henikoff 2002). H3.3 interacts with the histone chaperone HIRA, and HIRA is essential for H3.3 deposition (Tagami *et al.* 2004). The interaction domain lies within a C-terminal domain that contains three out of the four amino acid exchanges between H3.2 and H3.3 (Fig. 1). H3.3 is enriched in covalent modifications linked to transcriptionally active genes (McKittrick *et al.* 2004). During transcription at the *hsp70* locus, H3.3 initially replaces H3, and then is turned over constantly until the transcription of the gene seizes (Schwartz and Ahmad 2005). In differentiated, non-dividing cells, H3.3 can contribute to over 25% of the total H3 in chromatin, supporting that it is mainly incorporated in a replication-independent fashion. A deregulation of H3.3 turnover can have severe consequences. In the nerve cells of patients suffering from the X chromosome-linked neurodegenerative Rett syndrome, an excess accumulation

Table 1. Histone variants and their function

Histone	Yeast	*Drosophila*	Mammals	Function	Deposition
H3	–	–	H3.1	Canonical	RD
	–	H3.2	H3.2	Canonical	RD
	H3.3	H3.3	H3.3	Transcription	RD/RI
	Cse4	CID	CENP-A	Centromer identifier; Kinetochore assembly	RD/RI
H2A	–	H2A	H2A	Canonical	RD
	H2A(.X)	H2Av	H2A.X	Phosphorylation in DNA damage, Meiotic recombination, Mammals: Meiotic Sex chromosome condensation	RD/RI
	Htz1	H2Av	H2A.Z	Various functions in different organisms in euchromatin (transcription, boundary function, repression); maintenance or heterochromatin (subtelomeric, pericentric, and other)	RD/RI
	–	–	macroH2A	Inactive X chromosomes in female vertebrates	?
	–	–	$H2A^{Bbd}$	Associated with active euchromatin in female mammals	?
H2B	H2B	H2B	H2B	Canonical	RD
	–	–	TH2B	Testis-specific	?
H4	H4	H4	H4	Canonical	RD
H1/H5	Hho1p	His1.1-3	H1.1-5/H1°	Canonical; Successively replace germline-specific histones during gastrulation and specification	RD/RI?
	–	–	H1t/HILS	Testis-specific; Mid-pachytene stage of meiosis	RI?
	–	–	H1foo	Oocytes; In the zygote up to midblastula transition; Replaces somatic H1 during somatic cell nuclear transfer	RI?

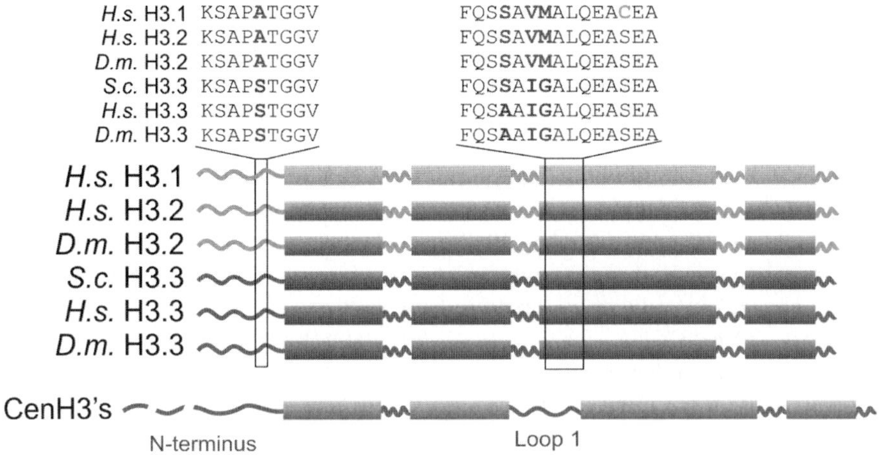

Figure 1. Variants of the histones H3 from yeast (*Saccharomyces cerevisiae; S.c.*), fruit fly (*Drosophila melanogaster; D.m.*), and human (*Homo sapiens; H.s.*). H3.1 is identical to H3.2 with the exception of a serine to cysteine exchange (top). H3.3 differs from H3.1/H3.2 only in four amino acid positions. Centromer-specific histones (CenH3's) have an amino terminus of variable length (between 20 and 200 residues). They also possess an extended loop 1 region in the histone fold domain

of H3.3 leads to a progression of the neurological deterioration. Rett syndrome is usually associated with the loss of the methyl-CpG binding protein MeCP2. The lack of MeCP2 causes reduced levels of H3 methylation at lysine 9, and leads to impaired recruitment of histone deacetylases to chromatin. These defects eventually cause hyperacetylation of H3-relatives within the cells (Shahbazian *et al.* 2002). An excessive incorporation of the transcription-associated variant H3.3 might further contribute to the loss of the epigenetic silencing by DNA methylation and associated chromatin modifications.

More recent studies in *Drosophila* revealed that H3.3 also plays an important role in the male pronucleus after fertilization (Loppin *et al.* 2005). The loss of the H3.3 chaperone HIRA impairs the replacement of paternal non-histone proteins from the sperm nucleus with maternally provided histones including H3.3, while the maternal genome exclusively contains the canonical H3. Thus, H3.3 and its deposition factor HIRA function in early fertilization events and might have a role in imprinting in higher eukaryotes.

2.3. Centromer-specific H3 Variants

Centromer-specific H3 variants (CenH3's) are known as CENP-A in mammals, Cid in flies, Cse4 in yeast (Fig. 1; Table 1; reviewed in Smith 2002). CenH3's are incorporated into nucleosomes independently of centromeric DNA replication. CenH3-containing nucleosomes associate with a number of factors with important roles in centromere structure and kinetochore assembly. Recently, the deposition complex of human CENP-A has been identified (Foltz *et al.* 2006; Okada *et al.* 2006). It contains a number of centromere proteins, the chaperone nucleophosmin, and subunits of the FACT (facilitates chromatin transcription) complex. These findings further illustrate that different histone chaperones are specifically associated with distinct H3 variants as shown for H3.1 and H3.3 (see above). CenH3's are atypical H3 variants with no sequence similarity to H3.1/H3.2 or H3.3 in the N-terminal tail (Fig. 1). This region varies between 20 and 200 amino acids in different species. The N-terminal tail is involved in binding of kinetochore components and other proteins. It is possible that the high variability within the N-terminus of CenH3's is due to changes in centromere structure between these organisms. In their histone fold domain, CenH3's also are only about 50% homologous to other H3 relatives, and they have a longer Loop 1 region with atypical amino acid composition compared to other H3 relatives (Fig. 1). This loop is one of the histone-DNA interaction regions, and the extended surface might contribute to a higher DNA-binding specificity.

3. VARIANTS OF THE HISTONE H2A

Variants of histone H2A are most common in higher eukaryotes. Thus far, five H2A-type histones have been described, of which two are found in all eukaryotes from yeast to mammals (Table 1). These are the histones H2A.X, and H2A.Z (Thatcher and Gorovsky 1994). While all other eukaryotes possess a canonical H2A, *S. cerevisiae* utilizes H2A.X as general, replication-dependent H2A form. Vertebrates possess an additional H2A variant named macroH2A, while the fifth known H2A variant H2ABBd (Barr body-deficient), is only conserved for mammals (Chow and Brown 2003; Gautier *et al.* 2004). Besides the most abundant canonical H2A, which is deposited into chromatin during DNA synthesis, other H2A variants also are synthesized outside of the S phase. Like specialized variants of H3, these proteins also are available for incorporation into chromatin independent of DNA replication.

3.1. Histone H2A.Z

H2A.Z is common for all eukaryotes from yeast to mammals, suggesting that this histone variant has important functions in chromatin-related processes. H2A.Z variants are known as the yeast Htz1, *Plasmodium* H2A.Z, *Tetrahymena* hv1, *Drosophila* H2Av, sea urchin H2AF.Z, avian H2AF, and the mammalian H2A.Z (also see Fig. 2; Table 1). While H2A.Z relatives only share about 60% homology to the canonical H2A, their amino acid sequences are highly conserved suggesting that this variant arose very early during evolution (Thatcher and Gorovsky 1994). H2A.Z contributes to an estimated 5–10% of all nucleosomal H2A variants (Redon *et al.* 2002). Although H2A.Z has been identified in the 1980s, its role within chromatin is not yet fully understood due to its complex biology.

The crystal structure of nucleosomes core particles containing H2A.Z has been resolved and indicates that H2A.Z confers unique structural features to a nucleosome compared to H2A (Suto *et al.* 2000). The most prominent difference exists in the region of interaction between H2A and the $(H3-H4)_2$ tetramer, which also is called docking-domain (amino acids 81–119; also see Fig. 2). Within the docking domain, three amino acid substitutions are mainly responsible for unique structural changes. These exchanges are believed to cause a partial destabilization of the H2A.Z-tetramer interaction. In addition, an extended acidic path on the surface of the H2A.Z-containing histone octamers might change the interaction with the N-terminal tail of H4 or generate a surface for the interaction with other non-histone proteins.

Various studies support the crystallographic analyses and confirmed that nucleosomes containing H2A.Z are more salt-labile than canonical nucleosomes (Abbott *et al.* 2001; Flaus *et al.* 2004; Zhang *et al.* 2005). In fluorescence resonance energy transfer analyses, however, H2A.Z-containing nucleosomes exhibited slower dissociation kinetics compared to their canonical counterparts (Park *et al.* 2004). The different observations made in these studies could be due to the different sources of chromatin as well as the methodology used to compare the stability of H2A.Z-containing nucleosomes. Future studies will be necessary to determine whether the incorporation of H2A.Z into a nucleosome has positive or negative impact on its stability under *in vivo* conditions.

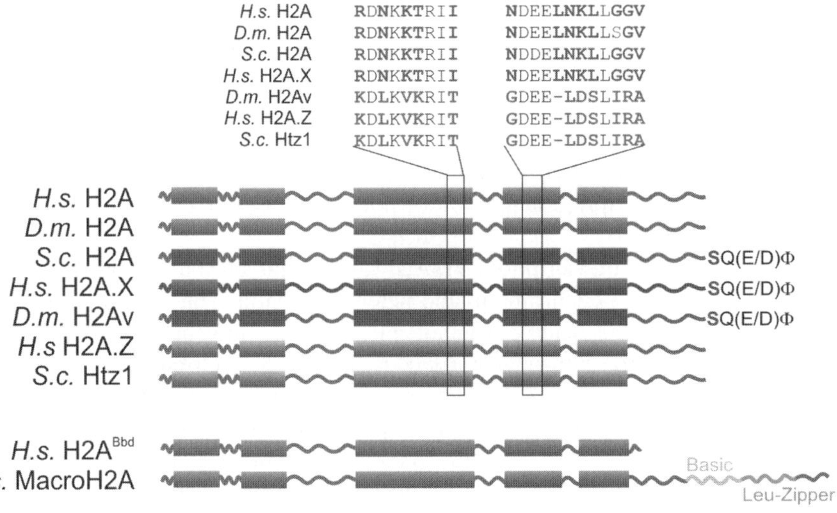

Figure 2. Histone H2A variants from yeast (*Saccharomyces cerevisiae; S.c.*), fruit fly (*Drosophila melanogaster; D.m.*), and human (*Homo sapiens; H.s.*). Two conserved domains distinguish H2A.Z-relatives (boxed regions; amino acid sequences in the top). H2A.X possesses a conserved C-terminal stretch of four amino acids. The serine (red) becomes phosphorylated at sites of DNA damage. $H2A^{Bbd}$ ('Barr body-deficient') and marcoH2A are present in mammals

The genome-wide incorporation of H2A.Z into chromatin is not completely random as suggested by its replication-independent deposition. However, its deposition is not strictly associated to a specific cellular response and also seems to vary between different species. In *Tetrahymena*, the H2A.Z-varaint hv1 is predominantly associated with the transcriptionally active macronucleus (Stargell *et al.* 1993). This early findings suggested that H2A.Z mainly functions in transcriptional regulation in *Tetrahymena*. Genome-wide gene expression studies in *S. cerevisiae* supported this notion. Htz1 is required for the proper expression of about 200 genes (Mizuguchi *et al.* 2004; Kobor *et al.* 2004). Recent studies demonstrated that H2A.Z is deposited into the nucleosome directly upstream of promoters of the majority of gene promoters in yeast (Guillemette *et al.* 2005; Raisner *et al.* 2005; Zhang *et al.* 2005). The positioning of H2A.Z upstream of promoters is remarkably specific. In general, H2A.Z is present in two nucleosomes flanking a nucleosome-free region of about 22 base pairs. This region corresponds to the transcription initiation site of most genes. The 22 base pair sequence is sufficient to promote the deposition of the two H2A.Z-containing nucleosomes, but at the same time remains nucleosome-free (Raisner *et al.* 2005). These observations suggest that the promoter-proximal regions of most genes in yeast share a DNA sequence that mediates the deposition of H2A.Z. Indeed, the sequences contain a highly conserved binding site for the Myb-domain factor Reb1 flanked by an AT-rich tract, both of which are important for positioning of the two H2A.Z-containing nucleosomes. Taken together, the deposition of H2A.Z in promoter-proximal regions is likely to depend on a specific DNA sequence in yeast.

In mammalian cells, H2A.Z becomes removed from open reading frames upon transcriptional activation, suggesting that it might be inhibitory to transcription in higher eukaryotes (Farris *et al.* 2005). The *Drosophila* H2A.Z-relative H2Av also is involved in developmental gene silencing and functions as Polycomb Group protein (Swaminathan *et al.* 2005). Although yeast Htz1 also appears to be enriched upstream of inactive genes, it has not been clarified whether it becomes mobilized upon transcription activation or not (Li *et al.* 2005b; Zhang *et al.* 2005; Millar *et al.* 2006).

Besides its role in transcription, H2A.Z also might play an important role at the boundaries between euchromatin and heterochromatin. The loss of Htz1 causes ectopic spreading of the heterochromatic factors Sir2, Sir3, and Sir4 (Meneghini *et al.* 2003). In chicken erythrocytes, the 5′ region of the β-globin locus is enriched for H2A.Z; However, H2A.Z is also more abundant at the heterochromatin-euchromatin boundary at nearby insulator (Bruce *et al.* 2005), suggesting that H2A.Z an insulator function of H2A.Z at the barrier between eu- and heterochromatin. In mammalian cell lines, H2A.Z is enriched in heterochromatic foci, and it must be assumed that H2A.Z has more functions in these cells than in the definition of eu- versus heterochromatin. Intriguingly, the RNAi-mediated knockdown of H2A.Z causes genomic instability and leads to chromosome breaks, and it likely acts as a guardian of genome integrity (Rangasamy *et al.* 2004). In yeast, the loss of Htz1 causes chromosome segregation defects (Krogan *et al.* 2004). *Drosophila* H2Av is not only present in euchromatin, but also is enriched in the chromocenters, pericentric chromatin and other heterochromatic regions (Leach *et al.* 2000). In flies, H2Av has important roles in heterochromatin formation as demonstrated by the effects of its mutation on Position Effect Variegation, and genome integrity control (Kusch *et al.* 2004; Swaminathan *et al.* 2005).

More recent studies indicate that H2A.Z deposition is dynamic and changes during development. In early mouse embryos, H2.A.Z is concentrated at pericentric heterochromatin, while it a little later becomes depleted from the inactive X chromosome. In cell lines from even later developmental stages, H2A.Z is not detectable at centromeres, but becomes enriched in other heterochromatin in the chromosome arms (Rangasamy *et al.* 2004). H2A.Z is also not detectable in totipotent cells prior to their differentiation, suggesting that the histone variant is not involved in the regulation of early developmental transcriptional programs (Rangasamy *et al.* 2003).

First studies in yeast demonstrated that H2A.Z is deposited by a complex containing 13 proteins including the ATPase Swr1p (Krogan *et al.* 2003; Kobor *et al.* 2004; Mizuguchi *et al.* 2004). This complex exchanges the canonical H2A from nucleosomes with H2A.Z in an ATP-dependent manner. While the Swr complex is mainly responsible for genome-wide deposition, residual chromosomal deposition independent of Swr1 has recently been reported (Wu *et al.* 2005). The subunits of this complex are conserved from yeast to mammals (Kusch *et al.* 2004; Cai *et al.* 2005). In mammals, the Swr1 homologue, SRCAP, catalyzes a similar histone exchange reaction. Intriguingly, mammals possess a second Swr1-type protein called p400. This factor is component of the Tip60 complex (Ikura *et al.* 2000), which is comprised of subunits of the Swr1 complex as well as the yeast NuA4 histone acetyltransferase complex. In *Drosophila*, a homologous complex has been demonstrated to be sufficient to catalyze the replication-independent incorporation of H2Av into nucleosomes (Kusch *et al.* 2004). Taken together, the Tip60 complex from higher eukaryotes is likely to combine the features of the Swr1 and NuA4 complexes.

Intriguingly, the acetyltransferase activity of the fly dTip60 complex is essential for efficient exchange, and recently it has been demonstrated that the NuA4 complex transiently acetylates Htz1 at lysine 14 (Millar *et al.* 2006). This acetylation depends on the transcriptional activity of the affected gene and is redundantly regulated by the NuA4 and SAGA histone acetyltransferase complexes. The latter is also highly conserved between yeast, flies and mammals, and future studies will reveal whether this complex also has a role in the posttranslational modification of H2A.Z in other organisms (Ogryzko *et al.* 1996; Kusch *et al.* 2003). In *Tetrahymena*, the deletion of all six lysines in the N-terminus of hv1 is lethal (Ren and Gorovsky 2001). In *Drosophila*, acetylation of lysine 5 of H2Av by the dTip60 complex is linked to its ATP-dependent exchange during DNA repair (Kusch *et al.* 2004).

The striking similarities between yeast and higher eukaryotes in H2A.Z metabolism suggest that the fundamental molecular mechanisms of chromatin regulation by this variant are mostly conserved. However, the readout of H2A.Z in both euchromatin and heterochromatin appears to have undergone some variation in different species. One must assume that both the targeted deposition as well as differential posttranslational modifications of nucleosomes containing H2A.Z are responsible for the dynamic and somewhat paradoxical roles of this variant in the definition of domains within eu- as well as heterochromatin. It is also likely that the role and regulation of H2A.Z incorporation into chromatin has changed during evolution. Further studies will be necessary to understand the function of H2A.Z-relatives in its entirety.

3.2. Histone H2A.X

H2A and its variant H2A.X are very similar in their amino acid sequences aside from their C-terminal region. Homologues of H2A.X are found in all eukaryotes including fungi, plants, protostomes, and deuterostomes (Malik and Henikoff 2003). Their similarity to the canonical H2A makes it difficult to trace their evolutionary links by comparative analyses. The C-terminal amino acid sequence contains a conserved serine residue (Fig. 2). This serine becomes phosphorylated in chromatin flanking sites of DNA lesions within minutes after the damage has occurred (reviewed in Li *et al.* 2005a). Phosphorylated H2A.X (also known as γ-H2A.X) functions in the control of DNA repair and has other functions in genome integrity control as outlined below. The incorporation of H2A.X into chromatin is rather random and is not restricted to certain chromatin domains. The general deposition of H2A.X also does not seem to occur outside of DNA replication as it has been reported for H2A.Z. Factors that specifically deposit H2A.X/H2B heterodimers have not been identified thus far.

The phosphatidylinositol 3 (PI3)-kinases, ATM, ATR, and DNA-PK, phosphorylate the C-terminal serine of H2A.X within minutes after DNA damage occurs (reviewed in Sedelnikova *et al.* 2003). Members of this kinase family are found in all eukaryotes and they play highly conserved roles in the regulation of DNA damage response pathways. The addition of the bulky, negatively charged phosphate in γ-H2A.X might have some impact on the decondensation of the nucleosome, but it appears more likely that γ-H2A.X serves signaling purposes for a number of reasons. Only about 10% of all H2A.X-containing nucleosomes become phosphorylated at sites of DNA damage. The phosphorylation spreads for several kilobases from the site of DNA damage, while 1–2 kilobases immediately adjacent to the site of the lesion are free of γ-H2A.X (Petersen *et al.* 2001). A mutation mimicking the phosphorylation of H2A.X (serine to glutamate) rescues DNA damage sensitivity of yeast cells lacking PI3 kinases (Downs *et al.* 2000). In mammalian cells, H2A.X is not essential for the initial formation of DNA repair foci, but for their stabilization (Fernandez-Capetillo *et al.* 2003a). In addition, γ-H2A.X recruits cohesins to DNA flanking sites of damage, presumably to tether the broken DNA ends or to allow repair by homologous recombination (Unal *et al.* 2004).

Knockout studies in mice revealed that the loss of H2A.X causes an increased frequency of chromosomal rearrangements, which often result in oncogenic transformations (Celeste *et al.* 2002). In addition, H2A.X seems to be important in functions associated with telomere maintenance and meiotic recombination (Fernandez-Capetillo *et al.* 2003b). In male mice, H2AX functions in the silencing of the sex chromosomes as well as in meiosis and its loss results in infertility (Fernandez-Capetillo *et al.* 2003c). Intriguingly, H2A.X is also necessary for the incorporation of the H2A variant macroH2A1.2 (see below) into the inactivated sex chromosomes. The underlying mechanism and the link between these H2A variants are not yet understood. H2A.X also mediates apoptosis and somatic recombination events (Petersen *et al.* 2001).

In yeast, H2A.X is dephosphorylated by the phosphatase PPH3p after it has been released from chromatin. The release of γ-H2A.X from repaired chromatin is independent of DNA replication, and it therefore must be assumed that a chromatin remodeling complex actively exchanges γ-H2A.X from nucleosomes. Thus far, several candidate remodeling complexes have been identified that specifically target nucleosomes containing γ-H2A.X.

A complex containing the ATPase Ino80p is recruited to sites of DNA double strand breaks and its deletion renders cells sensitive to genotoxic stresses (Downs *et al.* 2004; Morrison *et al.* 2004; van Attikum *et al.* 2004). Although Ino80 shares similarities in the ATPase domain with the H2A/H2A.Z-exchange factor Swr1p, it has not yet been demonstrated that the Ino80 complex indeed can catalyze the removal of γ-H2A.X from chromatin. Like for the INO80 complex, mutants for subunits of the SWR1 complex also exhibit DNA-damage sensitivity. Both complexes contain Arp4, a protein that preferentially interacts with γ-H2A.X over its unmodified counterpart (Downs *et al.* 2004), and it therefore cannot be excluded that both complexes function in γ-H2A.X turnover at sites of DNA damage. Arp4-relatives are also present in another complex with a conserved role in DNA repair: The Tip60-type complexes from yeast, flies, and mammals contain Arp4-relatives and their histone acetyltransferase activity is important for DNA double strand break repair (Ikura *et al.* 2000; Allard *et al.* 2004; Kusch *et al.* 2004). Indeed, the *Drosophila* dTip60 complex is capable of exchanging γ-H2Av with unmodified H2Av *in vitro* (Kusch *et al.* 2004). Mutations of subunits of the fly dTip60 complex severely impair the clearance of γ-H2Av from chromatin during DNA repair. Since the dTip60 complex is a combination of the yeast NuA4 and SWR1 complexes, it appears highly likely that the mechanisms of γ-H2A.X clearance are conserved in eukaryotes. The role of Ino80 in this process remains somewhat unclear. Recent studies in *Drosophila* suggest that Ino80 is involved in repression and might contribute to the generation of more condensed chromatin domains (Klymenko *et al.* 2006). Indeed, the loss of Ino80 complex subunits leads to a precocious decrease of γ-H2Av levels in *Drosophila* after DNA damage, suggesting that Ino80 might protect γ-H2A.X from premature release from damaged chromatin (T.K., unpublished). This might provide a mechanism by which damaged chromatin is remodeled to facilitate access of the damaged DNA to repair proteins without losing an important checkpoint for damage-dependent cell cycle arrest.

3.3. Histone MacroH2A

MacroH2A is an unusual vertebrate-specific H2A lineage. MacroH2A is enriched in the inactive female X chromosome ('Barr body') and in the inactivated male sex chromosomes during meiosis (Costanzi and Pehrson 1998; Hoyer-Fender *et al.* 2000). Its role in X chromosome inactivation might be a rather recent evolutionary event, since macroH2A is also present in birds, which do not regulate dosage compensation by inactivation of one sex chromosome (Ellegren 2002). Two distinct macroH2A variants have been identified in both mice and humans. Although they are closely related, their amino acid sequence deviates in the macroH2A-specific extended C-terminal stretch of about 200 amino acids. This C-terminus has similarity to nucleic acid binding domains, a putative leucine zipper domain, and can bind histone deacetylases (Chakravarthy *et al.* 2005). The C-terminal domain of macroH2A also bears strong homology to proteins with phosphoesterase activity found in RNA viruses, eubacteria, archea, and eukaryotes (Pehrson and Fuji 1998). This suggests that the C-terminal domain might carry out an enzymatic function; however, such an enzymatic activity and a potential substrate of macroH2A have not yet been described. Several reports support a role of macroH2A in transcriptional silencing (Angelov *et al.* 2003; Doyen *et al.* 2006). Nucleosomal arrays containing macroH2A are not accessible to transcription factors and are resistant to mobilization by chromatin remodeling complexes. In addition, macroH2A interferes with P300-dependent histone acetylation and RNA polymerase II passage through the nucleosomes *in vitro*. Taken together, macroH2A appears to confer a tighter packaging of nucleosomes. Little is known about the incorporation of macroH2A into chromosomes, and no macroH2A-specific deposition factors have been identified thus far.

3.4. Histone $H2A^{Bbd}$

H2A 'Barr body-deficient' (Bbd) is an evolutionary relatively 'young' histone variant sharing only about 48% amino acid sequence similarity to H2A. This histone variant appears to be specific for mammals (Chadwick and Willard 2001). As indicated by the name, the transcriptionally inactive and highly condensed X chromosome in female mammals (also known as 'Barr body') is depleted for $H2A^{Bbd}$, while this variant is detectable in autosomes and the active sex chromosomes. This observation suggested that $H2A^{Bbd}$ is linked to transcriptionally active euchromatin. $H2A^{Bbd}$ cofractionates in sedimentation centrifugation with hyperacetylated histone H4, further corroborating that it associates with transcriptionally active euchromatin.

Nucleosomes core particles containing $H2A^{Bbd}$ only have 118 base pairs of DNA incorporated compared to the canonical nucleosomes protecting about 147 base pairs from micrococcal nuclease (Bao *et al.* 2004). These nucleosomes are more flexible in structure and might facilitate passage of RNA polymerase II. However, the function of this histone variant in mammalian cells is not fully understood. As for macroH2A, it is also not yet known how and when H2A^{Bbd} becomes deposited into chromatin, and which chaperones associate with this H2A variant.

4. VARIANTS OF OTHER HISTONES

4.1. Histone H4

Histone H4 seems to be the most slowly evolving histones, very few amino acid exchanges have been reported for fungi, plants, and metazoans (Malik and Henikoff 2003). One explanation for this could be the extensive protein-protein interactions between H4 and other histones within the nucleosome. Amino acid substitutions within H4 would be possible only in a few positions without disrupting the structure or the nucleosome. The N-terminal tail of H4 is also conserved, since it is extensively posttranslationally modified by methyltransferases, kinases, acetyltransferases, and other enzymes. Thus far, no H4 gene with expression outside of S-phase has been reported, supporting that H4 has not undergone any functional specialization during evolution.

4.2. Histone H2B

A number of variants of histone H2B have been identified, but they do not appear to have evolved significantly different functions. H2B interacts with nucleosomal DNA in the histone fold domain and in the C-terminus. This helix lies on the outer plane of the nucleosomal 'disc' and could also help the stacking of nucleosomes into higher order structures. Two notable exceptions for H2B lineages have gonadal functions (Table 1). Vertebrates possess a testis-specific histone variant TH2B, which is detectable in chromatin during early spermatogenesis until mid-late pachytene (Thatcher and Gorovsky 1994). In lilies, a pollen-specific H2B variant has been described. Lilies also possess pollen-specific H2A and H3 variants (Ueda *et al.* 2000). The function and deposition of germline-specific H2B histone variants has yet to be further defined.

4.3. Linker Histone Variants

In mammals, at least six somatic H1/H5 variants have been identified (see Fig. 3; Table 1). These are H1.1 to H1.5, and H1°. In addition, an oocyte-specific linker, H1oo/H1foo, and two testis-specific forms, H1t and HILS1, were described (reviewed in Khochbin *et al.* 2001; Woodcock *et al.* 2006).

In the mid-pachytene stage of meiosis, H1t becomes preferentially enriched in testicular cells and contributes to up to 50% of the total chromosomal linker histones (Cole *et al.* 1986). The deletion of H1t has no effect on viability and fertility of mice (Drabent *et al.* 2000). It is not quite clear whether other H1 variants are upregulated in pachytene spermatocytes of H1t-deficient animals and compensate for the loss of H1t (Drabent *et al.* 2003; Nayernia *et al.* 2005). Thus, very little is known about

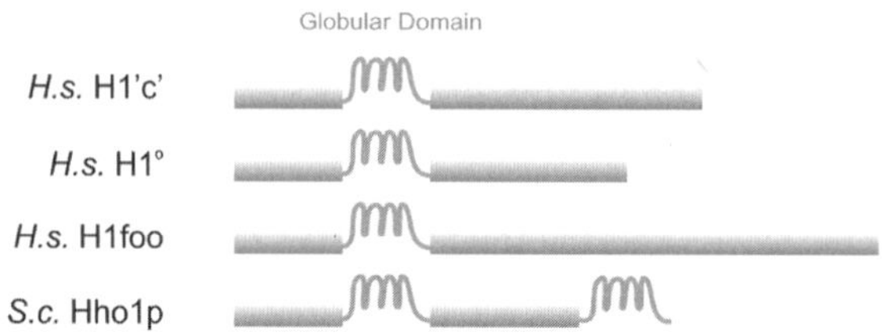

Figure 3. Representative linker histone variants from human (*Homo sapiens; H.s.*), and yeast (*Saccharomyces cerevisiae; S.c.*). The globular domain indicates the winged helix DNA binding domain that is common for all linker histones. H1c is a typical somatic linker histone common for most cells. H° is expressed in all somatic cells, but is more abundant in terminally differentiated tissues. H1foo is an oocyte-specific linker histone. The structure of *S. cerevisiae* Hho1p with two globular domains is a notable deviation

the function, deposition, and turnover of H1t as well as the other testis-specific variant HILS1. Thus far, no linker histone-specific deposition complexes have been identified either.

The oocyte-specific H1 variant, H1Foo (also known as H1oo), becomes incorporated into chromatin in the growing and maturing oocyte as well as in the zygote prior to midblastula transition (Tanaka *et al.* 2001, 2005). Upon fertilization, H1Foo rapidly associates with the introduced paternal genome and replaces sperm-specific histone-like proteins (Becker *et al.* 2005). In contrast to H1, only H1Foo can properly associate with chromatin in developing oocytes. This specificity is conferred by the N-terminal and globular domains of H1Foo. Intriguingly, H1 is replaced by H1Foo in a similar fashion during somatic cell nuclear transfer. This strongly suggests that H1Foo associates with a germline-specific histone deposition and exchange machinery, which has not yet been characterized. After midblastula transition, H1Foo is gradually replaced by somatic linker histones. In *Xenopus*, this replacement is completed at the end of gastrulation (Steinbach *et al.* 1997). The accumulation of somatic H1 is rate limiting to the loss of mesodermal competence and leads to the selective silencing of mesoderm-specifying transcriptional regulators. Future studies will be necessary to shed more light on the function and structure of H1Foo, which apparently plays a key role in the regulation of early developmental transcription programs.

A series of recent studies has questioned the role of H1 as general repressor of transcription and as stoichiometric unit of chromatin (reviewed in Catez *et al.* 2006). Live imaging studies have demonstrated that H1 is highly mobile in living cells, suggesting that it might not necessarily package chromatin into a static higher order structure as suggested from earlier studies. These findings also suggest that H1 might be subjected to active removal from chromosomes in large proportions. In addition, the ratio of H1 to nucleosomes is not one to one, and it is possible that H1 deposition also might be restricted to certain regions of chromatin. In mouse embryonic stem cells, a twofold reduction of H1 levels leads to shortening of the nucleosomal repeat length; however, the expression of less than 30 genes was affected in these cells. Yet, this two-fold reduction led to severe developmental defects at midgestation stage, causing an early embryonic death of the affected animals.

In general, the differences in expression and structure of H1 variants strongly suggests that variants of linker histones have important roles in chromatin architecture, and might be essential players in the epigenetic control of developmental gene expression. Future studies will be necessary to identify factors that target, modify, and mobilize different linker histones.

5. CONCLUDING REMARKS

From our current albeit limited knowledge, the incorporation of histone variants into chromatin appears to have major impact on the organization of chromatin in cells. Most importantly, the turnover and the patterns of their posttranslational modifications are highly dynamic and undergo dramatic changes in developing as well as maturing and senescing cells. Thus, histone variant metabolism is likely to be one of the main determinants in chromatin architecture and eventually the regulation and maintenance of developmental gene expression programs. We only are beginning to understand how the manifestation and the progression of cancer and hereditary diseases depend on epigenetic aspects of chromatin regulation by histone variants. Deciphering the functions of histone variants, their modifications, and their interplay within chromatin will be an endeavor for many research laboratories in the upcoming years. The generation of epigenetic maps of the human genome is already on its way.

Along this line, a great challenge of equal importance will be the characterization of multiprotein machineries that are involved in histone variant exchange and thereby regulate the dynamic changes of epigenetic patterning within cells. The identification of these factors will unravel the molecular nature of cancers, infertility, mental disorders, ageing, and degenerative diseases and eventually will help develop new and effective diagnostics and therapeutics.

REFERENCES

Abbott DW, Ivanova VS, Wang X, Bonner WM, Ausio J (2001) Characterization of the stability and folding of H2A.Z chromatin particles: implications for transcriptional activation. J Biol Chem 276:41945–41949

Ahmad K, Henikoff S (2002) The histone variant H3.3 marks active chromatin by replication-independent nucleosome assembly. Mol Cell 9:1191–1200

Allard S, Masson JY, Cote J (2004) Chromatin remodeling and the maintenance of genome integrity. Biochim Biophys Acta 1677:158–64

Angelov D, Molla A, Perche PY, Hans F, Cote J, Khochbin S, Bouvet P, Dimitrov S (2003) The histone variant macroH2A interferes with transcription factor binding and SWI/SNF nucleosome remodeling. Mol Cell 11:1033–1041

Bao Y, Konesky K, Park YJ, Rosu S, Dyer PN, Rangasamy D, Tremethick DJ, Laybourn PJ, Luger K (2004) Nucleosomes containing the histone variant H2A.Bbd organize only 118 base pairs of DNA. Embo J 23:3314–3324

Bates DL, Thomas JO (1981) Histones H1 and H5: one or two molecules per nucleosome? Nucleic Acids Res 9:5883–5894

Becker M, Becker A, Miyara F, Han Z, Kihara M, Brown DT, Hager GL, Latham K, Adashi EY, Misteli T (2005) Differential in vivo binding dynamics of somatic and oocyte-specific linker histones in oocytes and during ES cell nuclear transfer. Mol Biol Cell 16:3887–3895

Bruce K, Myers FA, Mantouvalou E, Lefevre P, Greaves I, Bonifer C, Tremethick DJ, Thorne AW, Crane C-Robinson (2005) The replacement histone H2A.Z in a hyperacetylated form is a feature of active genes in the chicken. Nucleic Acids Res 33:5633–5639

Cai Y, Jin J, Florens L, Swanson SK, Kusch T, Li B, Workman JL, Washburn MP, Conaway RC, Conaway JW (2005) The mammalian YL1 protein is a shared subunit of the TRRAP/TIP60 histone acetyltransferase and SRCAP complexes. J Biol Chem 280:13665–13670

Catez F, Ueda T, Bustin M (2006) Determinants of histone H1 mobility and chromatin binding in living cells. Nat Struct Mol Biol 13:305–310

Celeste A, Petersen S, Romanienko PJ, Fernandez-Capetillo O, Chen HT, Sedelnikova OA, Reina-San-Martin B, Coppola V, Meffre E, Difilippantonio MJ, Redon C, Pilch DR, Olaru A, Eckhaus M, Camerini-Otero RD, Tessarollo L, Livak F, Manova K, Bonner WM, Nussenzweig MC, Nussenzweig A (2002) Genomic instability in mice lacking histone H2AX. Science 296:922–927

Chadwick BP, Willard HF (2001) A novel chromatin protein, distantly related to histone H2A, is largely excluded from the inactive X chromosome. J Cell Biol 152:375–384

Chakravarthy S, Gundimella SK, Caron C, Perche PY, Pehrson JR, Khochbin S, Luger K (2005) Structural characterization of the histone variant macroH2A. Mol Cell Biol 25:7616–7624

Chow JC, Brown CJ (2003) Forming facultative heterochromatin: silencing of an X chromosome in mammalian females. Cell Mol Life Sci 60:2586–2603

Cole KD, Kandala JC, Kistler WS (1986) Isolation of the gene for the testis-specific H1 histone variant H1t. J Biol Chem 261:7178–7183

Costanzi C, Pehrson JR (1998) Histone macroH2A1 is concentrated in the inactive X chromosome of female mammals. Nature 393:599–601

Downs JA, Lowndes NF, Jackson SP (2000) A role for Saccharomyces cerevisiae histone H2A in DNA repair. Nature 408:1001–1004

Downs JA, Allard S, Jobin-Robitaille O, Javaheri A, Auger A, Bouchard N, Kron SJ, Jackson SP, Cote J (2004) Binding of chromatin-modifying activities to phosphorylated histone H2A at DNA damage sites. Mol Cell 16:979–990

Doyen CM, An W, Angelov D, Bondarenko V, Mietton F, Studitsky VM, Hamiche A, Roeder RG, Bouvet P, Dimitrov S (2006) Mechanism of polymerase II transcription repression by the histone variant macroH2A. Mol Cell Biol 26:1156–1164

Drabent B, Saftig P, Bode C, Doenecke D (2000) Spermatogenesis proceeds normally in mice without linker histone H1t. Histochem Cell Biol 113:433–442

Drabent B, Benavente R, Hoyer-Fender S (2003) Histone H1t is not replaced by H1.1 or H1.2 in pachytene spermatocytes or spermatids of H1t-deficient mice. Cytogenet Genome Res 103:307–313

Ehrenhofer-Murray AE (2004) Chromatin dynamics at DNA replication, transcription and repair. Eur J Biochem 271:2335–2349

Ellegren H (2002) Dosage compensation: do birds do it as well? Trends Genet 18:25–28

Farris SD, Rubio ED, Moon JJ, Gombert WM, Nelson BH, Krumm A (2005) Transcription-induced chromatin remodeling at the c-myc gene involves the local exchange of histone H2A.Z. J Biol Chem 280:25298–25303

Fernandez-Capetillo O, Celeste A, Nussenzweig A (2003a) Focusing on foci: H2AX and the recruitment of DNA-damage response factors. Cell Cycle 2:426–427

Fernandez-Capetillo O, Liebe B, Scherthan H, Nussenzweig A (2003b) H2AX regulates meiotic telomere clustering. J Cell Biol 163:15–20

Fernandez-Capetillo O, Mahadevaiah SK, Celeste A, Romanienko PJ, Camerini-Otero RD, Bonner WM, Manova K, Burgoyne P, Nussenzweig A (2003c) H2AX is required for chromatin remodeling and inactivation of sex chromosomes in male mouse meiosis. Dev Cell 4:497–508

Flaus A, Rencurel C, Ferreira H, Wiechens N, Owen-Hughes T (2004) Sin mutations alter inherent nucleosome mobility. Embo J 23:343–353

Foltz DR, Jansen LE, Black BE, Bailey AO, Yates JR, 3rd, Cleveland DW (2006) The human CENP-A centromeric nucleosome-associated complex. Nat Cell Biol 8:458–469

Gautier T, Abbott DW, Molla A, Verdel A, Ausio J, Dimitrov S (2004) Histone variant H2ABbd confers lower stability to the nucleosome. EMBO Rep 5:715–720

Guillemette B, Bataille AR, Gevry N, Adam M, Blanchette M, Robert F, Gaudreau L (2005) Variant histone H2A.Z is globally localized to the promoters of inactive yeast genes and regulates nucleosome positioning. PLoS Biol 3:e384.

Hake SB, Garcia BA, Duncan EM, Kauer M, Dellaire G, Shabanowitz J, Bazett-Jones DP, Allis CD, Hunt DF (2006) Expression patterns and post-translational modifications associated with mammalian histone H3 variants. J Biol Chem 281:559–568

Hoyer-Fender S, Costanzi C, Pehrson JR (2000) Histone macroH2A1.2 is concentrated in the XY-body by the early pachytene stage of spermatogenesis. Exp Cell Res 258:254–260

Ikura T, Ogryzko VV, Grigoriev M, Groisman R, Wang J, Horikoshi M, Scully R, Qin J, Nakatani Y (2000) Involvement of the TIP60 histone acetylase complex in DNA repair and apoptosis. Cell 102:463–473

Isenberg I (1979) Histones. Annu Rev Biochem 48:159–191

Jin J, Cai Y, Li B, Conaway RC, Workman JL, Conaway JW, Kusch T (2005) In and out: histone variant exchange in chromatin. Trends Biochem Sci 30:680–687

Khochbin S (2001) Histone H1 diversity: bridging regulatory signals to linker histone function. Gene 271:1–12

Klymenko T, Papp B, Fischle W, Kocher T, Schelder M, Fritsch C, Wild B, Wilm M, Muller J (2006) A Polycomb group protein complex with sequence-specific DNA-binding and selective methyl-lysine-binding activities. Genes Dev 20:1110–1122

Kobor MS, Venkatasubrahmanyam S, Meneghini MD, Gin JW, Jennings JL, Link AJ, Madhani HD, Rine J (2004) A protein complex containing the conserved Swi2/Snf2-related ATPase Swr1p deposits histone variant H2A.Z into euchromatin. PLoS Biol 2:E131.

Krogan NJ, Keogh MC, Datta N, Sawa C, Ryan OW, Ding H, Haw RA, Pootoolal J, Tong A, Canadien V, Richards DP, Wu X, Emili A, Hughes TR, Buratowski S, Greenblatt JF (2003) A Snf2 family ATPase complex required for recruitment of the histone H2A variant Htz1. Mol Cell 12:1565–1576

Krogan NJ, Baetz K, Keogh MC, Datta N, Sawa C, Kwok TC, Thompson NJ, Davey MG, Pootoolal J, Hughes TR, Emili A, Buratowski S, Hieter P, Greenblatt JF (2004) Regulation of chromosome stability by the histone H2A variant Htz1, the Swr1 chromatin remodeling complex, and the histone acetyltransferase NuA4. Proc Natl Acad Sci U S A 101:13513–13518

Kusch T, Florens L, Macdonald WH, Swanson SK, Glaser RL, Yates JR, 3rd, Abmayr SM, Washburn MP, Workman JL (2004) Acetylation by Tip60 is required for selective histone variant exchange at DNA lesions. Science 306:2084–2087

Kusch T, Guelman S, Abmayr SM, Workman JL (2003) Two Drosophila Ada2 homologues function in different multiprotein complexes. Mol Cell Biol 23:3305–3319

Leach TJ, Mazzeo M, Chotkowski HL, Madigan JP, Wotring MG, Glaser RL (2000) Histone H2A.Z is widely but nonrandomly distributed in chromosomes of Drosophila melanogaster. J Biol Chem 275:23267–23272

Li A, Eirin-Lopez JM, Ausio J (2005a) H2AX: tailoring histone H2A for chromatin-dependent genomic integrity. Biochem Cell Biol 83:505–515

Li B, Pattenden SG, Lee D, Gutierrez J, Chen J, Seidel C, Gerton J, Workman JL (2005b) Preferential occupancy of histone variant H2AZ at inactive promoters influences local histone modifications and chromatin remodeling. Proc Natl Acad Sci U S A 102:18385–18390

Loppin B, Bonnefoy E, Anselme C, Laurencon A, Karr TL, Couble P (2005) The histone H3.3 chaperone HIRA is essential for chromatin assembly in the male pronucleus. Nature 437:1386–1390

Luger K, Richmond TJ (1998) DNA binding within the nucleosome core. Curr Opin Struct Biol 8:33–40

Malik HS, Henikoff S (2003) Phylogenomics of the nucleosome. Nat Struct Biol 10:882–891

McKittrick E, Gafken PR, Ahmad K, Henikoff S (2004) Histone H3.3 is enriched in covalent modifications associated with active chromatin. Proc Natl Acad Sci U S A 101:1525–1530
Meneghini MD, Wu M, Madhani HD (2003) Conserved histone variant H2A.Z protects euchromatin from the ectopic spread of silent heterochromatin. Cell 112:725–736
Millar CB, Xu F, Zhang K, Grunstein M (2006) Acetylation of H2AZ Lys 14 is associated with genome-wide gene activity in yeast. Genes Dev 20:711–722
Mizuguchi G, Shen X, Landry J, Wu WH, Sen S, Wu C (2004) ATP-driven exchange of histone H2AZ variant catalyzed by SWR1 chromatin remodeling complex. Science 303:343–348
Morrison AJ, Highland J, Krogan NJ, Arbel-Eden A, Greenblatt JF, Haber JE, Shen X (2004) INO80 and gamma-H2AX interaction links ATP-dependent chromatin remodeling to DNA damage repair. Cell 119:767–775
Nayernia K, Drabent B, Meinhardt A, Adham IM, Schwandt I, Muller C, Sancken U, Kleene KC, Engel W (2005) Triple knockouts reveal gene interactions affecting fertility of male mice. Mol Reprod Dev 70:406–416
Ogryzko VV, Schiltz RL, Russanova V, Howard BH, Nakatani Y (1996) The transcriptional coactivators p300 and CBP are histone acetyltransferases. Cell 87:953–959
Okada M, Cheeseman IM, Hori T, Okawa K, McLeod IX, Yates JR, 3rd, Desai A, Fukagawa T (2006) The CENP-H-I complex is required for the efficient incorporation of newly synthesized CENP-A into centromeres. Nat Cell Biol 8:446–457
Park YJ, Dyer PN, Tremethick DJ, Luger K (2004) A new fluorescence resonance energy transfer approach demonstrates that the histone variant H2AZ stabilizes the histone octamer within the nucleosome. J Biol Chem 279:24274–24282
Parseghian MH, Hamkalo BA (2001) A compendium of the histone H1 family of somatic subtypes: an elusive cast of characters and their characteristics. Biochem Cell Biol 79:289–304
Pehrson JR, Fuji RN (1998) Evolutionary conservation of histone macroH2A subtypes and domains. Nucleic Acids Res 26:2837–2842
Petersen S, Casellas R, Reina-San-Martin B, Chen HT, Difilippantonio MJ, Wilson PC, Hanitsch L, Celeste A, Muramatsu M, Pilch DR, Redon C, Ried T, Bonner WM, Honjo T, Nussenzweig MC, Nussenzweig A (2001) AID is required to initiate Nbs1/gamma-H2AX focus formation and mutations at sites of class switching. Nature 414:660–665
Raisner RM, Hartley PD, Meneghini MD, Bao MZ, Liu CL, Schreiber SL, Rando OJ, Madhani HD (2005) Histone variant H2A.Z marks the 5' ends of both active and inactive genes in euchromatin. Cell 123:233–248
Rangasamy D, Berven L, Ridgway P, Tremethick DJ (2003) Pericentric heterochromatin becomes enriched with H2A.Z during early mammalian development. Embo J 22:1599–1607
Rangasamy D, Greaves I, Tremethick DJ (2004) RNA interference demonstrates a novel role for H2A.Z in chromosome segregation. Nat Struct Mol Biol 11:650–655
Redon C, Pilch D, Rogakou E, Sedelnikova O, Newrock K, Bonner W (2002) Histone H2A variants H2AX and H2AZ. Curr Opin Genet Dev 12:162–169
Ren Q, Gorovsky MA (2001) Histone H2A.Z acetylation modulates an essential charge patch. Mol Cell 7:1329–1335
Schwartz BE, Ahmad K (2005) Transcriptional activation triggers deposition and removal of the histone variant H3.3. Genes Dev 19:804–814
Sedelnikova OA, Pilch DR, Redon C, Bonner WM (2003) Histone H2AX in DNA damage and repair. Cancer Biol Ther 2:233–235
Shahbazian M, Young J, Yuva-Paylor L, Spencer C, Antalffy B, Noebels J, Armstrong D, Paylor R, Zoghbi H (2002) Mice with truncated MeCP2 recapitulate many Rett syndrome features and display hyperacetylation of histone H3. Neuron 35:243–254
Smith MM (2002) Centromeres and variant histones: what, where, when and why? Curr Opin Cell Biol 14:279–285
Stargell LA, Bowen J, Dadd CA, Dedon PC, Davis M, Cook RG, Allis CD, Gorovsky MA (1993) Temporal and spatial association of histone H2A variant hv1 with transcriptionally competent chromatin during nuclear development in Tetrahymena thermophila. Genes Dev 7:2641–2651

Steinbach OC, Wolffe AP, Rupp RA (1997) Somatic linker histones cause loss of mesodermal competence in Xenopus. Nature 389:395–399

Suto RK, Clarkson MJ, Tremethick DJ, Luger K (2000) Crystal structure of a nucleosome core particle containing the variant histone H2A.Z. Nat Struct Biol 7:1121–1124

Swaminathan J, Baxter EM, Corces VG (2005) The role of histone H2Av variant replacement and histone H4 acetylation in the establishment of Drosophila heterochromatin. Genes Dev 19:65–76

Tagami H, Ray-Gallet D, Almouzni G, Nakatani Y (2004) Histone H3.1 and H3.3 complexes mediate nucleosome assembly pathways dependent or independent of DNA synthesis. Cell 116:51–61

Tanaka M, Hennebold JD, Macfarlane J, Adashi EY (2001) A mammalian oocyte-specific linker histone gene H1oo: homology with the genes for the oocyte-specific cleavage stage histone (cs-H1) of sea urchin and the B4/H1M histone of the frog. Development 128:655–664

Tanaka M, Kihara M, Hennebold JD, Eppig JJ, Viveiros MM, Emery BR, Carrell DT, Kirkman NJ, Meczekalski B, Zhou J, Bondy CA, Becker M, Schultz RM, Misteli T, De La Fuente R, King GJ, Adashi EY (2005) H1FOO is coupled to the initiation of oocytic growth. Biol Reprod 72:135–142

Thatcher TH, Gorovsky MA (1994) Phylogenetic analysis of the core histones H2A, H2B, H3, and H4. Nucleic Acids Res 22:174–179

Ueda K, Kinoshita Y, Xu ZJ, Ide N, Ono M, Akahori Y, Tanaka I, Inoue M (2000) Unusual core histones specifically expressed in male gametic cells of Lilium longiflorum. Chromosoma 108:491–500

Unal E, Arbel-Eden A, Sattler U, Shroff R, Lichten M, Haber JE, Koshland D (2004) DNA damage response pathway uses histone modification to assemble a double-strand break-specific cohesin domain. Mol Cell 16:991–1002

van Attikum H, Fritsch O, Hohn B, Gasser SM (2004) Recruitment of the INO80 complex by H2A phosphorylation links ATP-dependent chromatin remodeling with DNA double-strand break repair. Cell 119:777–788

Witt O, Albig W, Doenecke D (1996) Testis-specific expression of a novel human H3 histone gene. Exp Cell Res 229:301–306

Woodcock CL, Skoultchi AI, Fan Y (2006) Role of linker histone in chromatin structure and function: H1 stoichiometry and nucleosome repeat length. Chromosome Res 14:17–25

Wu WH, Alami S, Luk E, Wu CH, Sen S, Mizuguchi G, Wei D, Wu C (2005) Swc2 is a widely conserved H2AZ-binding module essential for ATP-dependent histone exchange. Nat Struct Mol Biol 12:1064–1071

Zhang H, Roberts DN, Cairns BR (2005) Genome-wide dynamics of Htz1, a histone H2A variant that poises repressed/basal promoters for activation through histone loss. Cell 123:219–231

CHAPTER 6

HISTONE CHAPERONES IN CHROMATIN DYNAMICS

Implications in disease manifestation

JAYASHA SHANDILYA*, SHRIKANTH GADAD*, V. SWAMINATHAN* AND TAPAS K. KUNDU†

Transcription and Disease Laboratory, Molecular Biology and Genetics Unit, Jawaharlal Nehru Centre for Advanced Scientific Research, Jakkur P O, Bangalore 560064, India

Abstract: Histone chaperones are the histone interacting factors that stimulate histone transfer reaction without being a part of the final product. They are involved in the histone storage, histone translocation to the nucleus, and histone exchange and histone deposition onto the DNA for replication dependent chromatin assembly. Interestingly, they have also been demonstrated to possess the histone removal activity. While the involvement of the histone chaperones in chromatin transcription is undisputed, the question of their local versus global involvement is under scrutiny. This review enumerates the role played by various histone chaperones in the establishment of chromatin structure and regulation of chromatin transcription. The role of histone chaperones in disease manifestation is not very clear, preliminary results with few histone chaperones suggest that expression and function of these factors dramatically alters in carcinogenesis. This review will also focus on the possible role of histone chaperones in cancer diagnosis and progression

1. INTRODUCTION

The genome of the eukaryotic cell is packaged in a topologically complex, fibrous superstructure known as chromatin. The nucleosome core particle is the fundamental building block of chromatin and contains 146 bp of DNA wrapped in roughly two super helical turns around an octamer of four core histones (H3, H2B, H2A and H4) resulting in a beads on a string structure. This 10 nm structure further folds and

* Equal Contribution

† E-mail: tapas@jncasr.ac.in

T.K. Kundu and D. Dasgupta (eds.), Chromatin and Disease, 111–124.

organizes into higher order of compaction through interaction of N-terminal histone tails, linker histone H1 and other chromatin associated proteins (CAPs). This facilitates the accommodation of vast expanse of eukaryotic genome into the limited confines of the nucleus. However, in spite of such higher degree of compaction, the eukaryotic chromatin is highly dynamic in nature with diversity being generated through local chromatin composition (presence of histone variants), covalent modifications of histones as well as methylation status of the DNA and the ATP dependent chromatin remodeling, which ensures the progress of various nuclear processes. The maintenance of chromatin integrity is crucial for the cellular life. During the S phase of cell cycle high fidelity duplication of chromatin organization occurs in proper coordination with DNA replication. Like chromatin assembly, disassembly or reorganization of chromatin is also essential for most of the cellular processes. Histone chaperones are among the key factors for this dynamic organization of chromatin template. They are involved in the histone storage, histone translocation to the nucleus, and histone exchange and histone deposition onto DNA for replication dependent chromatin assembly (Table 1) (Fig. 1). Chromatin assembly is a two-step process, where initial step is the deposition of H3-H4 tetramer on DNA (Worcel *et al.*, 1978; Smith and Stillman, 1991) followed by the deposition of a pair of H2A-H2B dimers (Smith and Stillman, 1991). Various histone chaperones and ATP dependent chromatin spacing and remodeling factors facilitate the organization of nucleosomes in a regularly spaced array (Haushalter and Kadonaga, 2003). Newly synthesized histones are acetylated which are recognized and bound by histone chaperones and loaded onto the DNA in a sequential manner to assemble a functional nucleosome. Lastly linker histone H1 is assembled onto each dyad axis of the nucleosomes (Bustin *et al.*, 2005) with the help of specific linker histone chaperone, which compacts it further into a 30 nm solenoid structure. Increasing number of reports suggest that histone chaperones have significant contribution in chromatin reorganization (variant exchange and/or histone removal) during transcription initiation as well as during elongation (Fig. 2) (Gamble *et al.*, 2005; Angelov *et al.*, 2006; Schwabish and Stuhl, 2006). Their involvement in both replication and transcription (also in repair) clearly indicates that dysfunction of these group of proteins would result in several disease manifestation. Emerging evidences show that indeed proteins having histone chaperone activity are responsible for viral pathogenesis (Van Leeuwen *et al.*, 2003) and cancer prone disorders (Verhaak *et al.*, 2005; Grisendi *et al.*, 2006).

2. HISTONE CHAPERONES IN CHROMATIN ASSEMBLY

Histone chaperones are the physiological chromatin assembly factors, which are crucial to regulate the deposition of histones onto DNA *in vivo*, both in replication dependent and independent manner. The chromatin assembly factor 1 (CAF-1) initiates the replication dependent chromatin assembly. The newly synthesized

Table 1. Functional Diversity among Histone Chaperones (Based on Haushalter and Kadonaga, 2003)

Histone Chaperone	Binding Preference	Function
CAF-1	H3-H4 (Preferably acetylated)	• Replication-coupled chromatin assembly • DNA repair • Silencing • Cell cycle progression
Asf1	H3-H4 (Preferably acetylated)	• Replication-coupled chromatin assembly(as RCAF complex) • DNA repair,Silencing • Global regulator of transcription • Histone sink during replicational stress
NAP1	H2A-H2B H3-H4	• Nuclear import of histones • Linker histone chaperone (in *Xenopus oocytes)*
HIR	H2A-H2B H3-H4 H3.3	• Histone gene regulation • Replication independent chromatin assembly • Assembly factor for histone variant H3.3 • Formation of heterochromatin foci
Nucleoplasmin	H2A-H2B H3-H4	• Storage of histones in oocytes • Chaperone activity during rapid replication in early embryogenesis • Sperm decondensation • Chromatin condensation during apoptosis
N1/N2	H3-H4	• Storage of histones in oocytes • Chaperone activity during rapid replication
Spt6	H3-H4	• Histone-transfer vehicle • Transcription-elongation factor (FACT)
Nucleophosmin	H2B H3-H4	• Activation of transcription in acetylation dependent manner
FKBP	H2A-H2B (dimer) H3-H4 (tetramer)	• Histone chaperone regulating rDNA silencing
JDP2	H2A-H2B H3-H4	• Regulates transcription via inhibition of p300 mediated histone acetylation
Nucleolin	H2A-H2B (preferentially)	• FACT like activity • Facilitates transcriptional elongation
Rtt106	H3-H4	• Heterochromatin mediated silencing
Hif1p	H3-H4 (Preferably acetylated)	• Chromatin assembly
NASP	H1	• Linker histone chaperone

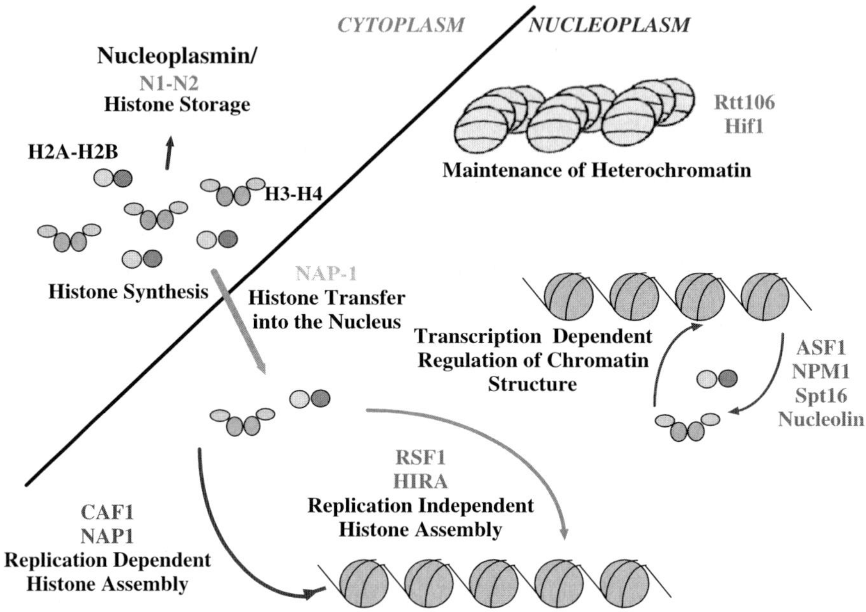

Figure 1. Different histone chaperones in the key histone metabolic pathways: Functions of histone chaperones range from the storage of newly synthesized histones in the cytoplasm, its transfer into the nucleus and in histone assembly into nucleosomes. Apart from this the histone chaperones are also involved in histone exchange, maintenance of heterochromatin and in the regulation of chromatin structure during transcription. (See Colour Plate 10.)

H3-H4 tetramers which are acetylated at their conserved lysine residues (K5, K8 and K12 of histone H4) by cytoplasmic histone acetyl transferase, Hat1 (Verreault *et al.*, 1998) are deposited first in a newly replicated DNA followed by addition of H2A-H2B dimers. Chromatin Assembly Factor1 (CAF-1) which is a heterotrimer that comprises of p150, p60, and p48 subunits is found to be generally complexed with newly synthesized H3 and H4. CAF-1 deposits histones on the replicating fork and is recruited there through its interaction with Proliferating Cell Nuclear Antigen (PCNA) (Smith and Stillman, 1991). The p48 subunit of CAF-1 is known to associate with a wide variety of histone deacetylases, which could deacetylate the freshly assembled histones. However, it was found that CAF-1 is not essential for yeast cell viability suggesting that there are several other pathways of chromatin assembly. Efficient nucleosome assembly also requires a replication-coupled assembly factor (RCAF), which comprises of anti-silencing factor1 (Asf1). RCAF acts synergistically with CAF-1 for chromatin assembly and interacts directly with CAF-1. Apart from its role in the replication coupled chromatin assembly, Asf1 is also involved in the replication independent chromatin assembly. It was originally identified as a protein involved in derepressing silenced chromatin upon over expression (Munakata *et al.*, 2000). More recently, Asf1 has been shown to act as a global chromatin disassembly factor (Adkins and Tyler, 2004). Rtt106p is another

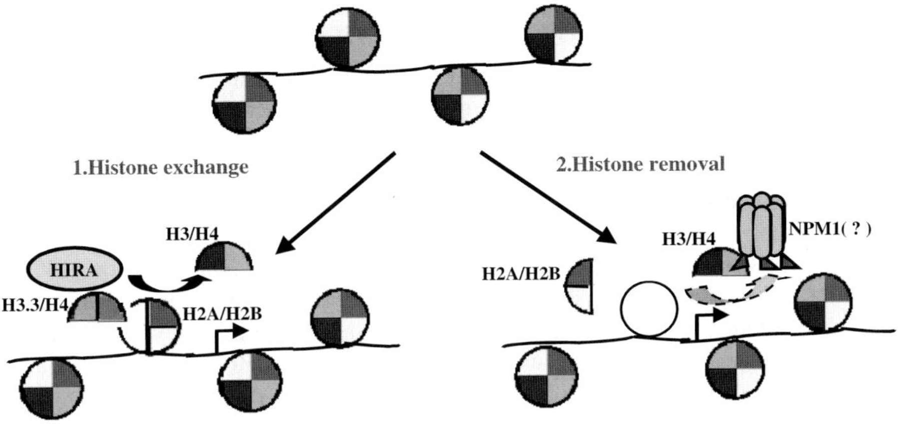

Figure 2. Histone chaperones facilitate favorable chromatin dynamics during transcriptional activation: Transcriptional competence of chromatin template is achieved by the replacement of histone variants and finally removal of histones. Histone chaperone may help in both the process in a replication independent manner. Acetylation of histone and also the chaperone may assist in this process. Recent evidence suggest that NPM1 may participate in these events globally or gene specific manner. (See Colour Plate 11.)

H3-H4 chaperone, which shares several properties with Asf1. It also interacts with CAF-1 and enhances the ability of CAF-1 to promote nucleosome formation onto replicated DNA. Similar to Asf1 mediated telomere silencing, Rtt106 is involved in maintaining silenced heterochromatin most probably through its interaction with Hir1p (Huang *et al.*, 2005). Nucleosome Assembly Protein1 (NAP1) is a histone chaperone that facilitates the random assembly of nucleosomes on the DNA. This histone chaperone preferentially binds to H2A-H2B, even its interaction with H3-H4 is also documented (McBryant *et al.*, 2003). NAP1 and its homolog NAP2 are known to translocate into the nucleus in the S phase and are involved in replication coupled deposition and mitotic progression (Ito *et al.*, 1996; Rodriguez *et al.*, 2000). NAP2 is a homolog of NAP1 that has neural tissue specific expression. NAP1 was found to have linker histone chaperone activity (Kepert *et al.*, 2005), which deposits histone H1 after the core particle assembly. However, it is not known whether this activity is replication dependent. Apart from NAP1, recently another linker histone chaperone, NASP (Nuclear Autoantigenic Sperm Protein) has been discovered (Richardson *et al.*, 2006). NASP is present in a multichaperone complex in the cell, which also contains Asf1 and CAF-1. It transports histone H1 into nuclei and exchanges H1 histones with DNA and is required for normal cell cycle progression and development (Richardson *et al.*, 2006). Hif1p is a histone chaperone that selectively interacts with H3 and H4 in yeast. Hif1p associates with nuclear Hat1p/Hat2p complex, which is bound to, acetylated histone H4 and histone H3. Hif1p is also a chromatin assembly factor and it promotes deposition of histones onto DNA (Ai and Parthum, 2004). Hif1p and NASP share sequence similarities with the *Xenopus* protein N1/N2 (Ai and Parthum, 2004). Consequently NASP appears to be a conserved and a critical member of multichaperone complexes that participate in nucleosome remodeling.

The replication independent chromatin assembly is essential for the regulation of transcription and DNA repair. Presumably, HIRA (Histone Regulator A) plays a major role in the deposition of H3-H4 tetramer in this process (Polo and Almouzni, 2005). Spt6 was identified as histone chaperone through genetic studies that indicated that mutations in spt6 could overcome transcription defects in the swi/snf defective yeast strains. Furthermore, spt6 could alter chromatin structure primarily through its interaction with histone H3 (Bortivin and Winston, 1996). The H2A-H2B chaperone may be common for both the replication dependent and independent chromatin assembly. The mechanism of replication independent chromatin organization and its functional consequences will be discussed in the following sections.

3. HISTONE STORAGE AND TRANSPORT BY HISTONE CHAPERONE

Apart from histone chaperones such as NAP1 and Asf1, which are important in nucleosome assembly at replication fork, others have prominent roles in oogenesis, decondensation of sperm chromatin after fertilization and nucleosome assembly in early embryonic cells. Nucleoplasmin and N1/N2 represent distinct chaperone families that bind to H2A-H2B dimers and H3-H4 tetramer, respectively (Dilworth *et al.*, 1987). These chaperones are the most abundant proteins in the nuclei of *Xenopus* oocytes. Thus, nucleoplasmin and N1/N2 may function as a repository for excess core histones in oocytes and eggs, which are required in order to assemble chromatin during the rapid rounds of replication, characteristic of the early cleavage divisions in frog embryo. Nucleoplasmin has been found as histone storage molecule in egg cytoplasm bound to early embryonic pool histone H2A and H2B. *Xenopus* sperm chromatin has a markedly reduced content of H2A and H2B relative to H3 and H4, which is replaced by sperm specific basic proteins (SSBP). Nucleoplasmin has higher affinity for SSBP than for the H2A-H2B dimer, thus upon sperm decondensation nucleoplasmin binds and removes the SSBP while depositing an H2A-H2B dimer. Furthermore, both nucleoplasmin and N1/N2 can act together to assemble nucleosomes onto naked double-stranded DNA in absence of replication (Kleinschmidt *et al.*, 1990; Dilworth *et al.*, 1987). Physiological significance of this observation, however, is yet to be shown. The versatile histone chaperone, NAP1 is also involved in the nuclear transport of histone H2A-H2B dimer from cytoplasm.

4. TRANSCRIPTIONAL REGULATION BY HISTONE CHAPERONES

The four canonical histones H2A, H2B, H3 and H4 of nucleosomes are often replaced by specific variants that help in specialized chromatin function, for example, transcription (Henikoff *et al.*, 2000). The notable instance is the presence of histone H3 variant, H3.3 in the transcriptionally active loci. H3.3 get deposited onto the DNA in replication independent manner (Ahmad and Henikoff, 2002). Apart from this, covalent modifications such as acetylation, methylation and phosphorylation of histones further influences composition and active state of particular gene or chromatin territories (Jenuwein and Allis, 2001). The nucleosomes are also mobilized by complexes termed as remodelers, which can either, facilitate or impede processes, such as transcription (Owen-Hughes, 2003). Prominent among the non-enzymatic machineries that are involved in the alteration of chromatin structure are replication independent histone chaperones. The transcriptional activation brought about by these chaperones occurs through multiple ways. The prominent mechanisms being either nucleosome depletion at the promoter region or histone variant exchange at the active loci. In addition, they might also help in the recruitment of components of basal polymerase machinery at the initiation site. In yeast, Anti silencing function1 (Asf1) is a highly conserved histone H3-H4 chaperone. It was originally identified as a protein facilitating Chromatin Assembly Factor1 (CAF-1), mediated chromatin assembly on newly replicated DNA (Tyler *et al.*, 1999). In yeast, Asf1 plays a significant role in transcriptional activation of PHO5 and PHO8 gene by the removal of nucleosomes from their promoter region. Physical and genetic interactions between Asf1 and yeast specific transcription factors like Bdf1 and Bdf2 also place Asf1 at the site of transcription (Chimura *et al.*, 2002). Similarly *Drosophila* Asf1 was also localized at the transcriptionally active intergenic bands of polytene chromosome (Tyler *et al.*, 2001). In higher eukaryotes, the existence of Asf1 in a protein complex with histone variant H3.3 further strengthens the hypothesis that the replication independent chromatin assembly function of Asf1 is involved in reorganizing the chromatin template for transcriptional activation (Tagami *et al.*, 2004). In vertebrates, HIRA has recently been shown to be critical for a nucleosome assembly pathway independent of DNA synthesis that specifically involves the H3.3 histone variant. Nucleosomes containing H3.3 and not H3 are specifically assembled in paternal Drosophila chromatin before the first round of DNA replication. The exclusive marking of paternal chromosomes with H3.3 represents a primary epigenetic distinction between parental genomes in the zygote, and underlines an important consequence of the critical and highly specialized function of HIRA at fertilization (Loppin *et al.*, 2005) (Fig. 2).

Histone chaperone NAP1 is involved in transcriptional regulation through chromatin remodeling (LeRoy *et al.*, 2000). NAP1 acts synergistically with ATP dependent chromatin remodeling factors such as NURF and ACF to remodel the chromatin structure (Jiang *et al.*, 2000). NAP1 and NAP2 are found to interact with p300 and stimulate transcription from p21 and E2F promoters (Ito *et al.*, 2000). Nucleoplasmin and NAP1 co-operate with SWI/SNF complex in chromatin remodeling and facilitate transcription factor binding to the nucleosomal DNA (Chen *et al.*, 1994). TAF1/SET (Template Activating Factor/patient SE translocation) a member of Nucleosome Assembly Protein (NAP) family of histone chaperones was recently identified as a factor generally required for chromatin transcription. TAF1 plays distinct roles in chromatin assembly and transcription. Although TAF1 acts after chromatin is assembled, it is required prior to the elongation phase of the transcription cycle distinguishing it mechanistically from FACT a factor required for RNA polymerase II to overcome nucleosome barrier in elongation. TAF1 is likely to act in concert with other transcription factors or chromatin remodelers to render chromatin transcriptionally competent. The other H2A-H2B histone chaperone complex is the FACT complex, required for the RNA polymerase II mediated transcriptional activation from the chromatin template. FACT is mainly involved in efficient transcript elongation through removal of H2A-H2B in front of the advancing polymerase (Belotserkovskaya *et al.*, 2003). Among the subunits of FACT a human homologue of yeast spt16/cdc68 protein is involved in cell cycle regulation and transcription. More recently spt6 was found to be involved in maintenance of chromatin structure during transcriptional elongation (Saunders *et al.*, 2003). The human spt6 has been shown to stimulate transcriptional elongation also *in vitro* transcription assays (Endoh *et al.*, 2004).

Recently, we have shown that the multifunctional nucleolar protein nucleophosmin (NPM1) is a histone chaperone (Okuwaki *et al.*, 2001; Swaminathan *et al.*, 2005) and activates transcription from chromatin template in an acetylation dependent manner (Swaminathan *et al.*, 2005) (Fig. 3). Interestingly, it is found that NPM1 gets acetylated *in vitro* and *in vivo*, specifically by p300. The enhancement of transcription requires prior acetylation of NPM1. This would be the first report of acetylation of histone chaperone. Presumably, reversible acetylation of histone chaperones may act as a regulatory mechanism for the chaperone function. NPM1 is known to interact with both histone H3-H4 tetramer and H2A-H2B dimer. Acetylation of histones and as well as the chaperone further strengthens the histone and chaperone interaction. Possibly higher affinity of acetylated NPM1 to the hyperacetylated histones induces the removal of promoter proximal histones, which further facilitates the transcription (Swaminathan *et al.*, 2005). More recently, another member of histone chaperone family, nucleolin has been found to facilitate transcription through displacement of H2A-H2B dimer, which resembles the activity of FACT complex. Interestingly, nucleolin also increases the efficiency of nucleosomal remodeling activity of ACF and SWI/SNF on nucleosomes composed of canonical histones and macroH2A nucleosomes, but not H2ABbd nucleosomes (Angelov *et al.*, 2006). These data suggest that composition of nucleosomes (presence of particular variants) may add another regulatory point for the histone chaperone function.

5. INVOLVEMENT OF HISTONE CHAPERONES IN DNA REPAIR

Sensing DNA damage is essential for maintenance of genomic integrity and cell cycle progression. DNA repair is the second major site of DNA synthesis in a cell after replication and it involves chromatin assembly after efficient repair process. In yeast, the histone chaperone CAF-1 (chromatin Assembly factor1) mediates histone H3-H4 assembly on to newly replicated DNA and also mediates nucleotide

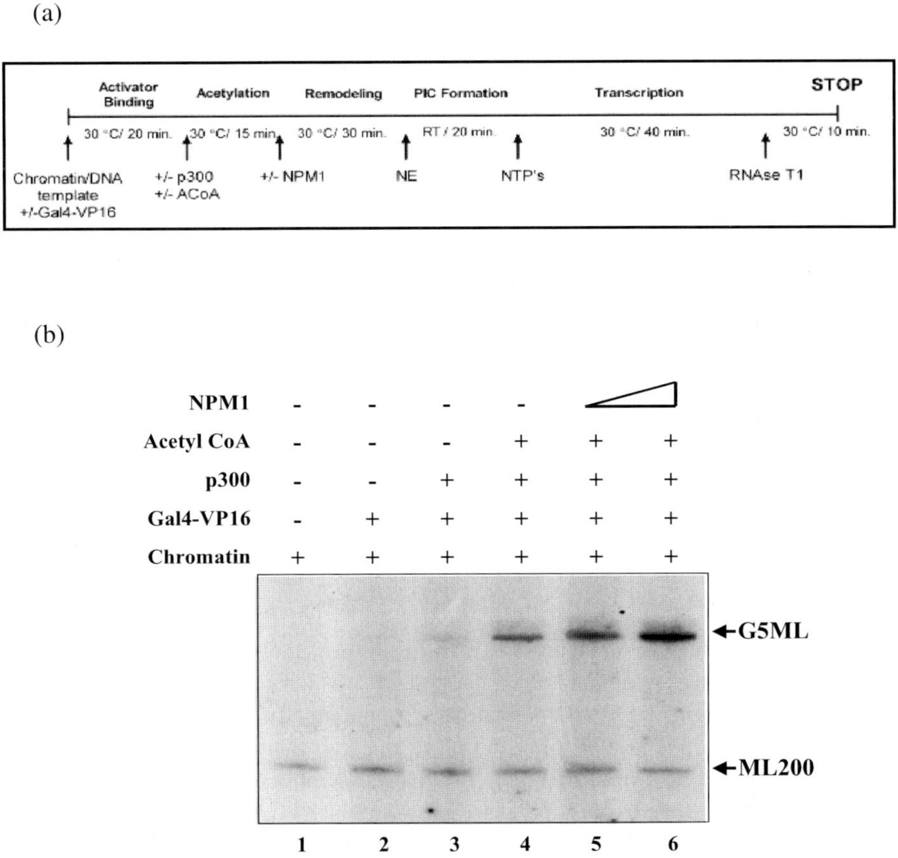

Figure 3. (a) Scheme of transcription. (b) Histone chaperone nucleophosmin/NPM1 enhance acetylation dependent chromatin transcription: NPM1 stimulates chromatin transcription in a dose dependent manner. Lane 1,without activator; lanes 2–6, with the activator Gal-VP16 (50 ng); lanes 3–6, with p300 (25 ng); lanes 4–6, with acetyl CoA (1.5 μM); lane 5, 1 pmol, lane 6, 10 pmols of full length NPM1

excision repair *in vitro* (Gaillard *et al.*, 1996). In the absence of CAF-1 yeast cells are highly sensitive to double-strand DNA-damaging agents and sensitive to UV irradiation. CAF-1 is found to play a role in both homologous recombination and non-homologous end-joining pathways. (Emili *et al.*, 2001). The largest subunit (p150) of this factor is known to interact with PCNA (Proliferating Cell Nuclear Antigen), which is essential for DNA damage processing and check point control. (Moggs *et al.*, 2000). The function of CAF-1 during double-strand break repair is distinct from that of another histone H3-H4 chaperone, Anti-silencing function 1 (Asf1). Asf1 also plays a role in the response to UV irradiation that is apparent in the absence of CAF-1 (Tyler *et al.*, 1999). In absence of DNA damage or replicative stress, nearly all of the soluble pool of Rad53, a DNA damage checkpoint protein in yeast is bound to Asf1 (Linger and Tyler, 2005). Activation of Rad53 by the DNA damage checkpoint kinases releases Asf1 into the vicinity of the DNA lesion in order to bind to histones and assemble chromatin. The human homologs of Asf1 (hAsf1) are shown to function synergistically with human CAF-1 (hCAF-1) to assemble nucleosomes during nucleotide excision repair *in vitro*. Asf1 proteins can interact directly with the p60 subunit of CAF-1. In contrast, to hCAF-1 p60, the nuclear hAsf1 proteins are not significantly associated with chromatin in cells before or after the induction of DNA damage. A transient physical interaction between Asf1 and CAF-1 occurs for allowing histone delivery from Asf1 to CAF-1, which facilitates nucleosome formation during DNA repair (Mello *et al.*, 2002). In yeast Hif1, which is a H3-H4 chaperone, exists as a complex with Hat1p/Hat2p and was found to be required for telomeric silencing and DNA double strand break repair (Qin and Parthun, 2006). These evidences clearly indicate at the increasing involvement of histone chaperones in repair of a wide spectrum of DNA damage in the chromatin context.

6. HISTONE CHAPERONE AND DISEASE

Role of histone chaperone in disease manifestation is not clearly understood yet. However, emerging evidences strongly suggest that soon these factors will be considered as diagnostic markers and also as targets for therapeutics. Here, we shall discuss about a few available examples. The significant requirement of CAF-1 complex for the replication dependent chromatin assembly and S phase progression, clearly suggests that the expression as well as function of CAF-1 would be strikingly different between proliferating and quiescent cells. In agreement with this hypothesis it was found that indeed CAF-1 expression is dramatically down regulated in quiescent cells (Polo and Almouzni, 2005). On the other hand CAF-1 was found to be overexpressed in breast cancer cell line as compared to the cell line derived from the normal cells. This observation raises the possibility of using CAF-1 as a prognostic marker for breast cancer and eventually also as a therapeutic target.

Multifunctional histone chaperone proteins like NPM1 are involved in regulating diverse cellular processes. It is more abundant in tumor cells than in normal resting cells (Chan *et al.*, 1989). Mutations in NPM1 are implicated in the disease phenotype of Acute Myelocytic Leukemia (AML). Probably cytoplasmic localization of NPM1 instead of nucleolar is one of the causes of this malignancy. Acute promyelocytic leukemia (APL) is uniquely associated with chromosomal translocations that disrupt the gene encoding the retinoic acid receptor. The presence of cytoplasmic NPM1 correlated with the clinical and biological features of the disease. NPM1 has also been reported to get upregulated in Glioblastoma multiforme (GBM), the most malignant class of glial neoplasm (grade IV), which carries the worst clinical prognosis among primary brain tumors in adults. However, it is not known whether the histone chaperone activity of NPM1 is involved during these disease manifestations. NPM1 directly interacts with the tumor suppressor protein p53 and confers to its stability (Colombo *et al.*, 2002). We have shown that NPM1 activates the p53 responsive gene expression (Colombo *et al.*, 2002, Swaminathan *et al.*, 2005). Taken together these reports suggest that NPM1 would be actively involved in tumor suppressor activity of p53 and also the stress response. Furthermore, NPM1 also interacts with several viral proteins with varied functional consequences. It binds to the human immunodeficiency virus protein Tat (Transactivating regulatory protein) via the nucleolar localization motif of Tat. Presumably, NPM1 acts as a human factor for the nucleolar localization of Tat (Li., 1997). NPM1 also interacts with another HIV protein Rev (Miyazak, and Nosaka Hatanaka 1996). It would be interesting to find out whether, the histone chaperone activity of NPM1 is exploited by the retroviruses like HIV for establishment of the pathogenesis by altering the integrated chromatinized viral genome in humans.

Recently, it was discovered that Nucleolin, a major multifunctional nuclear phosphoprotein also possesses histone chaperone activity (Angelov *et al.*, 2006). It is phosphorylated by Cdc2 kinase during mitosis (Dranovsky *et al.*, 2001). In patients with Alzheimer's disease (AD), Cdc2 phosphorylated nucleolin was present in neurofibrillary tangles (NFT) (Dranovsky *et al.*, 2001). In the brain, nucleolin was localized not only to nuclei but also to neuronal cytoplasm, and it is a marker for early NFT. These findings suggest that phosphorylation of nucleolin by Cdc2 kinase is a critical event and is the point of convergence of the two distinct pathways, mitosis and neurodegeneration (Dranovsky *et al.*, 2001). Role of nucleolin in disease manifestation has been discussed in details in a separate chapter of this book.

7. FUTURE PERSPECTIVES

Though histone chaperone, nucleoplasmin is known since past two decades, an active effort to understand its role in chromatin function is rather quite recent. Molecular events of replication dependent chromatin assembly and function of histone chaperones have been elucidated to a great extent, but very little is known about the regulation of chaperone function. Possibly post-translational modifications of histones and also the chaperones would play crucial role in regulating their functions. Recent reports suggest that histone chaperone play equally important role in transcription and DNA repair. Further analysis of replication independent functions like involvement in histone transfer or removal should be done, in the context of histone variants. Histone chaperones are probably recruited in the form of a large complex. Specific variants may be associated with specific complex. The exact mechanism of recruitment of the chaperone complex to the promoter needs to be elucidated. Posttranslational modifications, especially acetylation of the chaperone seems to be important for chromatin-mediated transcriptional activation. It would be interesting to find out whether other epigenetic modifications and chromatin remodeling are also involved in the process of activation. Since, epigenetic markers dramatically alter under various pathological state (viral infection or malignancy/neoplastic development), the better understanding of the functional regulation of histone chaperones in disease manifestation would be helpful in targeting them for therapeutics.

ACKNOWLEDGEMENT

Research work in authors' laboratory was supported by Jawaharlal Nehru Centre for Advanced Scientific Research, Jakkur, Bangalore and Council for Scientific and Industrial Research (CSIR), Government of India. JS and SG are the Junior Research Fellow of CSIR.

REFERENCES

Adkins MW, Tyler JK (2004) The histone chaperone Asf1p mediates chromatin disassembly *in vivo*. J Biol.Chem 279:52069–52074

Ahmad K, Henikoff S (2002) The histone variant H3.3 marks active chromatin by replication independent nucleosome assembly.Mol.Cell9:1191–1200

Ai X, ParthumMR (2004) The nuclear Hat1p/Hat2p complex: A molecular link between type B histone acetyltransferases and chromatin assembly.Mol.Cell14:195–205

Angelov D, Bondarenko VA, Almargo S, Menoni H, Mongelard F, Mietton F, Studitsky VM, Hamiche A, Dimitrov S,Bouvet P (2006) Nucleolin is a histone chaperone with FACT-like activity and assists remodeling of nucleosomes. EMBO J 25:1669–1679

Belotserkovskaya R, Oh S, Bondarenko VA, Orphanides G, Studitsky VM,Reinberg D (2003) FACT facilitates transcription–dependent nucleosome alteration. Science 301:1090–1093

Bortivin A, Winston F (1996) Evidence that Spt6p controls chromatin structure by a direct interaction with histones. Science272:1473–1476

Bustin M, Catez F, Lim, JH (2005). The dynamics of H1 function in chromatin.Mol.Cell17:617–620

Chan W-Y, Liu Q-R, Borjigin, J, Busch H, Rennert OM, Tease LA,and ChanPK(1989) Characterization of the cDNA encoding human nucleophosmin and studies of its role in normal and abnormal growth. Biochemistry 28:1033–1039

Chen H, Li B, Workman JL (1994) A histone-binding protein, nucleoplasmin, stimulates transcription factor binding to nucleosomes and factor-induced chromatin disassembly. EMBO J13:380–390

Chimura T, Kuzuhara T,Horikoshi M (2002) Identification and characterization of CIA/ASF1 as an interacter of bromodomains associated with TFIID. Proc.Nat.Aca.Sci. 9:9334–9339

Colombo E, Marine JC, Danovi D, Falini B Pelicci PG (2002) Nucleophosmin regulates the stability and transcriptional activity of p53. Nat. Cell. Biol. 4:529–533

Dilworth SM, Black SJ, Laskey RA (1987) Two complexes that contain histones are required for nucleosome assembly *in vitro*:role of nucleoplasmin and N1/N2 in *Xenopus* egg extracts. Cell 51:1009–1018

Dranovsky A, Vincent I, Gregori L, Schwarzman A, Colflesh D, Enghild J, Strittmatter W, Davies P Goldgaber D (2001) Cdc2 phosphorylation of nucleolin demarcates mitotic stages and Alzheimer's disease pathology. Neurobiol. Aging 22:517–528

Emili A, Schieltz DM, Yates JR, Hartwell LH (2001). Dynamic interaction of DNA damage checkpoint protein Rad53 with chromatin assembly factor Asf1.Mol.Cell 7:13–20

Endoh M, Zhu W, Hasegawa J, Wantanabe H, Kim DK, Aida M, Inukai N, Narita T, Yamada T, Furuva A, Sato H, Yamaguchi Y, Mandal SS, Reinberg D, Wada T,Handa H (2004) Human Spt6 stimulates transcription elongation by RNA polymerase II *in vitro*. Mol.Cell.Biol 24:3324–3336

Gaillard PH, Martini EM, Kaufman PD, Stillman B, Moustacchi E, Almouzni G (1996) Chromatin assembly coupled to DNA repair: a new role for chromatin assembly factor 1. Cell 86:887–896

Gamble MJ, Erdjument-Bromage H, Tempest P, Freedman LP, Fisher R P (2005) The histone chaperone TAF-1/SET/INHAT is required for transcription in vitro of chromatin templates. Mol.Cell.Biol 25:797–807

Grisendi S, Mecucci C, Falini B, Pandolfi PP (2006) Nucleophosmin and cancer. Nat. Rev. Can 6:493–505

Haushalter K A, Kadonaga JT (2003) Chromatin assembly by DNA-translocating motors. Nat.Rev.Mol.Cell.Biol.4:613–620

Henikoff S, Ahmad K, Platero JS, van Steensel B (2000) Heterochromatic deposition of centromeric histone H3-like proteins. Proc.Natl.Acad.Sci.U S A 18:716–721

Huang S, Zhou H, Katzmann D, Hochstrasser M, Atanasova E, Zhang Z (2005) Rtt106p is a histone chaperone involved in heterochromatin-mediated silencing. Proc.Natl.Acad.Sci 102:13410–13415

Ito T, Ikehara T, Nagakawa T, Kraus WL, Muramatsu M (2000) p300-mediated acetylation facilitates the transfer of histone H2A-H2B dimers from nucleosomes to a histone chaperone 14:1899–1907

Ito T, Tyler JK, Bugler M, Kobayashi R Kadonaga JT (1996) ATP-facilitated chromatin assembly with a nucleoplasmin-like protein from *Drosophila melanogaster*.J.Biol.Chem 271:25041–25048

Jenuwein T, Allis CD (2001) Translating histone code.Science 293:1074–1080

Jiang W, Nordeen SK, Kadonaga JT (2000) Transcriptional analysis of chromatin assembled with purified ACF and dNAP1 reveals that acetyl-CoA is required for preinitiation complex assembly. J.Biol.Chem 22:39819–39822

Kepert J.F Mazurkiewicz J, Heuvelman GL, Toth KF, Rippe K (2005) NAP1 modulates binding of linker histone H1 to chromatin and induces an extended chromatin fiber conformation. J. Biol. Chem 280:34063–34072

Kleinschmidt J.A, Seiter A Zentgraf H (1990) Nucleosome assembly *in vitro*: separate histone transfer and synergistic interaction of native histone complexes purified from nuclei of *Xenopus laevis* oocytes. EMBO J 9:1309–1318

LeRoy G, Loyla A, Lane W Reinberg D (2000). Purification and characterization of a human factor that assembles and remodels chromatin. J.Biol.Chem 275:14787–14790

Li YP (1997) Protein B23 is an important human factor for the nucleolar localization of the human immunodeficiency virus protein Tat. J Virol 71:4098–4102

Linger J, Tyler JK (2005) The yeast histone chaperone chromatin assembly factor 1 protects against double-strand DNA-damaging agents. Genetics 171:1513–1522

Loppin B, Bonnefoy E, Anselme C, Laurengon A, Karr T L, Couble P (2005) The histone H3.3 chaperone HIRA is essential for chromatin assembly in the male pronucleus.Nature 437:1386–1390

McBryant SJ, Park Y-J, Abernathy SM, Laybourn PJ, Nyborg JK,Luger K (2003) Preferential binding of the histone $(H3\text{-}H4)_2$tetramer by NAP1 is mediated by the amino-terminal histone tails. J. Biol Chem 278:44574–44583

Mello JA, Sillje HH, Roche DM, Kirschner D B, Nigg EA, Almouzni G (2002). Human Asf1 and CAF-1 interact and synergize in a repair coupled nucleosome assembly pathway. EMBO Rep.3:329–334

Miyazak Y, Nosaka T, Hatanaka M (1996) The post-transcriptional regulator Rev of HIV: Implications for its interaction with the nucleolar protein B23.78:1081–1086

Moggs JG, Grandi P, Quivy JP, Jonsson Z O, Hubscher U, Becker PB, Almouzni G (2000) A CAF-1-PCNA-mediated chromatin assembly pathway triggered by sensing DNA damage. Mol.Cell.Biol 20:1206–1218

Munakata T, Adachi N, Yokoyama N, Kuzuhara T, Horikoshi M (2000) A human homolog of yeast anti-silencing factor has histone chaperone activity. Genes to Cells 5:221–233

Okuwaki M, Matsumoto K, Tsujimoto M, Nagata K (2001) Function of nucleophosmin/B23, a nuclear acidic protein, as a histone chaperone. FEBS Lett 506:272–276

Owen-Hughes T (2003) Colworth memorial lecture. Pathways for remodeling chromatin. Biochem. Soc.Trans 31:893–905

Polo SE, Almouzni G (2005). Histone metabolic pathways and chromatin assembly factors as proliferation markers Cancer Lett 220:1–9

Qin S, Parthun MR (2006) Recruitment of the type B histone acetyltransferase Hat1p to chromatin is linked to DNA double-strand breaks. Mol.Cell.Biol 26:3649–3658

Richardson RT, Alekseev OM, Grossman G, Widgren EE, Thresher R, Wagner EJ, Sullivan KD, Marzluff WF, O'rand MG (2006). NASP, a linker histone chaperone that is required for cell proliferation. J. Biol. Chem 281:21526–21534

Rodriguez P, Pelletier J, Price GB, Zannis-Hadjopoulos M (2000) NAP2: histone chaperone function and phosphorylation state through the cell cycle. J.Mol.Biol 298:225–238

Saunders A,Werner J, Andrulis ED, Nakayama T, Hirose S, Reinberg D, Lis JT (2003) Tracking FACT and the RNA polymerase II elongation complex through chromatin *in vivo*. Science 301:1094–1096

Schwabish MA, Stuhl K (2006) Asf1 mediates histone eviction and deposition during elongation by RNA polymerase II. Mol.Cell 22:415–422

Smith S, Stillman B (1991) Stepwise assembly of chromatin during DNA replication *in vitro*. EMBO J 10:971–980

Swaminathan V, Kishore AH, Febitha KK, Kundu TK (2005) Human histone chaperone nucleophosmin enhances acetylation-dependent chromatin transcription. Mol.Cell.Biol 25:7534–7545

Tagami H, Ray-Gallet D, Almouzni G, Nakatani Y (2004) Histone H3.1 and H3.3 complexes mediate nucleosome assembly pathways dependent or independent of DNA synthesis Cell 116:51–61

Tyler JK, Adams CR, Chen SR, Kobayashi R, Kamakaka RT, Kadonaga JT (1999) The RCAF complex mediates chromatin assembly during DNA replication and repair. Nature 402:555–560

Tyler JK, Collins K A, Prasad-Sinha J, Amiott E, Bugler M, Harte P J, Kobayashi R Kadonaga JT (2001) Interaction between the Drosophila CAF-1 and Asf1 chromatin assembly factors. Mol.Cell.Biol 21:6574–6584

Van Leeuwen H, Okuwaki M, Hong R, Chakravarti D, Nagata K, O'Hare P (2003) Herpes simplex virus type 1 tegument protein VP22 interacts with TAF-1 proteins and inhibits nucleosome assembly but not regulation of histone acetylation by INHAT. J. Gen.Virol 84:2501–2510

Verhaak RG, Goudswaard CS, van Putten W, Bijl MA, Sanders MA, Hugens W, Uitterlinden AG, Erpelinck CA, Delwel R, Lowenberg B, Valk PJ (2005) Mutations in nucleophosmin (NPM1) in acute myeloid leukemia (AML): association with other gene abnormalities and previously established gene expression signatures and their favorable prognostic significance. Blood 106:3747–3754

Verreault A, Kaufman P D, Kobayashi R, Stillman B(1998) Nucleosomal DNA regulates the core-histone-binding subunit of the human Hat1 acetyltransferase. Curr.Biol 15:96–108

Worcel A, Hans S, Wong ML (1978) Assembly of newly replicated chromatin. Cell 15:969–977

CHAPTER 7

FUNCTIONS OF THE HISTONE CHAPERONE NUCLEOLIN IN DISEASES

SÉBASTIEN STORCK[1,2], MANU SHUKLA[1,3], STEFAN DIMITROV[1,3] AND PHILIPPE BOUVET[1,2]

[1]*Laboratoire Joliot-Curie, Ecole Normale Supérieure de Lyon, 46 Allée d'Italie, 69007 Lyon, France*
[2]*Laboratoire de Biologie Moléculaire de la Cellule, CNRS-UMR 5161/INRA 1237/IFR128 Biosciences Lyon-Gerland Ecole Normale Supérieure de Lyon, 46 Allée d'Italie, 69007 Lyon, France*
[3]*Institut Albert Bonniot, INSERM U309, 38706 La Tronche cedex, France*

Abstract: Alteration of nuclear morphology is often used by pathologist as diagnostic marker for malignancies like cancer. In particular, the staining of cells by the silver staining methods (AgNOR) has been proved to be an important tool for predicting the clinical outcome of some cancer diseases. Two major argyrophilic proteins responsible for the strong staining of cells in interphase are the nucleophosmin (B23) and the nucleolin (C23) nucleolar proteins. Interestingly these two proteins have been described as chromatin associated proteins with histone chaperone activities and also as proteins able to regulate chromatin transcription. Nucleolin seems to be over-expressed in highly proliferative cells and is involved in many aspect of gene expression: chromatin remodeling, DNA recombination and replication, RNA transcription by RNA polymerase I and II, rRNA processing, mRNA stabilisation, cytokinesis and apoptosis. Interestingly, nucleolin is also found on the cell surface in a wide range of cancer cells, a property which is being used as a marker for the diagnosis of cancer and for the development of anti-cancer drugs to inhibit proliferation of cancer cells. In addition to its implication in cancer, nucleolin has been described not only as a marker or as a protein being involved in many diseases like viral infections, autoimmune diseases, Alzheimer's disease pathology but also in drug resistance. In this review we will focus on the chromatin associated functions of nucleolin and discuss the functions of nucleolin or its use as diagnostic marker and as a target for therapy

INTRODUCTION

Hypertrophy of the nucleolus is one of the most distinctive cytological features of cancer cells (Derenzini *et al.*, 2000). The larger nucleolus displayed in malignant cells compared to benign cells is a key diagnostic feature of high grade prostatic intraepithelial neoplasia, an early stage that appears to be the precursor to the

*E-mail: pbouvet@ens-lyon.fr

T.K. Kundu and D. Dasgupta (eds.), Chromatin and Disease, 125–144.

majority of invasive prostate cancer (Fischer *et al.*, 2004) and reflects the deregulation of biological functions accomplished by the nucleolar compartment. This structural modification is associated with a deregulated expression of several nucleolar proteins like nucleolin, nucleophosmin (B23) and P120.

Besides the well known role of the nucleolus in ribosome biogenesis, nucleoli play important roles in the regulation of many fundamental cellular processes, including cell cycle regulation, apoptosis, telomerase production, RNA processing (Pederson , 2002) and therefore it is not surprising that many nucleolar proteins appear to be multifunctional proteins. Nucleolin, one of the most abundant non-ribosomal proteins of the nucleolus (Bugler *et al.*, 1982) has been the focus of many studies since it was first described 33 years ago (Orrick *et al.*, 1973). In addition to its role in RNA polymerase I transcription and pre-ribosomal processing and assembly in pre-ribosomes (Ginisty *et al.*, 1999), nucleolin seems to be involved in many aspects of DNA metabolism, chromatin regulation and appeared to be a good pharmalogical target for drug development.

Nucleolin, a 77 KDa protein is highly conserved in higher eukaryotes (Bourbon *et al.*, 1988; Caizergues-Ferrer *et al.*, 1989; Maridor and Nigg, 1990; Srivastava *et al.*, 1989) and analogous proteins could be found in plants and yeast (for a review see (Ginisty *et al.*, 1999)). Nucleolin is composed of three main domains: a N-terminal domain rich in acidic residues; a central domain containing 4 RNA recognition motifs (RBD or RRM motif) and a C-terminal domain rich in arginine and glycine residues (RGG or GAR domain). The exact contribution of the N-terminal domain for nucleolin function is not known, but since it is the site of numerous phosphorylation by CK2 and CDK1 (Belenguer *et al.*, 1990; Caizergues-Ferrer *et al.*, 1987; Peter *et al.*, 1990), it might play an important role in the regulation of nucleolin function during cell cycle.

The central domain has been the focus of many studies. Biochemical, genetic and NMR structural studies have shown that this domain interacts specifically with RNA (Allain *et al.*, 2000a,b; Bouvet *et al.*, 2001; Ghisolfi-Nieto *et al.*, 1996; Serin *et al.*, 1997). Through its first two RBDs, nucleolin interacts with a stem-loop structure found in pre-ribosomal RNA (Serin *et al.*, 1996, 1997). This interaction is very specific and mutation of a single residue in the RNA loop or in the active binding site of the protein almost completely abolishes the interaction. It is believed that by binding to this RNA structure during pre-rRNA transcription, nucleolin plays a role of chaperone, helping pre-rRNA to fold correctly (Allain *et al.*, 2000a). Interestingly, all four RBDs of nucleolin are required for binding to another single-stranded RNA motif (Ginisty *et al.*, 2001), and this binding is required for the first processing step of pre-rRNA (Ginisty *et al.*, 1998, 2000). These RBDs also seem to be involved in binding other RNA and DNA targets although the specificity of these interactions has not been always convincingly demonstrated. The RGG domain is the site of N^G, N^G-dimethylarginines modifications (Lischwe *et al.*, 1982) and interacts non specifically with nucleic acids (Ghisolfi *et al.*, 1992a,b) which could facilitate the specific binding of the RBD with targets located within large RNA. RGG domains have also been described as protein-protein interaction domains (Bouvet *et al.*, 1998; Sicard *et al.*, 1998).

1. NUCLEOLIN AS A PROTO-ONCOGENE THAT STIMULATES CELL PROLIFERATION

It has been shown that in cancer cell lines characterized by different proliferation rates, the transcriptional activity of RNA polymerase I and the expression of the major nucleolar proteins involved in the control of rRNA transcription and processing are directly related to nucleolar size and to the rapidity of cell proliferation (Derenzini *et al.*, 1998). Several observations suggest that nucleolin is a major actor in promoting cell proliferation. First of all, its amount is correlated with the proliferative status of the cell: nucleolin level is higher in tumours and actively dividing cells (Derenzini *et al.*, 1995; Mehes and Pajor, 1995; Roussel *et al.*, 1994; Sirri *et al.*, 1995, 1997) and is widely used as a marker of cell proliferation. Secondly, over expression of nucleolin cooperates with oncogenic mutant Ras in a rat embryonic fibroblast transformation assay (Takagi *et al.*, 2005). Nucleolin can therefore be considered a *bona fide* proto-oncogene. Thirdly, transfection of several tumour cell lines with G-rich oligonucleotides (GROs) arrests cells in S phase (Xu *et al.*, 2001) and the anti-proliferative activity of GROs is perfectly correlated with their ability to bind nucleolin (Bates *et al.*, 1999; Dapic *et al.*, 2003). The exact mechanism of action of GROs is not unravelled yet, but one likely explanation is that they sequester nucleolin, thereby preventing its binding to DNA (see below).

If these results suggest a positive role for nucleolin on proliferation, they do not indicate which of its activities are responsible for it. One hypothesis is that the stimulation of ribosome biogenesis by nucleolin is indispensable for active cell division. Indeed, a direct link between protein translation and cancer is clearly emerging (Ruggero and Pandolfi, 2003). However the studies mentioned above suggest a more direct impact of nucleolin on cell division. If part of its effects could come from its capacity to repress p53 mRNA translation (Takagi *et al.*, 2005), a more direct role of nucleolin in DNA replication can also be considered.

1.1. Nucleolin and RNA Polymerase I Transcription

Although the higher level of nucleolin in tumours and actively dividing cells suggests that nucleolin could play a positive role in the production of ribosomal RNA, most of the studies rather show a repressive effect of nucleolin on RNA polymerase I transcription.

A strong link between the phosphorylation of nucleolin, its proteolysis and the production of ribosomal RNA has been observed (Bouche *et al.*, 1984; Bourbon *et al.*, 1983; Warrener and Petryshyn, 1991). The inhibition of proteolysis using leupeptin leads to a lower rRNA transcription in an *in vitro* transcription system (Bouche *et al.*, 1984). In another series of experiments, the injection of nucleolin antiserum leads to 2–3.5 fold stimulation of pre-rRNA synthesis in *Chironomus tentans* salivary glands (Egyhazi *et al.*, 1988), although it was not clearly demonstrated that these antibodies blocked specifically the homolog of nucleolin in this species. A model was proposed based on these observations where nucleolin was suggested to be involved in regulating elongation of rRNA transcripts. In this model, nucleolin could interact with the nascent rRNA through RNA binding domains (RBDs) and with the RNA pol I machinery through its N terminal. This interaction was envisaged as a stalling mechanism, and proteolysis of nucleolin after its phosphorylation could release the transcriptional block (Ginisty *et al.*, 1999). However, this model was challenged by recent experiments using the xenopus oocyte system (Roger *et al.*, 2002). The injection of a 2–4 fold excess of nucleolin in *Xenopus laevis* stage IV oocytes leads to a significant reduction in accumulation of 40 S pre-RNA (Roger *et al.*, 2002). This repression was specific for RNA polymerase I, and could be obtained with a minimal pol I promoter independently of the nature of the RNA sequence that is transcribed, ruling out that the specific interaction of nucleolin with nascent pre-rRNA was required to regulate transcription elongation.

A recent study in the carp (*Cyprinus carpio*) shows that level of nucleolin is up regulated in cold-acclimatized carp with a concomitant nucleolar segregation and depression in rRNA transcription (Alvarez *et al.*, 2003).

On the other side, it has been demonstrated that nucleolin phosphorylation and rRNA transcription go hand in hand. Nucleolin phosphorylation could be triggered by a variety of stimuli like androgens and growth factors (Bonnet *et al.*, 1996; Bouche *et al.*, 1987; Issinger *et al.*, 1988; Suzuki *et al.*, 1985, 1991; Tawfic *et al.*, 1994) and the phosphorylation is invariably accompanied by increased rRNA transcription and cell proliferation. All these observations suggest that indeed nucleolin could have regulatory role in RNA polymerase I transcription. However, further experiments are clearly required to clarify the function of nucleolin in rRNA transcription.

1.2. Nucleolin and RNA Polymerase II Transcription

Several reports indicate that nucleolin is also involved in some aspect of RNA polymerase II transcription. A multiprotein complex containing YY1, HMGB2 and nucleolin binds the D4Z4 repeats and could regulate the transcription of genes located on chromosome 4q35 involved in Facioscapulohumeral muscular dystrophy (FSHD) (Gabellini *et al*, 2002). However, these proteins do not seem to be implicated in the FSHD cases not linked to 4q35 (FSHD1B) (Bastress *et al.*, 2005).

Cell transformation induced by the Human Papillomavirus (HPV18) could also involve nucleolin (Grinstein *et al.*, 2002). Nucleolin binds in a sequence-specific manner to the HPV18 enhancer. Antisense inactivation of nucleolin blocks E6 and E7 oncogene transcription and selectively decreases HPV18(+) cervical cancer cell growth. Nucleolin might be involved in opening the chromatin structure of the HPV18 enhancer suggesting that through this mechanism, nucleolin functions as a regulator of HPV18 oncogene transcription and HPV18-induced cervical carcinogenesis.

Nucleolin has also been implicated in the regulation of transcription of the Krüppel-like Factor 2 (KLF2) transcription factor. KLF2 is required for developmental and cellular functions in several distinct tissue types. Its transcription involves several transacting factors, chromatin modifications, and signaling pathways. Nucleolin has been identified as a protein that binds to a palindromic response region in the KLF2 promoter. Co-immunoprecipitation experiments indicated that nucleolin interacts with additional factors involved in KLF2 gene regulation (p85 and hnRNP-D) and small interfering RNAs targeting the nucleolin sequence selectively reduced nucleolin expression and were sufficient to block the induction of KLF2.

Nucleolin together with HnRNP D has been shown to form the LR1 transcription factor. LR1 is a B cell-specific, sequence-specific DNA binding activity that regulates transcription in activated B cells (Hanakahi *et al.*, 1997; Hanakahi and Maizels, 2000). DNA bending induced by nucleolin and hnRNP D might regulate the transcriptional activation by LR1 (Hanakahi and Maizels, 2000).

Finally, nucleolin has also been identified as a repressor of polymerase II transcription (Yang *et al.*, 1994). Biochemical and functional studies further established that nucleolin is a transcription repressor for regulation of alpha-1 acid glycoprotein (AGP)

These different reports indicate that nucleolin could influence transcription positively or negatively. These seemingly contradictory data might be explained if one considers that nucleolin does not act as a real transcriptional activator or repressor, but rather indirectly through the regulation of chromatin structure.

Indeed, since it has been discovered, it is well known that nucleolin binds to chromatin (Olson *et al.*, 1975; Olson and Thompson, 1983). Furthermore, it interacts with histone H1 and seems to modulate chromatin structure (Erard *et al.*, 1988, 1990).

This is only recently that a clear role for nucleolin in chromatin dynamic has been reported (Angelov *et al.*, 2006) (Fig. 1). It was shown that nucleolin possesses a histone chaperone activity and assists chromatin remodelers like SWI/SNF and ACF in their functions. In particular, nucleosome SWI/SNF dependent sliding and remodeling are greatly enhanced in the presence of nucleolin (Fig. 1). The presence of the acidic region in the N-terminal domain of nucleolin is a characteristic of many proteins with histone chaperone activity (Philpott *et al.*, 2000). The acidic domain is required for nucleolin chaperone activity, but is not sufficient to explain this activity since a nucleolin truncated protein containing only this acidic region does not carry the chaperone or co-remodeling activity (Angelov *et al.*, 2006). Nucleolin is able to destabilise the histone octamer, promoting H2A-H2B dimer displacement. Interestingly, passage of RNA polymerase II through the nucleosome is greatly facilitated in presence of nucleolin (Angelov *et al.*, 2006). H2A-H2B dimer displacement and facilitated transcription of pol II through the nucleosome are reminiscent of the activity of FACT complex (Orphanides *et al.*, 1998). The dual properties of nucleolin to bind non-specifically to DNA sequences and to interact with histones through its highly acidic domain recapitulate the characteristics of FACT subunits (Spt16 and SSRP1) in a single polypeptide. As mentioned above, nucleolin could exert a positive or a repressive effect on transcription. Such seemingly contradictory results could be explained if one takes into account that

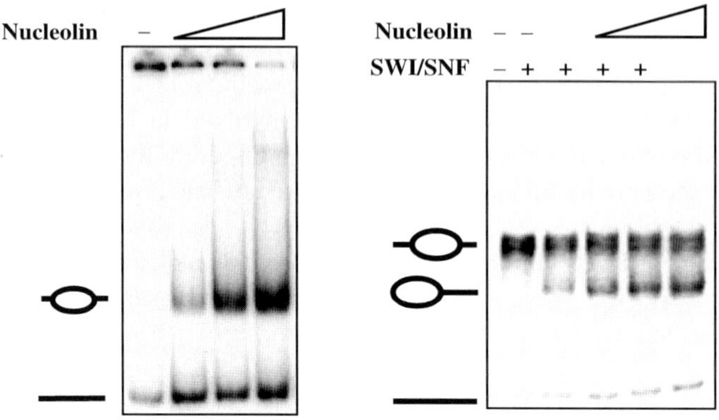

Figure 1. Nucleolin acts as a histone chaperone and boosts the activity of chromatin remodeler SWI/SNF. (a) *In vitro* histone deposition assay shows the histone chaperone activity of nucleolin. Recombinant core histones were incubated with 5S 147 bp DNA in presence or absence of nucleolin. Note the disappearance of aggregates in the wells and the formation of nucleosomal particles in presence of increasing amount of nucleolin. (b) Sliding assay performed on centrally positioned 601 nucleosomes (250 bp). The amount of SWI/SNF used here is insufficient to slide the nucleosomes. Note that in presence of nucleolin even the suboptimal quantity of SWI/SNF is able to slide the nucleosomes

nucleolin assists remodelers like SWI/SNF in their functions, and these remodelers are involved in both activation and repression of transcription (Martens and Winston, 2003).

2. NUCLEOLIN, DNA METABOLISM AND CANCER

DNA metabolism, i.e. DNA replication, recombination and repair, is a central process in oncogenesis. Proteins that ensure an accurate replication of the genome and protect it from genotoxic agents are "caretaker" tumor suppressor (Kinzler and Vogelstein, 1997): their loss of function leads to the rapid accumulation of deleterious mutations or chromosomal abnormalities. In parallel, proteins that stop cell proliferation when the integrity of the genome is offended are "gatekeeper tumor suppressors" (Kinzler and Vogelstein, 1997). On the other side, proteins that promote entry into S phase are potential proto-oncogenes whose gain of function enhances transformation and immortalization. It should be noted that the effects of the loss of function of many DNA replication and repair enzymes are by far not restricted to carcinogenesis. For example, mutations in several RecQ helicases cause various syndromes characterized by quite different defects (Hickson, 2003): premature ageing in Werner's syndrome, dwarfism, immunodeficiency and infertility in Bloom's syndrome or skin atrophy and skeletal abnormalities in Rothmund-Thomson syndrome.

A role for nucleolin in DNA replication, recombination and repair is suggested by its ability to bind directly both DNA and proteins involved in these processes (see below). As we shall see now, nucleolin could function either as a "caretaker", a "gatekeeper" or a proto-oncogene, depending on the cellular context.

2.1. A Positive Role for Nucleolin in DNA Replication?

Nucleolin has been shown to bind non specifically to denatured single-stranded DNA (Sapp *et al.*, 1986), particularly to G-rich sequences, such as those found in non-transcribed spacer region of ribosomal DNA (Olson *et al.*, 1983), telomeric DNA (Ishikawa *et al.*, 1993; Pollice *et al.*, 2000) or switch regions of immunoglobulin genes (Hanakahi *et al.*, 1997). This probably reflects a specific affinity for G quartets (Bates *et al.*, 1999; Hanakahi *et al.*, 1999), a conformation adopted by G-rich single strands through Hoogsteen bonding between guanines. However, this protein is also able to bind single-stranded DNA devoid of G-rich sequences such as MVMp DNA (Barrijal *et al.*, 1992), but this target also adopts a complex secondary structure.

Even though nucleolin has no or very little affinity for double-stranded DNA, it binds efficiently Matrix-Attachment Regions (MARs) (Dickinson and Kohwi-Shigematsu, 1995). Nevertheless, MARs, which mediate attachment of chromosomes to the nuclear matrix contain a sequence with an unusually high base unpairing potential (Galande, 2002), and nucleolin binds far more efficiently the T-rich single strand of MARs (Dickinson and Kohwi-Shigematsu, 1995).

Nucleolin lacks a characteristic DNA binding domain (Ginisty *et al.*, 1999). Its non-specific affinity for DNA is conferred by two different domains: its four RNA binding domains, particularly the 3rd and the 4th ones, and its C-terminal GAR domain (Hanakahi *et al.*, 1999; Sapp *et al.*, 1989). Of importance, these properties were determined *in vitro* with the native protein purified from cell extracts or recombinant truncated proteins; they are likely to be altered *in vivo* by interaction with other DNA binding factors (Dempsey *et al.*, 1998) and/or by post-translational modifications.

What could be the function of nucleolin in DNA replication? An intriguing possibility is that nucleolin may be a component of the DNA replication machinery as it was found associated with a DNA synthesome (Applegren *et al.*, 1995). Some authors attributed a helicase activity to human nucleolin and identified it as human helicase IV (Tuteja *et al.*, 1995). A possible ortholog of nucleolin in Pea displayed a similar activity which mapped to its GAR domain, even though it has no sequence homology with any other helicase (Nasirudin *et al.*, 2005). These results should, however, be interpreted cautiously as the activity described was far less efficient than that of typical helicases, while other groups simply did not detect it (Ginisty *et al.*, 1999). Moreover, two other studies demonstrated an antagonistic activity of nucleolin that favors complementary strand annealing (Hanakahi *et al.*, 2000; Sapp *et al.*, 1986). The reason for the discrepancy is not clear: either nucleolin does have antagonistic activities depending on its post-translational modifications, or some activities reported are conferred by contaminants.

However, one recent paper reports a genuine activity of nucleolin in replication (Seinsoth *et al.*, 2003). In this study, the authors demonstrate that nucleolin forms a ternary complex with the SV40 helicase T-antigene and endogenous topoisomerase I. Nucleolin could mediate the cohesion of this bipartite holoenzyme helicase complex, thus enhancing plasmid unwinding. It is not known at present whether nucleolin could play a more general role in endogenous replication initiation complexes or replication forks, through its helicase activity, its histone chaperone properties (Angelov *et al.*, 2006) or some yet unknown functions. Nevertheless, this notion is supported by the fact that nucleolin is a MAR binding protein (Dickinson and Kohwi-Shigematsu, 1995; Olson *et al.*, 1983) and a *bona fide* component of nuclear and nucleolar matrix (Dickinson and Kohwi-Shigematsu, 1995; Gotzmann *et al.*, 1997) where DNA replication is believed to occur (Cook, 1999; Falaschi, 2000). It is noteworthy that MAR-binding proteins are frequently involved in cancer (Galande, 2002), but one has to keep in mind that MARs and nuclear matrix likely play equally important roles in other cellular processes, namely regulation of transcription (Falaschi, 2000).

At last, nucleolin might play a specific role in telomeric replication and maintenance, as suggested by two types of data. First, it binds telomeric repeat (TTAGGG)n *in vitro* (Ishikawa *et al.*, 1993; Pollice *et al.*, 2000), with a marked preference for the single-stranded form. Secondly, it interacts *in vitro* and *in vivo* with hTERT (Khurts *et al.*, 2004), the protein catalytic component of human telomerase. This interaction takes place both in the cytoplasm and in the nucleolus, where it could promote the assembly of hTERT with the RNA subunit hTERC. As a conclusion, many data regarding the involvement of nucleolin in DNA replication are indirect and an experimental demonstration through knockdown or knockout studies is still awaited.

2.2. Nucleolin as a Stress-Responsive Tumour Suppressor?

An unexpected role of nucleolin in an S-phase checkpoint triggered by various stresses was recently uncovered. The story began with the discovery that heat shock causes a dramatic redistribution of nucleolin from the nucleolus to the nucleoplasm in HeLa cells (Daniely and Borowiec, 2000). This relocalization is very quick, since it starts as early as 5 minutes after a 15 minute heat shock at 44°C and is only transient, lasting around 1 hour. These results were subsequently confirmed in other cell lines and extended to other cell stresses such as exposition to gamma-irradiation or camptothecine, a topoisomerase I inhibitor (Daniely and Borowiec, 2000; Wang *et al.*, 2001). This phenomenon is nevertheless stress-selective, since it is not triggered by UV irradiation or hydroxurea exposure.

How specific stresses provoke nucleolin redistribution remains an open question. A block of ribosomal transcription cannot account for the whole phenomenon as shutdown of RNA polymerase I activity by actinomycin D induces rather a nucleolar-to-cytoplasmic change in nucleolin localization (Daniely and Borowiec, 2000). Interestingly, the relocalization is dependent on p53 since it does not occur in cell lines that lack p53 (Daniely *et al.*, 2002). Activated p53 interacts transiently with nucleolin after heat shock or gamma irradiation, through its last 30 C-terminal amino acids (Daniely *et al.*, 2002) known to bind several DNA repair proteins. The precise cell compartment where this interaction takes place is not known for the moment as nucleolin shuttles between the cytoplasm, the nucleoplasm and the nucleolus (Borer *et al.*, 1989). However, stress-activated nucleoplasmic p53 most likely binds nucleolin and prevents its import into the nucleolus, resulting in its accumulation onto the nuclear matrix (Daniely *et al.*, 2002). What induces the transient interaction in either compartment is also not known at present, but it is highly probable that nucleolin itself undergoes specific post-translational modification after stress, such as serine phosphorylation by casein kinase II (Kim *et al.*, 2005).

What is the biological meaning of this transient relocalization? Heat shock causes a transient arrest of DNA replication and kills cells preferentially in S-phase (Wang *et al.*, 2001). Elevated temperature has pleitotropic effects, including inhibition of origin firing, elongation step, histone deposition into chromatin and ligation of replication intermediates. These effects cannot be explained by mere alterations of the chromatine substrate, as lysates of heated cells fail to replicate exogenous plasmids carrying the SV40 origin of replication (Wang *et al.*, 2001): this *in vitro* assay accurately reproduces natural DNA replication, requiring only the addition of SV40 T-antigen. One of the main components affected by heat shock is RPA, the primary single-stranded DNA binding protein of eukaryotic cells which is necessary for both initiation and elongation steps of chromosomal DNA replication (Iftode *et al.*, 1999). Indeed, addition of recombinant hRPA to the cell lysates reverses the inhibition (Wang *et al.*, 1998). Interestingly, nucleolin was shown to interact *in vitro* with hRPA via its GAR domain and this interaction is induced *in vivo* upon heat shock (Daniely and Borowiec, 2000; Kim *et al.*, 2005). Since there is a striking correlation between the kinetics of replication inhibition, nucleolin mobilization and formation of nucleolin-RPA complex, it is tempting to build the following model: heat shock would provoke p53-dependent nucleolin redistribution to the nucleoplasm where it could sequester RPA and thereby inhibit origin unwinding and replication elongation (Daniely *et al.*, 2002). Importantly, nucleolin inhibits hRPA without affecting its single-stranded DNA binding activity (Daniely and Borowiec, 2000), which suggests that it sterically prevents interaction of hRPA with another factor. This model is strengthened by the observation that nucleolin-RPA complexes are preferentially located outside replicating regions (Daniely and Borowiec, 2000).

However, although this model is very attractive, it is still a matter of debate (Kim *et al.*, 2005). First of all, co-immunoprecipation experiments proved that the formation of a nucleolin-RPA complex occurs in the nucleolus as well as in the nucleoplasm (Kim *et al.*, 2005). Secondly, the interaction can be detected in a cell line which lacks p53 expression. Thirdly, hydroxyurea also induces RPA-nucleolin interaction without mobilizing nucleolin. On the other hand, mutant forms of Nucleolin that are constitutively mislocalized outside the nucleolus also constitutively interact with RPA, provided they retain the GAR domain (Kim *et al.*, 2005).

Moreover, overexpression of the same mutants inhibit DNA replication and block the cells at the G1/S-phase transition (Kim *et al.*, 2005), emphasizing the potential role of nucleolin mobilization. It is therefore highly probable that two different processes help the formation of RPA-nucleolin complexes after a genotoxic stress: a post-transcriptional modification of nucleolin that renders the GAR domain of nucleolin accessible to RPA, and its p53-dependent relocalization to the nucleoplasm where a higher amount of RPA is available. Of importance, nucleolin relocalization is transient and lasts far less than replication inhibition (Daniely and Borowiec, 2000). This means that nucleolin-RPA interaction is only an initial event and that other mechanisms account for prolonged replication inhibition.

As a conclusion, vertebrate nucleolin has acquired new functions in cell cycle control, compared to its yeast homologs GAR2 and NSR1. Nucleolin is a genuine stress-responsive protein which functions in a new S-phase checkpoint. This could relate to a more general function of the nucleolus as a stress sensor that enables coupling of cell metabolism (through ribosome biogenesis and therefore protein translation) and proper cell division (through control of DNA replication). Though paradoxical, this function is not incompatible with other results describing nucleolin as a proto-oncogene that promotes cell replication in normal conditions: depending on its post-translational modifications triggered by the phase of the cell cycle or the environment, nucleolin could play antagonistic roles. Of interest, other nucleolar proteins involved in ribosomal biogenesis display an additional role in cell cycle control. For example, nucleophosmin/B23 is rearranged or mutated in a number of hematological malignancies. Its inactivation in mouse leads to embryonic lethality and carcinogenesis, caused by unrestricted centrosome duplication and genome instability (Grisendi *et al.*, 2005).

2.3. A Function for Nucleolin in DNA Repair and Recombination?

Several enzymatic properties of nucleolin support a role for this protein in DNA repair and recombination: it is able to bend DNA (Hanakahi and Maizels, 2000) and can either unwind double-stranded DNA or enhance annealing of complementary DNA strands (Hanakahi and Maizels, 2000; Nasirudin *et al.*, 2005; Sapp *et al.*, 1986; Tuteja *et al.*, 1995).

Consistent with this idea, nucleolin was found in a B-cell specific recombination complex (Borggrefe *et al.*, 1998). It also forms a complex with hRNP-D that binds switch regions of immunoglobulin genes in B lymphocytes which undergo class-switch recombination (Hanakahi *et al.*, 1997). If nucleolin is clearly not the endonuclease of this recombination process, it still could have an accessory function such as structuring the switch regions or recruiting other factors.

Interestingly, nucleolin interacts with several key DNA repair and recombination proteins. First of all, it interacts with topoisomerase I through its N terminus (Bharti *et al.*, 1996; Edwards *et al.*, 2000). This interaction does not modify the enzymatic activity of topoisomerase I *per se*, but it could play important role in its predominantly nucleolar localization as it was demonstrated in yeast (Edwards *et al.*, 2000). Nucleolin could thus modulate rDNA recombination, an important (but not unique) function of topoisomerase I. Nucleolin also interacts with p53 (Daniely *et al.*, 2002), YB-1 (Gaudreault *et al.*, 2004) and RPA (Daniely and Borowiec, 2000), three proteins notably involved in DNA repair. At last, a recent paper describes a surprising interaction between nucleolin and Rad51 (De *et al.*, 2006), a RecA homolog required for homologous recombination. Most strikingly, the authors of this study show that inhibition of nucleolin by electroporation of a blocking antibody impairs intra-plasmid homologous recombination activity and sensitizes the cells to a topoisomerase II inhibitor, a phenotype quite similar to that observed after Rad51 inactivation by the same approach (De *et al.*, 2006).

Nucleolin could therefore contribute to genome stability in two different ways: by modulating the activities of several DNA repair enzymes through physical interaction and by enhancing the translation of some of these proteins after a genotoxic stress (Yang *et al.*, 2002).

2.4. Nucleolin as a Cell Surface Receptor

Although nucleolin is found almost exclusively within the nucleolus, increasing evidence indicates its presence at the surface of various cell types. While it can be argued that the detected protein could be a closely related antigen in some reports, several laboratories unequivocally identified nucleolin, using a combination of western blot, immunofluorescence, immunogold labeling, FACS analysis and micro-sequencing of the purified protein (Callebaut *et al.*, 1998b; Dumler *et al.*, 1999; Hirano *et al.*, 2005).

Although the presence of nucleolin at the cell surface could appear at first surprising, there are now a growing number of nuclear proteins involved in DNA metabolism and chromatin structure that have been shown to be present on the cell surface. Examples include the Ku protein which is involved in multiple nuclear processes (Dalziel *et al.*, 1992), nucleophosmin (Brandt *et al.*, 2004), Nopp140 (Kubler, 2001) and the chromosomal associated protein HMGB1 which was recently discovered to be a crucial cytokine that mediates the response to infection, injury and inflammation (Lotze and Tracey, 2005). Interestingly, auto-antibodies to human nucleolin (Valdez *et al.*, 1995), nucleophosmin (Chung and Utz, 2004), Ku proteins (Kelavkar *et al.*, 2002; Reeves *et al.*, 1991) and even to nucleosomal particles (Amoura *et al.*, 2000; Ghirardello *et al.*, 2004) were found associated with auto immune diseases like the Systemic lupus erythematosus.

Difficulties in detecting nucleolin on the cell surface could be explained by its very low concentration in this compartment (Hovanessian *et al.*, 2000). Moreover, this cell surface expressed nucleolin protein has a different isoelectric point and it is recognized by only one monoclonal antibody (mAb D3) in its native conformation (Hovanessian *et al.*, 2000). This probably reflects specific post-translational modifications undergone by nucleolin on the cell surface. Consistent with this hypothesis, extracellular nucleolin is a substrate of ecto-protein kinases including casein kinase II (Dumler *et al.*, 1999; Jordan *et al.*, 1994). Interestingly, indirect evidence suggests that nucleolin shutlles back and forth between the nucleus, the cytoplasm and the cell surface. Indeed, mAb D3 anti-nucleolin antibodies are quickly internalized by living cells (Hovanessian *et al.*, 2000) and eventually gain access to the nucleolus (Deng *et al.*, 1996).

These observations are quite amazing since nucleolin lacks a hydrophobic signal sequence (Ginisty *et al.*, 1999) They raise two questions: how is nucleolin attached to the cell surface and how is it translocated across the membranes? First of all, nucleolin can be dissociated from the membrane at high salt concentrations (Harms *et al.*, 2001). Thus, it is likely to bind integral membrane or GPI-anchored proteins through electrostatic interactions. The mechanisms of its externalization and subsequent re-internalization remain completely obscure for the moment. Externalization of nucleolin occurs through a mysterious alternative secretory pathway which is insensitive to inhibitors of the transport within the ER-Golgi network but requires physiological temperatures (Hovanessian *et al.*, 2000). Of note, such alternative pathways have already been described for several proteins, namely interleukin-1β and FGF-2 (Nickel, 2005). After internalization, nucleolin is detected within cytoplasmic smooth vesicles (Dumler *et al.*, 1999; Hovanessian *et al.*, 2000) and colocalizes with EEA1, a marker specifically associated with clathrin in early endosomes (Legrand *et al.*, 2004). Nucleolin may therefore use the classical endocytic pathway in an initial step owing to its association with the actin cytoskeleton (Hovanessian *et al.*, 2000) and unidentified membrane-anchored proteins. It would subsequently penetrate the cytosol through an unknown mechanism.

What could be the function of this protein at the cell surface? Independent affinity chromatography experiments suggested that nucleolin is a receptor for plenty of ligands, including several growth factors and adhesion molecules. It is doubtful that a single nucleolar protein is endowed with so many activities and the physiological relevance of these results awaits further confirmation. However, several experimental data support the idea that nucleolin is a genuine surface receptor in at least a few cases. First of all, a few putative ligands induce clustering of nucleolin at the cell surface: midkine (Said *et al.*, 2002), HIV particles (Nisole *et al.*, 2002a,b), lactoferrin (Legrand *et al.*, 2004) and pleiotrophin (Said *et al.*, 2005). Secondly, the same molecules and two others (urokinase (Dumler *et al.*, 1999) and F3 peptide (Christian *et al.*, 2003)) are co-internalized with nucleolin and this process can be blocked by anti-nucleolin antibodies. Nucleolin could thus act as a bridge between the cell surface and the nucleus, guiding growth factors such as midkine, lactoferrin and FGF-2 directly to the effector compartment where they would modulate the expression of ribosomal genes for example. This would explain why many tumour cells and angiogenic endothelial cells display an apparent accumulation of nucleolin at the cell surface (Christian *et al.*, 2003): abnormal relocalization of nucleolin would confer a competitive advantage by increasing the access to growth factors. At the same time, this observation makes nucleolin an attractive target for cancer therapy. Indeed, F3, a tumor homing peptide that binds specifically to nucleolin, and mAb D3 antibodies injected intravenously selectively accumulate in tumour and angiogenic vessels (Christian *et al.*, 2003; Joo *et al.*, 2005).

Cell surface expression of nucleolin could also have a pathophysiological significance as it has been shown to interact with several viruses and enteropathogenic bacteria. Depending on the micro organism, nucleolin may promote either their initial attachment to the cell surface (Callebaut *et al.*, 1998a,b; de Verdugo *et al.*, 1995; Sinclair *et al.*, 2006; Sinclair and O'Brien, 2002, 2004) or their entry in the host cell (Bose *et al.*, 2004; Nisole *et al.*, 2002a). A well documented example is the relationship between nucleolin and HIV. Nucleolin binds to V3 loop of gp120 and to a pseudo-peptide, HB-19, which mimics its structure (Callebaut *et al.*, 1998a). As a consequence, HB-19 (Nisole *et al.*, 2002a) as well as other putative ligands of nucleolin such as midkine (Callebaut *et al.*, 2001), lactoferrin (Legrand *et al.*, 2004) and pleitrophin (Said *et al.*, 2005) inhibits HIV attachment to the cell surface of mammalian cells and subsequent viral entry. Therefore, nucleolin could be a target of interest to design new antiviral molecules, but this would require first a deeper understanding of the physiological role of nucleolin at the cell surface.

3. CONCLUDING REMARKS: NUCLEOLIN AS A NEW PHARMACOLOGICAL TARGET

The recent findings of the localisation of nucleolin at the cell surface of cancer cells and its involvement in DNA replication have promoted the development of anti-cancer drugs targeting nucleolin and the use of nucleolin as a marker for the diagnosis of cancer.

For many years, the detection of AgNOR proteins by immuno cytochemistry has been used by pathologists to predict the clinical outcome of some cancer diseases. Nucleolin, together with nucleophosmin is mostly responsible for this specific silver staining and is, therefore, an interesting marker for diagnosis of cancer. More recently, the apparent preferential expression of nucleolin on the cell surface of cancer cells, is leading to the development of new tools for diagnosis of cancer cells. In particular detection of nucleolin at the cell surface using a monoclonal antibody was used in the Nucleolin OncoMarker kit (Assay Designs).

Several molecules with potential pharmaceutical activities targeting nucleolin function have been also developed. One can mention for example the development of the aptamer AS1411 (also known as AGRO100) by the company Antisoma (England). These aptamers are G-rich oligonucleotides, which can form G-quartet structures. They affect cell proliferation by inhibiting DNA replication (Xu *et al.*, 2001). It is claimed that these aptamers bind nucleolin on the cell surface of cancer cells, and then inhibit nucleolin function in cell proliferation by inhibiting DNA replication or another unknown process. Interestingly, these aptamers AS1411 reduced tumour growth in xenograft models of both renal and lung cancers and are currently being tested in phase I study with patients with kidney and lung cancers.

By targeting nucleolin on the cell surface, HB-19 pseudo-peptides (Nisole *et al.*, 2002a) inhibit HIV attachment to the cell surface of mammalian cells and subsequent viral entry, demonstrating that nucleolin could also be an interesting pharmaceutical target to prevent viral infection.

These examples demonstrate that nucleolin could be a therapeutic target for the development of innovative molecules against cancer and virus infection. The newly described function of nucleolin in chromatin dynamics and gene expression is potentially another function of nucleolin that could be targeted for the search of specific inhibitors of cell proliferation.

ACKNOWLEDGEMENTS

The work in authors laboratory is supported by CNRS, ATIP Plus to PB, Région Rhône-Alpes and grants from the Ministère de la Recherche: BCM0070; DRAB 04 2 136; ANR Project n° NT05-1_41978 and the ARCUS-INDIA program.

REFERENCES

Allain FH, Bouvet P, Dieckmann T, Feigon J (2000a) Molecular basis of sequence-specific recognition of pre-ribosomal RNA by nucleolin. Embo J 19:6870–6881

Allain FH, Gilbert DE, Bouvet P, Feigon J (2000b) Solution structure of the two N-terminal RNA-binding domains of nucleolin and NMR study of the interaction with its RNA target [In Process Citation]. J Mol Biol 303:227–241

Alvarez M, Quezada C, Navarro C, Molina A, Bouvet P, Krauskopf M, Vera MI (2003) An increased expression of nucleolin is associated with a physiological nucleolar segregation. Biochem Biophys Res Commun 301:152–158

Amoura Z, Koutouzov S, Piette JC (2000) The role of nucleosomes in lupus. Curr Opin Rheumatol 12:369–373

Angelov D, Bondarenko VA, Almagro S, Menoni H, Mongelard F, Hans F, Mietton F, Studitsky VM, Hamiche A, Dimitrov S, Bouvet P (2006) Nucleolin is a histone chaperone with FACT-like activity and assists remodeling of nucleosomes. Embo J 25:1669–1679

Applegren N, Hickey RJ, Kleinschmidt AM, Zhou Q, Coll J, Wills P, Swaby R, Wei Y, Quan JY, Lee MY et al. (1995) Further characterization of the human cell multiprotein DNA replication complex. J Cell Biochem 59:91–107

Barrijal S, Perros M, Gu Z, Avalosse BL, Belenguer P, Amalric F, Rommelaere J (1992) Nucleolin forms a specific complex with a fragment of the viral (minus) strand of minute virus of mice DNA. Nucleic Acids Res 20:5053–5060

Bastress KL, Stajich JM, Speer MC, Gilbert JR (2005) The genes encoding for D4Z4 binding proteins HMGB2, YY1, NCL, and MYOD1 are excluded as candidate genes for FSHD1B. Neuromuscul Disord 15:316–320

Bates PJ, Kahlon JB, Thomas SD, Trent JO, Miller DM (1999) Antiproliferative activity of G-rich oligonucleotides correlates with protein binding. J Biol Chem 274:26369–26377

Belenguer P, Caizergues-Ferrer M, Labbe JC, Doree M, Amalric F (1990) Mitosis-specific phosphorylation of nucleolin by p34cdc2 protein kinase. Mol Cell Biol 10:3607–3618

Bharti AK, Olson MO, Kufe DW, Rubin EH (1996) Identification of a nucleolin binding site in human topoisomerase I. J Biol Chem 271:1993–1997

Bonnet H, Filhol O, Truchet I, Brethenou P, Cochet C, Amalric F, Bouche G (1996) Fibroblast growth factor-2 binds to the regulatory beta subunit of CK2 and directly stimulates CK2 activity toward nucleolin. J Biol Chem 271(40):24781–24787

Borer RA, Lehner CF, Eppenberger HM, Nigg EA (1989) Major nucleolar proteins shuttle between nucleus and cytoplasm. Cell 56:379–390

Borggrefe T, Wabl M, Akhmedov AT, Jessberger R (1998) A B-cell-specific DNA recombination complex. J Biol Chem 273:17025–17035

Bose S, Basu M, Banerjee AK (2004) Role of nucleolin in human parainfluenza virus type 3 infection of human lung epithelial cells. J Virol 78:8146–8158

Bouche G, Caizergues-Ferrer M, Bugler B, Amalric F (1984) Interrelations between the maturation of a 100 kDa nucleolar protein and pre rRNA synthesis in CHO cells. Nucleic Acids Res 12:3025–3035

Bouche G, Gas N, Prats H, Baldin V, Tauber JP, Teissie J, Amalric F (1987) Basic fibroblast growth factor enters the nucleolus and stimulates the transcription of ribosomal genes in ABAE cells undergoing G0-G1 transition. Proc Natl Acad Sci U S A 84(19):6770–6774.

Bourbon H, Bugler B, Caizergues-Ferrer M, Amalric F (1983) Role of phosphorylation on the maturation pathways of a 100 kDa nucleolar protein. FEBS Lett 155:218–222

Bourbon HM, Lapeyre B, Amalric F (1988) Structure of the mouse nucleolin gene. The complete sequence reveals that each RNA binding domain is encoded by two independent exons. J Mol Biol 200:627–638

Bouvet P, Allain FH, Finger LD, Dieckmann T, Feigon J (2001) Recognition of pre-formed and flexible elements of an RNA stem-loop by nucleolin. J Mol Biol 309:763–775

Bouvet P, Diaz JJ, Kindbeiter K, Madjar JJ, Amalric F (1998) Nucleolin interacts with several ribosomal proteins through its RGG domain. J Biol Chem 273:19025–19029

Brandt R, Nawka M, Kellermann J, Salazar R, Becher D, Krantz S (2004) Nucleophosmin is a component of the fructoselysine-specific receptor in cell membranes of Mono Mac 6 and U937 monocyte-like cells. Biochim Biophys Acta 1670:132–136

Bugler B, Caizergues-Ferrer M, Bouche G, Bourbon H, Amalric F (1982) Detection and localization of a class of proteins immunologically related to a 100-kDa nucleolar protein. Eur J Biochem 128: 475–480

Caizergues-Ferrer M, Belenguer P, Lapeyre B, Amalric F, Wallace MO, Olson MO (1987) Phosphorylation of nucleolin by a nucleolar type NII protein kinase. Biochemistry 26:7876–7883

Caizergues-Ferrer M, Mariottini P, Curie C, Lapeyre B, Gas N, Amalric F, Amaldi F (1989) Nucleolin from Xenopus laevis: cDNA cloning and expression during development. Genes Dev 3:324–333

Callebaut C, Blanco J, Benkirane N, Krust B, Jacotot E, Guichard G, Seddiki N, Svab J, Dam E, Muller S, Briand JP, Hovanessian AG (1998a) Identification of V3 loop-binding proteins as potential receptors implicated in the binding of HIV particles to CD4(+) cells. J Biol Chem 273:21988–21997

Callebaut C, Jacotot E, Blanco J, Krust B, Hovanessian AG (1998b) Increased rate of HIV-1 entry and its cytopathic effect in CD4+/CXCR4+ T cells expressing relatively high levels of CD26. Exp Cell Res 241:352–362

Callebaut C, Nisole S, Briand JP, Krust B, Hovanessian AG (2001) Inhibition of HIV infection by the cytokine midkine. Virology 281:248–264

Christian S, Pilch J, Akerman ME, Porkka K, Laakkonen P, Ruoslahti E (2003) Nucleolin expressed at the cell surface is a marker of endothelial cells in angiogenic blood vessels. J Cell Biol 163:871–878

Chung L, Utz PJ (2004) Antibodies in scleroderma: direct pathogenicity and phenotypic associations. Curr Rheumatol Rep 6:156–163

Cook PR (1999) The organization of replication and transcription. Science 284:1790–1795

Dalziel RG, Mendelson SC, Quinn JP (1992) The nuclear autoimmune antigen Ku is also present on the cell surface. Autoimmunity 13:265–267

Daniely Y, Borowiec JA (2000) Formation of a complex between nucleolin and replication protein A after cell stress prevents initiation of DNA replication. J Cell Biol 149:799–810

Daniely Y, Dimitrova DD, Borowiec JA (2002) Stress-dependent nucleolin mobilization mediated by p53-nucleolin complex formation. Mol Cell Biol 22:6014–6022

Dapic V, Abdomerovic V, Marrington R, Peberdy J, Rodger A, Trent JO, Bates PJ (2003) Biophysical and biological properties of quadruplex oligodeoxyribonucleotides. Nucleic Acids Res 31:2097–2107

De A, Donahue SL, Tabah A, Castro NE, Mraz N, Cruise JL, Campbell C (2006) A novel interation of nucleolin with Rad51. Biochem Biophys Res Commun 344:206–213

de Verdugo UR, Selinka HC, Huber M, Kramer B, Kellermann J, Hofschneider PH, Kandolf R (1995) Characterization of a 100-kilodalton binding protein for the six serotypes of coxsackie B viruses. J Virol 69:6751–6757

Dempsey LA, Hanakahi LA, Maizels N (1998) A specific isoform of hnRNP D interacts with DNA in the LR1 heterodimer: canonical RNA binding motifs in a sequence-specific duplex DNA binding protein. J Biol Chem 273:29224–29229

Deng JS, Ballou B, Hofmeister JK (1996) Internalization of anti-nucleolin antibody into viable HEp-2 cells. Mol Biol Rep 23:191–195

Derenzini M, Sirri V, Trere D, Ochs RL (1995) The quantity of nucleolar proteins nucleolin and protein B23 is related to cell doubling time in human cancer cells. Lab Invest 73:497–502

Derenzini M, Trere D, Pession A, Govoni M, Sirri V, Chieco P (2000) Nucleolar size indicates the rapidity of cell proliferation in cancer tissues. J Pathol 191:181–186

Derenzini M, Trere D, Pession A, Montanaro L, Sirri V, Ochs RL (1998) Nucleolar function and size in cancer cells. Am J Pathol 152:1291–1297

Dickinson LA, Kohwi-Shigematsu T (1995) Nucleolin is a matrix attachment region DNA-binding protein that specifically recognizes a region with high base-unpairing potential. Mol Cell Biol 15:456–465

Dumler I, Stepanova V, Jerke U, Mayboroda OA, Vogel F, Bouvet P, Tkachuk V, Haller H, Gulba DC (1999) Urokinase-induced mitogenesis is mediated by casein kinase 2 and nucleolin. Curr Biol 9:1468–1476

Edwards TK, Saleem A, Shaman JA, Dennis T, Gerigk C, Oliveros E, Gartenberg MR, Rubin EH (2000) Role for nucleolin/Nsr1 in the cellular localization of topoisomerase I. J Biol Chem 275:36181–36188

Egyhazi E, Pigon A, Chang JH, Ghaffari SH, Dreesen TD, Wellman SE, Case ST, Olson MO (1988) Effects of anti-C23 (nucleolin) antibody on transcription of ribosomal DNA in Chironomus salivary gland cells. Exp Cell Res 178:264–272

Erard M, Lakhdar-Ghazal F, Amalric F (1990) Repeat peptide motifs which contain beta-turns and modulate DNA condensation in chromatin. Eur J Biochem 191:19–26

Erard MS, Belenguer P, Caizergues-Ferrer M, Pantaloni A, Amalric F (1988) A major nucleolar protein, nucleolin, induces chromatin decondensation by binding to histone H1. Eur J Biochem 175:525–530

Falaschi A (2000) Eukaryotic DNA replication: a model for a fixed double replisome. Trends Genet 16:88–92

Fischer AH, Bardarov S Jr, Jiang Z (2004) Molecular aspects of diagnostic nucleolar and nuclear envelope changes in prostate cancer. J Cell Biochem 91:170–184

Gabellini D, Green MR, Tupler R (2002) Inappropriate gene activation in FSHD: a repressor complex binds a chromosomal repeat deleted in dystrophic muscle. Cell 110:339–348

Galande S (2002) Chromatin (dis)organization and cancer: BUR-binding proteins as biomarkers for cancer. Curr Cancer Drug Targets 2:157–190

Gaudreault I, Guay D, Lebel M (2004) YB-1 promotes strand separation in vitro of duplex DNA containing either mispaired bases or cisplatin modifications, exhibits endonucleolytic activities and binds several DNA repair proteins. Nucleic Acids Res 32:316–327

Ghirardello A, Doria A, Zampieri S, Tarricone E, Tozzoli R, Villalta D, Bizzaro N, Piccoli A, Gambari PF (2004) Antinucleosome antibodies in SLE: a two-year follow-up study of 101 patients. J Autoimmun 22:235–240

Ghisolfi L, Joseph G, Amalric F, Erard M (1992a) The glycine-rich domain of nucleolin has an unusual supersecondary structure responsible for its RNA-helix-destabilizing properties. J Biol Chem 267:2955–2959

Ghisolfi L, Kharrat A, Joseph G, Amalric F, Erard M (1992b) Concerted activities of the RNA recognition and the glycine-rich C- terminal domains of nucleolin are required for efficient complex formation with pre-ribosomal RNA. Eur J Biochem 209:541–548

Ghisolfi-Nieto L, Joseph G, Puvion-Dutilleul F, Amalric F, Bouvet P (1996) Nucleolin is a sequence-specific RNA-binding protein: characterization of targets on pre-ribosomal RNA. J Mol Biol 260:34–53

Ginisty H, Amalric F, Bouvet P (1998) Nucleolin functions in the first step of ribosomal RNA processing. Embo J 17:1476–1486

Ginisty H, Amalric F, Bouvet P (2001) Two different combinations of RNA-binding domains determine the RNA binding specificity of nucleolin. J Biol Chem 276:14338–14343

Ginisty H, Serin G, Ghisolfi-Nieto L, Roger B, Libante V, Amalric F, Bouvet P (2000) Interaction of Nucleolin with an Evolutionarily Conserved Pre-ribosomal RNA Sequence Is Required for the Assembly of the Primary Processing Complex. J Biol Chem 275:18845–18850

Ginisty H, Sicard H, Roger B, Bouvet P (1999) Structure and functions of nucleolin. J Cell Sci 112:761–772

Gotzmann J, Eger A, Meissner M, Grimm R, Gerner C, Sauermann G, Foisner R (1997) Two-dimensional electrophoresis reveals a nuclear matrix-associated nucleolin complex of basic isoelectric point [In Process Citation]. Electrophoresis 18:2645–2653

Grinstein E, Wernet P, Snijders PJ, Rosl F, Weinert I, Jia W, Kraft R, Schewe C, Schwabe M, Hauptmann S, Dietel M, Meijer CJ, Royer HD (2002) Nucleolin as activator of human papillomavirus type 18 oncogene transcription in cervical cancer. J Exp Med 196:1067–1078

Grisendi S, Bernardi R, Rossi M, Cheng K, Khandker L, Manova K, Pandolfi PP (2005) Role of nucleophosmin in embryonic development and tumorigenesis. Nature 437:147–153

Hanakahi LA, Bu Z, Maizels N (2000) The C-terminal domain of nucleolin accelerates nucleic acid annealing. Biochemistry 39:15493–15499

Hanakahi LA, Dempsey LA, Li MJ, Maizels N (1997) Nucleolin is one component of the B cell-specific transcription factor and switch region binding protein, LR1. Proc Natl Acad Sci USA 94: 3605–3610

Hanakahi LA, Maizels N (2000) Transcriptional activation by LR1 at the Emu enhancer and switch region sites. Nucleic Acids Res 28:2651–2657

Hanakahi LA, Sun H, Maizels N (1999) High affinity interactions of nucleolin with G-G-paired rDNA. J Biol Chem 274:15908–15912

Harms G, Kraft R, Grelle G, Volz B, Dernedde J, Tauber R (2001) Identification of nucleolin as a new L-selectin ligand. Biochem J 360:531–538

Hickson ID (2003) RecQ helicases: caretakers of the genome. Nat Rev Cancer 3:169–178

Hirano K, Miki Y, Hirai Y, Sato R, Itoh T, Hayashi A, Yamanaka M, Eda S, Beppu M (2005) A multifunctional shuttling protein nucleolin is a macrophage receptor for apoptotic cells. J Biol Chem 280:39284–39293

Hovanessian AG, Puvion-Dutilleul F, Nisole S, Svab J, Perret E, Deng JS, Krust B (2000) The cell-surface-expressed nucleolin is associated with the actin cytoskeleton. Exp Cell Res 261:312–328

Iftode C, Daniely Y, Borowiec JA (1999) Replication protein A (RPA): the eukaryotic SSB. Crit Rev Biochem Mol Biol 34:141–180

Ishikawa F, Matunis MJ, Dreyfuss G, Cech TR (1993) Nuclear proteins that bind the pre-mRNA 3' splice site sequence r(UUAG/G) and the human telomeric DNA sequence d(TTAGGG)n. Mol Cell Biol 13:4301–4310

Issinger OG, Martin T, Richter WW, Olson M and Fujiki H (1988) Hyperphosphorylation of N-60, a protein structurally and immunologically related to nucleolin after tumour-promoter treatment. EMBO J. 7:1621–1626

Joo EJ, ten Dam GB, van Kuppevelt TH, Toida T, Linhardt RJ, Kim YS (2005) Nucleolin: acharan sulfate-binding protein on the surface of cancer cells. Glycobiology 15:1–9

Jordan P, Heid H, Kinzel V, Kubler D (1994) Major cell surface-located protein substrates of an ecto-protein kinase are homologs of known nuclear proteins. Biochemistry 33:14696–14706

Kelavkar U, Wang S, Badr K (2002) KU 70/80 lupus autoantigen is the transcription factor induced by interleukins (IL)-13 and -4 leading to induction of 15-lipoxygenase (15-LO) in human cells. Adv Exp Med Biol 507:469–481

Khurts S, Masutomi K, Delgermaa L, Arai K, Oishi N, Mizuno H, Hayashi N, Hahn WC, Murakami S (2004) Nucleolin interacts with telomerase. J Biol Chem 279:51508–51515

Kim K, Dimitrova DD, Carta KM, Saxena A, Daras M, Borowiec JA (2005) Novel checkpoint response to genotoxic stress mediated by nucleolin-replication protein a complex formation. Mol Cell Biol 25:2463–2474

Kinzler KW, Vogelstein B (1997) Cancer-susceptibility genes. Gatekeepers and caretakers. Nature 386:761, 763

Kubler D (2001) Ecto-protein kinase substrate p120 revealed as the cell-surface-expressed nucleolar phosphoprotein Nopp140: a candidate protein for extracellular Ca2+-sensing. Biochem J 360:579–587

Legrand D, Vigie K, Said EA, Elass E, Masson M, Slomianny MC, Carpentier M, Briand JP, Mazurier J, Hovanessian AG (2004) Surface nucleolin participates in both the binding and endocytosis of lactoferrin in target cells. Eur J Biochem 271:303–317

Lischwe MA, Roberts KD, Yeoman LC, Busch H (1982) Nucleolar specific acidic phosphoprotein C23 is highly methylated. J Biol Chem 257:14600–14602

Lotze MT, Tracey KJ (2005) High-mobility group box 1 protein (HMGB1): nuclear weapon in the immune arsenal. Nat Rev Immunol 5:331–342

Maridor G, Nigg EA (1990) cDna sequences of chicken nucleolin/C23 and No38/B23, two major nucleolar proteins. Nucleic Acids Res 18:1286

Martens JA, Winston F (2003) Recent advances in understanding chromatin remodeling by Swi/Snf complexes. Curr Opin Genet Dev 13:136–142

Mehes G, Pajor L (1995) Nucleolin and fibrillarin expression in stimulated lymphocytes and differentiating HL-60 cells. A flow cytometric assay. Cell Prolif 28:329–336

Nasirudin KM, Ehtesham NZ, Tuteja R, Sopory SK, Tuteja N (2005) The Gly-Arg-rich C-terminal domain of pea nucleolin is a DNA helicase that catalytically translocates in the 5'- to 3'-direction. Arch Biochem Biophys 434:306–315

Nickel W (2005) Unconventional secretory routes: direct protein export across the plasma membrane of mammalian cells. Traffic 6:607–614

Nisole S, Krust B, Hovanessian AG (2002a) Anchorage of HIV on Permissive Cells Leads to Coaggregation of Viral Particles with Surface Nucleolin at Membrane Raft Microdomains. Exp Cell Res 276:155–173

Nisole S, Said EA, Mische C, Prevost MC, Krust B, Bouvet P, Bianco A, Briand JP, Hovanessian AG (2002b) The anti-HIV pentameric pseudopeptide HB-19 binds the C-terminal end of nucleolin and prevents anchorage of virus particles in the plasma membrane of target cells. J Biol Chem 277:20877–20886

Olson MO, Ezrailson EG, Guetzow K, Busch H (1975) Localization and phosphorylation of nuclear, nucleolar and extranucleolar non-histone proteins of Novikoff hepatoma ascites cells. J Mol Biol 97:611–619

Olson MO, Rivers ZM, Thompson BA, Kao WY, Case ST (1983) Interaction of nucleolar phosphoprotein C23 with cloned segments of rat ribosomal deoxyribonucleic acid. Biochemistry 22:3345–3351

Olson MO, Thompson BA (1983) Distribution of proteins among chromatin components of nucleoli. Biochemistry 22:3187–3193

Orphanides G, LeRoy G, Chang CH, Luse DS, Reinberg D (1998) FACT, a factor that facilitates transcript elongation through nucleosomes. Cell 92:105–116

Orrick LR, Olson MO, Busch H (1973) Comparison of nucleolar proteins of normal rat liver and Novikoff hepatoma ascites cells by two-dimensional polyacrylamide gel electrophoresis. Proc Natl Acad Sci USA 70:1316–1320

Pederson T (2002) Proteomics of the nucleolus: more proteins, more functions? Trends Biochem Sci 27:111–112

Peter M, Nakagawa J, Doree M, Labbe JC, Nigg EA (1990) Identification of major nucleolar proteins as candidate mitotic substrates of cdc2 kinase. Cell 60:791–801

Philpott A, Krude T, Laskey RA (2000) Nuclear chaperones. Semin Cell Dev Biol 11:7–14

Pollice A, Zibella MP, Bilaud T, Laroche T, Pulitzer JF, Gilson E (2000) In vitro binding of nucleolin to double-stranded telomeric DNA. Biochem Biophys Res Commun 268:909–915

Reeves WH, Pierani A, Chou CH, Ng T, Nicastri C, Roeder RG, Sthoeger ZM (1991) Epitopes of the p70 and p80 (Ku) lupus autoantigens. J Immunol 146:2678–2686

Roger B, Moisand A, Amalric F, Bouvet P (2002) Repression of RNA polymerase I transcription by nucleolin is independent of the RNA sequence that is transcribed. J Biol Chem 277:10209–10219

Roussel P, Sirri V, Hernandez-Verdun D (1994) Quantification of Ag-NOR proteins using Ag-NOR staining on western blots. J Histochem Cytochem 42:1513–1517

Ruggero D, Pandolfi PP (2003) Does the ribosome translate cancer? Nat Rev Cancer 3:179–192

Said EA, Courty J, Svab J, Delbe J, Krust B, Hovanessian AG (2005) Pleiotrophin inhibits HIV infection by binding the cell surface-expressed nucleolin. Febs J 272:4646–4659

Said EA, Krust B, Nisole S, Svab J, Briand JP, Hovanessian AG (2002) The anti-HIV cytokine midkine binds the cell surface-expressed nucleolin as a low affinity receptor. J Biol Chem 277:37492–37502

Sapp M, Knippers R, Richter A (1986) DNA binding properties of a 110 kDa nucleolar protein. Nucleic Acids Res 14:6803–6820

Sapp M, Richter A, Weisshart K, Caizergues-Ferrer M, Amalric F, Wallace MO, Kirstein MN, Olson MO (1989) Characterization of a 48-kDa nucleic-acid-binding fragment of nucleolin. Eur J Biochem 179:541–548

Seinsoth S, Uhlmann-Schiffler H, Stahl H (2003) Bidirectional DNA unwinding by a ternary complex of T antigen, nucleolin and topoisomerase I. EMBO Rep 4:263–268

Serin G, Joseph G, Faucher C, Ghisolfi L, Bouche G, Amalric F, Bouvet P (1996) Localization of nucleolin binding sites on human and mouse pre- ribosomal RNA. Biochimie 78:530–538

Serin G, Joseph G, Ghisolfi L, Bauzan M, Erard M, Amalric F, Bouvet P (1997) Two RNA-binding domains determine the RNA-binding specificity of nucleolin. J Biol Chem 272:13109–13116

Sicard H, Faubladier M, Noaillac-Depeyre J, Leger-Silvestre I, Gas N, Caizergues-Ferrer M (1998) The role of the Schizosaccharomyces pombe gar2 protein in nucleolar structure and function depends on the concerted action of its highly charged N terminus and its RNA-binding domains. Mol Biol Cell 9:2011–2023

Sinclair JF, Dean-Nystrom EA, O'Brien AD (2006) The established intimin receptor Tir and the putative eucaryotic intimin receptors nucleolin and beta1 integrin localize at or near the site of enterohemorrhagic Escherichia coli O157:H7 adherence to enterocytes in vivo. Infect Immun 74:1255–1265

Sinclair JF, O'Brien AD (2002) Cell surface-localized nucleolin is a eukaryotic receptor for the adhesin intimin-gamma of enterohemorrhagic Escherichia coli O157:H7. J Biol Chem 277:2876–2885

Sinclair JF, O'Brien AD (2004) Intimin types alpha, beta, and gamma bind to nucleolin with equivalent affinity but lower avidity than to the translocated intimin receptor. J Biol Chem 279:33751–33758

Sirri V, Roussel P, Gendron MC, Hernandez-Verdun D (1997) Amount of the two major Ag-NOR proteins, nucleolin, and protein B23 is cell-cycle dependent. Cytometry 28:147–156

Sirri V, Roussel P, Trere D, Derenzini M, Hernandez-Verdun D (1995) Amount variability of total and individual Ag-NOR proteins in cells stimulated to proliferate. J Histochem Cytochem 43:887–893

Srivastava M, Fleming PJ, Pollard HB, Burns AL (1989) Cloning and sequencing of the human nucleolin cDNA. FEBS Lett 250:99–105

Suzuki N, Kobayashi M, Sakara K, Suzuki T, Hosoya T (1991) Synergistic stimulatory effect of glucocorticoid, EGF and insulin on the synthesis of ribosomal RNA and phosphorylation of nucleolin in primary cultured rat hepatocytes. Biochim Biophys Acta 1092:367–375

Suzuki N, Matsui H, Hosoya T (1985) Effects of androgen and polyamines on the phosphorylation of nucleolar proteins from rat ventral prostates with particular reference to 110-kDa phosphoprotein. J Biol Chem 260:8050–8055

Takagi M, Absalon MJ, McLure KG, Kastan MB (2005) Regulation of p53 translation and induction after DNA damage by ribosomal protein L26 and nucleolin. Cell 123:49–63

Tawfic S, Goueli SA, Olson MO, Ahmed K (1994) Androgenic regulation of phosphorylation and stability of nucleolar protein nucleolin in rat ventral prostate. Prostate 24:101–106.

Tuteja N, Huang NW, Skopac D, Tuteja R, Hrvatic S, Zhang J, Pongor S, Joseph G, Faucher C, Amalric F et al (1995) Human DNA helicase IV is nucleolin, an RNA helicase modulated by phosphorylation. Gene 160:143–148

Valdez BC, Henning D, Busch RK, Srivastava M, Busch H (1995) Immunodominant RNA recognition motifs of human nucleolin/C23. Mol Immunol 32:1207–1213

Wang Y, Guan J, Wang H, Leeper D, Iliakis G (2001) Regulation of dna replication vafter heat shock by replication protein a-nucleolin interactions. J Biol Chem 276:20579–20588

Wang Y, Perrault AR, Iliakis G (1998) Replication protein A as a potential regulator of DNA replication in cells exposed to hyperthermia. Radiat Res 149:284–293

Warrener P, Petryshyn R (1991) Phosphorylation and proteolytic degradation of nucleolin from 3T3-F442A cells. Biochem Biophys Res Commun 180:716–723

Xu X, Hamhouyia F, Thomas SD, Burke TJ, Girvan AC, McGregor WG, Trent JO, Miller DM, Bates PJ (2001) Inhibition of DNA replication and induction of S phase cell cycle arrest by G-rich oligonucleotides. J Biol Chem 276:43221–43230

Yang C, Maiguel DA, Carrier F (2002) Identification of nucleolin and nucleophosmin as genotoxic stress-responsive RNA-binding proteins. Nucleic Acids Res 30:2251–2260

Yang TH, Tsai WH, Lee YM, Lei HY, Lai MY, Chen DS, Yeh NH, Lee SC (1994) Purification and characterization of nucleolin and its identification as a transcription repressor. Mol Cell Biol 14:6068–6074

CHAPTER 8

CHROMATIN AS A TARGET FOR THE DNA-BINDING ANTICANCER DRUGS

PARIJAT MAJUMDER, SUMAN K. PRADHAN, PUKHRAMBAM GRIHANJALI DEVI, SUDIPTA PAL AND DIPAK DASGUPTA*

Biophysics Division, Saha Institute of Nuclear Physics, Block-AF, Sector-I, Bidhannagar, Kolkata – 700 064, India

Abstract: Chemotherapy has been a major approach to treat cancer. Both constituents of chromatin, chromosomal DNA and the associated chromosomal histone proteins are the molecular targets of the anticancer drugs. Small DNA binding ligands, which inhibit enzymatic processes with DNA substrate, are well known in cancer chemotherapy. These drugs inhibit the polymerase and topoisomerase activity. With the advent in the knowledge of chromatin chemistry and biology, attempts have shifted from studies of the structural basis of the association of these drugs or small ligands (with the potential of drugs) with DNA to their association with chromatin and nucleosome. These drugs often inhibit the expression of specific genes leading to a series of biochemical events. An overview will be given about the latest understanding of the molecular basis of their action. We shall restrict to those drugs, synthetic or natural, whose prime cellular targets are so far known to be chromosomal DNA

Keywords: Nucleosome, Chromosome, Intercalators, Groove-binders, DNA-cleaving agents, Cross-linkers, Chromatin condensation, Histone tails, Nucleosome phasing

The history of modern chemotherapy of cancer dates back to as early as 1946 when Goodman *et al.*, 1946, produced the first report of clinical results from 67 patients treated with nitrogen mustards for Hodgkin's disease, lymphosarcoma, and leukemia. From then onwards, the search for newer and better anticancer agents has continued with an ever-increasing pace. Drugs designed vary not only in their sources, chemical compositions and modes of action, but also in their targets. Whereas certain anticancer drugs target nucleic acids, others target specific

* E-mail: dipak.dasgupta@saha.ac.in

T.K. Kundu and D. Dasgupta (eds.), Chromatin and Disease, 145–189.

metabolic and signaling enzymes and even microtubules. Unfortunately, the major problem of anticancer drugs is their cytotoxicity. Lack of specificity of these compounds along with their chemically reactive centers augments their cytotoxic effects.

Chromatin is well recognized as a target for anticancer agents, in that both constituents of chromatin – chromosomal DNA and the associated chromosomal proteins have been utilized as drug targets for the past few decades. From the biochemical perspective, drugs have been designed that (1) may block the template property of DNA at the chromatin level, thereby inhibiting the action of replication or transcription machinery (Straney and Crothers, 1987; Gniazdowski and Czyz, 1999); (2) may inhibit the function of DNA associated proteins such as topoisomerase, DNA methyltransferase, high mobility group of proteins (HMGs), TATA binding proteins (TBPs), histone acetyltransferases (HATs), histone deacetylases (HDACs) (Schäfer and Jung, 2005); (3) may block the telomere function. All of the above processes lead to cell death.

In this review article we shall mostly confine ourselves to drugs that target the DNA, thereby falling into the first category mentioned above. The following figure depicts the target sites of the DNA-interactive anticancer drugs. Since the discovery of the double helical nature of DNA and the associated central dogma of molecular biology, there has been a plethora of studies to understand the structural basis of recognition of these drugs and the biological consequence of this recognition upon the nucleic acid template under *in vitro* condition. Interestingly, though many DNA-binding drugs, synthetic or from natural sources, are used to treat various types of cancer after extensive clinical trials, their mode of action inside the normal and neoplastic cells are still not very well known. In many cases there has been no extensive studies aimed at the target validation for these drugs. This is almost certainly due to the imperative social necessity to employ them to alleviate the sufferings and prolong the life expectancy of cancer-afflicted patients before a comprehensive understanding of their mode of action. In the last two or three decades efforts have been made to attain an incisive understanding of their interaction at the cellular level along with the validation of their targets at the level of cellular organelle.

More recently the concept of ADMET profile (**A**bsorption **D**istribution **M**etabolism **E**xcretion **T**oxicity profile) has further streamlined the molecular pharmacology aspects of these drugs with the ultimate objective of providing efficient target directed drugs with least toxicity.

Since chromatin targeted DNA-binding anticancer drugs take advantage of the dynamic nature of chromatin, so for the appreciation of the molecular and structural aspects of chromatin as target for these drugs, a brief introduction to the chromatin structure is essential. The nucleus houses over two meters of DNA, compacted to nearly one hundred thousandth of its dimension by a hierarchical scheme of folding, with an equal mass of proteins. This nucleoprotein complex is called 'chromatin' (Widom, 1998; Richmond and Widom, 2000). The term chromatin structure is open in the sense that it covers a wide range of phenomena and levels

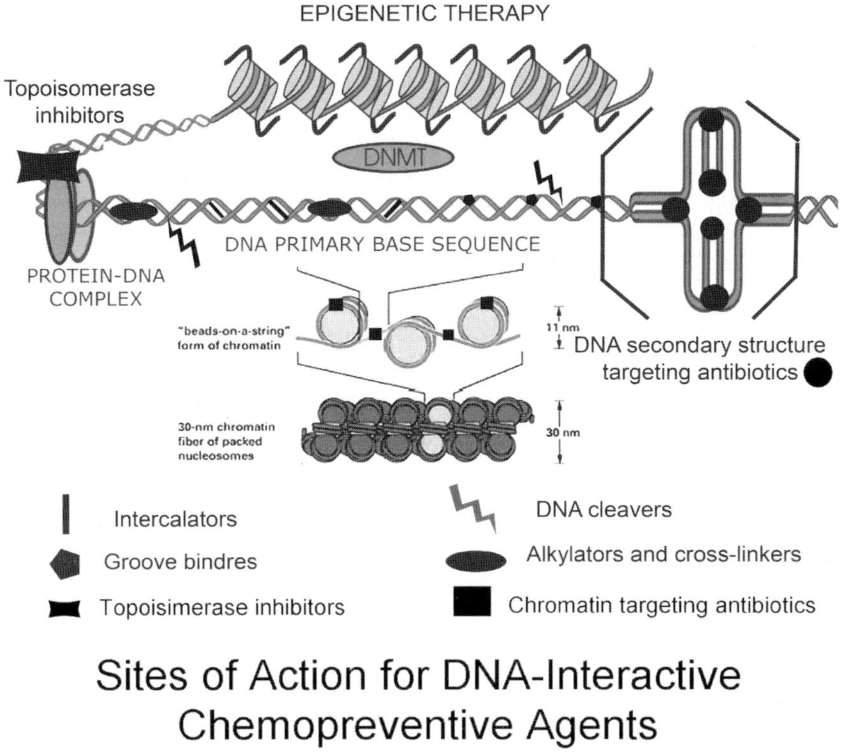

of organization, encompassing the atomic details of nucleosome architecture to the large-scale arrangements of interphase chromosomes that make up the nucleus.

It is now well established that the packaging of DNA in the eukaryotic nucleus, the principal target site of the DNA-binding antibiotics, involves several distinct hierarchical events. The first level of compaction occurs when DNA is wrapped around an octamer of core histones to form the repeating subunit nucleosome. An additional stretch of linker DNA connects adjacent nucleosomes. The linker DNA and nucleosome core are associated with linker histone H1 (Kornberg and Lorch, 1999). Each of the four core histones comprises a structured central domain, an amino-terminal tail, and in some cases a carboxyterminal tail (H2A and H3) (Van Holde *et al.*, 1995). N and C terminal tail sequences contribute nearly 28% of their mass. The high-resolution structure of the nucleosome core particle (Luger *et al.*, 1997, 2006; Davey *et al.*, 2002) has shown that two turns of the DNA superhelix in the nucleosome core are arranged in such a way that they create sufficient gaps for the amino-terminal tails of both H2B and H3 histones to pass through to the outside of the core particle. H2A and H4 tails pass across the superhelix on the flat faces of the particle to the outside as well. Histone tails are exposed. They are posttranslationally modified during eukaryotic transcription. The modifications alter the charge distribution pattern on the N-terminal tails. It is an essential requirement for the eukaryotic gene expression (Workman and Kingston, 1998). Along with sliding of the core particles along the linker (chromatin remodeling) the tail domains also play an important role in the access of transcription factors and other DNA-binding proteins to the nucleosomal templates. These tails have extreme biological significance as they play a vital role in regulating the nucleosomal spacing (i.e. in chromatin remodeling) and the state of condensation of chromatin. The inter-conversion of fluid interphase chromatin (Gasser, 2002) from transcriptionally blocked to transcriptionally active states is possibly tightly regulated by reversible modifications of selective amino acid residues in the tail parts of the histones, of other associated proteins, and of DNA (Jenuwein and Allis, 2001; Berger, 2002; Urnov, 2002). Trypsinized nucleosomes leading to chopped N-terminal tails have been found to be more accessible to transcription factors such as TFIIIA, DNase cleavage, and other sequence-specific DNA binding proteins (Ausio *et al.*, 1989; Juan *et al.*, 1994, Garcia-Ramirez *et al.*, 1992). Thus the tails could impede access of proteins and DNA-binding drugs to nucleosomal DNA (Workman and Kingston, 1998)

In fact, a nucleosome is a dynamic system that can take up many subtly different conformations and substructures at all organizational levels and thereby allow chromosomal DNA to be accessed while simultaneously being packaged into highly condensed chromosomal structures. The resulting accessibility of DNA has fundamental implications in all physiological processes that use DNA as substrate, such as transcription, replication, DNA repair and recombination. Nucleosome dynamics may be intrinsic or may be due to the action of protein-mediated pathways. Using stopped-flow fluorescence resonance energy transfer, (Li and Widom, 2004), have shown that the ends of nucleosomal DNA unwrap and rewrap rapidly (within 50–250 ms) from the histone surface. This timescale is sufficient to permit transcription factor and anti-cancer drug binding to nucleosomal DNA during its partially unwrapped stage and to allow access of the transcription machinery. Like the nucleosomes, the higher level of organization, the chromatin fiber is also capable of assembling and disassembling of superstructures by both intrinsic and protein mediated mechanisms. A recent report (Bucceri *et al.*, 2006) using yeast DNA repair by photolyase as a model system to monitor the rapid accessibility of nucleosomal DNA in yeast on a second time scale suggests that spontaneous unwrapping of nucleosomes rather than histone dissociation or chromatin remodeling provides DNA access to transcription factors, enzymes and probably small DNA-binding ligands. Thus, study of the kinetics might be relevant to understand the molecular approach of the anticancer drugs towards chromosomal DNA (Sischka *et al.*, 2005).

1. DNA AS TARGET FOR ANTICANCER DRUGS

The structure of the DNA molecule makes it an extremely versatile target for anticancer drugs. It has a negatively charged phosphate backbone, hydrogen accepting and donating functional groups from the bases in the major and minor grooves, phosphate oxygen atoms and aromatic hydrophobic components of the bases able to promote van der Waals interactions. DNA is polymorphic, and has been observed in several different conformations (e.g. A, B, Z etc.) that differ in the geometry of the double helix, including the depth and width of major and minor grooves (Saenger, 1984). The groove shapes and hydration patterns in a particular conformation are also sequence dependent to certain extent. Along with these, the molecule of DNA also has several levels of structural organization. The most basic level is the primary base sequence. The roll, tilt and twist from one base pair to the next in the double strand depend on the base, thereby giving rise to sequence dependent microheterogeneity in the DNA backbone. The bending of the backbone arising from stretches of A-bases is probably the extreme example of such feature. The small and large DNA binding ligands recognizes this structural variability along the double helix. It leads to sequence specific recognition. Higher levels of organization include the DNA secondary structures, such as hairpins, and Holliday junctions, triple helices and DNA quadruplexes. All of the above features have been exhaustively exploited in designing anticancer drugs.

2. MOLECULAR AND STRUCTURAL BASIS OF DNA-BINDING ANTICANCER DRUG ACTION

Anticancer drugs targeted to DNA bind to the same by either non-covalent forces or covalent interactions. The primary and most important step in the functioning of a DNA targeting drug is the base-specific recognition of DNA by the drug. This initial process of recognition is driven by size and shape complementarities between the drug and its DNA – binding site. Non-covalent forces such as the coulombic force, van der Waals interactions, hydrogen bonding and stacking interaction stabilize the primary complex formed. For drugs that interact with DNA via non-covalent forces, the active primary complex is responsible for its intracellular function; whereas, for drugs that interact with DNA via covalent forces, the covalent bond formation is preceded by the non-covalent reversible drug-DNA association that determines the specificity, because it positions the reactive part of the drug along the target site in the DNA. The general mechanism of action of DNA targeting anti-cancer drugs is shown below:

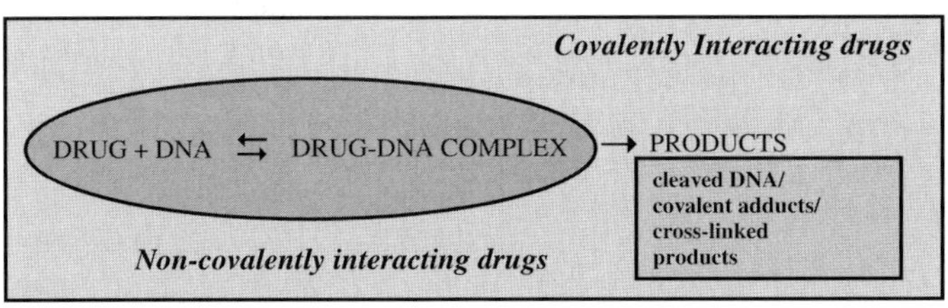

Chemical basis of action for DNA targeting anti-cancer drugs

3. GENERAL APPROACHES TO STUDY ASSOCIATION OF ANTICANCER DRUGS WITH DNA AT THE CHROMATIN LEVEL

There have been two basic approaches. First one involves isolation of the chromatin and nucleosome from the healthy and diseased cell line. The second approach is the reconstitution of the model target such as nucleosome followed by the association with the drug(s). The second approach has been extensively employed to identify the binding site in the protein-nucleic acid complex. A pre-knowledge about the components and their arrangements in the reconstituted system sometime makes it the preferred approach. Different biophysical, biochemical and genetic techniques have been employed to understand the mode of association and the effect of the drugs upon chromatin/nucleosome structure and function.

3.1. Non-covalently Interacting Anticancer Drugs

There are two principal modes by which drugs can bind non-covalently to DNA: intercalation and external groove binding.

3.1.1. Intercalators

Intercalating drugs consist of planar heteroaromatic ring systems, which have the potential to be inserted between two adjacent base pairs in a helix. The complex is stabilized by π-π stacking interactions between the drug molecule and the DNA base pair accommodating the ring. In general, intercalating drugs are characterized by the presence of one or more planar aromatic rings, parallely oriented and/or separated by linker regions of varying lengths. Structure of a typical intercalator, Actinomycin, is given below:

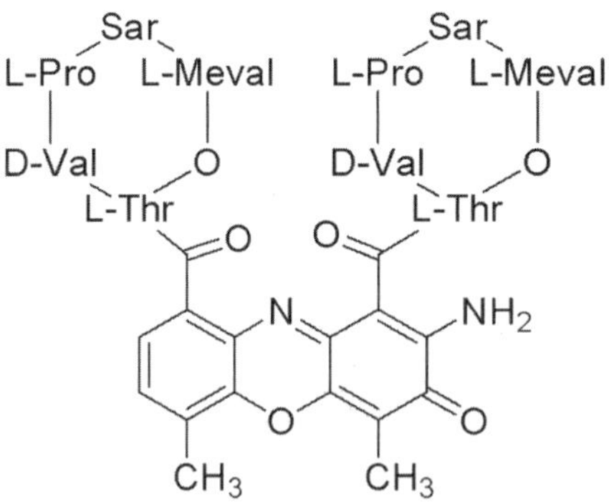

Actinomycin D

Intercalation results in structural perturbations in DNA so as to lengthen it by ~1 bp spacing (i.e. ~ 3.4 Å) and there is also some unwinding of the DNA helix. As a result, the DNA adopts a C3′-endo –(5′, 3′)-C2′-endo mixed puckering conformation around the intercalation site. Intercalators are known to bind the DNA via both the minor and major grooves.

Mono-alkylators (e.g. daunomycin and related anthracycline drugs) contain a planar chromophore group of 4 fused six-membered rings, substituted at different loci and containing an amino sugar residue. The chromophores are inserted in a 'head-on' fashion with the long axis of the aglycone moiety oriented at right angles to the long axis of adjacent base pairs. One of the cyclohexane rings protrudes into the minor groove leaving the amino sugar in the minor groove.

The anthracycline antibiotic daunomycin and its derivatives are antitumor drugs widely used in cancer chemotherapy to treat myelogenous leukemia and solid tumors (Gianni *et al.*, 1983, Myers *et al.*, 1988, Carrion *et al.*, 2004). Cellular DNA is the primary target for these drugs. Daunomycin acts by intercalation of its planar aglycon chromophore between DNA base pairs, and its amino sugar ring lies in the minor groove of the double helix (Barcello *et al.*, 1988; Chaires *et al.*, 1982, 1990; Chaires, 1996) leading to inhibition of replication and transcription.

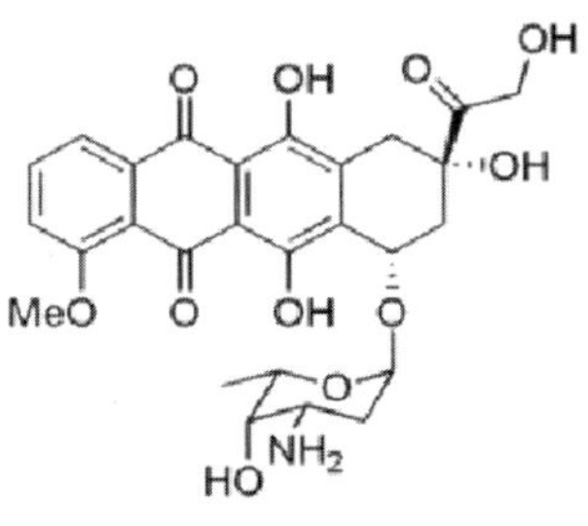

Doxorubicin

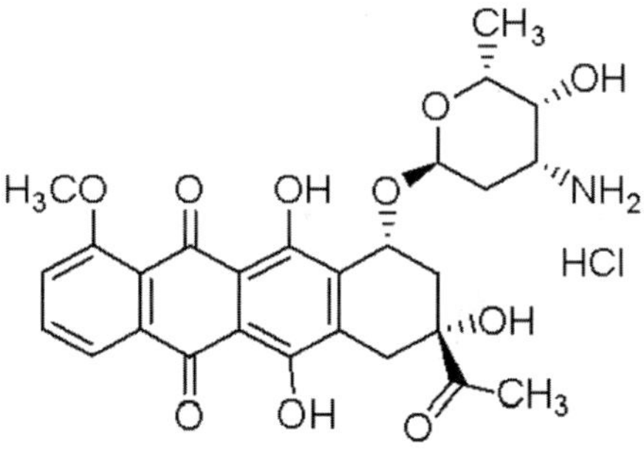

Daunomycin

These drugs were few of those whose association with chromatin and nucleosome was studied (Mizuno *et al.*, 1975; Zunino *et al.*, 1980; Chaires *et al.*, 1982, 1983, 1990; Ganguli *et al.*, 1983; Terasaki *et al.*, 1984; Fritzsche *et al.*, 1987; Simpkins *et al.*, 1984; Chaires, 1996; Chakrabarti *et al.*, 1996) when the molecular nature of the chromatin was established. The first detailed structural study (Chaires *et al.*, 1982) employed equilibrium, hydrodynamic, and electric dichroism studies of the complex of daunomycin with H1-depleted 175 bp nucleosomes, along with some comparative data for ethidium. The results showed that in contrast to ethidium, daunomycin binding to nucleosomes is strongly reduced relative to the affinity for free DNA. The salt concentration dependence of the binding constant suggested that approximately one Na^+ ion is released from both nucleosomes and free DNA upon daunomycin binding. The early melting transition of nucleosomes is preferentially stabilized by low levels of both drugs, but more markedly by ethidium.

Ethidium also stabilizes the second nucleosome melting transition, but daunomycin does not. Dichroism and rotational relaxation time measurements indicated that daunomycin unfolds nucleosomes. The data favored an unfolded structure in which the nucleosome elongates along the DNA superhelical axis. Higher concentration of the drug at a ratio more than 0.15 per DNA base pair promotes nucleosome aggregation. The authors suggested that the activity of daunomycin as an antitumor agent arises out of its special intercalation geometry that strongly prefers free DNA regions to the bent helices found in nucleosomes and chromatin. As a result of this preference there is an increased local concentration of the drug in the genetically active regions of nuclear DNA in which nucleosomal structure is less prevalent. Presumably the abundance of such regions in tumor cells makes them especially sensitive to daunomycin. There have been many other reports, which aimed to quantitatively define the affinity of daunomycin and related drugs with chromatin and nucleosome (van Helden and Wild, 1982). Fluorescent probes were used to examine the effect of adriamycin on supercoiled DNA and calf thymus nucleosomes.

Equilibrium dialysis and sedimentation velocity analysis (Rabbani *et al.*, 1999) were also employed to characterize the binding of the antitumor drug daunomycin to chicken erythrocyte chromatin before and after depletion of linker histones and to its constitutive DNA under several ionic strengths (5, 25, and 75 mM NaCl). The equilibrium dialysis experiments showed that the drug binds cooperatively to both the chromatin fractions and to the DNA counterpart within the range of ionic strength used in this study. A significant decrease in the binding affinity was reported at 75 mM NaCl. Binding of daunomycin to DNA does not significantly affect the sedimentation coefficient of the molecule in contrast to the effect noticed when the drug binds to chromatin and to its linker histone-depleted counterpart. In these instances, preferential binding of the drug to the linker DNA regions induces an unfolding of the chromatin fiber that is followed by aggregation. Transient electrical dichroism studies also supported the condensation of the chromatin fiber.

Earlier reports showed that treatment of adriamycin to Novikoff hepatoma nuclei enhanced the DNA fragmentaion by micrococcal nuclease (Ross *et al.*, 1978; Gyapay *et al.*, 1985). Exposure of murine thymocytes to doxorubicin triggered rapid DNA degradation, as indicated by the appearance of a major subdiploid population demonstrated by DNA flow cytometry (Zaleskis *et al.*, 1994). Electron microscopic comparison of samples with large subdiploid populations versus those with little or no such subset revealed significantly more cells with the characteristic features of apoptosis. Daunorubicin – induced internucleosomal DNA fragmentation was also reported in acute myeloid cell lines. A plausible mechanism of this DNA-degradation may be the chemical activation of the drug by an intracellular redox system, which leads to the production of active oxygen species (Akman *et al.*, 1992; Quillet-Mary *et al.*, 1996). There are reports of DNA base modifications induced in isolated human chromatin by NADH dehydrogenase- catalyzed reduction of doxorubicin. Induction of apoptosis via DNA cleavage has now been accepted as one of the actions of the drug and its related compounds.

In a recent study (Nair *et al.*, 2005) the transcription inhibitory effect upon a particular gene by the drug and the related compound WP361 was demonstrated with the urokinase receptor (uPAR). It is transcriptionally activated in several cancers and contributes to tumor progression by promoting cell migration and proteolysis. The bisanthracycline (WP631) represses uPAR gene expression and cell migration of RKO colon cancer cells by interfering with transcription factor binding to a chromatin-accessible -148/-124 promoter region. It was suggested from DNaseI hypersensitivity, genomic footprinting, and chromatin immunoprecipitation experiments that chromatinized -148/-124 regulatory region of the uPAR promoter is accessible to small molecules and that WP631, which disrupts the interaction of DNA binding proteins with this region, diminishes uPAR expression and function.

Recently, a novel observation has raised question about the DNA as the only target of daunomycin at the chromatin level inside the cell. It was shown using a compositionally defined chicken erythrocyte chromatin fraction that that the drug is also able to interact with chromatin-bound linker histones without any noticeable binding to core histones (Rabbani *et al.*, 2004). The drug can interact in an equal fashion with both histone H1 and H5 and to a greater extent with core histones H3/H4 and H2A/H2B as free proteins in solution. Binding of daunomycin to linker histones appears to primarily involve the trypsin-resistant (winged-helix) domain of these proteins.

Since anthracycline antibiotics play an important role in cancer chemotherapy, the necessity for an improvement of their therapeutic index has enthused an ongoing search for anthracycline analogues with improved pharmacological properties. Cardiac toxicity is a negative feature in their use as drugs. Along with DNA, the DNA topoisomerase II was recognised to be another prime cellular target. Several anthracyclines interfere with topoisomerase II functions by stabilizing a reaction intermediate in which DNA strands are cut and covalently linked to tyrosine residues of the enzyme (Binaschi *et al.*, 2001). Investigations on the sequence specificity of doxorubicin *in vitro* and in nuclear chromatin of living cell have led to a molecular model of drug receptor on the topoisomerase II-DNA complex. Anthracyclines are likely placed at the interface between the DNA cleavage site and the active site of the enzyme, forming a DNA-drug-enzyme ternary complex. Moreover, a quite detailed structure-function relationship has been established for anthracyclines. These studies have revealed that (a) drug intercalation is necessary but not sufficient for topoisomerase II poisoning; (b) the deletion of the 4-methoxy and 3′-amino substituents results in an enhancement of the drug activity; and (c) the 3′ substituent of the sugar moiety markedly influences the sequence selectivity of anthracycline-stimulated DNA cleavage. These relationships have been exploited during the last decade by several groups in the search for new anthracycline drugs with lower side effects and higher activity against resistant cancer cells. A scheme utilizing the DNA replication in *Xenopus* egg extract system to simultaneously evaluate DNA-interacting drugs as potential anti-cancer agents and gain insight into the mechanism of drug action has been proposed (Kumar *et al.*, 2004). According to the authors this system might be useful for large scale screening of DNA-interacting chemotherapeutic compounds in cellular milieu.

3.1.2. External groove binders

The grooves (minor and major) of the DNA molecule are of immense structural and chemical significance. The difference in the pattern of hydrogen bond donors and acceptors in the minor and major grooves allow for sequence readout by these group of ligands. Thus, the external groove binders search for the shape and size of DNA grooves, which are specifically recognized by them. The molecules may even induce structural changes in the DNA duplex for better structural complementarity between drug and DNA. The drug can also alter its own structure by 'induced fit' type of mechanism (Spolar *et al.*, 1994, Chaires, 1997).

3.1.2.1. Minor groove binders. Minor groove binders are typically composed of several aromatic rings such as pyrrole, furan, or benzene that are connected by bonds with torsional freedom. In all complexes of minor groove binders with DNA, the drug displaces the 'spine of hydration' and fits snugly into the minor groove. These drugs generally adopt a characteristic curved shape isohelical with the target groove. Van der Waals interactions, hydrophobic forces, and hydrogen bonds stabilize the resulting DNA–drug complex.

Minor groove binders have a binding preference for A•T base pairs, probably because there are favorable hydrophobic contacts between the adenine C2 hydrogen atoms and the aromatic rings in the drug. Besides, the A•T base pairs possess hydrogen bond acceptors such as C2 carbonyl oxygen of thymine or N3 nitrogen of adenine that can readily interact with any hydrogen bond donors. Although similar hydrogen bonding opportunities exist at G•C base pairs, the amino group of guanine may be a steric block to hydrogen bonds involving either guanine N3 or cytosine C2 atoms. The favorable curvature of some tailor made ligands can overcome this and can bind to G•C base pair as well.

However, with the discovery of a new motif – the side-by-side pyrrole-imidazole amino acid pairing, or 2:1 (ligand: DNA) complex, it is possible to distinguish all four Watson-Crick base pairs (G•C, C•G, A•T and T•A) (Dervan *et al.*, 1999). Using the 2:1 motif for recognition of the minor groove, a 6-bp sequence can be read even at subnanomolar concentrations (Tranger *et al.*, 1996). Minor groove binding drugs are comprehensively reviewed in Zimmer (1986).

Aureolic acid group of antibiotics: We have given below an extended summary of the work done with this group of antibiotics, because we have been actively involved in understanding the mode of action of these DNA-binding antibiotics. The summary also illustrates the model example of how different biophysical and biochemical approaches were undertaken to characterize the association of these drugs with chromosomal DNA and the sequential effect of this binding upon the chromatin structure.

Aureolic acid group of antitumor antibiotics, mithramycin (MTR) and chromomycin A_3 (CHR), from *Streptomyces plicatus* and *Streptomyces griseus*, respectively, are clinically employed for testicular carcinoma and Paget's disease (Calabresi and Chabner, 1991). With a gross structural similarity, MTR and CHR have difference in the nature of sugar rings connected to aglycone ring via O-glycosidic bond. Antitumor properties of MTR and CHR in experimental tumors have been ascribed to their inhibitory roles in replication and transcription processes during macro molecular biosyntheses (Wohlert *et al.*, 1999). They inhibit the expression of proto-oncogenes like *c-myc*, that have an important role in the regulation of cell proliferation and differentiation (Synder *et al.*, 1991). A potential use of these antibiotics is as neurological therapeutics for the treatment of diseases associated with aberrant activation of apoptosis (Chatterjee *et al.*, 2001).

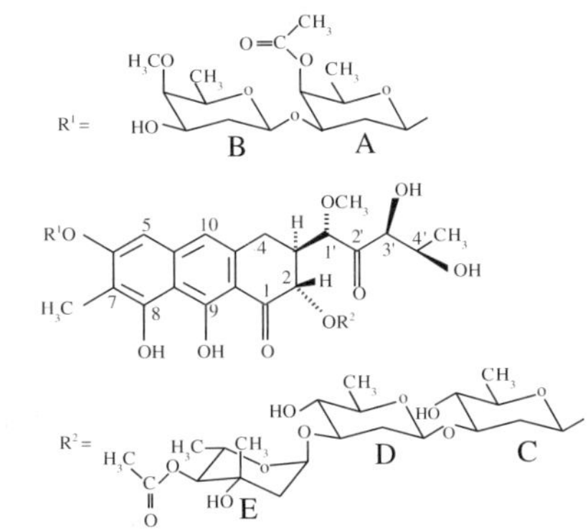

Chromomycin A_3

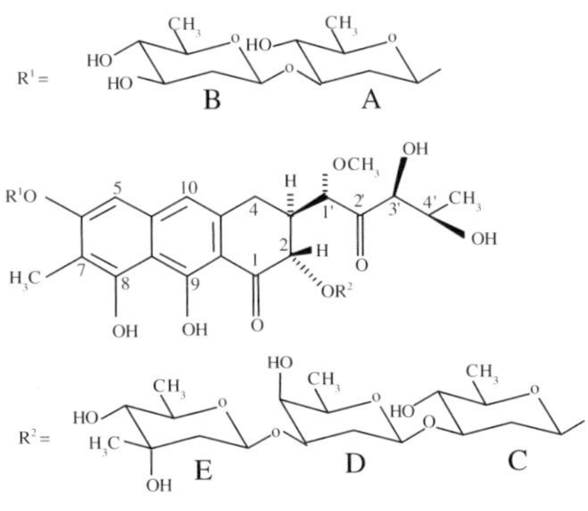

Mithramycin

Prime cellular target of these antibiotics is DNA. A bivalent cation such as Mg^{2+} is an essential requirement for their association with DNA at and above physiological pH (Dimaraco *et al.*, 1975). We have shown that in absence of DNA these antibiotics bind to Mg^{2+} and form two different types of complexes, complex I (1:1 in terms of antibiotic : Mg^{2+}) and complex II (2:1 in terms of antibiotic : Mg^{2+}) (Dimaraco *et al.*, 1975; Aich and Dasgupta, 1990; Aich *et al.*, 1992a, b). The equilibria associated with the reversible association along with the affinity constants are given below:

$$\text{Drug} + Mg^{2+} \leftrightarrows \text{Drug} : Mg^{2+} \ \textbf{(Complex I)}$$

$$\text{Drug} : Mg^{2+} + \text{Drug} \leftrightarrows (\text{Drug})_2 : Mg^{2+} \ \textbf{(Complex II)}$$

Affinity parameters for the formation of antibiotic: Mg^{2+} complexes

Antibiotic	Type of complex	Apparent association constant (M^{-1})	Stoichiometry
Mithramycin	I	1.8×10^4	1.1
	II	1.6×10^3	2:1
Chromomycin	I	1.9×10^4	1:1
	II	5.8×10^2	2:1

Since the second complex contains two molecules of drug, therefore, we refer to it as dimer complex. Keeping in view the milli molar concentration of the metal ion present in the cell, possibility of the formation of dimer complex is more under *in vivo* conditions. However, in certain cases of cancer the metal ion concentration goes down to micro molar range. Under these unusual conditions, complex I is formed. Recently we have shown that mithramycin forms only dimer complex with Zn^{2+}, another metal ion playing an important role as cofactor in many enzymes and DNA binding proteins like transcription factors.

These complexes are DNA binding ligands at and above physiological pH and bind to DNA via minor groove (Dimaraco *et al.*, 1975; Cons and Fox, 1989; Aich and Dasgupta, 1990; Aich *et al.*, 1992a, b; Aich and Dasgupta, 1995). It was established in our laboratory from spectroscopic and thermodynamic studies that the modes of binding of the two ligands with natural DNA, polynucleotides and oligomeric duplexes are different (Dimaraco *et al.*, 1975; Aich and Dasgupta, 1990; Aich *et al.*, 1992a, b; Aich and Dasgupta, 1995). We also illustrated the role of DNA minor groove size and the accessibility of the 2-amino group in the minor groove of guanosine in drug–DNA interaction using designed nucleotide sequences (Aich and Dasgupta, 1995; Majee *et al.*, 1997; Chakrabarti *et al.*, 2000–2001, 2002). Detailed NMR studies from other laboratories have helped to understand how the bulky complex II is accommodated at the cost of a considerable widening of the minor groove in B-DNA type structure (Keniry *et al.*, 1993; Sastry *et al.*, 1995). In our laboratory we have shown from a detailed thermodynamic analysis of the association of the dimer complex with different DNAs, natural and oligonucleoides, with defined sequences, that B to A type transition in the groove leads to a positive change in enthalpy. This is compensated by a positive change in entropy arising from the release of bound water in the minor groove (Aich and Dasgupta, 1995; Majee *et al.*, 1997; Chakrabarti *et al.*, 2000–2001, 2002). Sugars present in the antibiotics play a significant role during the association with nucleic acids (Majee *et al.*, 1997; Chakrabarti *et al.*, 2000–2001, 2002). Absence of substituents like acetoxy group in the sugar moieties of mithramycin imparts conformational flexibility of greater degree than chromomycin. Therefore, the drug dimer of mithramycin has been found to have a better conformational plasticity than chromomycin when it binds to the minor groove of DNA.

Two approaches are usually taken to study the effect of the association of DNA binding anticancer drugs upon the structure of chromatin and nucleosome. The first one is reconstitution of the model nucleosome in the presence of the drugs. This has been reported earlier in the case of mithramycin (Fox and Cons, 1993; Carpenter *et al.*, 1993). In our laboratory, so far we have taken the second approach of comparing the association of the anticancer drugs with isolated chromatin at various levels.

Spectroscopic studies such as absorbance, fluorescence and CD have demonstrated directly the association of the above complexes with chromatin and its components under different conditions. We made a comparison of the affinity parameters, apparent dissociation constant and binding stoichiometry, in order to throw light on the nature of the association. The reduced binding affinity of the antibiotic: Mg^{2+} complexes to nucleosome or chromatin might be a consequence of bending of double helix or, additionally, of unusual DNA conformations induced by the histone binding. Alternatively, one can say that histone-DNA contacts and N-terminal tail domains of individual core proteins in nucleosome core particle reduce the accessibility of nucleosomal DNA to antibiotic: Mg^{2+} complexes. In the chromatin, presence of linker H1 further reduces the binding potential of the ligand. Structural and thermodynamic studies are mutually complementary and both are necessary for understanding molecular basis of the binding process. Therefore we made a comparison of the associated energetics. Linear van't Hoff plot characteristic of the association processes reported here implies the small value of heat capacity changes. It can be attributed to the absence of any major conformational alterations involving histones and DNA. In developing correlation between thermodynamic and structural data, it is essential to consider that enthalpy-entropy compensation leads to the observed free energy change. We resolved the enthalpy contributions from three plausible sources: (a) the molecular interactions between bound ligand and polymer binding site, (b) conformational changes in either DNA or drug molecule or the complex.

Antibiotic:Mg^{2+} complex induced alteration in the ultrastructural changes in the native and H1 depleted chromatin were monitored by thermal melting analysis, polyacrylamide gel mobility assay, dynamic light scattering experiments and transmission electron microscopic studies. Micrococcal nuclease digestion is the biochemical probe to assess the accessibility of the antibiotic: Mg^{2+} complexes to nucleosomal DNA.

Quantitative characterization of the binding of antibiotic: Mg^{2+} complexes with different levels of chromatin

Antibiotic	Type of complex	System	K_d (μM)	n (base/drug)	ΔG (Kcal mol^{-1})	ΔH (Kcal mol^{-1})	ΔS (eu)
Mithramycin	I	Native chromatin	107	14	−5.4	−12.0	−22.6
		H1 depleted chromatin	85	13	−5.4	−10.8	−18.4
		Core particle	154	18	−5.1	−9.8	−16.0
		Chopped core particle[a]	72	13	−5.5	−5.9	−1.4
		Naked DNA	33	5	−5.9	−7.5	−5.5
	II	Native chromatin	184	33	−5.1	2.1	24.3
		H1 depleted chromatin	153	24	−5.2	3.9	30.9
		Core particle	201	38	−4.9	4.6	32.6
		Chopped core particle	85	15	−5.3	5.2	35.7
		Naked DNA	32	7	−6.0	3.5	32.3
Chromomycin	I	Native chromatin	110	13	−5.4	−9.9	−15.4
		H1 depleted chromatin	85	11	−5.4	−7.8	−8.3
		Core particle	116	18	−5.3	−7.7	−8.1
		Chopped core particle	85	14	−5.5	−4.8	2.2
		Naked DNA	54	6	−5.7	−5.2	1.7
	II	Native chromatin	nd[b]	nd	nd	nd	nd
		H1 depleted chromatin	nd	nd	nd	nd	nd
		Core particle	210	32	−4.9	2.2	24.2
		Chopped core particle	85	15	−5.5	6.0	39.2
		Naked DNA	20	7	−6.3	7.0	45.6

[a]Chopped core particle means nucleosome core particle with the N-terminal tails of core histones removed by tryptic digestion.
[b]nd: not determined because aggregation of native chromatin at $[Mg^{2+}] > 3$ mM has confined our studies to complex II of mithramycin.

The antibiotic: Mg^{2+} complexes bind to nucleosomal DNA in presence of histones none of which bind to them. Role of the histones is probably limited to steric hindrance for the access of these complexes to the minor groove of DNA, though the possibility of noncovalent interactions between the DNA bound ligand and the potential hydrogen bonding site(s) in the histones can not be overlooked. In general we have noticed that the presence of histones leads to an increase in the dissociation constant and binding site size compared to naked DNA. Further increase in these parameters for nucleosome core particle provides indirect support that they bind to both core and linker DNA. Nuclease digestion pattern of the chromatin in presence of the antibiotics also favors the above proposition. Presence of the antibiotics reduces the accessibility of the nuclease to the potential cleavage sites in the linker region. This observation implies that presence of the anticancer drugs bound to the chromatin is a potential obstruction for the entry of the transcription factor(s) at the target promoter sites in the gene. Stabilization of the chromatin structure as indicated from the increase in transition temperatures as a sequel to the binding of the ligands would make the RNA polymerase induced opening of the duplex energetically costly during transcription. Furthermore, during nucleosome tracking, another key step in eukaryotic transcription, histone-DNA contacts in the nucleosome region need to be ruptured.

Depletion of histone H1 after covalent modification from chromatin is a key step in eukaryotic transcription (Lee *et al.*, 1993; Juan *et al.*, 1994; Rice and Allis, 2001). A comparison of the association of the antibiotic: Mg^{2+} complexes with the normal and H1 depleted chromatin suggests that smaller ligands, like anticancer drugs, have better accessibility for H1 depleted chromatin compared to native chromatin. H1 depleted chromatin is also more prone to aggregation upon association with the complex I of the antibiotic: Mg^{2+} complexes. It is also less accessible to micrococcal nuclease. We propose that H1 depleted chromatin is a better target of these antibiotics compared to native chromatin. This observation is particularly significant in case of neoplastic cells where most of the cell nuclei are transcriptionally active, and, therefore, contain H1 depleted chromatin.

We modified the nucleosome by chopping the N-terminal tails sticking out of the core particle. From a scrutiny of the spectroscopic features of two bound complexes (for mithramycin and chromomycin) and comparison of the binding and associated thermodynamic parameters, we noticed the following features. N-terminal and intact and chopped core particles interact differentially with the same antibiotic: Mg^{2+} complex. Tryptic removal of the N-terminals tail domains of core histones enhances the binding potential and access of the antibiotic: Mg^{2+} complexes to the nucleosomal DNA. The association of the DNA-binding anticancer drugs with nucleosome core particle leads to a slow release of the nucleosomal DNA. Such disruption of histone-DNA interaction might be one of the mechanisms of the transcription inhibitory potential of these drugs. Compared to the N-terminal intact nucleosome, the N-terminal chopped nucleosome is more susceptible to disruption (shown in a carton below). Release of free DNA has also been reported when another groove binder DAPI interacts with reconstituted nucleosome (Fitzgerald and Anderson, 1999).

We extended this study to include daunomycin, which is widely used as an anticancer drug, in order to examine whether mode of binding to DNA influences the association to normal and N-terminal chopped nucleosome. The extent of the above effects is more pronounced in case of the intercalator, daunomycin. N-terminal tail domains protect the eukaryotic genome from external agents, such as anticancer drugs and the degree of protection is dependent upon the mode of binding to DNA. Although N-terminal tails are structureless entities of the nucleosome (Van Holde *et al.*, 1995), our studies indicate that they have a part in maintaining the structural integrity of the nucleosome. The DNA release has been noted for other drugs also (unpublished observations from our laboratory) and may be a general feature at least with mononucleosome.

Notwithstanding the importance of in-depth analysis of the structural features of the association of the anti-gene transcription inhibitors with DNA, it is clear that structural studies on the association of these compounds with chromatin and its components like nucleosome and H1 histone under different conditions are essential to get a molecular picture of how they function *in vivo*. To the best of our knowledge very few extensive structural studies at the level of chromatin as presented above have yet been done with any other anticancer drugs (or synthetic ligand) working via the inhibition of transcription (Gottesfeld *et al.*, 2002). It is now well known that chromatin of the cell under neoplastic conditions is highly transcriptionally active. Even in a normal cell the process of transcription is dynamic in terms of the chromatin structure. Remodeling of the chromatin is a key factor in this process (Tsukiyama and Wu, 1997; Kadam and Emerson, 2002). Our future goal is to ascertain the molecular role of these small transcription inhibitors in this dynamic process.

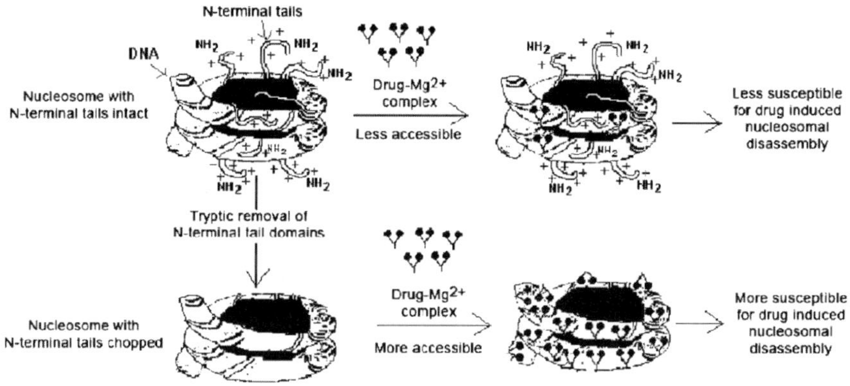

Role of N-terminal tails of nucleosome core histones to the accessibility of small ligands

Distamycin and other minor groove binders of the same class: In order to detect the binding loci of this class of groove binders in the nucleosomal DNA, hydroxyl radical and DNase I footprinting studies were carried out on the complexes of four AT-selective minor groove binding ligands (Hoechst, distamycin, netropsin and berenil) with DNA fragments which have been reconstituted with nucleosome core particles (Brown and Fox, 1996). Hydroxyl radical footprints of reconstituted *tyr*T DNA show that all four ligands induce changes in the phased cleavage pattern, consistent with the suggestion that they cause the DNA to rotate by 180° on the nucleosome surface. Regions to which the ligands are bound are turned away from the protein surface, thereby minimising electrostatic repulsion between the cationic charges on the ligand and protein.

In one of the earlier reports hydroxyl radical footprinting was employed to analyze the interaction of distamycin and actinomycin with the **5s** ribosomal RNA genes of *Xenopus* (Churchill *et al.*, 1990). The two drugs showed different hydroxyl radical footprints. Distamycin gives a conventional (albeit high-resolution) footprint, while actinomycin does not protect DNA from hydroxyl radical attack, but instead induces discrete sites of hyper reactivity. The results suggest that the shape of the DNA rather than the specific sequences leads to the recognition by distamycin. Similar trends were noticed from the footprinting studies of the reconstitution of two fragments of Xenopus borealis DNA 135 and 189 base-pairs long with chicken erythrocyte histones, after incubation with echinomycin (a bisintercalating antitumour antibiotic) or distamycin (Low *et al.*, 1986, Portugal and Waring, 1987). Controlled digestion of these defined sequence core particles using DNAase I revealed new cleavage products, indicative of a change in orientation of the DNA molecule on the surface of the nucleosome. This new rotational setting of DNA within the core particle appears to be practically independent of DNA sequence (Brown and Fox, 1996). This study has shown that minor groove binding ligands like distamycin, netropsin and berenil alter the rotational positioning of DNA fragments on nucleosome core particles.

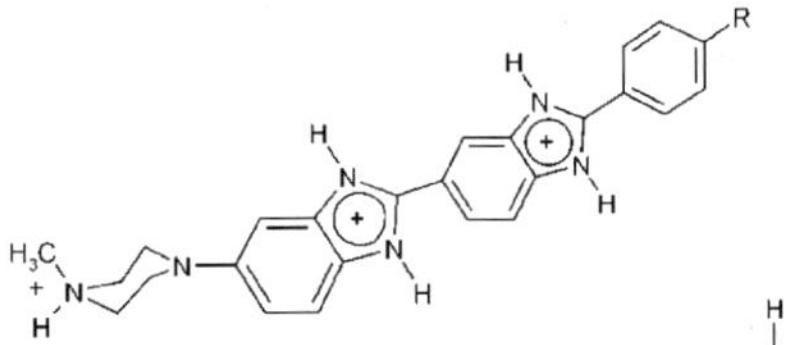

Benzimides:
Hoescht 33258 R= OH
Hoescht 33342 R= OCH_2CH_3

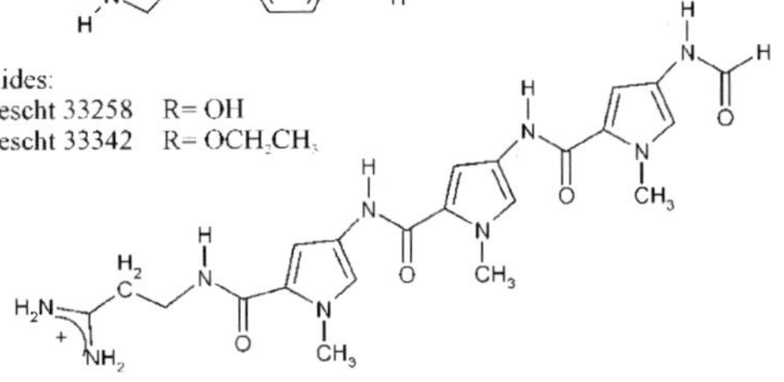

Distamycin A

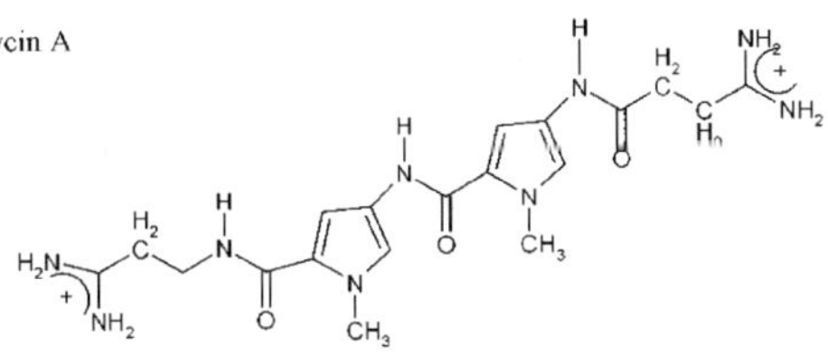

Netropsin

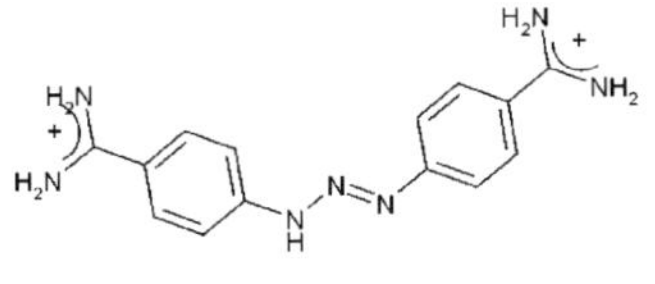

Berenil

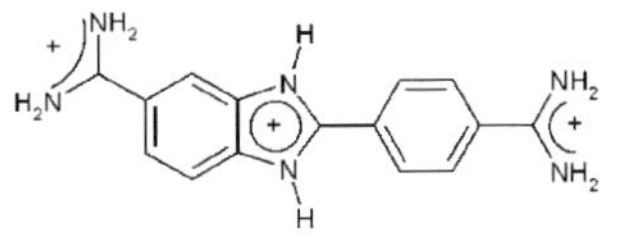

DAPI

Combilexins are a group of synthetic DNA ligands having a sequence-specific minor groove-binding element combined with an intercalating chromophore, which stabilizes the DNA complex and can interfere with topoisomerases (Pindur *et al.*, 2005 and related references therein). Collectively, the structural and kinetic data concur that the conjugate threads through the DNA double helix so that its acridine chromophore could intercalate, leaving the netropsin moiety and the methanesulfonanilino group positioned within the minor and major grooves of the double helix, respectively. The hybrid maintains the AT selectivity conferred by the netropsin moiety. The threading-type intercalation process, evidenced by stopped-flow measurements, is affected when the DNA is wrapped around histones. The composite drug can bind to both the DNA linker segments and the nucleosomal cores in chromatin. In contrast to its constituents, it antagonizes the salt-induced condensation of chromatin.

The (A•T)-selective recognition of these group of DNA binding ligands have been used to selectively block these regions during the experiments to understand the activation or repression of a gene by transcription factor. One such typical example is the following report (Kas *et al.*, 1989, 1993). Scaffold-associated regions (SARs) are A+T-rich sequences defined by their specific interaction with the nuclear scaffold. The interaction of distamycin with SAR sequences leads to a complete suppression of binding to either scaffolds or histone H1, implying that $(dA.dT)_n$ tracts play a direct role in mediating these specific interactions and that histone H1 and nuclear scaffold proteins may recognize a characteristic minor groove width or conformation. The effect of distamycin on these specific DNA–protein interactions *in vitro* also proposes that binding of SARs to the nuclear scaffold and SAR-dependent nucleation of H1 assembly might be important targets of the drug *in vivo* (Sumer *et al.*, 2004).

In fact much progress has been made in recent years in developing small molecules that target the minor groove of DNA. Synthesis of molecules that recognize specific DNA sequences with affinities comparable to those of eukaryotic transcription factors is a major stride in this direction. This makes it feasible to modulate or inhibit DNA/protein interactions *in vivo*, a major step towards the development of general strategies of anti-gene therapy. Examples from anti-parasitic drugs also suggest that synthetic molecules can affect a variety of cellular fractions crucial to cell viability by more generally targeting vast portions of genomes based on their biased base composition. Approaches based on selective interactions with broad genomic targets such as satellite repeats, essential for cellular proliferation employ synthetic polyamides or diamidines that bind the DNA minor groove. These highly selective agents are capable of interfering with specific protein/DNA interactions that occur in A+T-rich repeated sequences that constitute a significant portion of eukaryotic genomes. The satellite localization of cellular proteins that bind the minor groove of DNA via domains such as the AT hook motif is highly sensitive to these molecules (Susbielle *et al.*, 2005).

3.1.2.2. Major groove binders. Mainly proteins recognize the major grooves of DNA. Non-peptidyl compounds have a tendency to bind to the minor groove, potentially allowing simultaneous major groove recognition by proteins. However, oligomers called triplex forming oligonucleotides (TFOs) (Thuong and Hélène, 1993) can bind to polypurine–polypyrimidine duplex sequences in the major groove to form hydrogen bond with bases on the purine strand. TFOs bind within the existing major groove of DNA and the orientation of the third strand relative to the duplex is dependent on the sequence.

Peptide nucleic acids (PNAs with a peptide like backbone) also bind via the major groove (Neilsen, 1999). PNAs form a triplex (Lohse *et al.*, 1999), which then result in the displacement of the non-complementary oligopyrimidine DNA strand. This has been extensively reviewed by Hurley (2002).

3.2. Covalently Interacting Anticancer Drugs

Anticancer drugs that interact with DNA covalently do so by a primary non-covalent interaction, which is immediately followed by covalent interaction with either the phosphate backbone, or DNA bases, or deoxyribose sugar moieties. Covalent interactions lead to any of the following effects: cleavage of DNA strand, cross-linking, alkylation and adduct formation and base release. These events are mutually interconnected. Mono-alkylating agents alkylate the nucleophilic sites within the double helix. Bifunctional alkylating agents cross-link the two strands of DNA and as a result, the integrity of DNA, as a template is lost.

3.2.1. DNA adduct forming or alkylating agents

Alkylating agents were the first drugs to be used to treat highly proliferating cancers. They are mainly of two types: the relatively non-specific nitrogen mustards and the even less selective nitrosoureas.

Nitrogen mustards form strong electrophiles through the formation of carbonium ion intermediates. Most common sites of attack are the N7 atoms of adenine and guanine bases in the major groove. N2 and N3 atoms are the other potential sites. Chloroethyl side chains are the main structural features of nitrogen mustards such as *Chlorambucil, Melphalan*, and *Cyclophosphamide*. Chlorine is a good leaving group, therefore, it facilitates nucleophilic attack of nitrogen to form an imminium ion in a strained ring system. This readily undergoes alkylation at N7 of guanine in the major groove to form a monoalkylation product. Bifunctional alkylating agents can undergo a second cyclization of the second side chain and form a covalent bond with another nucleophilic group, possibly an N7 of another guanine or some other nucleophilic moiety. This results in the cross-linking of the two complementary strands of DNA, primarily at 5’GPuC sequences (Hansson *et al.*, 1987). Cyclophosphamide requires activation by cellular mixed function oxidases. It is a non-specific pro-drug of the active metabolite phosphoramide mustard and the most widely used alkylating agent.

Nitrosoureas encompass a class of compounds that breakdown to very unstable intermediates leading to indiscriminate reactions. Among this class are included *Streptozotocin*, which is used in the treatment of Hodgkin's disease (Schein *et al.*, 1974), triazenes such as *Decarbazene*, widely used for malignant melanoma (Cohen *et al.*, 1998), and the more recent *Temozolomide*, used for gliomas (Friedman *et al.*, 2000). Temozolomide is a monoalkylating drug that methylates guanine residues in DNA following a DNA facilitated rearrangement. Figure shows the chemical structures of some DNA alkylating agents.

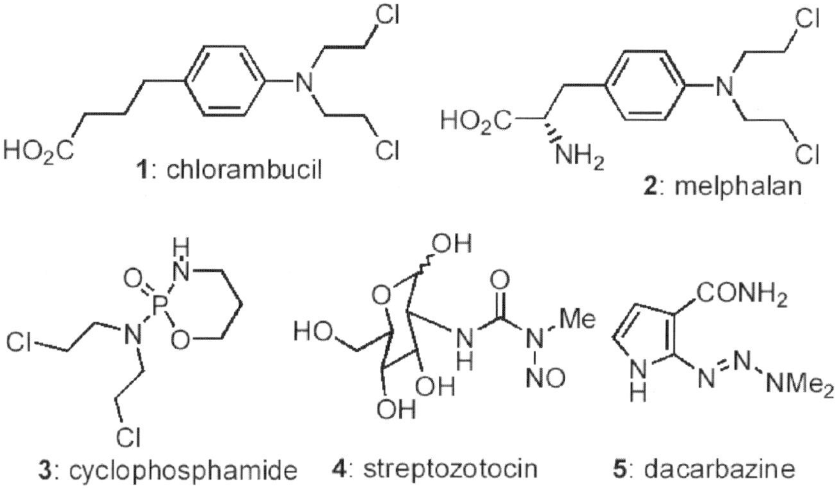

Chemical structures of some DNA alkylating anticancer drugs.

3.3. DNA Cleaving Anticancer Drugs

This class consists of potent antitumor agents that mainly exhibit a radical mechanism. Among the members of this class, *esperamicin, calicheamicin, dynemicin* and *neocarzinostatin* function by an endiyne mechanism. They undergo an inducible chloroaromatization to an aryl or indenyl diradical, which abstracts hydrogens from proximate deoxyribosyl sites, leading to DNA scission (Smith *et al.*, 1994). *Bleomycin*, on the contrary, has a different mechanism of action. The amino terminal tripeptide of Bleomycin molecule seemingly intercalates between guanine-cytosine base pairs of DNA. The opposite end of the Bleomycin peptide binds Fe(II) and serves as a ferrous oxidase, able to catalyze the reduction of molecular oxygen to superoxide or hydroxyl radicals responsible for DNA strand scission (Takeshita *et al.*, 1978; Giloni *et al.*, 1981). Recently, a new class of drugs is being developed – pyrazolo-triazoles that are DNA cleaving agents upon

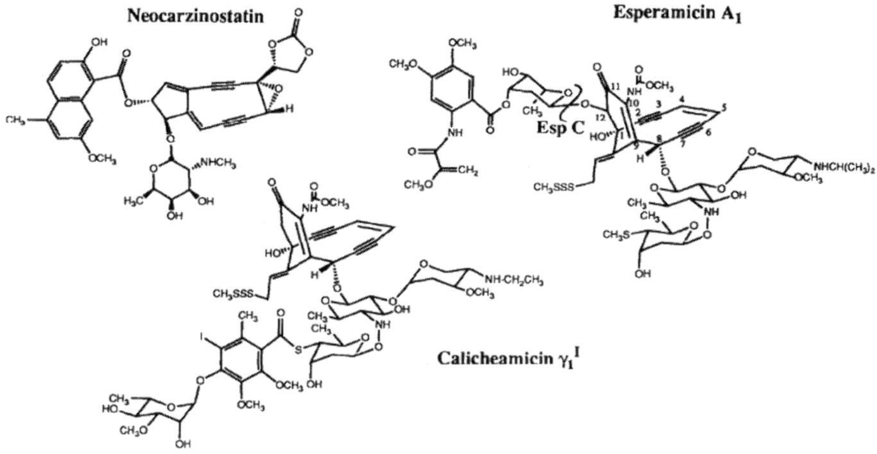

Chemical structures of DNA cleaving antibiotics

photoactivation (Manfredini *et al.*, 2000). The reaction proceeds from the lowest excited singlet state to an azoimine that converts thermally or photochemically to an intermediate. The intermediate in the singlet state reacts directly or undergoes intersystem crossing to the triplet. It is capable of hydrogen abstraction followed by DNA cleavage.

Constraints on the structure of nucleosomal and the linker DNA induced by the associated histones and the non-histone chromosomal proteins will play an important role in the loci of DNA-damage in the nucleus by this group of antibiotics. One of the initial reports employed Chinese Hamster cell nuclei (Kuo and Samy, 1978). They were reacted with neocarzinostatin and its DNA was analyzed on non-denaturing agarose gel. A series of bands with a multiplicity of 175 bp was obtained. A similar result was also obtained when the DNA samples were electrophoresed under denaturing gels. Later, a detailed study reported the DNA damage in HeLa nuclei and isolated nucleosome core particles with several more members of the enediyne family of antitumor antibiotics such as calicheamicin γl (CAL), esperamicin Al (ESP Al), esperamicin C (ESP C), and neocarzinostatin (NCS). All three enediyne antitumor antibiotics produce DNA damage in He La nuclei that was modulated at the level of the nucleosome (Yu *et al.*,1994; Smith *et al.*, 1996). DNA damage induced by ESPA1 and NCS was limited to the linker DNA. On the other hand, the damage produced by CAL and ESP C also occurred in the nucleosome core with a 10-nucleotide periodicity. The differences in the site of the damage have been ascribed to the structural differences between the enediynes. Distinctive features of drug structure that may limit damage to the nucleosome core include the presence of substituents on both sides of the CAL/ESP-type core, and the presence of an intercalating moiety, such as the naphthoate of NCS and possibly the anthranilate of ESP Al. These observations further emphasize the point that DNA-damage by the covalently interacting drugs depend on the geometry of the drug–DNA complex resulting from the first step of the non-covalent association between them. Keeping in view the fact that these antigene compounds are potent cytotoxic agents, there is scope to understand how these drugs might switch off the active genes for transcription. As proposed earlier it is possible that the conversion of nucleosomes to nuclease hypersensitive sites is a mechanism of transcription inhibition for the active genes. However, the chemically active nature of these drugs also make them possible players in any one or more steps in the chromatin remodeling, an essential pre-requisite for transcriptional activity of the active chromatin.

3.3.1. Bleomycin

The antitumor antibiotic bleomycin (structure shown below) is a glycopeptide that binds to the DNA minor groove and induces sequence specific single and double strand breaks in DNA by a free radical mechanism. It forms a coordination complex with Fe (II) and in the complexed state it combines with oxygen to produce a highly reactive species, which specifically abstracts hydrogen from the C4' of deoxyribose sugar moiety. This leads to strand breakage or production of abasic sites. In the process of formation of double strand breaks, bleomycin primarily induces single strand breaks at pyrimidines of G-C and G-T sequences which is followed by secondary cleavage on the opposite strand (Povirk and Goldberg, 1987; Steighner and Povirk, 1990).

Biochemical and cytological studies of bleomycin actions on chromatin and chromosomes have revealed that bleomycin interacts with nuclei isolated from a variety of mammalian cells to release nucleosomes. i.e. bleomycin cleaves chromatin at linker regions (Kuo and Samy, 1978). Moreover, the ability of bleomycin to induce DNA lesions depends on nucleosome repeat length (Lonn *et al.*, 1990), the number of DNA lesions created being lower for salt-incubated nuclei with short average nucleosome repeat length (140–145 bp) compared to nuclei with longer (190–195 bp) repeat length. bleomycin induced DNA cleavage is also asymmetric towards the periphery of nucleosome bound DNA (Smith *et al.*, 1994), where marked inhibition of cleavage is observed toward the upstream side, but negligible inhibition occurs towards the downstream side for chromatin, reconstituted with *Xenopus laevis* 5S rRNA gene.

Another factor, which is of prime importance in the context of bleomycin activity on chromatin, is the degree of chromatin compaction. Restriction enzyme digestion of DNA from drug-treated nuclei along with Southern blotting procedures has shown that bleomycin preferentially cleaves the chromatin at actively transcribing regions (Kuo, 1981). In fact, the DNA sensitivity to bleomycin is inversely correlated with the degree of chromatin coiling (Lopez-Larraza and Bianchi, 1993). CHO cells with decondensed chromatin show higher DNA sensitivity to bleomycin than CHO cells with maximal chromatin compactness. Furthermore, a comparative study of the response of mosquito (ATC-15) and mammalian (CHO) cells to bleomycin reveals that ATC-15 cells, which have higher chromatin condensation compared to

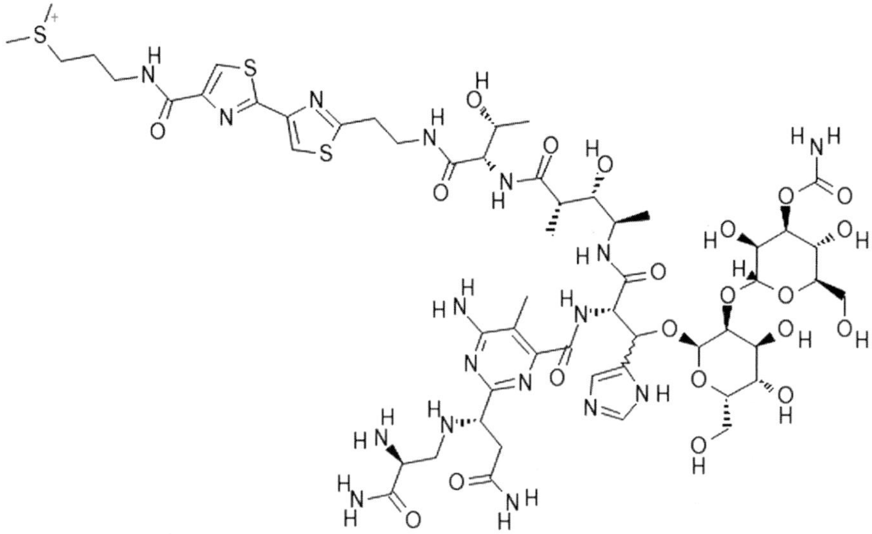

Bleomycin

CHO cells, also show a lower sensitivity to bleomycin (Lopez-Larraza *et al.*, 2006). The ATC-15 cells exhibit satisfactory growth at bleomycin doses that produce a permanent growth arrest of CHO cells, thereby suggesting that mosquito cells might have linker DNA shorter than that of mammalian cells.

Bleomycin induced chromosomal damage in Chinese hamster bone marrow gives rise to micronuclei by means of lagging chromatin; main and micronuclei eventually become asynchronous in consecutive cell cycles and mitosing main nuclei induce premature chromosome condensation in the micronuclei (Kurten and Obe, 1975).

The cells respond to bleomycin induced chromatin damage by activating nuclear poly (ADP-ribose) polymerase (PARP), which regulates chromatin structure and DNA repair. Closely associated with PARP, is the activity of cellular integrins. A fluorescence microscopy based study using wild type and PARP knockout mouse lung endothelial cells and the PARP inhibitor, 3-aminobenzamide showed that integrin clustering protect wild type cells from DNA breakage and 3-aminobenzamide and PARP knockout inhibit this protection. Hence, the antigenotoxic effect of integrin activation requires PARP, but at the same time, integrins alter chromatin structure by both PARP-dependent and independent mechanisms (Sidik and Smerdon, 1990).

Ultimately the DNA lesion is repaired by a 'short patch' repair mechanism, which, in linker regions of nucleosomes or open regions of chromatin (where lesions are generally concentrated) is associated with minimal nucleosome rearrangement (Jones *et al.*, 2001).

3.4. DNA Cross-linkers: Cisplatin and Mitomycin C

The most widely used drug in this category is cisplatin. Therfeore, we have summarized below its action at the chromatin level. The anti-tumour drug cis-diamminedichloroplatinum(II) (cisplatin) is employed for the treatment of ovarian and testicular carcinomas, as well as solid tumours (Loehrer and Einhorn, 1984; Zamble and Lippard; 1995).

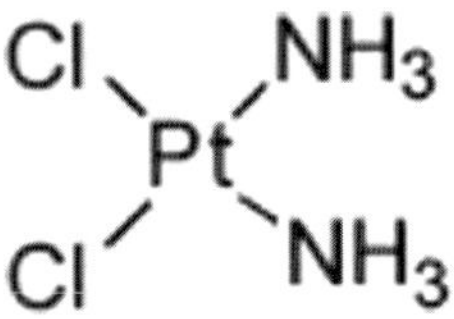

Cisplatin

The covalent binding of cisplatin to cellular DNA mediates the cytotoxicity of the anti-cancer agent (Roberts and Thomson, 1979; Bruhn *et al.*, 1992; reviewed in Jamieson and Lippard, 1999). The reaction of cisplatin with DNA results in covalent cisplatin-DNA adducts that can inhibit DNA replication. The most prevalent covalent adduct is an intra-strand cross-link formed between the N-7 of two adjacent guanine residues (Dabrowiak and Bradner, 1987; Bruhn *et al.*, 1992). Inter-strand DNA crosslinks, as well as some DNA–protein cross-links, also occur. An interesting observation is that linker histone H1 binds preferentially to cisplatin damaged DNA (Yaneva *et al.*, 1997). The sequence specificity of cisplatin DNA damage using a polymerase stop assay (Temple *et al.*, 2000) has shown that runs of consecutive guanines is the most prevalent with lesser damage at AG, GA and GC dinucleotides. A similar DNA sequence selectivity has also been found in intact human cells (Murray *et al.*, 1998; Davies *et al.*, 2000). Footprinting techniques, utilising bleomycin and DNase I as the damaging agents, were employed to establish the precise location of positioned nucleosomes with respect to the DNA sequence in reconstituted chromatin (Galea and Murray, 2002). Reconstituted nucleosomal DNA was treated with cisplatin and drug-induced DNA adduct formation was quantitatively analysed with a polymerase stop assay using Taq DNA polymerase. The results from the studies show that the preferred site of cisplatin DNA binding was in the linker region of the nucleosome. The effect of chromatin structure upon cisplatin damage has also been studied in the intact human cells using epsilon-globin promoter as the DNA target. The study had shown that chromatin structure has a large impact upon the degree of damage, particularly, the binding of a transcription factor resulted in an enhancement of the DNA damage. Protein induced distortion of the DNA could lead to the formation of novel adduct that could evade the normal repair pathway thereby leading to the anti-tumor activity of the drug. In the cell, exposure of the damaged

DNA duplex to the exterior of the nucleosome is necessary for the damage repair of the platinum lesion by the appropriate machinery. The rotational setting of the nucleosomal DNA on the surface of the histone octamer decides the solvent accessibility of the nucleotides (Danford *et al.*, 2005 and related references therein). Enzymatic digestion by exonuclease III of the nucleosome substrates suggested that the platinum cross-link affects the translational positioning of the DNA, forcing it into an asymmetric arrangement with respect to the core histone proteins. These phasing phenomena might play an important role in the recognition and processing of platinum- DNA adducts in cancer cells treated with these drugs.

Mitomycin C (structure shown below) is a naturally occurring antitumor antibiotic, used in cancer chemotherapy, particularly for the treatment of bladder cancer (Bradner, 2001). The cytotoxicity of this drug arises due to inter and intra-strand DNA crosslinking, following adduct formation. Mitomycin C induces the formation of monoadduct at guanine N2. Similarly, two guanine residues in proximity in the DNA minor groove may cross-link with each other through their respective 2-amino groups (Iyer and Szybalski, 1963; Dorr *et al.*, 1985; Keyes *et al.*, 1991). Crosslinking renders the DNA unsuitable as a template for replication or transcription to occur, and if left unrepaired, are highly cytotoxic.

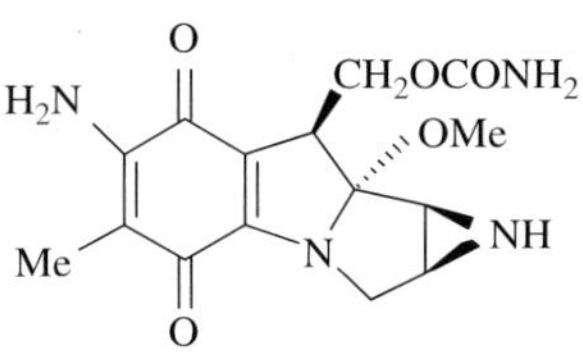

Mitomycin C

It has been found that mitomycin C treatment initiates certain active cellular processes, which result in non-random chromatid interchanges. The frequency of exchange between homologous chromosomes by far exceeds what is expected by chance (Shaw and Cohen, 1965; Morad *et al.*, 1973). Quadriradical formation predominantly occurs with human chromosomes1, 9, and 16 and exchange breakpoints appear within the C bands of these chromosomes. Furthermore, the frequency of exchange events in these chromosomes tends to be directly correlated with the size of their paracentromeric heterochromatic bands (C-bands). Abdel-Halim *et al.*, (2005) have reported that majority of exchange breakpoints of chromosome 9 are located within the paracentromeric heterochromatin and that over 70% of the exchanges occur between its homologues in G_0/G_1 and S-phase cells. It therefore implies that mitomycin C treatment induces heterochromatin-specific pairing and formation of exchanges.

Mitomycin C treatment is often followed by the induction of cross-link repair *in vivo*. The initiation of cross-link repair possibly occurs when the DNA replication or transcription machinery are stalled at the damaged site. Single strand DNA foci appear as an outcome of cross-link repair and these single strand DNA foci may occur in S-phase cells (Rothfuss and Grompe, 2004). A detailed *in vivo* analysis of mitomycin C induced DNA damage and repair (Lee *et al.*, 2006) identifies the involvement of *Xeroderma pigmentosum* (XP) proteins. Induction of single strand DNA patches by mitomycin C is completely dependent on the presence of XPG and XPE proteins, as human mutant cells, defective in XPG and XPE fail to form single strand DNA foci on treatment with mitomycin C. Moreover, mitomycin C induced cross-link causes XPG to localize exclusively in the nucleus and to associate with chromatin. However, treatment of XPF deficient cells with mitomycin C results in a strong reduction of chromatid interchange frequency (Abdel-Halim *et al.*, 2005). Lack of XPF also significantly delays the formation of mitomycin C induced single strand DNA foci *in vivo*. Thus, XPG, XPE and XPF, all play unique roles in the repair of mitomycin C induced DNA damage.

3.5. Anticancer Drugs Targeting the Structural Organization of DNA

Different levels of structural organization of DNA are targets for anticancer drugs. The first, most basic level is the chemical interaction of the drug with the DNA double helix. The DNA primary base sequence may be targeted in a non-sequence-specific (global) manner or the same may be targeted at repetitive DNA sequences such as AT rich regions (ORIs and MARs) or specific DNA sequences (e.g. oncogenes). Similarly, the secondary DNA structures such as DNA quadruplexes (telomeres), hairpins, Holliday junctions and triple helices also serve as potential targets for anticancer drugs.

3.6. Targeting the DNA Primary Base Sequence

Non-sequence-specific or global DNA damage is generally caused by random DNA alkylation, cross-linking or strand scission, the mechanisms and consequences of which have been previously mentioned.

3.7. Targeting the Repetitive DNA Sequences

Clusters of repetitive DNA sequences are present over vast areas of the human genome. In some cases, these sequences may provide important regulatory functions. Repetitive DNA often has the ability to take on non-B-form DNA conformations, which might recruit certain regulatory proteins that participate in control of gene expression. AT-islands, containing nearly 85–100% AT sequences may function as matrix attachment regions (MARs) that organize DNA loops on the nuclear matrix and coordinate nuclear activities such as DNA replication, transcription and mitosis (Woynarowski, 2002). Moreover, origins of replication and certain promoters contain AT rich sequences. Unfortunately, these repetitive sequences are unstable. Polymerase slippage or unequal recombination (Debrauwere *et al.*, 1997) may cause expansion or deletion of the repetitive elements and this is often associated with disease state. Therefore, selective damage to AT-rich DNA might be an important mechanism of drug action since binding to these sequences affects specific gene expression by preventing transcription factor binding, increasing the affinity of a transcription factor for its sequence, or creating unnatural binding sites (Gniazdowski *et al.*, 2005).

Efforts are being made to design anticancer drugs that will specifically interact with AT-rich sequences and interfere directly with the metabolic processes therein. In fact, certain cyclopropyindoline compounds such as CC-1065, adozelesin, and bizelesin have been studied that alkylate the N3 of adenine in the minor groove of AT-rich DNA sequences. These compounds have been shown to inhibit DNA replication in cell free and cell based yeast and mammalian systems (McHugh *et al.*, 1994, 1999; Cobuzzi *et al.*, 1996; Woynarowski and Beerman, 1997; Weinberg *et al.*, 1999; Wang *et al.*, 2001). In addition, certain aminoindoline compounds have been found to target AT-rich sites located within the c-MYC gene *in vitro* (Nelson *et al.*, 2005).

3.8. Targeting Specific DNA Sequences or Oncogenes

Certain drugs have been designed that span DNA and recognize a limited number of specific sequences. The pyrrole-imidazole polyamides (Dickinson *et al.*, 2004) are the most discriminatory sequence selective DNA binding agents that inhibit transcription factor binding *in vitro*. However, these hairpin polyamides have not been found effective *in vivo* (Dudouet *et al.*, 2003). Chlorambucil conjugated polyamides have been designed that cause cell cycle arrest in G2/M. Molecules such as polyamide 1-CBI (1-chloromethyl)-5-hydroxyl-1,2-dihydro-3H-(benz[e]indole) conjugate exhibit sequence specific DNA alkylation in mammalian cells. 1-CBI is able to damage encapsidated SV40 DNA by penetrating the virions (Philips *et al.*, 2005). Moreover, the differences in sequence specificities of DNA alkylation of these conjugated polyamides lead to marked differences in biological activities (Shinohara *et al.*, 2006).

Ecteinascidin (Et-743) is a minor groove-alkylating agent, currently in clinical development. The drug alkylates N2 of the central guanine of the DNA binding triplet and causes a conformational change in DNA, with the minor groove widening and the double helix bending towards the major groove (Pommier *et al.*, 1996; Garcia-Nieto *et al.*, 2000; Hurley *et al.*, 2001; Zewail-Foote and Hurley, 1999, 2001). This compound demonstrates a unique potential to alter gene expression of discrete loci based on the presence of GC boxes in the promoter regions.

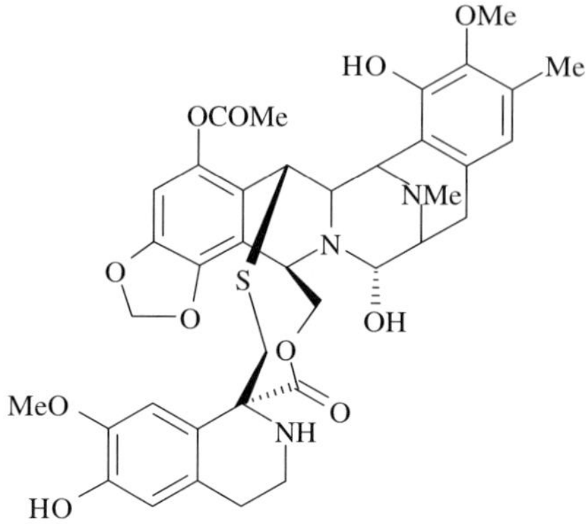

Et-743

4. DNA SECONDARY STRUCTURES AS TARGETS FOR ANTICANCER DRUGS

DNA secondary structures containing stretches of single stranded DNA are present in the human genome and are involved in the regulation of crucial processes such as transcription. Hairpins or cruciforms are the potential recognition sites for binding of transcription factors. Of equal importance are telomeres, the DNA-protein complexes marking the ends of chromosomes. Facile interconversion between double and single stranded DNA and G-quadruplex at physiological conditions renders these secondary DNA structures attractive candidates for biological signaling molecules and consequently, potential targets for anticancer agents.

4.1. Targeting DNA Quadruplexes

Telomeres are DNA protein complexes at the ends of eukaryotic chromosomes that protect the linear DNA ends from erosion over multiple replication cycles and also from being recognized as double strand breaks and subsequent repair by exonucleolytic trimming and end-to-end fusion. Chromosomal telomeres contain 3′ G-rich overhang of 150—200 bp that forms a G-quartet structure. It is in the form of stacked tetrads of guanines in a cyclic Hoogsteen hydrogen bonding arrangement. G-quartets can be stabilized by sodium and potassium ions and this stabilization can inhibit telomerase activity. A number of small molecules have been identified that interact with G-quartets. For instance, 2,6-diamidoanthraquinone BSU1051 has been found to interact with and stabilize the G-quartet, and thereby inhibit telomerase activity (Sun *et al.*, 1997). A 3,6,9-trisubstituted acridine is also a potent inhibitor of telomerase. Apart from these, certain cationic porphyrins such as $TMPyP_4$ are another class of agents that bind to G-tetrads by interactive stacking. $TMPyP_4$ has selectivity for intermolecular G-quadruplex structures (Liu *et al.*, 2005). Telomestatin is a natural G-quadruplex intercalating agent, which holds greater promise compared to previously studied G-quadruplex targeted molecules. On treatment of multiple myeloma cells with Telomestatin, inhibition of telomerase activity occurs along with reduction in telomere length followed by cell growth inhibition (Shammas *et al.*, 2004). A third class of G-tetrad interacting compounds comprises of perylenetetracarboxylic diimide PIPER which shows binding characteristics, similar to the porphyrins. These compounds not only bind to G-quadruplexes, but also induce their formation in cells (Han *et al.*, 1999).

Apart from telomeres, G-quadruplexes are also present in the upstream promoter regions of certain oncogenes. G-quadruplex targeted molecules may interact at these sites as well. In fact, the cationic prophyrin, $TMPyP_4$ and the core modified expanded prophyrin analogue 5,10,15,20-[tetra(N-methyl-3-pyridyl)]-26,28-diselenasapphyrin chloride (Se2SAP) have been found to cause repression of transcriptional activation of c-MYC in cells by G-quadruplex stabilization (Seenisamy *et al.*, 2005).

4.2. Targeting Hairpins and Holliday Junctions

Hairpins, cruciforms or single strand DNA containing secondary structures play an important role in the regulation of transcription. A molecule that can selectively bind to hairpins has the potential to block transcription by interfering with protein recognition of that specific site. The transcription inhibitor Actinomycin D and its analogues bind nearly 10 fold more tightly to the hairpin conformation formed from the single stranded DNA $5'$-$A_7TAGT_4A_3TAT_7$-$3'$ than to the same strand in fully duplexed form.

4.3. Targeting DNA Triplexes

Specific targeting of triplex DNA structures is one of the most important strategies of 'antigene' based chemotherapy (Jenkins, 2000). The main aim is to target individual gene sequences at the DNA duplex level to modulate their expression or interactions with DNA binding proteins or to interfere with the vital template processes. A genomic DNA duplex is targeted using either a triplex forming oligonucleotide (TFO) or peptide nucleic acid analogue (PNA) to produce local DNA triplex structures that can inhibit the transcription of specific genes. The specificity of these triple helical structures stems from the Z—X · Y base triplets formed by Hoogsteen or reverse Hoogsteen hydrogen bonding arrangements involving pyrimidine or purine bases (Z) in the third strand and the purine strand (X) of the host DNA duplex. The triple helical structures formed have low thermal and thermodynamic stability and this poses a problem for effective targeting. So DNA triplex-targeted drugs mainly aim at stabilization of DNA triplexes.

Significant increase in triplex stability can be achieved by using intercalating agents, either by conjugate attachment to the third strand TFO or as a separate adjunct ligand, if binding of this residue can preferentially stabilize a triple stranded DNA. Various fused and unfused heterocycles have been studied. They include naphthylquinoline, BePI (benzo[e]pyridoindole), imidazothioxanthone, acridine derivatives, coralyne, and the phenothiazinium dye, methylene blue. These molecules have (i) an extended planar aromatic system to maximize π-overlap with successive base triplet planes, (ii) a cationic charge on the intercalated chromophore and (iii) a pendant side-chain terminating in an amine residue that can be protonated. Due to these characteristics, the compound is easily delivered to the target DNA and it anchors the bound ligand through interaction with the grooves or the anionic phosphodiester backbone.

Efforts have also been made to design groove directed drugs for triplex stabilization. Ideally, these agents should have little or no inherent affinity for the underlying duplex in order to prevent intergroove cross talk and consequent binding induced displacement of the third strand. However, most of the compounds studied are mainly A/T specific with high affinity for DNA duplexes rather than DNA triplexes.

5. PROTEIN–DNA COMPLEXES AS MOLECULAR TARGETS FOR ANTICANCER DRUGS

Chromosomal DNA is the scaffold on which DNA binding proteins assemble and regulate vital cellular processes such as DNA packaging, transcription, replication, recombination and repair. Since these proteins use DNA as template for carrying out their respective functions, they are found associated with DNA some time or the other in the cell cycle. Several drugs have been designed that target the protein-DNA complexed state and thereby inhibit the enzyme function. Among this class of drugs, the best-studied are the topoisomerase poisons.

Topoisomerases (I & II) are enzymes that modify the DNA topology by a complex catalytic cycle involving DNA strand cleavage, strand passage and religation (Osheroff *et al.*, 1991). The necessity of topoisomerases in the cell arises from the fact that the DNA double helix is normally stored in a highly supercoiled complexed state in chromatin which has to be unwound for processing. The cytotoxicity of topoisomerase poisons is due to stabilization of the enzyme-DNA covalent 'cleavage complex' (Nelson *et al.*, 1984). Topoisomerase I inhibitors stabilize a covalent bond between a tyrosine residue on the protein and the 3′ phosphoryl end of the single strand it breaks (Hsiang *et al.*, 1985), while topoisomerase II inhibitors stabilize a covalent bond between a tyrosine residue on the protein and the 5′ phosphoryl end of each broken strand of DNA with a 4 bp stagger between cleavage sites on complementary strands. In either case, it leaves the topoisomerase molecule covalently bound to DNA, masking the cleavage site.

The major class of topoisomerase I inhibitors comprise of the camptothecins, while topoisomerase II inhibitors fall into several classes – anthracyclines (e.g. doxorubicin), anthracenediones (mitoxantrone), anthrapyrazoles (bianthrazole), actinomycins (actinomycin D), acridines (m-AMSA), ellipticines (9-hydroxyellipticine) and epidophyllotoxins (Etoposide (VP-16) and VM-26). The chemical

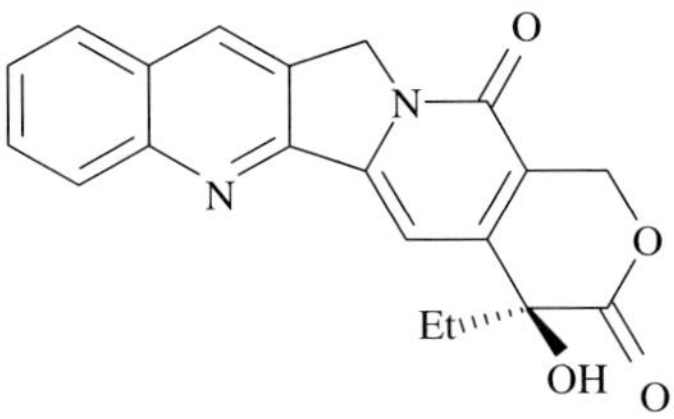

Camptothecin

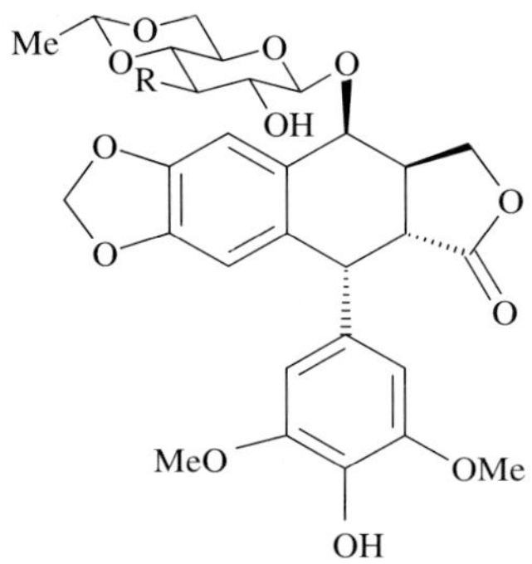

Etoposide

Chemical structures of some topoisomerase inhibitors

structures of some topoisomerase inhibitors are shown below. Except for the camptothecins and epidophyllotoxins, the drugs bind to DNA through intercalation and then form a ternary drug/DNA/enzyme complex that inhibits the DNA resealing activity of the enzyme, stabilizing the cleavage complex and resulting in DNA double strand breaks. Camptothecins and epidophyllotoxins are believed to bind primarily to the protein followed by ternary complex formation.

However, topoisomerase poisons show a limited sequence preference. Especially the topoisomerase II targeted drugs trigger random double stranded breaks throughout the genome, which at times induce chromosomal translocations and in turn cause secondary leukaemia. With the intention to impart greater specificity to topoisomerase II poisons, Duca *et al.*, 2006 have designed derivatives of VP16 conjugated to triplex forming oligonucleotides. These molecules induce cleavage 13—14 bp from the triplex end where the drug was attached. Hence, the molecules are expected to offer great promise in cancer chemotherapy.

Apart from the topoisomerase poisons, other drugs that trap proteins in ternary complexes include Cisplatin, and Et-743.

Cisplatin (cis-diamminedichloroplatinum (II)) initially binds covalently to adjacent guanines in the major groove of DNA and bends DNA in the direction of the major groove (Bellon *et al.*, 1991). This distortion facilitates protein binding in the minor groove and stabilizes the interactions between DNA and its binding proteins such as TATA box binding protein (TBP), High mobility group 1 (HMG1), High mobility group 2 (HMG2), human upstream binding factor (hUBP) and sex-determining region Y protein (SRY) (Gniazdowski and Czyz, 1999). These proteins bind to DNA and enforce bending of DNA. So, trapping of these proteins, either at their natural or unnatural binding sites imposes architectural changes that are eventually hazardous to the cell. The above review is a glimpse of the activities with the platinum containing anticancer drugs. There has been some excellent reviews in this area in the last year (Sedletska *et al.*, 2005; Wheate and Collins, 2005).

Ecteinascidin 743 (Et-743) is a complex natural product, which also seems to work by trapping DNA-binding proteins at sites where structural distortion of the DNA is recognized. The chemical structures of Cisplatin and Et-743 are shown earlier in this review.

Blocking an essential enzyme in its DNA bound state may be an intriguing mechanism to cause cell death, but in reality this feat has rarely been achieved. The drugs designed to target DNA/protein complexes have high cytotoxicity and low clinical efficacy. So, efforts are being made to handle these problems as far as possible and at the same time, newer classes of drugs, the 'epigenetic' drugs are coming up with greater promise.

6. EPIGENETIC THERAPY OF CANCER

'Epigenetic' is a term used to describe a state of gene expression that is mitotically and meiotically inherited without any change in the sequence of DNA. Epigenetic mechanisms are mainly of two classes: (1) the DNA may be modified by the covalent attachment of a moiety that is then perpetuated. (2) a self-perpetuating protein state may be established (Zelent *et al.*, 2004). The two most studied epigenetic phenomena are DNA methylation and histone tail modifications (Mai *et al.*, 2005).

Methylation is the most commonly occurring epigenetic modification of human DNA. Under normal conditions, it helps to maintain transcriptional silence in non-expressed or non-coding regions of the genome. Methylation results from the activity of a family of DNA methyl transferases (DNMTs) that catalyze the addition of a methyl group to the 5-position (C5) of the cytosine ring, almost exclusively in the context of CpG dinucleotides. Low levels of DNA methylation in the promoter region of genes are linked to active gene expression. On the contrary, methylation near the transcription start site stand in the way of gene expression.

Apart from gene silencing, other effects of DNA methylation include spontaneous deamination, enhanced DNA binding of carcinogens and increased UV absorption by DNA, all of which increase the rate of mutations, DNA adduct formation and subsequent gene inactivation

DNA methylation is one such epigenetic phenomenon, which is abnormal in tumor cells (Szyf, 2003; Lund and van Lohuizen, 2004). Methylation of CpG islands, a feature of cancer cells, occurs rarely in normal tissue. Hence, methylation provides a tumor specific therapeutic target. *De novo* methylation as well as maintenance of methylation is carried out by DNMTs. There are 5 known human DNMTs – DNMT1, DNMT2, DNMT3a, DNMT3b and DNMT3L. Apart from DNMT2 and DNMT3L, which lack the amino terminal regulatory domain and the catalytic domain respectively (Goll and Bestor, 2005), the remaining DNMTs all have enzymatic function. Inactivation of DNMTs is the most effective means of inhibiting DNA methylation and restoring normal patterns of methylation. However, targeting the methyl transferase enzyme leads to loss of specificity and hypomethylation of the genome. The overall decrease in methylation level may even activate the potentially deleterious oncogenes (Szyf *et al.*, 2004). DNA methylation inhibitors are of two broad classes: nucleoside and non-nucleoside analogues.

7. THERMODYNAMICS AND CRYSTALLOGRAPHY OF DRUG–DNA INTERACTION IN CHROMATIN AND NUCLEOSOME

An elucidation of the energetics of association of the non-covalently binding drugs to DNA as an integral part of chromatin and nucleosome is required to understand the structural basis of their association. It tells about the alteration in the state of the DNA in the chromatin (or nucleosome) as a result of the association with the drug. The scenario will be more complex for the covalently binding drugs. However, there has been very few studies aimed at understanding the energetics of drug–chromatin (nucleosome) interaction.

Haq, 2002, has described the overall observed drug–DNA binding free energy as being composed of at least five component free energy terms: ΔG_{conf} (the free energy contribution arising from conformational changes in DNA and drug); ΔG_{r+t} (the unfavorable contribution to free energy arising from losses in translational and rotational degrees of freedom upon complex formation); ΔG_{hyd} (free energy for the hydrophobic transfer of the drug from bulk solution to the DNA binding site); ΔG_{pe} (the polyelectrolyte contribution, mainly due to coupled polyelectrolyte effects, the most important of which is the release of condensed counterions from the DNA helix upon drug binding); ΔG_{mol} (the contribution to free energy from weak non-covalent interactions such as hydrogen bond formation, van der Waals interactions, specific electrostatic bond formation, dipole–dipole interactions, etc. between the drug and DNA). In the case of chromosomal DNA, the additional contributions from the reshuffling of histone–DNA interactions and the potential contribution from histones need to be considered.

The general trend for minor groove binders is that binding is driven largely by hydrophobic effect and favorable entropy is derived from the release of bound water and counterions from DNA and drug upon complex formation. Intercalators, on the other hand, have slightly lower affinities for DNA than minor groove binders. As with minor groove recognition the interaction of intercalators with DNA results from a delicate balance of opposing energetic factors. There is a free energy cost associated with deforming the DNA lattice in order to accommodate the intercalating chromophore. There are also energetic penalties that arise from losses in rotational and translational freedom in the DNA and drug upon complex formation. However, in case of intercalation, drug – binding results in additional van der Waals interactions between drug and adjacent bases that are capable of providing a significant favorable contribution to the overall free energy.

Detailed reports of the energetics of binding of an antibiotic to nucleosomal DNA under different conditions have been done only with aureolic acid group of antibiotics from our laboratory (Mir and Dasgupta, 2001a, b; Mir and Dasgupta, 2003) . The thermodynamic parameters were evaluated from temperature dependence of the affinity constants with native chromatin, nucleosomal DNA and naked DNA. The results showed that there is a gradual reduction in the free energy as we go down the series-naked DNA, native chromatin and nucleosomal core particle. It originates from an alteration in the extent of enthalpy-entropy compensation for each system. A continuation of the studies with the same group of antibiotics showed that chopping of the N-terminal tails leads to a decrease in the free energy of association of the drugs with nucleosome. The effect of lowering in free energy is more pronounced in the case of intercalator like daunomycin (Mir *et al.*, 2004). While these studies have thrown light upon the energetics of association, the isothermal titration calorimetric approach will provide an incisive picture. There is a report of calorimetric investigation of ethidium and netropsin binding to chicken erythrocyte chromatin (Taquet *et al.*, 1998). They demonstrated that the presence of histones on DNA still allows the ligand binding that takes place according to a simple one single-site model. The results show the enthalpic origin of the association with little variation of heat capacity change with temperature. Knowledge about the variation of heat capacity change with temperature for the DNA will be a necessary input for understanding the structural basis of drug-chromatin/nucleosome association. Such studies are in progress in our laboratory. Melting studies and differential scanning calorimetry are other potential thermodynamic tools to get an idea about the effect of the drugs upon the energetics of histone–DNA interactions at the nucleomal and linker level. There have been many differential scanning calorimetric studies to understand the fine structure of nuclei, chromatin and nucleosome. Differential scanning calorimetry of nuclei was used as test for the effects of strand breakers like bleomycin on human chromatin (Almagor and Cole, 1989). We have employed melting studies to examine the effect of aureolic acid group of antibiotics upon DNA in the chromatin, nucleosome and H1-depleted chromatin (Mir and Dasgupta, 2003).

Evaluation of crystal structure will provide insight to the site-specific nature of the drug-DNA interaction at the nucleosomal level. In this regard, a recent report on the crystal structures of three nucleosome core particles in complex with site-specific DNA-binding ligands, the pyrrole-imidazole polyamides has been the first study (Suto *et al.*, 2003). The notable feature emerging from this study shows that the minor groove of nucleosomal DNA is capable (within limits) of adjusting its parameters to allow recognition and binding of small ligands while retaining a full complement of histone–DNA interactions. However, nucleosomal DNA undergoes significant structural changes at the ligand-binding sites and in adjacent regions to accommodate the ligands. Notwithstanding the importance of this result, appropriate control studies should be done to check that association of the ligands does not lead to disruption of the nucleosome.

8. EFFECT OF REVERSIBLE DNA-BINDING OF ANTIBIOTICS UPON CHROMATIN CONDENSATION

The ability of DNA-binding drugs to induce chromatin condensation calls for in-depth studies, because the biological consequence of chromatin condensation is its loss of transcriptional ability. Therefore it might be a mechanism of transcription inhibition by the reversible DNA binders. The first report in this line (Sen and Crothers, 1986) employed transient electric dichroism to study the ability of the drugs to induce folding of chromatin from the 10- to 30-nm fiber either by

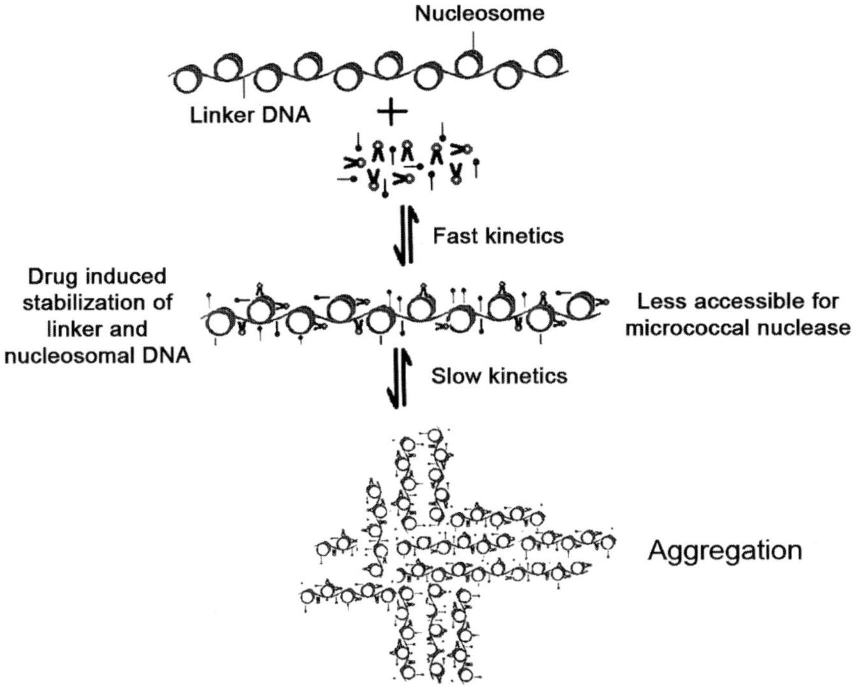

Effect of Drugs upon Chromatin Structure

themselves or in conjunction with multivalent cations. It was found that charge on the drug is a factor for its condensation potential. Interestingly these drugs, irrespective of the groove binder nature or intercalative nature, inhibit compaction of chromatin; they rather induce condensation. Physiscochemical studies with metal complexes of aureolic acid group of antibiotics, chromomycin A3 and mithramycin, and some other intercalators like daunomycin from our laboratory also suggest that these drugs induce condensation of the chromatin. Based on these results we propose a model for the effect of these drugs upon chromatin structure. At present we are examining the validity of this mechanism for other reversible binders to DNA. However, studies need to be done to examine whether the condensation ability differs for normal and neoplastic cells.

9. CONCLUDING REMARKS

There have been excellent review articles related to the present topics (e.g. Muenchen and Pienta, 1999; Nelson *et al.*, 2004, Inche and La Thangue, 2006). Our review has provided an overview of the research activities to understand the molecular basis of the function of the various classes of DNA-targeting drugs at the chromatin level. By no means, the review has covered all reports in this area. We have emphasized upon the salient features.

The rationale that has driven cancer drug design over the years is the specific targeting of tumorigenous cells, leading to induction of cell death. In that run, the DNA targeting drugs have occupied an important position. The earliest anticancer agents were the DNA alkylators/cross-linkers which modify the strands of DNA and thereby inhibit their templating properties. Later, utilizing this same principle, the DNA strand cleaving molecules were developed. Along with them reversible binders of DNA were also examined and employed as drugs (Martinez and Chacon-Garcia, 2005).

DNA secondary structures and protein-DNA complexes were subsequently targeted for specific drug design. The modern era of chemotherapeutics holds prospects for the DNA secondary structures and protein-DNA complexes as anticancer drug targets. This not only imparts high specificity to the drug, but also establishes the importance of the anti-gene approach and the drug target.

At present, cancer treatment schemes mostly use combination chemotherapy which (1) kills maximum number of cells within the range of toxicity tolerated by the host for each drug; (2) offers a broader range of coverage of resistant cell lines in a heterogeneous tumor population; and (3) prevents or slows the development of new drug-resistant cell lines. This strategy will perhaps remain the method of choice in future as well. Since, DNA damage and subsequent apoptosis induction is physiological effect of many DNA-binding anticancer drugs, therefore, nontoxic amplification of DNA-cleaving activity of anticancer drugs would effectively reduce drug dose and side effects, leading to development of improved chemotherapy. In a recent survey (Kawanishi and Hiraku, 2004), the enhancing effects of DNA-binding ligands ('amplifiers'), especially minor groove binders and intercalators, on anticancer drug-induced apoptosis and DNA cleavage were made using human cultured cells and (32)P-labeled DNA fragments obtained from the human genes. The mechanism of amplification of DNA cleavage has been ascribed to the fact that binding of amplifier changes the DNA conformation to allow anticancer drug to interact more appropriately with the specific sequences, resulting in enhancement of anticancer effect. This study on amplifiers of anticancer agents shows a novel approach to the potentially effective anticancer therapy.

A major objective of the present-day synthetic, chemical and structural biology and molecular medicine is to find natural or synthetic small molecules with the DNA – binding potential, so that their site-specific binding potential to DNA can be utilized to regulate the DNA-templated biological processes. The major problem confronting the above task is the inability to define their full range of specific targets in the cell nucleus (Kim *et al.*, 2003) and therefore to predict the target sites and their effectiveness to kill specifically the target cells afflicted with neoplasia. A recent report (Warren *et al.*, 2006) has been a pathfinder to address this problem at a global level. The group has developed *a comprehensive high-throughput platform that can rapidly and reliably identifies the cognate sites of DNA-binding molecules. This platform provides an unbiased analysis because it consists of a double-stranded DNA array that displays the entire sequence space represented by 8 bp (all possible permutations equal 32,896 molecules) and can currently be extended to as many as 10 variable base pair positions.* This report has featured the importance of chemical genomics approach as one of the tools to achieve the target (Jung *et al.*, 2003). Delivery of the drugs to the appropriate site is another challenge for chemical biology. There has been considerable progress in this area too.

The focus of research interest in my laboratory has been to elaborate the effect of DNA-binding class of anticancer drugs upon the chromatin structure related to its function during the process of gene expression. As the literature survey shows, there has been a lacuna of information in this area. A judicious combination of biophysical, biochemical and genetic approaches would definitely unfold the intricacies behind the different steps responsible for the mode of action of these drugs. In order to focus upon the site of action, the expression of genes and proteins, which are suppressed or enhanced as a consequence to the use of the drugs, also need to be identified in order to assess their efficacy as anticancer drugs. However, from the knowledge accrued from the research in the last decade on chromatin structure and function it is becoming progressively clear that effect may not be confined to a singular locus. A network of steps might be responsible for the ultimate cellular function of these DNA-binding drugs. In this connection identification and validation of a single (or more) target is an absolute prerequisite to understand the molecular pharmacology of the drugs.

ACKNOWLEDGEMENTS

DD wants to acknowledge Dr. Sangita Sen, Dr. Mohd.Ayoub Mir and Dr. Suman Das for their excellent contributions to the work on drug-chromatin interaction in his laboratory during their tenure. The financial support has come from intramural fund of SINP.

REFERENCES

Abdel-Halim HI, Natarajan AT, Mullenders LH, Boei JJ (2005) Mitomycin C-induced pairing of heterochromatin reflects initiation of DNA repair and chromatid exchange formation. J Cell Sci 118 (Pt 8):1757–1767

Aich P, Dasgupta D (1990) Role of Mg(II) in the Mithramycin-DNA interaction: evidence for two types of Mithramycin-Mg(II) complex. Biochem Biophys Res Commun 173:689–696

Aich P, Dasgupta D (1995) Role of magnesium ion in mithramycin-DNA interaction: binding of mithramycin-Mg^{2+} complexes with DNA. Biochemistr 34: 1376–1385

Aich P, Sen R, Dasgupta D (1992a) Interaction between antitumor antibiotic Chromomycin A3 and Mg(II) I. Evidence for the formation of two types of Chromomycin A3-Mg(II) complexes. Chem Biol Interact 83: 23–33

Aich P, Sen R, Dasgupta D (1992b) Role of magnesium ion in the interaction between chromomycin A_3 and DNA: binding of chromomycin $A_3 - Mg^{2+}$ complexes with DNA. Biochemistry 31:2988–2997

Akman SA, Doroshow JJH, Thomas G, Burke J, Dizdaroglus M (1992) DNA base modifications induced in isolated human chromatin by NADH de hydrogenase-cata1yzed reduction of doxorubicin. Biochemistry, 31:3500–3506

Almagor M, Cole RD (1989) Differential scanning calorimetry of nuclei as a test for the effects of anticancer drugs on human chromatin. Cancer Res 49(20):5561–5566

Ausio J, Dong F, van Holde KE (1989) Use of selectively trypsinized nucleosome coreparticles to analyze the role of the histone "tails" in the stabilization of the nucleosome. J Mol Biol 206(3):451–463

Barceló F, Martorell J, Gavilanes F, González-Ros JM (1988) Equilibrium binding of daunomycin and adriamycin to calf thymus DNA. Temperature and ionic strength dependence of thermodynamic parameters. Biochem Pharmacol 37:2133–2138

Bellon SF, Coleman JH, Lippard SJ (1991) DNA unwinding produced by site-specific intrastrand cross-links of the antitumor drug cis-diamminedichloroplatinum(II). Biochemistry 30(32):8026–8035

Berger SL (2002) Histone modifications in transcriptional regulation. Curr Opin Genet Dev 12:142–148

Binaschi M, Bigioni M, Cipollone A, Rossi C, Goso C, Maggi CA, Capranico G, Animati F (2001) Anthracyclines: selected new developments. Curr Med Chem Anticancer Agents 1(2):113–130

Bradner WT (2001) Mitomycin C: a clinical update. Cancer Treat Rev 27(1):35–50

Brown PM, Fox KR (1996) Minor groove binding ligands alter the rotational positioning of DNA fragments on nucleosome core particles. J Mol Biol 262(5):671–685

Bruhn SL, Pil PM, Essigmann JM, Housman DE, Lippard SJ (1992) Isolation and characterization of human cDNA clones encoding a high mobility group box protein that recognizes structural distortions to DNA caused by binding of the anticancer agent cisplatin. Proc Natl Acad Sci U S A 89(6):2307–2311

Bucceri A, Kapitza K, Thoma, F (2006) Rapid accessibility of nucleosomal DNA in yeast on a second time scale. EMBO J 15 June; (Epub ahead of print)

Calabresi P, Chabner BA (1991) The pharmacological basis of therapeutics (Goodman, Gilmann, eds). Macmillan Publishing Co., NY, pp1209–1263

Carpenter ML, Marks JN, Fox KR (1993) DNA-sequence binding preference of the GC-selective ligand mithramycin. Deoxyribonuclease-I/deoxyribonuclease-II and hydroxy-radical footprinting at CCCG, CCGC, CGGC, GCCC and GGGG flanked by (AT)n and An.Tn. Eur J Biochem 215:561–566

Carrión C, de Madariaga MA, Domingo JC (2004), *In vitro* cytotoxic study of immunoliposomal doxorubicin targeted to human CD34(+) leukemic cells. Life Sci 75:313–328

Chaires JB (1996) Molecular recognition of DNA. In: Hurley LH, Chaires JB (eds) Advances in DNA sequence-specific agents, vol 2. JAI Press, Greenwich CT, pp141–167

Chaires JB (1997) Energetics of drug-DNA interactions. Biopolymers 44(3):201–215

Chaires JB, Dattagupta ND, Crothers DM (1982) Studies on interaction of anthracycline antibiotics and DNA: Equilibrium binding studies on interaction of daunomycin with DNA. Biochemistry 21:3933–3940

Chaires JB, Dattagupta ND, Crothers DM (1983) Binding of daunomycin to calf thymus nucleosomes. Biochemistry 22(2):284–292

Chaires JB, Herrera JE, Waring MJ (1990) Preferential binding of daunomycin to 5-ATCG and 5-ATGC sequences revealed footprinting, titration experiment. Biochemistry 29:6145–6153

Chakrabarti S, Mahmood A, Kassis AI, Bump EA, Jones AG, Makrigiorgos GM (1996) Generation of hydroxyl radicals by nucleohistone-bound metal-adriamycin complexes. Free Radic Res 25(3):207–220

Chakrabarti S, Bhattacharyya D, Dasgupta D (2000–2001) Structural basis of DNA recognition by anticancer antibiotics, chromomycin A3 and mithramycin: roles of minor groove width and ligand flexibility. Biopolymers 56(2):85–95

Chakrabarti S, Bhattacharyya B, Dasgupta D (2002) Interaction of mithramycin and chromomycin A_3 with $d(TAGCTAGCTA)_2$: Role of sugars in antibiotic-DNA recognition. J Phys Chem 106:6947–6953

Chatterjee S, Zaman K, Ryu H, Conforto A, Ratan RR (2001) Sequence-selective DNA binding drugs mithramycin A and chromomycin A_3 are potent inhibitors of neuronal apoptosis induced by oxidative stress and DNA damage in cortical neurons. Ann Neurol 49:345–354

Churchill ME, Hayes JJ, Tullius TD (1990) Detection of drug binding to DNA by hydroxyl radical footprinting. Relationship of distamycin binding sites to DNA structure and positioned nucleosomes on 5S RNA genes of Xenopus. Biochemistry 29(25):6043–6050

Cobuzzi RJ, Burhans WC, Beerman TA (1996) Inhibition of initiation of simian virus 40 DNA replication in infected BSC-1 cells by the DNA alkylating drug adozelesin. J Biol Chem 271:19852–19859

Cohen GL, Falkson CI (1998) Current treatment options for malignant melanoma. Drugs 55:791–799
Cons BM, Fox KR (1989) High resolution hydroxyl radical footprinting of the binding of mithramycin and related antibiotics to DNA. Nucl Acids Res 17:5447–5459
Dabrowiak JC, Bradner WT (1987) Platinum antitumour agents. Prog Med Chem 24:129–158
Danford AJ, Wang D, Wang Q, Tullius TD, Lippard SJ (2005) Platinum anticancer drug damage enforces a particular rotational setting of DNA in nucleosomes. Proc Natl Acad Sci U S A 102(35):12311–12316
Davey CA, Sargent DF, Luger K, Maeder AW, Richmond TJ (2002) Solvent mediated interactions in the structure of the nucleosome core particle at 1.9Å resolution. J Mol Biol 319:1097–1113
Davies NP, Hardman LC, Murray V (2000) The effect of chromatin structure on cisplatin damage in intact human cells. Nucleic Acids Res 28(15):2954–2958
Debrauwere H, Gendrel CG, Lechat S, Dutreix M (1997) Differences and similarities between various tandem repeat sequences: minisatellites and microsatellites. Biochimie 79:577–586
Dervan PB, Brli RW (1999) Sequence specific DNA recognition by polyamides. Curr Opin Biol 3:688–693
Dickinson LA, Burnett R, Melander C, Edelson BS, Arora PS, Dervan PB, Gottesfeld JM (2004) Arresting cancer proliferation by small-molecule gene regulation. Chem Biol 11(11):1583–1594
Dimaraco A, Arcamone F, Zunino F (1975) Antibiotics, mechanism of action of antimicrobial & antitumor agents (Corcoran JW, Hahn FE, eds). Springler-Verlag, Berlin, pp101–128
Dorr RT, Bowden GT, Alberts DS, Liddil JD (1985) Interactions of mitomycin C with mammalian DNA detected by alkaline elution. Cancer Res 45(8):3510–3516
Duca M, Guianvarc'h, D, Oussedik K, Halby L, Garbesi A, Dauzonne D, Monneret C, Osheroff N, Giovannangeli C, Arimondo PB (2006) Molecular basis of the targeting of topoisomerase II-mediated DNA cleavage by VP16 derivatives conjugated to triplex-forming oligonucleotides. Nucleic Acids Res 34(6):1900–1911
Dudouet B, Burnett R, Dickinson LA, Wood MR, Melander C, Belitsky JM, Edelson B, Wurtz N, Briehn C, Dervan PB, Gottesfeld JM (2003) Accessibility of nuclear chromatin by DNA binding polyamides. Chem Biol 10:859–867
Fitzgerald DJ, Anderson JN (1999) Selective nucleosome disruption by drugs that bind in the minor groove of DNA. J Biol Chem 274(38):27128–27138
Fox KR, Cons BM (1993) Interaction of mithramycin with DNA fragments complexed with nucleosome core particles: comparison with distamycin and echinomycin. Biochemistry. 32(28):7162–7171
Friedman HS, Kerby T, Calvert, H (2000) Temozolomide and treatment of malignant gliomas. Clin Cancer Res 6:2585–2597
Fritzsche H, Wahnert U, Chaires JB, Dattagupta N, Schlessinger FB, Crothers DM (1987) Anthracycline antibiotics. Interaction with DNA and nucleosomes and inhibition of DNA synthesis. Biochemistry 26(7):1996–2000
Galea AM, Murray V (2002) The interaction of cisplatin and analogues with DNA in reconstituted chromatin. Biochim Biophys Acta 1579(2–3):142–152
Ganguli R, Chowdhury K, Chakraborty B, Neogy RK (1983) Interaction of nogalamycin with chromatin. Chem Biol Interact 46(3):347–352
Garcia-Nieto R, Manzanares I, Cuevas C, Gago F (2000) Increased DNA binding specificity for antitumor ecteinascidin 743 through protein-DNA interactions? J Med Chem 43:4367–4369
Garcio-Ramirez M, Dong F, Ausio J (1992) Role of the histone "tails" in the folding of oligonucleosomes depleted of histone H1. J Biol Chem 267(27):19587–19595
Gasser SM (2002) Visualizing chromatin dynamics in interphase nuclei. Science 296(5572):1412–1416
Gianni L, Cordon BJ, Myers CE (1983) The biochemical bases of anthracycline toxicity and antitumor action, In: Hodgson E, Bend JR, Philipot RM (eds) Reviews in biochemical toxicology, vol 5. Elsevier, Amsterdam, The Netherlands, pp1–82
Giloni L, Takeshita M, Johnson F, Iden C, Grollman AP (1981) Bleomycin induced strand-scission of DNA. Mechanism of deoxyribose cleavage. J Biol Chem 256(16):8608–8615
Gniazdowski M, Czyz M (1999) Transcription factors as targets of anticancer drugs. Acta Biochim Pol 46(2):255–262

Gniazdowski M, Denny WA, Nelson SM, Czyz M (2005) Effects of anticancer drugs on transcription factor-DNA interactions. Expert Opin Ther Targets 9(3):471–489

Goll MG, Bestor TH (2005) Eukaryotic cytosine methyltransferases. Annu Rev Biochem74:481–514

Goodman LS, Wintrobe MM, Dameshek W, Goodman MJ, Gilman A, McLennan MT (1946) A Nitrogen mustard therapy. Use of methyl-bis(beta-chloroethyl)amine hydrochloride and tris(beta-chloroethyl)amine hydrochloride for Hodgkin's disease, lymphosarcoma, leukemia and certain allied and miscellaneous disorders. JAMA 132:126–132

Gottesfeld JM, Belitsky JM, Melander C, Dervan PB, Luger K (2002) Blocking transcription through a nucleosome with synthetic DNA ligands. J Mol Biol 321:249–263

Gyapay G, Jeney A, Lapis K (1985) Drug action and chromatin structure. I. Adriamycin binding to core particle of nucleosomes and subsequent enhanced DNA fragmentation induced by micrococcal nuclease. Neoplasma 32(5):521–528

Han H, Cliff CL, Hurley LH (1999) Accelerated assembly of G-quadruplex structures by a small molecule. Biochemistry 38:6981–6986

Hansson J, Lewensohn R, Ringborg U, Nilsson B (1987) Formation and removal of DNA cross-links induced by melphalan and nitrogen mustard in relation to drug-induced cytotoxicity in human melanoma cells. Cancer Res 47:2631–2637

Haq I (2002) Thermodynamics of drug-DNA interactions. Arch Biochem Biophys 403:1–15

Hsiang YH, Hertzberg R, Hecht S, Liu LF (1985) Camptothecin induces protein-linked DNA breaks via mammalian DNA topoisomerase I. J Biol Chem 260(27):14873–14878

Hurley LH (2002) DNA and its associated processes as targets for cancer therapy. Nat Rev cancer 2:188–200

Hurley LH, Zewail-Foote, M (2001) The antitumor agent ecteinascidin 743: characterization of its covalent DNA adducts and chemical stability. Adv Exp Med Biol 500:289–299

Inche AG, La Thangue NB (2006) Chromatin control and cancer-drug discovery: realizing the promise. Drug Discov Today 11(3–4):97–109

Iyer VN, Szybalski W (1963) A molecular mechanism of mitomycin action: Linking of complementary DNA strands. Proc Natl Acad Sci U S A 50:355–362

Jamieson ER, Lippard SJ (1999) Structure, recognition, and processing of cisplatin-DNA adducts. Chem Rev 99(9):2467–2498

Jenkins T (2000) Targeting multi-stranded DNA structures. Curr Med Chem 7:99–115

Jenuwein T, Allis CD (2001) Translating the histone code. Science 293(5532):1074–1080

Juan LJ, Utley RT, Adams CC, Vettese-Dadey M, Workman JL (1994) Differential repression of transcription factor binding by histone H1 is regulated by the core histone amino termini. EMBO J 15:6031–6040

Jones CB, McIntosh J, Huang H, Graytock A, Hoyt DG (2001) Regulation of bleomycin-induced DNA breakage and chromatin structure in lung endothelial cells by integrins and poly(ADP-ribose) polymerase. Mol Pharmacol 59(1):69–75

Jung M, Kim H, Kim M (2003) Chemical genomics strategy for the discovery of new anticancer agents. Curr Med Chem 10(9):757–762

Kadam S, Emerson BM (2002) Mechanism of chromatin assembly and transcription. Curr Opin Cell Biol 14:262–268

Kas E, Izaurralde E, Laemmli UK (1989) Specific inhibition of DNA binding to nuclear scaffolds and histone H1 by distamycin. The role of oligo(dA).oligo(dT) tracts. J Mol Biol 210(3):587–599

Kas E, Poljak L, Adachi Y, Laemmli UK (1993) A model for chromatin opening: stimulation of topoisomerase II and restriction enzyme cleavage of chromatin by distamycin. EMBO J 12(1): 15–126

Kawanishi S, Hiraku Y (2004) Amplification of anticancer drug-induced DNA damage and apoptosis by DNA-binding compounds. 4(5):415–419

Keniry MA, Banville DL, Simmonds PM, Shafer RH (1993) Nuclear magnetic resonance comparison of the binding sites of mithramycin and chromomycin on the self-complementary oligonucleotide d(ACCCGGGT)2. Evidence that the saccharide chains have a role in sequence specificity. J Mol Biol 31:753–767

Keyes SR, Loomis R, DiGiovanna MP, Pritsos CA, Rockwell S, Sartorelli AC (1991) Cytotoxicity and DNA crosslinks produced by mitomycin analogs in aerobic and hypoxic EMT6 cells. Cancer Commun 3(10–11):351–356

Kim MS, Blake M, Baek JH, Kohlhagen G, Pommier Y, Carrier F (2003) Inhibition of histone deacetylase increases cytotoxicity to anticancer drugs targeting DNA. Cancer Res 63(21):7291–7300

Kornberg RD and Lorch Y (1999) Twenty-Five years of the nucleosome, fundamental particle of the eukaryote chromosome. Cell 98(3):285–294

Kumar A, Brown DT, Leno GH (2004) DNA intercalators differentially affect chromatin structure and DNA replication in Xenopus egg extract. Anticancer Drugs 15(6):633–639

Kuo MT (1981) Preferential damage of active chromatin by bleomycin. Cancer Res 41(6):2439–2443

Kuo MT, Samy TS (1978) Effects of neocarzinostatin on mammalian nuclei: release of nucleosomes. Biochim Biophys Acta 518(1):186–190

Kurten S, Obe, G (1975) Premature chromosome condensation in the bone marrow of Chinese hamster after application of bleomycin in vivo. Mutat Res 27(2):285–294

Lee DY, Hayes JJ, Pruss D, Wolffe AP (1993) A positive role for histone acetylation in transcription factor access to nucleosomal DNA. Cell 72:73–84

Lee YJ, Park SJ, Ciccone SL, Kim CR, Lee SH (2006) An in vivo analysis of MMC-induced DNA damage and its repair. Carcinogenesis 27(3):446–453

Loehrer PJ, Einhorn LH (1984) Drugs five years later. Cisplatin. Ann Intern Med 100(5):704–713

Li G, Widom J (2004) Nucleosomes facilitate their own invasion. Nat Struct Mol Biol 11:763–769

Liu W, Sun D, Hurley LH (2005) Binding of G-quadruplex interactive agents to distinct G-quadruplexes induces different biological effects in MiaPaca cells. Nucleosides Nucleotides Nucleic Acids 24(10–12):1801–1815

Lohse J, Dahl O, Neilsen PE (1999) Double duplex invasion by peptide nucleic acid: a general principle for sequence-specific targeting of double stranded DNA. Proc Natl Acad Sci U S A 96:11804–11808

Lonn U, Lonn S, Nylen U, Windblad G (1990) Bleomycin-induced DNA lesions are dependent on nucleosome repeat length. Biochem Pharmacol 39(1):101–107

Lopez-Larraza DM, Bianchi NO (1993) DNA response to bleomycin in mammalian cells with variable degrees of chromatin condensation. Environ Mol Mutagen 21(3):258–264

Lopez-Larraza DM, Padron J, Ronci NE, Vidal Rioja LA (2006) Chromatin condensation and differential sensitivity of mammalian and insect cells to DNA strand breaks induced by bleomycin. Mutat Res 16 April; [Epub ahead of print]

Low CM, Drew HR, Waring MJ (1986) Echinomycin and distamycin induce rotation of nucleosome core DNA. Nucleic Acids Res 14(17):6785–6801

Luger K (2006) Dynamic nucleosomes. Chromosome Res 14(1):5–16

Luger K, Mader AW, Richmond RK, Sargent DF, Richmond TJ (1997) Crystal structure of the nucleosome core particle at 2.8 A resolution. Nature 389(6648):251–260

Lund AH, and van Lohuizen M (2004) Epigenetics and cancer. Genes Dev 18(19):2315–2335

Mai A, Massa S, Rotili D, Cerbara I, Valente S, Pezzi R, Simeoni S, Ragno R (2005) Histone deacetylation in epigenetics: an attractive target for anticancer therapy. Med Res Rev 25(3):261–309

Majee S, Sen R, Guha S, Bhattacharya D, Dasgupta D (1997) Differential interactions of the Mg^{2+} complexes of chromomycin A_3 and mithramycin with poly(dG-dC) x poly(dC-dG) and poly(dG) x poly(dC). Biochemistry 36:2291–2299

Manfredini S, Vicentini CB, Manfrini M, Bianchi N, Rutigliano C, Mischiati C, Gambari R (2000) Pyrazolo-triazoles as light activable DNA cleaving agents. Bioorg Med Chem 8:2343–2346

Martinez R, Chacon-Garcia L (2005) The search of DNA-intercalators as antitumoral drugs: what it worked and what did not work. Curr Med Chem12(2):127–151

McHugh MM, Woynarowski JM, Mitchell MA, Gawron LS, Welland KL, Beerman TA (1994) CC-1065 bonding to intracellular and purified SV40 DNA: site specificity and functional effects. Biochemistry 33:9158–9168

McHugh MM, Kuo SR, Walsh-O'Beirne MH, Liu JS, Melendy T, Beerman TA (1999) Bizelesin, a bifunctional cyclopropylpyrroloindole alkylating agent, inhibits simian virus 40 replication in trans by induction of an inhibitor. Biochemistry 38(35):11508–11515

Mir MA, Dasgupta D (2001a) Interaction of antitumor drug, mithramycin, with chromatin, Biochem. Biophys Res Commun 280:68–74

Mir MA, Dasgupta D (2001b) Association of the anticancer antibiotic chromomycin A_3 with the nucleosome: role of core histone tail domains in the binding process. Biochemistry 40:11578–11585

Mir MA, Dasgupta D (2003) Association of anticancer drug mithramycin with H1-depleted chromatin: a comparison with native chromatin, J Inorg Biochem 94(1–2):72–77

Mir MA, Das S, Dasgupta D (2004) N-terminal tail domains of core histones in nucleosome block the access of anticancer drugs, Mithramycin and Daunomycin, to the nucleosomal DNA. Biophys Chem 109:121–135

Mizuno NS, Zakis B, Decker RW (1975) Binding of daunomycin to DNA and the inhibition of RNA and DNA synthesis. Cancer Res 35(6):1542–1546

Muenchen HJ, Pienta KJ (1999) The role of the nuclear matrix in cancer chemotherapy. Crit Rev Eukaryot Gene Expr 9(3–4):337–343

Morad M, Jonasson J, Lindsten J (1973) Distribution of mitomycin C induced breaks on human chromosomes. Hereditas 74(2):273–281

Murray V, Whittaker J, McFadyen WD (1998) DNA sequence selectivity of cisplatin analogues in intact human cells. Chem Biol Interact 110(1–2):27–37

Myers CE, Mimnaugh EG, Yeh GC, Sinha BK (1988) Anthracycline and anthracenedione-based anticancer agents (Lown JW, ed). Elsevier, Amsterdam, The Netherlands, pp527–569

Nair RR, Wang H, Jamaluddin MS, Fokt I, Priebe W, Boyd DD (2005) A bisanthracycline (WP631) represses uPAR gene expression and cell migration of RKO colon cancer cells by interfering with transcription factor binding to a chromatin-accessible -148/-124 promoter region. Oncol Res 15(5):265–279

Neilsen PE (1999) Peptide nucleic acids as therapeutic agents. Curr Opin Struct Biol 9:353–357

Nelson EM, Tewey KM, Liu LF (1984) Mechanism of antitumor drug action: poisoning of mammalian DNA topoisomerase II on DNA by 4′-(9-acridinylamino)-methanesulfon-m-anisidide. Proc Natl Acad Sci U S A 81(5):1361–1365

Nelson SM, Ferguson LR, Denny WA (2004) DNA and the chromosome – varied targets for chemotherapy. Cell Chromosome 3(1):2

Nelson SM, Ferguson LR, Denny WA (2005) Demonstration by real time polymerase chain reaction that cellular DNA alkylation by novel aminoindoline compounds affects expression of the protooncogene c-myc. Chem Res Toxicol 18(2):239–248

Osheroff N, Zechiedrich EL, Gale KC (1991) Catalytic function of DNA topoisomerase II. Bioessays 13(6):269–273

Philips BJ, Chang AY, Dervan PB, Beerman TA (2005) DNA damage effects of a polyamide-CBI conjugate in SV40 virions. Mol Pharmacol 67(3):877–882

Pindur U, Jansen M, Lemster T (2005) Advances in DNA-ligands with groove binding, intercalating and/or alkylating activity: chemistry, DNA-binding and biology. Curr Med Chem 12(24):2805–2847

Pommier Y, Kohlhagen G, Bailly C, Waring M, Mazumder A, Kohn KW (1996) DNA sequence and structure selective alkylation of guanine N2 in the DNA minor groove by ecteinascidin 743, a potent antitumor compound from the Caribbean tunicate Ecteinascidia turbinata. Biochemistry 35:13303–13309

Portugal J, Waring MJ (1987) Interaction of nucleosome core particles with distamycin and echinomycin: analysis of the effect of DNA sequences. Nucleic Acids Res 15(3):885–903

Povirk LF, Goldberg IH (1987) A role of oxidative DNA sugar damage in mutagenesis by neocarzinostatin and bleomycin. Biochimie 69(8):815–823

Quillet-Mary A, Mansat V, Duchayne E, Come MG, Allouche M, Bailly JD, Bordier C, Laurent G (1996) Daunorubicin-induced internucleosomal DNA fragmentation in acute myeloid cell lines. Leukemia 10(3):417–425

Rabbani A, Iskandar M, Ausio J (1999) Daunomycin-induced unfolding and aggregation of chromatin. J Biol Chem 274(26):18401–18406

Rabbani A, Finn RF, Thambirajah AA, Ausio J (2004) Binding of antitumor antibiotic daunomycin to histones in chromatin and in solution. Biochemistry 43:16497–16504

Rice JC, Allis CD (2001) Histone methylation versus histone acetylation: new insights into epigenetic regulation. Curr Opin Cell Biol 13:263–273

Richmond TJ, Widom J (2000) Nucleosome and chromatin structure. Oxford University Press, Oxford

Roberts JJ, Thomson AJ (1979) The mechanism of action of antitumor platinum compounds. Prog Nucleic Acid Res Mol Biol 22:71–133

Ross WE, Glaubiger DL, Kohn KW (1978) Protein-associated DNA breaks in cells treated with adriamycin or ellipticine. Biochim Biophys Acta 519(1):23–30

Rothfuss A, Grompe M (2004) Repair kinetics of genomic interstrand DNA cross-links: evidence for DNA double-strand break-dependent activation of the Fanconi anemia/BRCA pathway. Mol Cell Biol 24(1):123–134

Saenger W (1984) Principles of nucleic acid structure. Springer-Verlag, New York

Sastry M, Fiala R, Patel DJ (1995) Solution structure of mithramycin dimers bound to partially overlapping sites on DNA. J Mol Biol 251:674–689

Schäfer S, Jung M (2005) Chromatin modifications as targets for new anticancer drugs. Arch Pharm (Weinheim) 338(8):347–357

Schein PS, O'Connell MJ, Blom J, Hubbard S, Magrath IT, Bergevin P, Wiernik PH, Zeigler JL, DeVia VT (1974) Clinical antitumor activity and toxicity of streptozotocin (NSC-85998). Cancer 34:993–1000

Sedletska Y, Giraud-Panis MJ, Malinge JM (2005) Cisplatin is a DNA-damaging antitumour compound triggering multifactorial biochemical responses in cancer cells: importance of apoptotic pathways. Curr Med Chem Anticancer Agents 5(3):251–265

Seenisamy J, Bashyam S, Gokhale V, Vankayalapati H, Sun D, Siddiqui-Jain A, Streiner N, Shin-Ya K, White E, Wilson WD, Hurley LH (2005) Design and synthesis of an expanded porphyrin that has selectivity for the c-MYC G-quadruplex structure. J Am Chem Soc 127(9):2944–2959

Sen D, Crothers DM (1986) Influence of DNA-binding drugs on chromatin condensation. Biochemistry 25(7):1503–1509

Shammas MA, Reis RJS, Cheng L, Koley H, Hurley LH, Anedrson KC, Munshi NC (2004) Telomerase inhibition and cell growth arrest after telomestatin treatment in multiple myeloma. Clin Cancer Res 10:770–776

Shaw MW, Cohen MM (1965) Chromosome exchanges in human leukocytes induced by mitomycin C. Genetics 51:181–190

Shinohara K, Bando T, Sasaki S, Sakakibara Y, Minoshima M, Sugiyama H (2006) Antitumor activity of sequence-specific alkylating agents: pyrrole-imidazole CBI conjugates with indole linker. Cancer Sci 97(3):219–225

Sidik K, Smerdon MJ (1990) Nucleosome rearrangement in human cells following short patch repair of DNA damaged by Bleomycin. Biochemistry 29(32):7501–7511

Simpkins H, Pearlman LF, Thompson LM (1984) Effects of adriamycin on supercoiled DNA and calf thymus nucleosomes studied with fluorescent probes. Cancer Res 44(2):613–618

Sischka A, Toensing K, Eckel R, Wilking SD, Sewald N, Ros R, Anselmetti D (2005) Molecular mechanisms and kinetics between DNA and DNA binding ligands. Biophys J 88(1):404–411

Smith AL, Nicolaou KC (1996) The enediyne antibiotics. J Med Chem 39(11):2103–2117

Smith BL, Bauer GB, Povirk LF (1994) DNA damage induced by bleomycin, neocarzinostatin, and melphalan in a precisely positioned nucleosome. Asymmetry in protection at the periphery of nucleosome-bound DNA. J Biol Chem 269(48):30587–30594

Spolar RS, Record MT (1994) Coupling of local folding to site-specific binding of proteins to DNA. Science 263:777–784

Steighner RJ, Povirk LF (1990) Bleomycin-induced DNA lesions at mutational hot spots: implications for the mechanism of double-strand cleavage. Proc Natl Acad Sci U S A 87(21): 8350–8354

Straney DC, Crothers DM (1987) Effect of drug-DNA interactions upon transcription initiation at the lac promoter. Biochemistry 26(7): 1987–1995

Sumer H, Saffery R, Wong N, Craig JM, Choo KH (2004) Effects of scaffold/matrix alteration on centromeric function and gene expression. J Biol Chem 279(36):37631–37639

Sun D, Thompson B, Cathers BE, Salazar M, Kerwin SM, Trent JO, Jenkins TC, Neidle S, Hurley LH (1997) Inhibition of human telomerase by a G-quadruplex-interactive compound. J Med Chem 40:2113–2116

Susbielle G, Blattes R, Brevet V, Monod C, Kas, E (2005) Target practice: aiming at satellite repeats with DNA minor groove binders. Curr Med Chem Anticancer Agents. 5(4):409–420

Suto RK, Edayathumangalam RK, White CL, Melander C, Gottesfeld JM, Dervan PB, Luger, K (2003) Crystal structures of nucleosome core particles in complex with minor groove DNA-binding ligands. J Mol Biol 326:371–380

Synder RC, Ray R, Blume S, Miller DM (1991) Mithramycin blocks transcriptional initiation of the c-myc P1 and P2 promoters. Biochemistry 30:4290–4297

Szyf M (2003) DNA methylation and cancer therapy. Drug Resist Updat 6(6):341–353

Szyf M, Pakneshan P, Rabbani SA (2004) DNA demethylation and cancer: therapeutic implications. Cancer Lett 211(2):133–143

Takeshita M, Grollman AP, Ohtsubo E, Ohtsubo H (1978) Interaction of bleomycin with DNA. Proc Natl Acad Sci U S A 75(12):5983–5987

Taquet A, Labarbe R, Houssier C (1998) Calorimetric investigation of ethidium and netropsin binding to chicken erythrocyte chromatin. Biochemistry 37(25):9119–9126

Temple MD, McFadyen WD, Holmes RJ, Denny WA, Murray V (2000) Interaction of cisplatin and DNA-targeted 9-aminoacridine platinum complexes with DNA. Biochemistry 39(18):5593–5599

Terasaki T, Iga T, Sugiyama Y, Hanano M (1984) Interaction of doxorubicin with nuclei isolated from rat liver and kidney. J Pharm Sci 73(4):524–528

Thuong N, Hélène C (1993) Sequence specific recognition and modification of double helical DNA by oligonucleotides. Angew. Chem Int Ed Engl 32:666–690

Tien Kuo M, Hsu TC (1978) Biochemical and cytological studies of bleomycin actions on chromatin and chromosomes. Chromosoma 68(3):229–240

Tranger JW, Baird EE, Mrksich M, Dervan PB (1996) Extension of sequence-specific recognition in the minor groove of DNA by pyrrole-imidazole polyamides to 9–13 base pairs. J Am Chem Soc 118:6160–6166

Tsukiyama T, Wu C (1997) Chromatin remodeling and transcription. Curr Opin In Genetics and Dev 7:182–191

Urnov FD (2002) Methylation and the genome: the power of a small amendment. J Nutr 132:2450S-2456S

Van Helden PD, Wild IJ (1982) Effects of adriamycin on heart and skeletal muscle chromatin. Biochem Pharmacol 31(6):973–977

Van Holde KE, Zlatanova J, Arents G, Moudrianakis E (1995) Chromatin structure and gene expression (Elign SCR, ed). Oxford University Press, Oxford, pp1–26

Wang Y, Beerman TA, Kowalski D (2001) Antitumor drug adozelesin differentially affects active and silent origins of DNA replication in yeast checkpoint kinase mutants. Cancer Res 61(9):3787–3794

Warren CL, Kratochvil NC, Hauschild KE, Foister S, Brezinski ML, Dervan PB, Phillips GN Jr., Ansari AZ (2006) Defining the sequence-recognition profile of DNA-binding molecules. Proc Natl Acad Sci USA 103(4):867–872

Weinberg M, Trabold PA, Lu M, Sharma K, Huberman JA, Burhans WC (1999) Induction by adozelesin and hydroxyurea of origin recognition complex-dependent DNA damage and DNA replication checkpoints in Saccharomyces cerevisiae. J Biol Chem 274:35975–35984

Wheate NJ, and Collins JG (2005) Multi-nuclear platinum drugs: a new paradigm in chemotherapy. Curr Med Chem Anticancer Agents. 5(3): 267–279

Widom J (1998) Structure, dynamics, and function of chromatin in vitro. Annu Rev Biophys Biomol Struct 27:285–327

Wohlert SE, Kunzel E, Machinek R, Mendez C, Salas JA, Rohr J (1999) The structure of mithramycin reinvestigated. J Nat Prod 62:119–121

Workman JL and Kingston RE (1998) Alteration of nucleosome structure as a mechanism of transcriptional regulation. Annu Rev Biochem 67:545–579

Woynarowski JM (2002) Targeting critical regions in genomic DNA with AT specific cancer drugs. Biochim Biophys Acta 1587:300–308

Woynarowski JM, Beerman TA (1997) Effects of bizelesin (U-77,779), a bifunctional alkylating minor groove binder, on replication of genomic and simian virus 40 DNA in BSC-1 cells. Biochim Biophys Acta 1353:50–60

Yaneva J, Leuba SH, van Holde K, Zlatanova J (1997) The major chromatin protein histone H1 binds preferentially to cis-platinum-damaged DNA. Proc Natl Acad Sci U S A 94(25):13448–13451

Yu L, Goldberg IH, Dedon PC (1994) Enediyne-mediated DNA damage in nuclei is modulated at the level of the nucleosome. J Biol Chem 269(6):4144–4151

Zaleskis G, Berleth E, Verstovsek S, Ehrke MJ, Mihich E (1994) Doxorubicin-induced DNA degradation in murine thymocytes. Mol Pharmacol 46(5):901–908

Zamble DB, Lippard SJ (1995) Cisplatin and DNA repair in cancer chemotherapy. Trends Biochem Sci 20(10):435–439

Zelent A, Waxman S, C arducci M, Wright J, Zweibel J, Gore SD (2004) State of the translational science: summary of Baltimore workshop on gene re-expression as a therapeutic target in cancer January 2003. Clin Cancer Res 10(14):4622–4629

Zewail-Foote M, Hurley LH (1999) Ecteinascidin 743: a minor groove alkylator that bends DNA toward the major groove. J Med Chem 42:2493–2497

Zewail-Foote M, Hurley LH (2001) Differential rates of reversibility of ecteinascidin 743-DNA covalent adducts from different sequences lead to migration to favored bonding sites. J Am Chem Soc 123:6485–6495

Zimmer C, Wälmert U (1986) Nonintercalating DNA-binding ligands: specificity of the interaction and their use as tools in biophysical, biochemical and biological investigations of the genetic material. Prog Biophys Mol Biol 47(1):31–112

Zunino F, Di Marco A, Zaccara A, Gambetta RA (1980) The interaction of daunorubicin and doxorubicin with DNA and chromatin. Biochim Biophys Acta 607(2):206–214

PART II

NONHISTONE PROTEINS, SITES FOR EPIGENETIC MARKS

Link to disease

CHAPTER 9

REVERSIBLE ACETYLATION OF NON HISTONE PROTEINS

Role in cellular function and disease

KIRAN BATTA, CHANDRIMA DAS, SHRIKANTH GADAD, JAYASHA SHANDILYA, TAPAS K. KUNDU*

Transcription and Disease Laboratory, Molecular Biology and Genetics Unit. Jawaharlal Nehru Centre for Advanced Scientific Research, Jakkur, Bangalore-64, India

Abstract: Post-translational modifications of nonhistone proteins play a significant role in regulating the chromatin structure, dynamics and thereby gene regulation. Among the different posttranslational modifications, reversible acetylation of non-histone proteins has profound functional implications on wide range of cellular processes. The acetylation status of these proteins is regulated by several cellular and non-cellular factors like viruses, physiological stresses, DNA damaging agents and ROS. Mutations found in the acetylation sites of these proteins and aberrant acetylation are related to imbalances in different cellular pathways and various diseases. Several factor acetyltransferases and deacetylases are known to regulate the acetylation of the nonhistone proteins. Modulators of these enzymes derived from natural as well as synthetic sources can thus have important therapeutic implications. Designing strategies to specifically target the acetylation of these proteins can be used as a valuable tool for new generation drugs

1. INTRODUCTION

Chromatin is a highly complex, dynamic nucleoprotein organization of eukaryotic genome, which play significant role in the regulation of cellular homeostasis. Physiologically, chromatin is not just DNA and histone complex, rather it is a dynamic organization of DNA associated with histone and histone interacting nonhistone proteins. The structure and function of chromatin is regulated by post-translational

* Transcription and Disease Laboratory, Molecular Biology and Genetics Unit. Jawaharlal Nehru Centre for Advanced Scientific Research, Jakkur, Bangalore-560064, INDIA.
E-mail: tapas@jncasr.ac.in

T.K. Kundu and D. Dasgupta (eds.), Chromatin and Disease, 193–212.

modifications of chromatin (histone and nonhistone) proteins and ATP- dependent remodeling of chromatin-DNA in the nucleosomes. Furthermore, post-translational modifications of different transcription factors also play crucial role in transcriptional regulation(Sterner and Berger, 2000; Yoon and Seger, 2006; Martin and Zhang, 2005; Bedford and Richard, 2005). Among the different post-translational modifications (reversible acetylation, phosphorylation, methylation, ubiquitination and sumoylation) acetylation and deacetylation has been extensively studied (Sterner and Berger, 2000). However, different post-translational modifications and ATP-dependent chromatin remodeling, function in a highly coordinated manner and thus set the finer regulation of transcription. Recent reports suggest that acetylation, methylation and phosphorylation of histones as well as nonhistone proteins are regulated through highly interconnected network (Sterner and Berger, 2000; Martin and Zhang, 2005; Bedford and Richard, 2005). In this review we shall focus on reversible acetylation of nonhistone proteins. Significantly, apart from the histones, the acetyltransferases have a broad range of substrate specificity which include, transcription factors, coactivators, hormone receptors, structural proteins, histone chaperones and hence are known as Factor Acetyl Transferases (FAT). Reversible acetylation of these factors can lead to an altered DNA binding as well as protein (histone)-protein interaction, thereby influencing the structure-function dynamics of chromatin. Interestingly, acetylated histones and nonhistone proteins can act as specific signaling platform promoting critical interactions with several factors leading to an open chromatin structure facilitating transcription (Davie and Hendzel, 1994). However, the interplay of factor acetylation, chromatin remodeling and histone chaperones coordinately regulates the transcription. Many components of basal transcriptional machinery get acetylated, although its functional significance has not been clearly established. Acetylation of histones is pre-requisite for transcriptional activation and histone chaperones helps in removal of acetylated histones thereby leading to transcriptional activation. Acetylation of several transcription factors subsequently affects transcriptional activation (Fig. 1a). Reversible acetylation also influences DNA replication. It has been reported that HBO1 can acetylate ORC1 leading to activation of DNA replication (Fig. 1b). Presumably, chromatin remodeling may precede histone acetylation whereby H3-H4 tetramer, which are initially in a state of transient acetylation gets deposited by Chromatin Assembly Factor 1 (CAF1) and gets deacetylated shortly after incorporation into chromatin (Smith and Stillman, 1991; Ruiz-Carrillo *et al.*, 1975; Jackson *et al.*, 1976). This is subsequently followed by deposition of H2A-H2B dimer by Nucleosome Assembly Protein 1 (NAP1). Efficient nucleosome assembly also requires Replication Coupled Assembly Factor (RCAF), which has Anti-silencing Factor 1 (Asf1) as a component. RCAF acts synergistically with CAF1 for efficient chromatin assembly (Fig. 1c). Thus reversible acetylation has diverse physiological consequences. We elucidate here the role of reversible acetylation of nonhistone proteins in diverse cellular phenomena like transcription, DNA repair, cell cycle progression, apoptosis and various signaling pathways. Further aberrant acetylation status of these proteins and consequent disease manifestation and therapeutic implications has also been highlighted in this review.

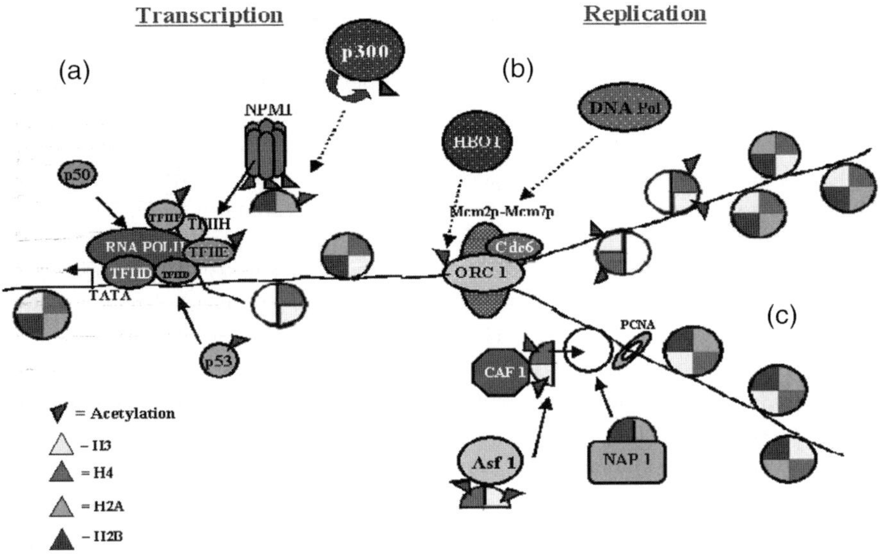

Figure 1. Mechanistic effect of acetylation/deacetylation of histones and nonhistones on chromatin structure.(a) Acetylation of non-histone proteins results in transcriptional activation (b) Acetylation of ORC1 by HBO1 is important for replication. (c) Acetylation of newly synthesized histones necessary for chromatin assembly. (See Colour Plate 12.)

2. ROLE OF NON HISTONE PROTEINS ACETYLATION ON CELLULAR FUNCTION

Reversible acetylation of histone and nonhistone proteins play key role in maintaining cellular homeostasis. In this following section we shall discuss about the physiological significances of acetylation and deacetylation of different classes of nonhistone proteins.

2.1. Chromatin Dynamics and Transcription

Nonhistone chromatin proteins are integral structural components of the chromatin. The dynamicity of the chromatin is maintained by chromatin compaction leading to higher order structure formation and decompaction leading to an open chromatin that is a prerequisite for gene expression. There is a vast repertoire of nonhistone structural components viz. HMGs, HP1, MeCP2, Polycomb group of proteins, MENT complex, PARPs (McBryant *et al.*, 2006), which are responsible for chromatin higher order structure formation. However, the acetylation dependent functional regulation is reported in few cases only. The HMGs are a set of nonhistone proteins that have direct influence in chromatin architecture. HMG group of proteins led to a decompaction of chromatin structure working in opposition to linker histone H1 (Catez *et al.*, 2003). Acetylation of HMGs can significantly alter their dynamic interaction with chromatin. Acetylation of HMGN1 by p300 or HMGN2 by p300 or PCAF has distinct functional consequence (Bergel *et al.*, 2000; Herrera *et al.*, 1999). Acetylation of HMGN2 leads to a weak nucleosome binding ability (Herrera *et al.*,1999). Binding of HMGN1/N2 to the nucleosomal core particle also inhibits the acetylation of nucleosomal H3 by PCAF (Lim *et al.*, 2005). However, HMGN (especially HMGN2) has been shown to act as chromatin specific transcriptional coactivator (Paranjape *et al.*, 1995). Interestingly, acetylation of HMGN2 by p300 inhibits this coactivator activity (Banerjee and Kundu unpublished data). Presumably HMGN2, binding to the chromatin assist in the activator binding, which recruits the HAT. The active-HAT (complex) then acetylates the activators, promoter proximal histones, as well as HMGNs. These events may lead to the opening of chromatin for activator dependent transcriptional activation and also the removal of acetylated HMGN.

Enhanceosomes are nucleoprotein complexes, where specific regulatory proteins and activators interact to bring about the activation of specific genes. HMGA proteins are found to be integral components of enhanceosomes. Acetylation of HMGA by p300/CBP leads to enhanceosome disruption and hence transcriptional repression (Bergel *et al.*, 2000). On the other hand acetylation of HMGA by PCAF leads to enhanceosome assembly and transcription activation (Herrera *et al.*, 1999). This differential functional consequence can be attributed to the change in protein-protein and DNA-protein interaction brought about by the HATs. PARP-1 functions both as a structural component of chromatin and a modulator of chromatin structure through its intrinsic enzymatic activity. PARP1 is acetylated by p300/ CBP *in vivo* and acetylation is induced in response to inflammatory responses (Hassa *et al.*, 2005). However PARPs acetylation and its effect on chromatin structure still needs to be understood.

Acetylation is known to regulate the function of several transcription factors by multiple ways, like effecting the DNA binding ability, protein protein interactions, protein half-life, and protein localization (Table 1). Acetylation causes enhanced sequence specific DNA binding for transcription factors like p53, E2F, EKLF, p50 and PC4 (Bode and Dong, 2004; Martínez-Balbás *et al.*, 2000; Marzio *et al.*, 2000; Chen and Bieker, 2004; Chen *et al.*, 2004; Kumar *et al.*, 2001, Table 1), where as it reduces DNA binding of certain factors like Foxo1, HMGI (Y), p65 (Matsuzaki *et al.*, 2005)(Table 1). The ability to activate or repress the DNA binding ability depends on the site of acetylation. If the acetylation sites lie in DNA binding domain it repress the DNA binding and if they are adjacent to DNA binding domain, then it activates DNA binding (see Table 1).

p53 is a sequence specific DNA binding transcription factor known to maintain the cellular homeostasis. p53 function is directly correlated with its sequence specific DNA binding to the promoters of its regulatory genes. p53 C-terminal domain binds to its core DNA binding domain and prevents p53 DNA binding. p53 was the first sequence specific transcription factor known to be acetylated by p300/CBP (Gu and Roeder, 1997). Acetylation at specific lysine residues

Table 1. Biochemical and cellular functions of non-histone protein acetylation

NON HISTONE PROTEIN ACETYLATION		
DNA Binding	Enhances	p53, E2F1, EKLF, p50, PC4, AML1, IRF, STAT3, SRY, ATA1, GATA1, GATA2
	Weakens	Foxo1, HMGI(Y), HMGB1, p65, YY1, MGA1
Protein stability	Stabilizes	p53, E2F, SMAD7, AR, ER- α, C myc
	Destabilizes	HIF1-α
Protein Protein interaction	Enhances	STAT3, AR, EKLF, Importin α
	Weakens	NF-kB, Ku70, HSP90
Protein localization	Nuclear Export	P53, HMGB1
	Nuclear import	STAT3, Importin , POP1
Transcriptional regulation	Activation	p53, E2F, MEf1, BETA2, AML1, CREB, IRF, NF-KB, GATA HMGA1, STAT3, AR, EKLF
	Repression	HMGI(Y), C myc, ER α, HIF1 α
DNA Repair	Activates	p53, E2F, TDG
	Inhibits	NEIL2, FEN1, Pol b

in the C-terminus of p53 consequently activate DNA binding by neutralizing the positive charge and preventing its interaction with core DNA binding domain (Luo *et al.*, 2004) (Fig. 2). While the acetyltransferases enhance p53 DNA binding ability, p53 dependant transactivation, deacetylases repress p53 dependant transcription. Mdm2 a well-known negative regulator of p53 suppresses p300-mediated acetylation by recruiting HDAC1 containing complex and that deacetylate p53 (Ito *et al.*, 2002; Jin *et al.*, 2002, Fig 2). However MDM2 is also a substrate of p300 and the acetylation of Mdm2 inhibits MDM2 mediated p53 degradation (Wang *et al.*, 2004). SirT1, a NAD dependant deacetylase, deacetylates p53 and suppresses p53 dependant function *in vivo* (Vaziri *et al.*, 2001, Fig 2). Inhibition of SirT1 by a small molecular inhibitor increases acetylation at Lysine 382 residue of p53 after different types of DNA damage in primary human mammary epithelial cells (Solomon *et al.*, 2006). Therefore, it is quite evident that reversible acetylation of p53 is a key regulator of the p53 activity as a transcriptional regulator. However, the effect of p53 acetylation on its activation by other proteins (like HMGB1 or PC4) is yet to be elucidated. The tumor suppressor, p53, represses more number of genes than it activates (Kinzler and Vogelstein, 1996). To date no information is available regarding the role of post-translational modifications of p53 on its transcription repression activity.

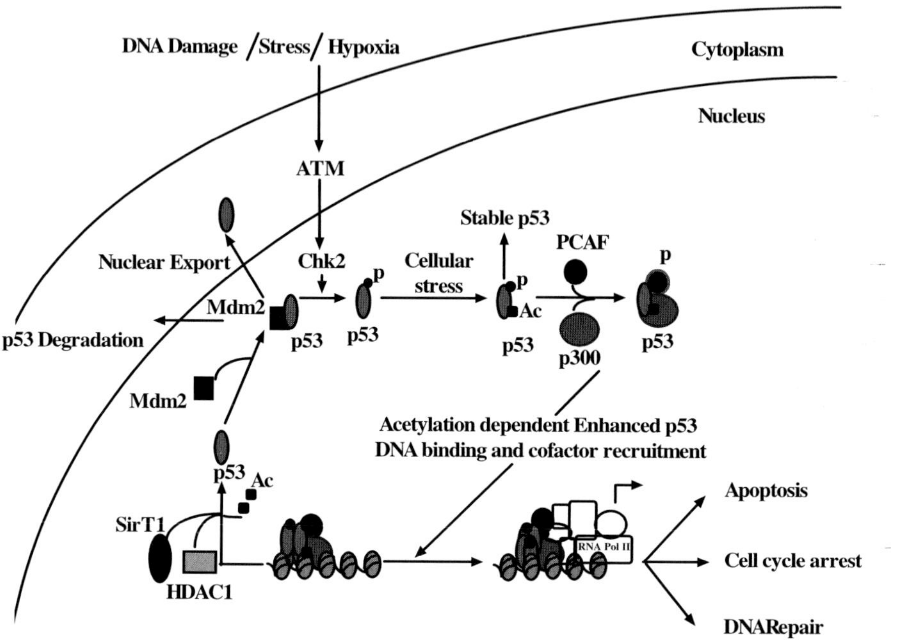

Figure 2. Regulation of p53 function by acetylation / deacetylation: Under stresses conditions p53 gets phosphorylated, acetylated and consequently gets stabilized. Acetylated p53 has enhanced transcriptional ability leading to the activation several p53 responsive genes, which plays important roles in diverse cellular processes. Decateylation of p53 by SirT1 and HDAC1 down regulates p53 activity by enabling interaction with MDM2 followed by nuclear export and p53 degradation. (See Colour Plate 13.)

Another important transcription factor that plays crucial role in cell cycle progression is E2F. p300/ CBP mediated acetylation at the N-terminus of E2F enhances its sequence specific DNA binding ability, which is correlated with the enhanced transcription from E2F responsive promoters. As expected the acetylation dependent effect could be reversed by HDAC1 mediated deacetylation (Martínez-Balbás *et al.*, 2000; Marzio *et al.*, 2000). There are several other non-histone factor acetyl transferase substrates like GATA1, CBP, Runx1 and HMGB1 etc. Prototype transcription factor GATA 1 regulates hematopoiesis and autoregulates its own expression by self-association. GATA1 interacts with p300/CBP and the acetylation of GATA1 increases the transcriptional activity (Nishikawa *et al.*, 2003). Mutations in acetylation sites present in GATA1 causes decreased transcriptional activity. Cyclic AMP response element binding protein is a substrate of CBP and the acetylation of CREB significantly elevates CREB/CBP mediated transcription. (Lu *et al.*, 2001). RUNX1 gene is mutated in many human leukemia. Acetylation of RUNX1 by p300 increases its DNA binding ability and transcriptional activation (Yamaguchi *et al.*, 2004).

Global transcriptional coactivator PC4 is known to get acetylated specifically by p300 but not PCAF and GCN5 *in vitro* and *in vivo* in humans (Kumar *et al.*, 2001, Das and Kundu unpublished data). Acetylation of PC4 at least in the two-lysine residues, enhances its DNA binding ability. In the cell PC4 is present predominantly as a phospho-protein. Phosphorylation of PC4 was found to negatively regulate the acetylation, which is a rare example for any protein, which harbor both the posttranslational modifications (Kumar *et al.*, 2001). Most of the cases, phosphorylation enhance (favor) the acetylation of proteins (Warnock *et al.*, 2005). PC4 directly interact with p53 *in vitro* and *in vivo* and enhance the sequence specific DNA binding of p53 (Banerjee 2004). Furthermore, it also induces the expression of p53 responsive genes and thereby enhances the p53 dependent apoptosis. Since both p53 and PC4 gets acetylated it would be interesting to find out the role of reversible acetylation of these two proteins in the regulation of cellular homeostasis. However, the deacetylation pathway of PC4 is not known yet. Initial data suggest that PC4 gets partially deacetylated by HDAC1 (Swaminathan and Kundu unpublished data).

Histone chaperones are also a group of nonhistone proteins, which directly interact with core or linker histones and actively participate in the histone metabolism pathway (also see chapter on Histone Chaperone in Chromatin Dynamics by Jayasha *et al.*). Emerging evidences suggest that the acetylation of histones as well as chaperones are the key regulator of histone chaperone function. The H2A-H2B histone chaperone, NAP1 though is not an *in vivo* substrate for p300 acetylation, it is a functional component of the p300 coactivator, which indicate, that NAP1 may be a point of connection between chromatin and transcriptional coactivators (Shikama *et al.*, 2000). Interestingly, it was found that indeed the acetylation of histones by p300 facilitates the transfer of H2A-H2B from nucleosomes to NAP1 (Ito *et al.*, 2000). Presumably, this would help in the alteration of chromatin structure for transcriptional activation. At this juncture it was important to know what would happen if any chaperone, which is required for transcriptional activation gets acetylated. Recently we have found that p300 specifically acetylates human histone chaperone, Nucleophosmin (NPM1). Acetylation enhances its histone chaperone function and chromatin transcriptional activity (Swaminathan *et al.*, 2005). NPM1 is a multifunctional nucleolar phosphoprotein. The level of nucleophosmin gets upregulated during various stress like UV rays, DNA damaging drug treatment, various disease conditions like acute myeloid leukemia, and glioblastoma multiforme (for further details see Chapter on Histone Chaperone in Chromatin Dynamics by Jayasha *et al.*).

Acetylation of p53, E2F and many other proteins not only activates their transcriptional activities but also half-life (Li *et al.*, 2002; Martínez-Balbás *et al.*, 2000). Apart from DNA binding, and protein half-life, acetylation also plays pivotal role in protein-protein interactions thereby cellular processes involved. However, the acetylation of pRb regulates the specific interaction with E2F-1 (Markham *et al.*, 2006). E2F activity is negatively regulated by its interaction with pRb, which recruits histone decetylase complexes and probably lead to the deacetylation of E2F.

CtBP (carboxyl-terminal binding protein) participates in regulating cellular development and differentiation by associating with a diverse array of transcriptional repressors. This family of protein plays crucial role in differentiation, apoptosis, oncogenesis and development (Corda *et al.*, 2006). CtBP interacts with p300 bromodomain and inhibits its transcriptional activity. It also interacts with nuclear hormone receptor corepressor RIP140 and the acetylation of RIP140 inhibits its interaction with CtBP and thereby acts negatively for transcriptional repression (Vo *et al.*, 2001). Acetylation of adenoviral protein E1A inhibits its interaction with CtBP and leading to alleviation of transcriptional repression mediated by CtBP. (Zhang *et al.*, 2000).

In general, retroviral gene expression is regulated by chromatin dynamics of the host genome. For example upon integration, HIV genome forms nucleosomal structure and depends upon the host chromatin modifications for the viral gene expression (for details see chapter on Chromatin modifying enzymes and HIV gene expression by Quivy *et al.*,). Furthermore, the HIV1 Tat protein, crucial for viral replication and transcription gets acetylated by both p300/CBP and PCAF. The p300-mediated acetylation of Tat, enhances transcriptional coactivation and also enhances its binding to core histones (Ott *et al.*, 1999; Deng *et al.*, 2000). Site-specific acetylation of Tat differentially regulates its interacting partners like PCAF and cyclin T1 (Tagami *et al.*, 2002). The multifunctional transcriptional activator NF-kB control the expression of several genes related to stress induced, immune, inflammatory responses and the HIV gene expression. The activity of NF-kB is negatively regulated by its interaction with IkB. Acetylation of NF-kB by p300/CBP inhibits its interaction with IkB and induces translocation of the factor to nucleus (Chen *et al.*, 2004). Acetylation of NF-kB is regulated by the prior phosphorylation. The phospho-acetylated forms of NF-kB display enhanced transcriptional activity. Histone decateylase 3 (HDAC3) deacetylates NF-kB enabling it to bind IKB and causing its translocation in to cytoplasm (Chen *et al.*, 2004).

Acetylation at lysine residues present in nuclear localization signal of proteins helps in nuclear retention. For example, acetylation regulates p53 subcellular localization by promoting p53 nuclear export (Kawaguchi *et al.*, 2006, Fig 2). The other important instance is in the case of multifunctional non-histone chromatin protein HMGB1, where p300 mediated acetylation causes nuclear export and cytosolic accumulation. POP1 is a transcription factor mediates Wnt signaling pathway and plays crucial role in embryogenesis. Acetylation of POP1 inhibits nuclear export and increases nuclear import (Gay *et al.*, 2003). HELA E box binding protein (HEB) has an important role in thymopoiesis and acetylation by HATs increases its nuclear retention and enhances its transcriptional activities. The orphan nuclear receptor SF-1 regulates the development and differentiation of steroidogenic tissues. Acetylation of SF1 by GCN5 regulates its transcriptional activity and stabilizes the protein. Inhibition of deacetylation using TSA increases SF1 mediated transcriptional activation and nuclear export of SF1 protein (Jacob *et al.*, 2001). These examples clearly establish the fact that acetylation of nonhistone proteins is crucial for cell signaling.

DNA viruses can regulate the activity of cellular acetyltransferases and to effect cell cycle progression in support of virus replication. Viral infection mediated assembly of enhanceosome at IFN-B promoter requires HMGI (Y). CBP, PCAF are recruited to this complex and CBP mediated acetylation of HMG I (Y) decreases its DNA binding ability and causes enhanceosome disruption (Munshi *et al.*, 2001). Autoacetylation of transcription factors is also one of the mechanisms in controlling the gene expression. Autoacetylation of acetyl transferases also regulates its activity. PCAF gets acetylated by itself and by p300 and the acetylation of PCAF enhances its acetyl transferase activity (Herrera *et al.*, 1997). Autoacetylation of general transcription factor TFIIB strengthens its interaction with TFIIF (Choi *et al.*, 2004) and thereby transcription suggesting the role of transcription factor auto acetylation in the regulation of transcription.

2.2. Cellular Processes

Cell cycle progression, apoptosis, DNA damage and DNA repair are cellular functions that are regulated by several mechanisms. One such important regulatory mechanism is posttranslational modification of histone and non-histone proteins. Myriad of reports have been shown that acetylation of non-histone proteins apart from histones, contributes in major to these processes.

2.2.1. Apoptosis

Apoptosis is physiological process of cell killing regulated by several families of proteins (divided as proapoptotic and antiapoptotic proteins) and their post translational modifications. Acetylation of p53 is very important for p53-mediated apoptosis. Acetylation of p53 at distinct sites regulates different cellular activities performed by p53. Acetylation at 373 position in p53 by p300/CBP leads to apoptosis where as acetylation at 320 residue by PCAF leads to cell cycle arrest (Knights *et al.*, 2006). Under stressed conditions acetylation levels of p53 increases, which results in active form of p53. Acetylation of p53 is also controlled by deacetylases such as HDAC1, HDCA3, and hSIRT1. hSIRT1, HDAC2 interact with p53. Inhibitors of hSIRT1 and HDAC2 enhance p53 acetylation and thereby p53 mediated apoptosis (Huang *et al.*, 2005; Olaharski *et al.*, 2005). p53 homologue, p73 also gets acetylated and activated in response to DNA damage and potentiates the p73 mediated apoptosis (Costanzo *et al.*, 2002). The DNA end joining protein Ku70 prevents apoptosis by sequestering a proapoptotic protein Bax from mitochondria (Cohen *et al.*, 2004). However, the acetylation of Ku70 disrupts its interaction with Bax and elevates Bax mediated apoptosis. On the other hand, Stat1 has been known to repress NF-kB mediated cell signaling. Acetylated Stat1 interacts with NF-kB thereby preventing its DNA binding ability, nuclear localization and finally expression of anti apoptotic genes (Kramer *et al.*, 2006). These examples clearly indicate that acetylation of nonhistone proteins may induce or inhibit the apoptosis depending upon the protein and physiological status. However, detailed information from this area of research, would help to design therapeutic molecules that would activate apoptosis in the malignant cells.

2.2.2. *Cell cycle progression*

Crucial stages of cell cycle are generally controlled through transcriptional regulation of a subset of genes, which in turn regulated by acetylation/deacetylation of histone and non histone proteins. One of the notable examples is the regulation of C-myc gene expression and cell cycle progression. C-myc regulates the expression of several genes involved in growth promotion by associating with its DNA binding partner max. p300 associates with C-myc and helps in C-myc stabilization independent of p300 mediated acetylation where as C-myc acetylation increases its turnover (Faiola *et al.*, 2005). Another cell cycle regulatory protein CyclinD1 plays key regulatory role during G1 phase and is over expressed in many cancers. Cyclin D1 interacts with PCAF and facilitates the association of ER and PCAF. Over expression of PCAF results in cyclinD1 dependent regulation of ER activity (McMahon *et al.*, 1999). Interestingly, Cyclin D1 expression is down regulated by HDAC1 complex recruited to its promoter by SMAR1, a matrix attachment region binding proteins (Rampalli *et al.*, 2005).

Nuclear receptor function is controlled by its acetylation similar to phosphorylation. The androgen receptor (AR) is a nuclear hormone receptor superfamily member that conveys both *trans* repression and ligand-dependent *trans*-activation function. Activation of the AR leads to diverse cellular processes like secondary sexual differentiation in males and the induction of apoptosis by the JNK kinase, MEKK1. p300 acetylates androgen receptor and acetylation governs ligand sensitivity, cofactor recruitment and growth properties of receptors. Point mutations of the AR acetylation motif that abrogate acetylation reduce trans-activation by p300 without affecting the trans-repression function of the AR. The AR acetylation mutant was also defective in MEKK1-induced apoptosis, suggesting that the conserved AR acetylation site contributes to a pathway governing prostate cancer cellular survival (Fu *et al.*, 2002). SirT1 physically interacts with AR and inhibits p300-mediated transactivation though its NAD dependent deacetylase activity. Estrogen receptor alfa regulates the ligand dependent and ligand independent transcription. p300 acetylates ER alfa and mutations in the acetylation sites dramatically enhanced the hormonal sensitivity and has no effect on MAPK signaling pathway suggesting acetylation may suppresses ligand sensitivity.

2.2.3. *Oxidative stress*

Oxidative Stress is a general term used to describe the steady state level of oxidative damage in a cell, tissue, or organ, caused by the reactive oxygen species (ROS) which are free radicals, reactive anions containing oxygen atoms. Oxidative stress and ROS have been implicated in disease states, such as Alzheimer's disease, Parkinson's disease, cancer, and aging. In addition to modulation of signaling processes and oxidation of cellular proteins and lipids, reactive oxygen species (ROS) induce multiple damages in both nuclear and mitochondrial genomes, most of which are repaired via the DNA base excision repair pathway. 8-Oxoguanine (8-oxoG), a major ROS product in the genome, is excised by 8-oxoG-DNA glycosylase (OGG1) and the resulting abasic (AP) site is cleaved by AP-endonuclease

(APE1) in the initial steps of repair (Szczesny *et al.*, 2004). OGG1 is acetylated by p300 *in vivo* predominantly at Lys338/Lys341. Acetylation significantly increases OGG1's activity in the presence of AP-endonuclease by reducing its affinity for the abasic (AP) site product. The enhanced rate of repair of 8-oxoG in the genome by wild-type OGG1 but not the unacetylable K338R/K341R mutant. oxidative stress increases the acetylation of OGG1 by about 2.5-fold after with no change at the polypeptide level. OGG1 interacts with class I histone deacetylases, which is responsible for its deacetylation (Bhakat *et al.*, 2004). This indicates towards a novel regulatory function of OGG1 acetylation in repair of its substrates in oxidatively stressed cells. Forkhead transcription factor, DAF-16, regulates genes that contribute both to longevity and resistance to various stresses in *C. elegans*. Members of the FOXO, mammalian homologs of DAF-16, also regulate genes related to stress resistance, such as GADD45. Acetylation of FOXO4, by the transcriptional coactivator p300, counteracted transcriptional activation of FOXO4 by p300 (Kobayashi *et al.*, 2005). In contrast, mammalian SIRT1 (a class III histone deacetylase) was found to bind to FOXO4, catalyze its deacetylation in an NAD-dependent manner, and thereby increase its transactivation activity (Horst *et al.*, 2004). In response to oxidative stress, FOXO accumulates within the nucleus and induces GADD45 expression. FOXO-mediated GADD45 induction is markedly impaired in the cell, which depleted SIRT1 expression by RNA-interference (Kobayashi *et al.*, 2005). These results indicate that mammalian SIRT1 plays a pivotal role for FOXO function via NAD-dependent deacetylation in response to oxidative stress, and thereby may contribute to cellular stress resistance and longevity.

Heat shock protein, Hsp90 is a well known stress-induced protein with a more general housekeeping function. The emerging significance of Hsp90 in both normal and oncogenic signaling highlights the need to understand how Hsp90 is regulated. It is known that Hsp90 becomes transiently acetylated upon GR activation after ligand stimulation, which might allow the conversion of GR-Hsp90 from a stable complex into a dynamic one thereby enabling GR to enter the nucleus for transcriptional activation. Subsequent deacetylation by HDAC6 would then allow Hsp90 to re-enter the productive chaperone complex (Murphy *et al.*, 2005). Acetylation of molecular chaperones might play a role in the dynamic reorganization of chaperone complexes in response to "oncogenic stress" which might be induced by high demand for "growth and proliferation"-associated signaling (Kovacs *et al.*, 2005). Under this scenario, HDAC6 may then be required for robust Hsp90 activity by regenerating deacetylated Hsp90 and thereby ensuring effective cell signaling.

2.2.4. *DNA damage and repair*

Acetylation levels of histone and non-histone proteins are regulated upon DNA damage. Retinoblastoma protein pRb controls G1-S phase transition and phosphorylation of pRb regulates it function. DNA damage induced acetylation of pRb near the phosphorylation sites prevents it phosphorylation and keep pRb in active form thereby leading to growth repression. DNA damage dependant association of E2F with p300 not only accumulates Ac-E2F but also subsequent ubiquitinated E2F. However p300 induced ubiquitination is not dependant on pRb (Galbiati *et al.*, 2005). In another case, DNA damage induced phosphorylation of p53 at ser 33 and 37 increases p53 affinity for p300 and PCAF thus promoting acetylation of carboxy-terminal sites including Lys-382 (by p300) and Lys-320 (by PCAF)(Sakaguchi *et al.*, 1998). Acetylation at both these sites causes conformational change and enhances sequence specific DNA binding ability of p53. p53 homologue p73 also gets accumulated during DNA damage and half-life of p73 is partially dependant on C-ABL kinase (Gong *et al.*, 1999). C-ABL is required for p300 mediated acetylation of p73 upon genotoxic insult and enhances p73 dependant apoptosis. Like p53 both phosphorylation and acetylation of p73 contributes to its function under DNA damaging condition (Costanzo *et al.*, 2002). The ATM protein kinase is a critical intermediate in a number of cellular responses to ionizing irradiation (IR) and possibly other stresses. DNA damage induces rapid acetylation of ATM by TIP60. Inhibition of Tip60 blocks the ATM dependant phosphorylation of p53 and Chk2 further sensitizes the cells to ionizing radiation (Sun *et al.*, 2005).

Werner's syndrome is a result of mutation in WRN gene, which encodes DNA helicase. The translocation of WRN from nucleolus into nucleoplasmic foci is significantly enhanced by its acetylation by p300 under DNA damaging condition (Blander *et al.*, 2002). DNA pol β is key protein involved in Base excision repair. p300 acetylates polβ and severely impairs its activity implying the role of acetylation in Base Excision Repair (Hasan *et al.*, 2002). Another DNA repair protein DNA glycosylase NEIL2 gets acetylated by p300 and the acetylation inhibits its repair activity (Bhakat *et al.*, 2004). Thymine DNA Glycolsylase (TDG) initiates repair of G/T and G/U mismatches by removing thymine and uracil residues. TDG associates with CBP/p300 and results in transcriptional activation by CBP. p300/CBP mediated acetylation of TDG leads to disruption of CBP from DNA ternary complex and also regulates recruitment of repair endonuclease APE (Lucey *et al.*, 2005). Above observation strongly argues for the potential regulatory role for protein acetylation in base mismatch repair and a role for CBP/p300 in maintaining genomic stability.

3. NON HISTONE PROTEIN ACETYLATION LINK TO DISEASES

Aberrant acetylation levels of non histone proteins not only lead to the dysfunction of these proteins but also cause disturbances in cellular processes thereby leading to several diseases (Fig. 3). There are several reports of aberrant acetylation in diseases like cancer, diabetes, cardiac hypertrophy and viral diseases.

Diabetes is hypothesized to cause cardiac protein acetylation and the acetylation alters the protein function (Fig. 3b). Hypoxia-inducible factor-1 (HIF1) is a transcription factor found in mammalian cells cultured under reduced oxygen tension that plays an essential role in cellular and systemic homeostatic responses to hypoxia. Diabetes interferes with cellular response to hypoxia. In hyperglycemic conditions HIF1 degradation is increased because of enhanced HIF1 acetylation by a factor acetyltransferase ARD1 that makes it susceptible to proteosomal degradation (Jeong *et al.*, 2002). Insulin receptor substrate proteins (IRS) are the key regulators of insulin action. IRS1 is acetylated and the acetylation enhances its phosphorylation, which mediates the down stream insulin dependent signaling. IRS1 is associated with HDAC2 and HDAC2 mediated deacetylation have the reverse effect. HDAC2 specific small molecular inhibitors may elevate insulin sensitivity in otherwise insulin resistant conditions (Kaiser and James, 2004). Hyperglycemia also activates p53 and p53-regulated genes leading to myocyte cell death (Fiordaliso *et al.*, 2001). p53 acetylation, site specific phosphorylation levels goes up in response to stress induced conditions like hypoxia, anti metabolites, nuclear export inhibitors thereby maintaining cellular homeostasis. p300/CBP mediated acetylation is negatively regulated by MDM2. MDM2 recruits HDAC1 and decatylates p53 making it susceptible for proteosomal degradation. Inhibitors of p53 deacetylases or MDM2, could further stabilize p53 and may act as anticancer drugs. Phosphorylation and acetylation on STAT3 are important events for STAT3 to stimulate cancer and metastasis. Small molecule inhibitors for both the events should be effective for cancer treatment. Histone acetyl transferase p300 is mutated in several cancers because the acetylation that is necessary for cell cycle regulation by this protein is disrupted, so cells proliferate incessantly and leads to cancer. There are several number of evidences suggest that p300 itself is a tumor suppressor (Imhof *et al.*, 1997).

Nuclear protein acetylation increases dramatically immediately after UV irradiation suggesting causal relationship exists between nuclear protein acetylation and nucleotide excision repair of DNA in human cells (Ramanathan and Smerdon, 1986). Over expression of oncoprotein PU.1 inhibits erythroid cell differentiation and contributes to the formation of murine erythroleukemias (MEL) by inhibiting CBP mediated acetylation of several nuclear proteins (Hong *et al.*, 2002) (Fig. 3b). Microtubules are formed from a protein called tubulin and have to be broken down for the cell to divide. Acetylation of tubulin leads to stabilizes microtubules (Maruta *et al.*, 1986) and therefore the inhibitors of tubulin deacetylases are important for anti cancer therapy (Fig. 3b). β-catenin is key component in Wnt signaling pathway, which is implicated in many human cancers (Clevers, 2004). CBP mediated acetylation of β-catenin differentially regulates its function in a promoter dependent manner. The site of acetylation found in β catenin is often found to be mutated in thyroid cancers (Ma *et al.*, 2005) (Fig. 3b). Small molecule inhibitors specifically inhibit β-catenin TCF signaling pathway by preventing CBP interaction with β-catenin and thereby acting as novel anticancer compounds.

Imbalances in acetylation/deacetylation is the primary cause for cellular pathogenesis in polyglutamine diseases such as Huntington disease (Sadri-Vakili *et al.*, 2006). Inhibitors of HDAC proteins could possible act as therapeutic reagents for the treatment of these devastating diseases. Steroid hormone receptors are essential cellular homeostasis. Over-expression or misregulation of receptors and their signaling pathways leads several malignancies. HSP90 dependent maturation

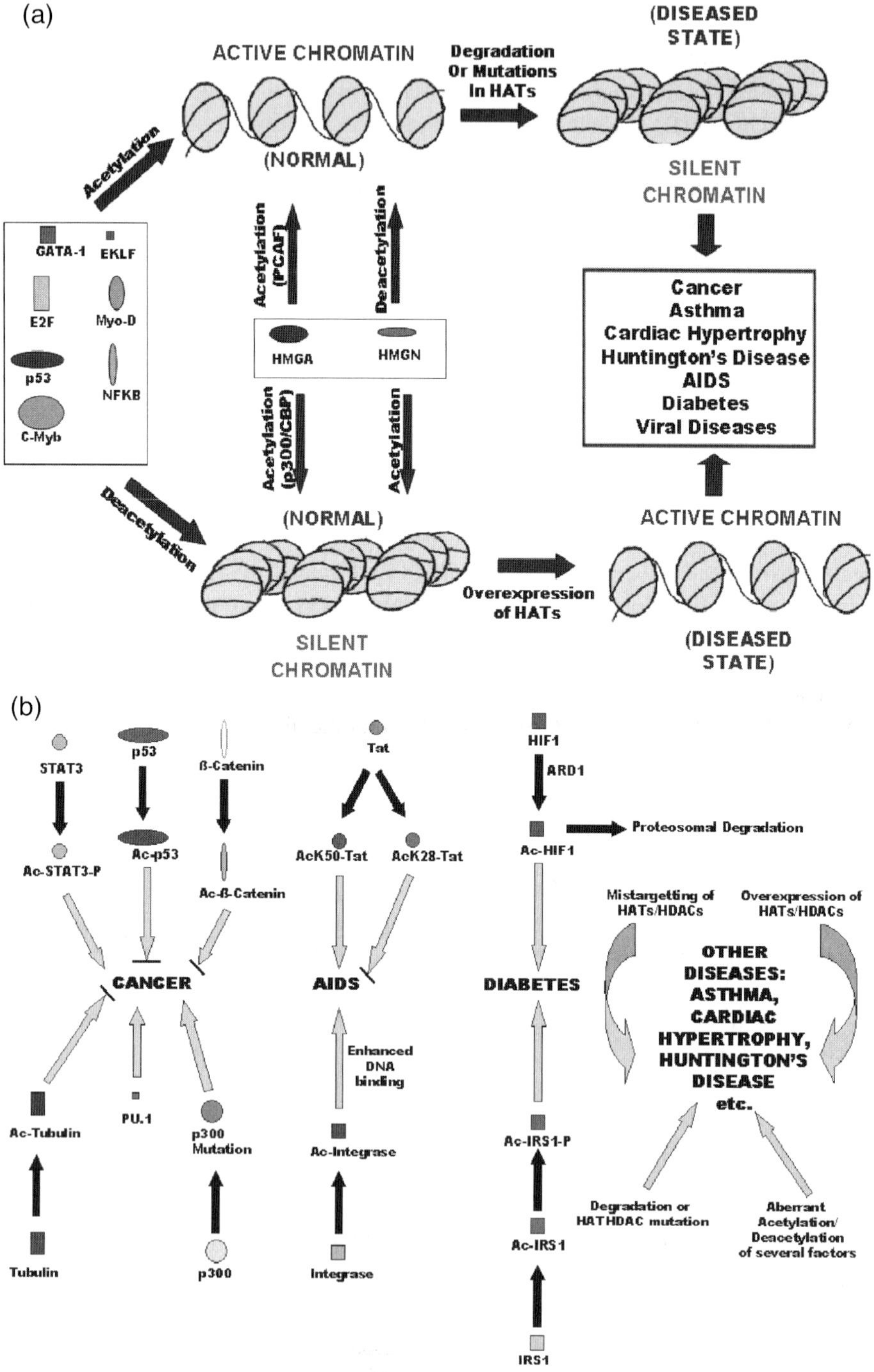
(a)
ACTIVE CHROMATIN
(NORMAL)
Degradation Or Mutations In HATs
(DISEASED STATE)
SILENT CHROMATIN
Acetylation
GATA-1
EKLF
E2F
Myo-D
p53
NFKB
C-Myb
Acetylation (PCAF)
Deacetylation
HMGA
HMGN
Acetylation (p300/CBP)
Acetylation
Cancer
Asthma
Cardiac Hypertrophy
Huntington's Disease
AIDS
Diabetes
Viral Diseases
Deacetylation
(NORMAL)
SILENT CHROMATIN
Overexpression of HATs
ACTIVE CHROMATIN
(DISEASED STATE)
(b)
STAT3
p53
ß-Catenin
Tat
HIF1
ARD1
Ac-STAT3-P
Ac-p53
Ac-ß-Catenin
AcK50-Tat
AcK28-Tat
Ac-HIF1
Proteosomal Degradation
Mistargetting of HATs/HDACs
Overexpression of HATs/HDACs
CANCER
AIDS
DIABETES
OTHER DISEASES: ASTHMA, CARDIAC HYPERTROPHY, HUNTINGTON'S DISEASE etc.
Enhanced DNA binding
Ac-Tubulin
PU.1
p300 Mutation
Ac-Integrase
Ac-IRS1-P
Degradation or HATHDAC mutation
Aberrant Acetylation/ Deacetylation of several factors
Tubulin
p300
Integrase
Ac-IRS1
IRS1

is critical for receptor ligand binding and activity. HDAC6 deacetylates HSP90 and inhibition of HDAC6 leads to accumulation of acetylated HSP90, which fails to form stable complex with receptors. Modulators of HSP90 acetylation can be used as effective therapeutic agents to abrogate aberrant steroid signaling pathways (Kovacs *et al.*, 2005).

HIV genome is composed of RNA, which is reverse transcribed to DNA by viral reverse transcriptase immediately after infecting a new host cell. The early phase of HIV infection is characterized by long latency period during which the viral DNA integrates into the host genome and do not express any virus specific proteins. This Integration is catalyzed by the viral protein integrase (IN) and it preferentially occurs near transcriptionally active genes. The catalytic activity of IN is regulated by p300 mediated acetylation. Direct interaction with p300 results in acetylation of three specific lysines (K264, K266, K273) in the carboxy-terminus DNA binding region of IN. Acetylation increases IN affinity to DNA, and promotes the DNA strand transfer activity of the protein. In the context of the viral replication cycle, point mutations in the IN acetylation sites abolish virus replication by specifically impairing its integration capacity (Cereseto *et al.*, 2005) (Fig. 3b). This is the first demonstration that HIV-1 IN activity is specifically regulated by post-translational modification. The late phase of viral life cycle begins with early synthesis of highly spliced mRNAs, HIV-1 transcription is predominantly dependent on viral transactivator Tat. p300 acetylates Tat at lysine 50 and 28. Acetylation of Tat at lysine 50 prevents its interaction with CyclinT1 and recruits PCAF to elongating polymerase. However acetylation at lysine 28 abrogates Tat and PCAF interaction. Moreover, Tat changes the conformation of CBP/p300 such that the altered p300/CBP can bind to other transcription factors with higher affinity, implying that Tat influences the transcription machinery by aiding CBP/p300 to acquire new partners and increase its functional repertoire (Deng *et al.*, 2001) (Fig. 3b). Other transcription factors which serves as a substrate for Tat and p300 HAT complex, is human TBP (Deng *et al.*, 2000). Acetylation of TBP increases its DNA binding to HIV-1 TATA box. The major obstacle in treating HIV is the virus developing drug resistance. Currently prevalent mode of treatment HAART (highly Active Antiretroviral therapy) targets multiple viral proteins at a time like inhibitors of viral Reverse transcriptase, protease and Integrase. Acetylation status of HIV Tat and Integrase is also crucial for viral maintenance and expression inside the host. Small molecule inhibitors, which could

---

Figure 3. Role of nonhistone protein acetylation in maintaining cellular homeostasis- mis-regulation and disease connection: (a) Acetylation of nonhistone proteins are associated with active or repressed chromatin architecture as guided by suitable cellular signals for maintenance of gene expression. Misregulation of HAT function leads to diseased state, where chromatin architecture is altered than under normal condition. In a parallel way the posttranslational modification status of these proteins may act as versatile tool to diagnose the various stages of disease manifestation e.g. probable involvement of acetylated NPM1 modulating its stress response function can lead us to use it as a marker for various disease states. (b) Acetylation of nonhistone proteins in connection to diseases like Cancer, AIDS, Diabetes and others. (See Colour Plate 14.)

specifically inhibit p300, mediated acetylation of Tat or IN would be useful in interrupting the progression of viral life cycle in host cell (Varier R and Kundu T K., 2006). SirT1 enhances HIV transcription by deacetylating Tat in a cyclinT1 dependent manner (Pagans *et al.*, 2005). So inhibitors of SirT1 could also be used for its efficacy against HIV infection.

4. FUTURE DIRECTIONS

Among various post translational modifications, reversible acetylation of non histone proteins play a very significant role in modulating their functions viz., transcriptional activation or repression, regulation of the cell cycle in spatial and temporal manner and apoptosis. Understanding of the molecular mechanisms of this reversible acetylation will help to design therapeutics in combating the diseases emerging from aberrant acetylation status. The problems arising from the aberrant acetylation can be countered either through targeting directly the non-histone proteins, or enzymes involved in the reversible acetylation. The most challenging task would be the target specificity of these molecules. Nanoparticle mediated drug delivery and substrate analogs are the promising directions we would look forward.

ACKNOWLEDGEMENT

Research work in author's laboratory was supported by Jawaharlal Nehru Centre for Advanced Scientific Research, Department of Science and Technology, and Council for Scientific and Industrial Research (CSIR), Government of India. KB is a CSIR senior research fellow and SG, JS are CSIR junior research fellows.

REFERENCES

Banerjee S, Kumar BRP, Kundu TK (2004) General transcriptional coactivator PC4 activates p53 function.Mol Cell Biol 24(5): 2052–2062

Bedford MT, Richard S (2005) Arginine methylation an emerging regulator of protein function. Mol Cell 18(3): 263–272

Bergel M, Herrera JE, Thatcher BJ, Prymakowska-Bosak M, Vassilev A, Nakatani Y, Martin B, Bustin M (2000) Acetylation of novel sites in the nucleosomal binding domain of chromosomal protein HMG-14 by p300 alters its interaction with nucleosomes. J Biol Chem 275(15): 11514–11520

Bhakat KK, Hazra TK, Mitra S (2004) Acetylation of the human DNA glycosylase NEIL2 and inhibition of its activity. Nucleic Acids Research 32(10): 3033–3039

Blander G, Zalle N, Daniely Y, Taplick J, Gray MD (2002). DNA Damage-induced Translocation of the Werner helicase is regulated by acetylation. J Biol Chem 277(52): 50934–50940

Bode AM, Dong Z (2004) Post-translational modification of p53 in tumorigenesis. Nat Rev Cancer 4(10): 793–805

Catez F, Lim JH, Hock R, Postnikov YV, Bustin M. (2003) HMGN dynamics and chromatin function. Biochem Cell Biol 81: 113–122

Cereseto A, Manganaro L, Gutierrez MI, Terreni M, Fittipaldi A, Lusic M, Marcello A, Giacca M (2005) Acetylation of HIV-1 integrase by p300 regulates viral integration. EMBOJ 24(17): 3070–3081

Chen LF, Greene WC (2004) Shaping the nuclear action of NF-kappaB. Nat Rev Mol Cell Biol 5(5): 392–401

Chen X, Bieker JJ (2004) Stage-specific repression by the EKLF transcriptional activator. Mol Cell Biol 24(23): 10416–10424

Choi CH, Burton ZF, Usheva A (2004) Auto-acetylation of transcription factors as a control mechanism in gene expression. Cell Cycle 3(2): 114–115

Clevers H (2004) Wnt breakers in colon cancer. Cancer Cell 5(1): 5–6

Cohen HY, Lavu S, Bitterman KJ, Hekking B, Imahiyerobo TA, Miller C, Frye R, Ploegh H, Kessler BM, Sinclair DA (2004) Acetylation of the C terminus of Ku70 by CBP and PCAF controls Bax-mediated apoptosis. Mol Cell 13(5): 627–638

Corda D, Colanzi A, Luini A (2006) The multiple activities of CtBP/BARS proteins: the Golgi view. Trends Cell Biol 16(3): 167–173

Costanzo A, Merlo P, Pediconi N, Fulco M, Sartorelli V, Cole PA, Fontemaggi G, Fanciulli M, Schiltz L, Blandino G, Balsano C, Levrero M (2002) DNA damage-dependent acetylation of p73 dictates the selective activation of apoptotic target genes. Mol Cell 9(1): 75–86

Davie JR, Hendzel MJ (1994) Multiple functions of dynamic histone acetylation. J Cell Biochem 55(1): 98–105

Deng L, de la Fuente C, Fu P, Wang L, Donnelly R, Wade JD, Lambert P, Li H, Lee CG, Kashanchi F (2000) Acetylation of HIV-1 Tat by CBP/P300 increases transcription of integrated HIV-1 genome and enhances binding to core histones. Virology 277(2): 278–295

Deng L, Wang D, de la Fuente C, Wang L, Li H, Lee CG, Donnelly R, Wade JD, Lambert P, Kashanchi F (2001) Enhancement of the p300 HAT activity by HIV-1 Tat on chromatin DNA. Virology 289(2): 312–326

Faiola F, Liu X, Lo S, Pan S, Zhang K, Lymar E, Farina A, Martinez E. (2005) Dual regulation of c-Myc by p300 via acetylation-dependent control of Myc protein turnover and coactivation of Myc-induced transcription. Mol Cell Biol 25(23): 10220–10234

Fiordaliso F, Leri A, Cesselli D, Limana F, Safai B, Nadal-Ginard B, Anversa P, Kajstura J (2001) Hyperglycemia activates p53 and p53-regulated genes leading to myocyte cell death. Diabetes 50(10): 2363–2375

Fu M, Wang C, Wang J, Zhang X, Sakamaki T, Yeung YG, Chang C, Hopp T, Fuqua SA, Jaffray E, Hay RT, Palvimo JJ, Janne OA, Pestell RG (2002) Androgen receptor acetylation governs trans activation and MEKK1-induced apoptosis without affecting *in vitro* sumoylation and trans-repression function. Mol Cell Biol 22(10): 3373–3388

Galbiati L, Mendoza-Maldonado R, Gutierrez MI, Giacca M (2005) Regulation of E2F-1 after DNA damage by p300-mediated acetylation and ubiquitination. Cell Cycle 4(7): 930–939

Gay F, Calvo D, LO MC, Ceron J, Maduro M, Lin R, Shi Y (2003) Acetylation regulates subcellular localization of the Wnt signaling nuclear effector POP-1. Genes Dev 17(6): 717–722

Gong JG, Costanzo A, Yang HQ, Melino G, Kaelin WG Jr, Levrero M, Wang JY (1999) The tyrosine kinase c-Abl regulates p73 in apoptotic response to cisplatin-induced DNA damage. Nature 399(6738): 734–735, 737

Gu W, Roeder RG (1997) Activation of p53 sequence-specific DNA binding by acetylation of the p53 C-terminal domain. Cell 90(4): 595–606

Hasan S, El-Andaloussi N, Hardeland U, Hassa PO, Burki C, Imhof R, Schar P, Hottiger MO (2002) Acetylation regulates the DNA end-trimming activity of DNA polymerase beta. Mol Cell 10(5): 1213–1222

Hassa PO, Haenni SS, Buerki C, Meier NI, Lane WS, Owen H, Gersbach M, Imhof R, Hottiger MO (2005) Acetylation of poly(ADP-ribose) polymerase-1 by p300/CREB-binding protein regulates coactivation of NF-kappaB-dependent transcription. J Biol Chem 280(49): 40450–40464

Herrera JE, Bergel M, Yang XJ Nakatani Y, Bustin M (1997) The histone acetyltransferase activity of human GCN5 and PCAF is stabilized by coenzymes. J Biol Chem 272(43): 27253–27258

Herrera JE, Sakaguchi K, Bergel M, Trieschmann L, Nakatani Y, Bustin M(1999) Specific acetylation of chromosomal protein HMG-17 by PCAF alters its interaction with nucleosomes. Mol Cell Biol 19(5): 3466–3473

Hong W, Kim AY, Ky S, Rakowski C, Seo SB, Chakravarti D, Atchison M, Blobel GA (2002) Inhibition of CBP-mediated protein acetylation by the Ets family oncoprotein PU.1. Mol Cell Biol 22(11): 3729–3743

Huang BH, Laban M, Leung CH, Lee L, Lee CK Salto-Tellez M, Raju GC, Hooi SC (2005) Inhibition of histone deacetylase increases apoptosis and p21(Cip1/WAF1) expression, independent of histone deacetylase. Cell death differ 12(4): 395–404
Imhof A, Yang XJ, Ogryzko VV, Nakatani Y, Wolffe AP, Ge H (1997) Acetylation of general transcription factors by histone acetyltransferases. Curr Biol 7(9): 689–692
Ito A, Kawaguchi Y, Lai CH, Kovacs JJ, Higashimoto Y, Appella E, Yao TP (2002) MDM2-HDAC1-mediated deacetylation of p53 is required for its degradation. EMBO 21(22): 6236–6245
Ito T, Ikehara T, Nakagawa T, Kraus WL, Muramatsu M (2000) p300-mediated acetylation facilitates the transfer of histone H2A-H2B dimers from nucleosomes to a histone chaperone. Genes Dev.14(15): 1899–1907
Jackson V, Shires A, Tanphaichitr N, Chalkley R (1976) Modifications to histones immediately after synthesis. J Mol Biol 104(2): 471–483
Jacob AL, Lund J, Martinez P, Hedin L (2001) Acetylation of steroidogenic factor 1 protein regulates its transcriptional activity and recruits the coactivator GCN5. J Biol Chem 276(40): 37659–37664
Jeong JW, Bae MK, Ahn MY, Kim SH, Sohn TK, Bae MH, Yoo MA, Song EJ, Lee KJ, Kim KW (2002) Regulation and destabilization of HIF-1alpha by ARD1-mediated acetylation. Cell 111(5): 709–720
Jin Y, Zeng SX, Dai MS, Yang XJ, Lu H (2002) MDM2 Inhibits PCAF (p300/CREB-binding Protein-associated Factor)-mediated p53 Acetylation. J Biol Chem 277(34): 30838–30843
Kaiser C, James SR (2004) Acetylation of insulin receptor substrate-1 is permissive for tyrosine phosphorylation. BMC Biol. 2:23
Kawaguchi Y, Ito A, Appella E, Yao TP (2006) Charge modification at multiple C-terminal lysine residues regulates p53 oligomerization and its nucleus-cytoplasm trafficking 281(3): 1394–1400
Kinzler KW, Vogelstein B (1996) Life (and death) in a malignant tumour. Nature 379(6560): 19–20
Keizo Nishikawa, Makoto Kobayashi, Atsuko Masumi, Susan E Lyons, Brant M Weinstein, Paul Liu P, Masayuki Yamamoto (2003) Self-association of gata1 enhances transcriptional activity *in vivo* in Zebra Fish embryos. Mol Cell Biol 23: 8295–8305
Knights CD, Catania J, Giovanni SD, Muratoglu S, Perez R, Swartzbeck A, Quong AA, Zhang X, Beerman T, Pestell RG, Avantaggiati ML (2006) Distinct p53 acetylation cassettes differentially influence gene-expression patterns and cell fate. J Cell Biol 173(4):533–544
Kobayashi Y, Furukawa-Hibi Y, Chen C, Horio Y, Isobe K, Ikeda K, Motoyama N (2005) SIRT1 is critical regulator of FOXO-mediated transcription in response to oxidative stress. Int J Mol Med 16(2): 237–243
Kovacs JJ Cohen TJ Yao TP (2005) Chaperoning steroid hormone signaling via reversible acetylation. Nucl Recept Signal. 3: e004
Kramer OH, Baus D, Knauer SK, Stein S, Jager E, Stauber RH, Grez M, Pfitzner E, Heinzel T (2006) Acetylation of Stat1 modulates NF-kappaB activity. Genes Dev 20(4): 473–485
Kumar BR, Swaminathan V, Banerjee S, Kundu TK (2001) p300-mediated acetylation of human transcriptional coactivator PC4 is inhibited by phosphorylation. J Biol Chem 276(20): 16804–16809
Lim JH, West KL, Rubinstein Y, Bergel M, Postnikov YV, Bustin M (2005) Chromosomal protein HMGN1 enhances the acetylation of lysine 14 in histone H3. EMBO J 24(17): 3038–3048
Li M, Luo J, Brooks CL, Gu W (2002) Acetylation of p53 inhibits its ubiquitination by Mdm2. J Biol Chem 277(52) 50607–50611
Lu Q, Hutchins AE, Doyle CM, Lundblad JR, Kwok RP (2003) Acetylation of CAMP-responsive element-binding protein (CREB) by CREB-binding protein enhances (REB-dependent Transcription. J Biol Chem. 278(18): 15727–15734
Lucey MJ, Chen D, Garcia JL, Hart SM, Phoenix F, Jehani RA, Alao JP, White R, Kindle RB, Losson R, Chambon P, Parker MG, Schär P, Heery DM, Buluwelan L, Ali S (2005) T:G mismatch-specific thymine-DNA glycosylase (TDG) as a coregulator of transcription interacts with SRC1 family members through a novel tyrosine repeat motif. Nucleic Acids Research 33(19): 6393–6404
Luo J, Li M, Tang Y, Laszkowska M, Roeder RG, Gu W (2004) Acetylation of p53 augments its site-specific DNA binding both *in vitro* and *in vivo*. PNAS 101(8) 2259–2264
Ma H, Nguyen C, Leė KS, Kahn M (2005) Differential roles for the coactivators CBP and p300 on TCF/beta-catenin-mediated survivin gene expression. Oncogene 24(22):3619–3631

Markham D, Munro S, Soloway J, O'Connor DP, La Thangue NB (2006) DNA-damage-responsive acetylation of pRb regulates binding to E2F-1. EMBO Rep.7(2): 192–198

Martin C, Zhang Y (2005) The diverse functions of histone lysine methylation. Nat Rev Mol Cell Biol. 6(11): 838–849

Martínez-Balbás MA, Bauer UM, Nielsen SJ, Brehm A, Kouzarides T (2000) Regulation of E2F1 activity by acetylation. EMBO J 19(4): 662–671

Maruta H, Greer K, Rosenbaum JL (1986) The acetylation of alpha-tubulin and its relationship to the assembly and disassembly of microtubules. J Cell Biol 103(2): 571–579

Marzio G, Wagener C, Gutierrez MI, Cartwright P, Helin K, Giacca M. (2000) E2F family members are differentially regulated by reversible acetylation. J Biol Chem 275(15): 10887–10892

Matsuzaki H, Daitoku H, Hatta M, Aoyama H, Yoshimochi K, Fukamizu A (2005) Acetylation of Foxol alters its DNA-binding ability and sensitivity to phosphorylation. PNAS 102(32): 11278–11283

McBryant SJ, Adams VH, Hansen JC (2006) Chromatin architectural proteins. Chromosome Res 14(1): 39–51

McMahon C, Suthiphongchai T, DiRenzo J, Ewen ME (1999) P/CAF associates with cyclin D1 and potentiates its activation of the estrogen receptor. PNAS 96(10) 5382–5387

Merika M, Thanos (2001) DEnhanceosomes. Curr Opin Genet Dev 11: 205–208

Munshi N, Agalioti T, Lomvardas S, Merika M, Chen G, Thanos D (2001) Coordination of a transcriptional switch by HMGI(Y) acetylation. Science 293: 1133–1136

Murphy PJ, Morishima Y, Kovacs JJ, Yao TP, Pratt WB (2005) Regulation of the dynamics of hsp90 action on the glucocorticoid receptor by acetylation/deacetylation of the chaperone. J Biol Chem 280(40): 33792–33799

Nishikawa K, Kobayashi M, Masumi A, Lyons SE, Weinstein BM, Liu PP, Yamamoto M(2003) Self-association of gatal enhances transcriptional activity *in vivo* in Zebra Fish embryos. Mol Cell Biol 23(22): 8295-8305

Olaharski AJ, Rine J, Marshall BL, Babiarz J, Zhang L, Verdin E, Smith MT (2005) The Flavoring agent dihydrocoumarin reverses pigenetic silencing and inhibits sirtuin deacetylases. PLoS Genet 1(6):e77

Ott M, Schnolzer M, Garnica J, Fischle W, Emiliani S, Rackwitz HR, Verdin E (1999) Acetylation of the HIV-1 Tat protein by p300 is important for its transcriptional activity. Curr Biol 16–30:9(24): 1489–1492

Pagans S, Pedal A, North BJ, Kaehlcke K, Marshall BL, Dorr A, Hetzer-Egger C, Henklein P, Frye R, McBurney MW, Hruby H, Jung M, Verdin E, Ott M (2005) SIRT1 regulates HIV transcription via Tat deacetylation. PLoS Biol 3(2):e41

Paranjape SM, Krumm A, Kadonaga JT (1995) HMG17 is a chromatin-specific transcriptional coactivator that increases the efficiency of transcription initiation. Genes Dev 9(16): 1978–1991

Qing Lu, Amanda E Hutchins, Colleen M Doyle, James R. Lundblad, Roland PS. Kwok (2001) Acetylation of cAMP-responsive element-binding protein (CREB) by CREB-binding protein enhances CREB-dependent Transcription. Mol Cell Biol 21(18): 6181–6188

Ramanathan B, Smerdon MJ (1986) Changes in nuclear protein acetylation in u.v.-damaged human cells. Carcinogenesis 7(7): 1087–1094

Rampalli S, Pavithra L, Bhatt A, Kundu TK, Chattopadhyay S (2005). Tumor suppressor SMAR1 mediates cyclin D1 repression by recruitment of the SIN3/histone deacetylase 1 complex. Mol Cell Biol 25(19) 8415–8429

Ruiz-Carrillo A, Wangh L, Allfrey V (1975) Assembly of newly replicated chromatin. Science 190(4210): 117–128

Sadri-Vakili G, Cha JH (2006) Mechanisms of Disease: Histone modifications in Huntington's disease. Nature Clinical Practice Neurology 2(6):330–338

Sakaguchi K, Herrera JE, Saito S, Miki T, Bustin M, Vassilev A, Anderson CW, Appella E (1998) DNA damage activates p53 through a phosphorylation-acetylation cascade. Genes Dev 12(18) 2831–2841

Shikama N, Chan HM, Krstic-Demonacos M, Smith L, Lee CW, Cairns W, La Thangue NB (2000). Functional interaction between nucleosome assembly proteins and p300/CREB-binding protein family coactivators. Mol Cell Biol 20(23): 8933–8943

Smith S, Stillman B (1991) Stepwise assembly of chromatin during DNA replication *in vitro*. EMBO J 10(4): 971–980

Solomon JM, Pasupuleti R, Xu L, McDonagh T, Curtis R, DiStefano PS, Huber LJ (2006) Inhibition of SIRT1 catalytic activity increases p53 acetylation but does not alter cell survival following DNA Damage. Mol cell Biol 26(1): 28–38

Sterner DE, Berger SL (2000) Acetylation of histones and transcription-related factors. Microbioloby and Molecular Biology Reviews 64(2): 435–459

Sun Y, Jiang X, Chen S, Fernandes N, Price BD (2005) A role for the Tip60 histone acetyltransferase in the acetylation and activation of ATM. PNAS 102(37): 13182–13187

Swaminathan V, Kishore AH, Febitha KK, Kundu TK (2005) Human histone chaperone nucleophosmin enhances acetylation-dependent chromatin transcription. Mol Cell Biol 25(17) 7534–7545

Szczesny B, Bhakat KK, Mitra S, Boldogh I (2004) Age-dependent modulation of DNA repair enzymes by covalent modification and subcellular distribution. Mech Ageing Dev 125(10–11): 755–765

Tagami H, Péloponèse JM, Loret E, Jeang KT, Nakatani y, Emiliani S, Benkirane M, Kiernan RE(2002) Differential acetylation of Tat coordinates its interaction with the co-activators cyclin T1 and PCAF. EMBO J 21(24): 6811–6819

Van der Horst A, Tertoolen LG, de Vries-Smits LM, Frye RA, Medema RH, Burgering BM (2004) FOXO4 is acetylated upon peroxide stress and deacetylated by the longevity protein hSir2(SIRT1). J Biol Chem 279(28) 28873–28879

Varier RA, Kundu TK (2006) Chromatin modifications (acetylation/ deacetylation/ methylation) as new targets for HIV therapy. Curr Pharm Des 12(16): 1975–1993

Vaziri H, Dessain SK, Ng Eaton E, Imai SI, Frye RA, Pandita TK, Guarente L, Weinberg RA (2001) hSIR2(SIRT1) functions as an NAD-dependent p53 deacetylase. Cell. Oct 19(107) 2149–159

Vo N, Fjeld C, Goodman RH. (2001) Acetylation of nuclear hormone receptor-interacting protein RIP140 regulates binding of the transcriptional corepressor CtBP. Mol cell Biol 21(18): 6181–6188

Wang X, Taplick J, Geva N, Oren M (2004) Inhibition of p53 degradation by Mdm2 acetylation. FEBS Lett 561(1–3): 195–201

Warnock LJ, Raines SA, Mee TR, Milner J (2005) Role of phosphorylation in p53 acetylation and PAb421 epitope recognition in baculoviral and mammalian expressed proteins. FEBS Journal 272(7): 1669–1675

Yamaguchi Y, Kurokawa M, Imai Y, Izutsu K, Asai T, Ichikawa M, Yamamoto G, Nitta E, Yamagata T, Sasaki K, Mitani K, Ogawa S, Chiba S, Hirai H (2004) AML1 is functionally regulated through p300-mediated acetylation on specific lysine residues. J Biol Chem 279(15): 15630–15638

Yoon S, Seger R (2006) The extracellular signal-regulated kinase: multiple substrates regulate diverse cellular functions. Growth Factors. 24(1): 21–44

Zhang Q, Yao H, Vo N, Goodman RH (2000) Acetylation of adenovirus EIA regulates binding of the transcriptional corepressor CtBP. PNAS 97(26): 14323–14328

CHAPTER 10

MARs AND MARBPs

Key modulators of gene regulation and disease manifestation

SAMIT CHATTOPADHYAY* AND LAKSHMINARASIMHAN PAVITHRA

National Center for Cell Science, Pune University Campus, Ganeshkhind, Pune 411007, India

Abstract: The DNA in eukaryotic genome is compartmentalized into various domains by a series of loops tethered onto the base of nuclear matrix. Scaffold/ Matrix attachment regions (S/MAR) punctuate these attachment sites and govern the nuclear architecture by establishing chromatin boundaries. In this context, specific proteins that interact with and bind to MAR sequences called MAR binding proteins (MARBPs), are of paramount importance, as these sequences spool the proteins that regulate transcription, replication, repair and recombination. Recent evidences also suggest a role for these *cis*-acting elements in viral integration, replication and transcription, thereby affecting host immune system. Owing to the complex nature of these nucleotide sequences, less is known about the MARBPs that bind to and bring about diverse effects on chromatin architecture and gene function. Several MARBPs have been identified and characterized so far and the list is growing. The fact that most the MARBPs exist in a co-repressor/ co-activator complex and bring about gene regulation makes them quintessential for cellular processes. This participation in gene regulation means that any perturbation in the regulation and levels of MARBPs could lead to disease conditions, particularly those caused by abnormal cell proliferation, like cancer. In the present chapter, we discuss the role of MARs and MARBPs in eukaryotic gene regulation, recombination, transcription and viral integration by altering the local chromatin structure and their dysregulation in disease manifestation

1. INTRODUCTION

The eukaryotic nuclei once referred to as "merely a bag of chromatin" has now been recognized to be a highly ordered structure or a 'hub' of cellular activities. The nucleus is seen as a three dimensional mosaic of nucleolus, inter-chromatin regions and condensed chromatin, dispersed in a nuclear ground substance

* E-mail: samit@nccs.res.in

T.K. Kundu and D. Dasgupta (eds.), Chromatin and Disease, 213–230.

traditionally called the "Nuclear Matrix" (NM). The NM is a dynamic fibro-granular structure postulated to contain chromatin and ribonucleoprotein domains (Berezney and Coffey, 1977; Smith *et al.*, 1984), serving as the structural milieu for gene function. The multitude of genomic functions occurring in nuclear matrix include gene replication (Smith and Berezney, 1980; Berezney and Buchholtz, 1981), transcription (Jackson and Cook, 1985) and RNA splicing and processing (Mariman *et al.*, 1982; Zeitlin *et al.*, 1987). All the cellular processes are highly coordinated and programmed and this demands that the genome be organized as a set of genes and gene clusters. Such an orderly arrangement of nuclear domains is brought about by anchorage of specific sequences to the matrix at interphase (Berezney and Coffey, 1974) and chromosomal scaffolds during mitosis (Mirkovitch *et al.*, 1984). These signature sequences known as S/MARs (Scaffold/ Matrix Attachment Regions) serve as boundary elements that punctuate chromosomal DNA into topologically restricted functional units, defining borders between chromatin domains (Breyne *et al.*, 1992; Chung *et al.*, 1993). Several studies also indicate the insulator nature of these elements, that together control single or multiple gene expression, by serving as Locus Control Regions (LCRs) (Grosveld *et al.*, 1987). MARs are also known to aid cell specific expression by their co-habitation with enhancers (Forrester *et al.*, 1994) and reduce the position effect variegation from local chromatin structures (Blasquez *et al.*, 1989). This is done by recruiting topoisomerase II (Razin *et al.*, 1991; Berrios *et al.*, 1985) and absorbing torsional stress by their base unwinding ability (Bode *et al.*, 1992). Thus they orchestrate topological organization of functional chromatin domains. Since MARs organize and govern the accessibility to local chromatin structures, they are also targets for viral integration and replication. Recent reports indicate the effect of MARs juxtaposed to retroviral integration sites on human genome. Several studies report that integration is favored near DNaseI hypersensitive sites or active genes (Mooslehner *et al.*, 1990). Small nuclear genome containing tumor viruses like HPV16, HBV, SV40, and HTLV-1 have also been shown to integrate near MARs (Shera *et al.*, 2001). Recent observations regarding the retroviral integrations reveal that HIV-1 and MoMuLV favor active genes for their integration (Schroder *et al.*, 2002; Mitchell *et al.*, 2004) Most of these integration sites (95%) have been shown to be flanked by S/MARs, around 1 Kb region of integration, that could serve as promoters (Kulkarni *et al.*, 2004; Johnson and Levy, 2005). In this context, specific proteins that bind to S/MARs, called MARBPs become significant, as they govern the chromatin accessibility of the region. For example the ratio of SATB1 and Cux/CDP, two well known MARBPs in various tissues for Mouse Mammary Tumor Virus region (MMTV) defines the transcriptional status of the virus. (Zhu *et al.*, 2000; Liu *et al.*, 1999). Similiarly, a study by Stunkel *et al.,* showed that Cux recruits HDAC1 to the LCR MAR of E6 promoter of Human Papilloma Virus (HPV) and represses this oncoprotein (2000).

MARBPs have been routinely isolated using high salt extraction of the matrix and their ability to have base unpairing potential (Galande and Kohwi-Shigematsu, 1999). Several studies report the ability of these MARBPs to control gene expression by binding to MAR sequences within the regulator regions of the gene and activate or repress the gene expression. They are often found in a complex with co-activators or co-repressors, modulating gene function by remodeling or covalently modifying the chromatin structure. Most of these MARBPs have been shown to be drastically affected upon malignant transformations. Since aberrant gene expression gives rise to malignancies and abnormal cell cycle progression, understanding the nature and order of these proteins assumes great importance in the current scenario of chromatin biology and disease manifestation.

2. DEFINING MARs

The available literature on MARs reveals that they can enhance the expression of reporter genes by forming a domain and insulating them from position variegation effect at the integration site (Breyne *et al.*, 1992; Allen *et al.*, 1993). Their main function is to bring together control elements like promoters and enhancers loaded with their transcriptional factors, creating a transcription factor and enzyme rich nuclear microenvironment.

While there is no stead and fast rule for identifying MARs, certain traits make the AT rich DNA elements function as MARs. The AT richness confers DNA unwinding potential, so that in their single stranded form specific DNA unwinding proteins bind to MARs, relieve the torsional stress and the energy is used to relax positive supercoiling generated ahead of transcription elongation point (Bode *et al.*, 1992). Studies by Amati *et al.*, (1990) and von Kries *et al.*, (1991) although suggest that this property of MARs might not be essential for matrix binding. A number of studies show that MARs can allow the induction of DNaseI hypersensitive sites in chromatin. This has been substantiated with several experimental observations that MAR sites overlap with DNase I hypersensitive sites. The proposed mechanism is that MARs become DNAse I hypersensitive at the time of their activation as a result of binding of single strand binding proteins (SSBPs) specifically during transcription and replication due to removal of nucleosomes that preludes ORI activation (Hsieh *et al.*, 1993).

Several triple helical or Z form, cruciform structure of MARs has also been proposed. The cruciform loop part is susceptible to DNaseI and hence characteristic nicking occurs. The role of MARs in chromatin dynamics has also been tested using an artificial MARBP called MATH 20 (Strick and Laemmli, 1995). This protein has numerous linked DNA binding domains called AT hooks, which preferentially bind to AT tracts. Since this protein could then bind MARs and associated chromatin specifically, this was used to study chromatin condensation in Xenopus oocytes. It was found that titration of MARs with MATH 20 specifically inhibited chromatid conversion without inhibiting condensation resulting in abortive mitotic structures. When MATH 20 was added to chromatids, it leads to a structural collapse and formation of chromatid balls. Thus, it is speculated that S/MARs could be target of binding proteins that mediate or facilitate formation and juxtapositioning of metaphase chromatin loops (Hart and Laemmli, 1998). Similarly, MAR sites have been found to be sites of illegitimate recombination, since these elements are found at the site of DNA insertion, deletion and translocation (Sperry *et al.*, 1989; Shapiro *et al.*, 1987).

List of well known MARBPs and functions

Name	Species	Tissue specificity	Function	Interactome
SATB1/L2a-P1/	*Homo sapiens, Mus musculus*	Predominantly thymus, minute amounts in brain	Repressor	CDP, HDAC1, ACF1, ISWI
Nucleolin	*Homo sapiens*	erythroleukemia cell line K562	Glycosaminoglycan stabilisation	Glycosaminoglycans
SAF-B	*Homo sapiens*	Ubiquitous	Repressor	RNA Pol II
SAF-A	*Homo sapiens*	HeLa, Embryonic kidney cell line 293, K562	Activator Organization of chromosomal DNA & packaging of hnRNA	p300, DNA PK
NFμNR/ Cux/ CDP	*Mus musculus*	Breast, Ubiquitous	Important in organ development	SATB1, HDAC1, SMAR1
p114	*Homo sapiens*	Infiltrating ductal carcinoma tissues, normal breast tissue, benign breast diseases	combined property of PARP and SAF-A (hnRNP-U)	PARP, Ku
Bright	*Mus musculus*	Ubiquitous	B cell regulator of immuno-globulin heavy-chain transcription	Sp100, LYSp100/p140
SP100	*Homo sapiens*	Ubiquitous	Repressor of Bright	Bright
Ku (subunit of PARP)	*Homo sapiens*	SKBR3 cell line	chromosome condensation, subunit of DNA-dependent protein kinase	PARP
DNA-PK	*Homo sapiens*	Breast cancer	downregulated during cellular senescence	PARP, SAF-A

LYSp100/p140	*Homo sapiens*	B cells	Co-activates Bright	Bright
SMAR1	*Homo sapiens, Mus musculus*	Ubiquitous but predominant in thymus	Represses transcription, downregulated in breast cancer cells	P53, HDAC1, mSin3a
MeCP2/ ARBP	*Homo sapiens*	Ubiquitous, particular high levels in neuron of the post natal brain	Mutations in the MECP2 gene cause Rett syndrome	mSin3a, HDAC1, HDAC2

3. ROLE OF MARBPS IN MODULATING MARS AS FUNCTIONAL ELEMENT

Gene expression status involves changing of attachment points of chromatin loops and hence implies the dynamicity of MARs and its association with nuclear matrix, regulated and governed by cell type (Liebich *et al.*, 2002). The intrinsic activity of MAR is not sufficient to bring about position effects or transcriptional regulation but may depend on the contribution of protein factors that specifically bind to these motifs. These are classified based on their recognition sites as: abundant multi-functional matrix proteins like High Mobility Group proteins (HMGs), transcription factors like H box, Y box and CAAT binding proteins. Some class of proteins like ARBP, MeCP2, SATB1, CDP and SMAR1 also interfere with the MAR- matrix productive effects on transcription, unlike Bright that activate transcription. The varied effects of MARBPs could involve changes in the chromatin structure and activity by the recruitment or interaction with chromatin remodeling complexes. For example MAR binding repressors like ARBP, MeCP2, SATB1, SMAR1, CDP and CRBP mediate transcriptional repression by recruitment of components of histone deacetylating pathways. This is in contrast with activator MARBPs like SAF-A and Bright that recruits SWI or p300 and affect the acetylation and remodeling of chromosomes. Apart from affecting chromatin remodeling, gene regulation is also achieved by the direct interaction of MARBPs like SAF-B with RNA polymerase II and/ or with RNA processing factor. Hence, MARBPs serve as molecular base to assemble transcriptome and chromatin remodeling network. (Fig. 1)

4. MARBPS WORK IN CONSORT

Proteins that mediate MAR function (MARBPs) appear to control numerous genes expressed in differentiated cells. SMAR1, one such MARBP, has been shown to be ubiquitously expressed in all tissues, predominantly in thymus and governs the transition of T cells from DN to DP stage. SATB1 is expressed predominantly in thymus but also in brain and several other organs, while CDP expression occurs

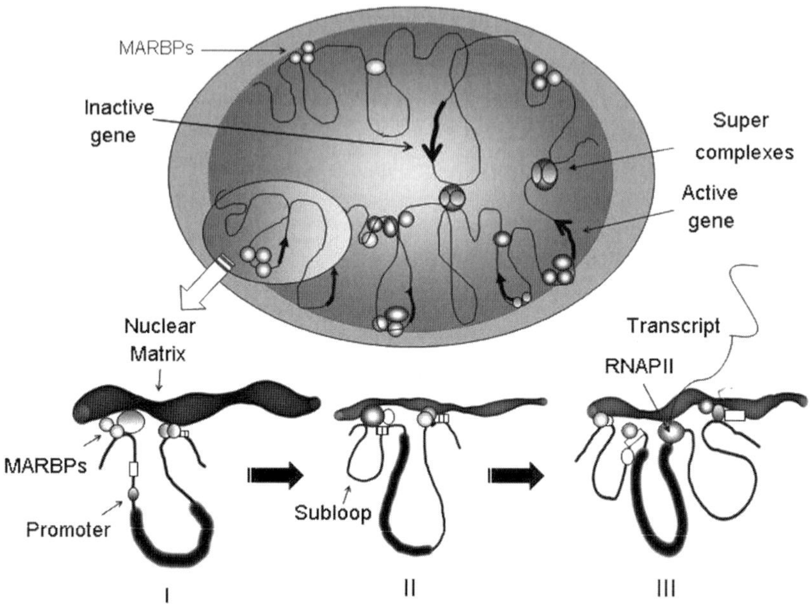

Figure 1. MARs serve to enrich the nuclear milieu by bringing in transcription factors and enzymes to regulate transcription

in all but terminally differentiated cells. Binding of SATB1 and CDP to regulatory elements has been associated with transcriptional repression of numerous genes expressed in differentiated cells. For instance, SATB1 regulates gene expression both spatially and temporally during T-cell development. Recently, SATB1 null-mice were generated, and genome-wide expression profiling analysis indicated that about 2% of all genes become significantly derepressed in thymocytes (Alvarez *et al.*, 2000). In contrast, the activator Bright is present in differentiated cells, where it may compete with repressors for MAR-binding sites. Thus, there would be a dynamic process of activation/ repression forming temporary sites of DNA attachment. Bright overexpression was found to enhance transgene expression, indicating that the normal activator protein level is limiting. In contrast, cotransfection of CDP abrogated Bright transactivation and reduced the basal expression level, indicating that CDP is able to override the DNA binding and/or transactivation capacity of Bright (Herrscher *et al.*, 1995). Some MARBPs, depending on the context, either function as transcriptional repressors or activators. MARBPs such as Cux/CDP and SATB1 can function as transcriptional repressors in non-B cells by interacting with their target MAR sequences flanking the IgH intronic enhancer (Alvarez *et al.*, 2000). On the other hand, the MARBP Bright acts as a transcriptional activator in B cells (Schubeler *et al.*, 1996) and this activation is context-dependent as it requires an intact IgH enhancer core (Zahn-Zabal *et al.*, 2001).

Different degrees of repression could be mediated by competition between repressors and activators for MAR binding. For example, MARβ, which resides 400 bp upstream of TCRβ enhancer (Eβ), is the docking site for three MARBPs: SMAR1, Cux and SATB1 (Chattopadhyay *et al.*, 2000). MARβ (HS1) is the major DNase I hypersensitive site induced during the TCR co-receptor $CD4^{-}CD8^{-}$ double negative (DN) to $CD8^{+}CD4^{+}$ double positive (DP) stage of thymocyte development. This induction is concomitant with halt of TCRβ V(D)J recombination in DP thymocytes. Studies by Chattopadhyay *et al.*, (1998) identified that SMAR1 binds to this region along with Cux and SATB1 (Fig. 2).

The 114 Kd protein PARP, exclusively found as a mixture with DNA-PK has been implicated in repair after DNA damage. Evidences show that PARP and Ku 70/86 interact together and synergistically enhance their binding to DNA (IgH MARs)

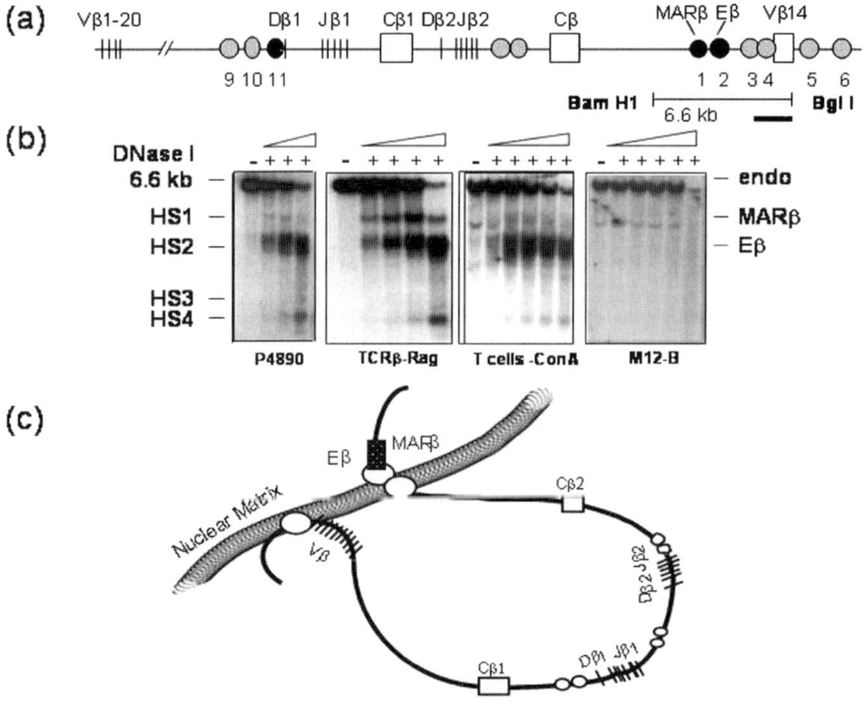

Figure 2. TCRβ locus showing eleven DNase1 hypersensitive sites. (a) HS1 and HS2 represent Eβ enhancer and MARβ sites respectively that are located upstream of Vβ14 segment. (b) DNaseI hypersensitivity assay showing four sites within a 6.6 kb BamH1-BglII genomic fragment. Panel 1: P4890 representing Rag mutant double negative (DN) lymphoma T cell line. Panel 2: DP cells of Rag mutant TCRβ transgenic mice. Panel 3: Mature single positive T cells from lymph node stimulated with ConA. Panel 4: Mature B cell line. No DNaseI hypersensitive sites were observed in B cells indicating that all four hypersensitive sites are specific for T cell and (c) A current model of possible role of MAR sites in maintaining cross-talk between Eβ enhancer, MARs and MARBPs that together control V(D)J recombination and transcription at TCRβ locus

(Galande and Kohwi-Shigematsu, 1999). Similarly, PARP and SAF-A contribute to the MAR binding ability of p114. Together, these observations suggest that the different MAR-binding proteins associate with diverse sets of proteins involved in transcriptional regulation and also cross-talk amongst each other, constituting a functional nuclear matrix.

5. MARBPs AND DISEASE MANIFESTATION

Since most cases of malignant transformations occur by means of deregulation of genes or viral integrations, the role of MARs and MARBPs become crucial. The composition of nuclear matrix has been reported to be altered during the course of transformation and the MARBPs that associate with transcriptional units might be involved in the progression/ cessation of disease. The uncontrolled cell proliferation and invasion of the cancer cells would naturally require an altered organization of the chromatin, to assure the differential expression of a subset of proteins compared to their normal counterparts. These genes have to be expressed or repressed ectopically. For example, the expression pattern of nuclear matrix proteins is considerably different between normal breast epithelial cells and malignant cells (Khanuja *et al.*, 1993).

The abnormal levels of MARBPs, apart from their altered cellular distribution, seems to govern the progression of proliferative diseases. Studies to identify the nuclear matrix binding proteins associated with aggressive cancer phenotype have lead to the identification of PARP, Ku, High mobility group proteins (I/Y), NMP, SAF-A/B that have binding affinity to double stranded BURs. The expression of these proteins is dramatically increased upon malignant transformation and marks the advanced cancer phenotype leading to metastasis. Unlike other MARBPs that are highly expressed at the onset of malignant transformation, SMAR1 level is downregulated in breast cancer derived lines, which may be explained in part to its regulation of Cyclin D1 gene, a hallmark of breast and prostate cancer (Rampalli *et al.*, 2005). In most cases of cancer, there is gene duplication or loss of transcriptional control of this gene that leads to uncontrolled cell proliferation, leading to cancer. We have shown that SMAR1 forms a co-repressor complex with HDAC1 and mSin3a, recruits this complex to the Cyclin D1 promoter on a segment that is rich in ATCs (putative MAR sequence as evaluated by MAR finder program MAR-WIZ). The recruited complex then deacetylates histones (H3K9 and H4K10) at the loci, leading to chromatin condensation and eventually shuts down transcription. (Fig. 3)

Several reports have highlighted the ability of such MARBPs to recruit chromatin modifying enzyme complexes to control gene transcription, showing that differential expression of these proteins is critical too. For example in breast cancer derived cell lines, SMAR1 levels are downregulated that correlates to an induced Cyclin D1 levels. Another example is HMG-I, an architectural chromatin protein that binds to minor groove of double stranded DNA. Recent reports suggest the role of HMG-I, C and Y proteins in cellular proliferation and neoplastic transformation

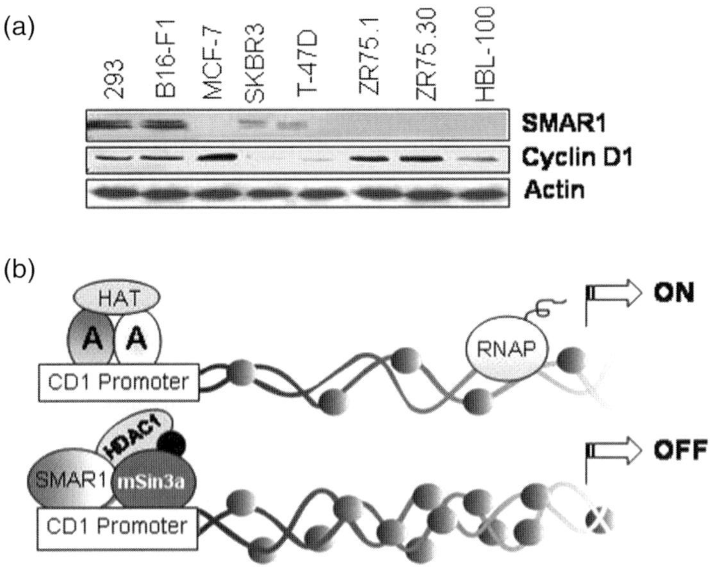

Figure 3. (a) Western blot analysis of breast cancer derived cell lines showing an induced Cyclin D1 levels that correlates to a decreased SMAR1 expression. (b) A schematic representation of SMAR1 recruitment of HDAC1 on CyclinD1 promoter, that switches off the transcription

(Tamimi *et al.*, 1993). HMG-I has been identified as a target gene for c-myc and is involved in c-myc mediated neoplastic transformation. Distant metastasis is reported in animals injected with HMG-I or -C (Wood *et al.*, 2000a, b). The mechanism of transformation has been attributed to the identification of HMG-C AT hooks in chimeric proteins associated with lipomas and mesenchymal tumors (Schoenmakers *et al.*, 1995; Rogalla *et al.*, 1997). These chimeras are thought to function by binding to DNA via the AT hooks and alter the gene regulation by transcriptional regulatory domains they acquire by rearrangement (Ashar *et al.*, 1995, Wunderlich and Bottger, 1997; Bustin, 1999). Apart from this, HMG-C, truncated transgenic mice display gigantism and lipomatosis while the null mice develop pygmy phenotype. (Battista *et al.*, 1999; Zhou *et al.*, 1995). This reveals that the levels of the MARBPs must be kept under a tight leash to avoid abnormal cellular proliferation and transformation.

Protein modifications of MARBPs also assume great importance as their binding properties can be altered. MARBPs like SATB1 and SAF-A get modified with special reference to a particular cellular process. For example, the cleavage of SATB1 by Caspase 6 disrupts the PDZ domain mediated dimerization, resulting in the formation of SATB1 monomers that do not have DNA binding ability. This causes detachment of the protein from chromatin that eventually leads to rapid and efficient disassembly of higher order chromatin structure and facilitates apoptosis in T cells, an important event in T cell receptor rearrangement (Galande *et al.*, 2001).

Likewise, the SAR binding domain of SAF-A loses its DNA binding potential upon proteolytic cleavage during apoptosis (Gohring and Fackelmayer, 1997). There are reports that nuclear matrix proteins associate with granular nuclear bodies and undergo modifications in cells that undergo apoptosis (Zweyer *et al.*, 1997). In most of the cases, the loss of DNA binding ability of MARBPs becomes central to the altered function. For example, Ku deficiency leads to extreme radiation sensitivity and high levels of chromosomal aberrations (Gu *et al.*, 1997). This is due to the fact that the Ku heterodimer (Ku 70/80) binds to DNA double strand breaks and facilitate repair by non-homologus end joining pathway (Walker *et al.*, 2001). In case of invasive breast cancer, a specific nuclear matrix binding protein called NMP has been identified, that recognizes a unconventional MAR in the promoter, stimulates the levels of NFκB, that in turn increases the DNA binding activity of NFκB observed in c-erb2 and BRCA1 positive human breast tumors, suggesting a role in breast cancer progression (Raziuddin *et al.*, 1997). HET/SAF-B was originally cloned as a nuclear matrix protein that binds to the MAR of hsp27 promoter and represses it in breast cancer cells (Oesterreich, 1997). In addition, it has been shown to bind to ER and function as its co-repressor (Townson *et al.*, 2000). There are also reports suggesting that the overexpression of HET causes growth inhibition in M phase of cell cycle and causes multicellularity. Consistent with this, it is also known to cause aneuploidy in breast tumor specimens and causes lower proliferation (Townson *et al.*, 2000). This suggests that these factors might have important and distinct roles in tumorigenesis independent of their transcriptional functions. The MARBPs have also been identified to play major role in normal tissue differentiation and organogenesis. For example, Cux has been shown to play role in specifying the identity of external sensory organs during peripheral nervous system development. It has also been implicated in controlling proliferation and differentiation (Nepveu, 2001). Cux knock out mice show retarded growth, curly whiskers, late development of lung epithelia, defective hair follicle development, infertile male progeny and less number of T and B cells and more myeloid cells (Sinclair *et al.*, 2001, Tufarelli *et al.*, 1998). On the other hand, Cux-1 transgenic mice show organomegaly and hyperplasia of different organs (Ledford *et al.*, 2002).

Certain novel isoforms of MARBPs have been suggested to play key role in disease manifestation. Two isoforms of Cux/CDP were previously known till Goulet *et al.*, showed that a novel isoform of Cux (p75) showed a higher binding affinity to DNA and represses the transcription of CDK inhibitor p21 and activated DNA polymerase a gene promoter. They also identified that a novel intronic transcription initiation was responsible for the expression of this isoform. Although this isoform is predominant in $CD4^+/CD8^+$ and $CD4^+$ T cells, expression was also activated in breast tumor cell lines and in primary human breast tumors. The stable lines expressing p75 isoform could not form tubule structures in collagen but developed as solid undifferentiated aggregates of cells. Some studies also indicate that Cux is a downstream target of Notch, activated in T cell leukemia. Cux could thus serve as a downstream target marker for Notch activation (Goulet *et al.*, 2002). Activation of CDP/Cux at the G_1/S transition involved the proteolytic processing of the protein to generate a shorter isoform in uterine leiomyomas (Moon *et al.*, 2002). Similarly, CUTL1 activity is associated with increased migration and invasiveness in numerous tumor cell lines, both *in vitro* and *in vivo* transcriptional target of transforming growth factor beta (TGFβ) and a mediator of its promigratory effects CUTL1 expression is significantly increased in high-grade carcinomas and is inversely correlated with survival in breast cancer. This suggests that CUTL1 plays a central role in coordinating a gene expression program associated with cell motility and tumor progression (Michl and Downward, 2006).

SMAR1 is shown to exist in two isoforms, the shorter form having a deletion of 117 bp at the N terminus (Chattopadhyay *et al.*, 2000). The 39 aa deleted shorter form has been shown to be more effective in regressing B16F-10 induced melanoma (Kaul *et al.*, 2003). SMAR1 transgenic mice showed abnormal V(D)J recombination and organomegaly of lymphoid organs with cellular infiltrations, suggesting a necessity for fine tuning of protein expression to continue cellular process (Kaul-Ghanekar *et al.*, 2005) (Fig. 4). This reveals that various isoforms of MARBPs may have different roles in the context of cellular functions.

Apart from serving as modulators of transcription, MARBPs alter cellular functions by their protein interactions. Nucleolin first described by Orrick *et al.*, (1973) is a major nucleolar protein involved in ribosome biogenesis. Recent reports also suggest that nucleolin binds to cell surface adhesion molecules like L-selectin expressed on leucocytes and hemopoeitic stem cell progenitors (Harms *et al.*, 2001). Modulation of hepatitis delta viral replication is also well documented (Lee *et al.*, 1998). This protein is also linked to actin cytoskeleton and inhibits HIV infection by cytokine midkine. Reports by Christian *et al.*, (2003) show that nucleolin on cell surface is a marker for endothelial cells and its interaction with Acharan sulfate (AS) may be a key in solving the mechanism of AS mediated inhibition of tumor growth. The identification of cancer cell specific markers led to the identification of a subset of nuclear matrix proteins (NMPs) that exist in prostrate, bladder, colon and renal cancer (Konety *et al.*, 1998). These NMPs can be detected in the serum samples of the patients because of these factors as tumor cells undergo degeneration and lysis. NMP 22 has been now routinely used to assess bladder cancer and similarly L4 is a candidate for identification of colon cancer (Brunagel *et al.*, 2002).

Several ubiquitous transcription factors like Lys 100/ p140, SP100 are also known to bind to MARs. While p140 is known to co-activate transcription by Bright, SP1 represses transcription. AP-1 family transcription factor is also known to associate with nuclear matrix (van Wijnen *et al.*, 1993) and allow the expression of Igκ gene in response to LPS stimulation (Schanke *et al.*, 1994). p53 another well known transcription factor, the function of which depends on its DNA binding ability to different promoters. Wild type p53 is known to be a tumor suppressor while mutant form exerts oncogenic functions of its own (Dittmer *et al.*, 1993; Levine *et al.*, 1995). The "gain of function" of mutant p53 has been demonstrated partly due to its high affinity to bind to MARs (Will *et al.*, 1998). These MARs have a typical AATATATTT unwinding motif, promoting structural alterations in

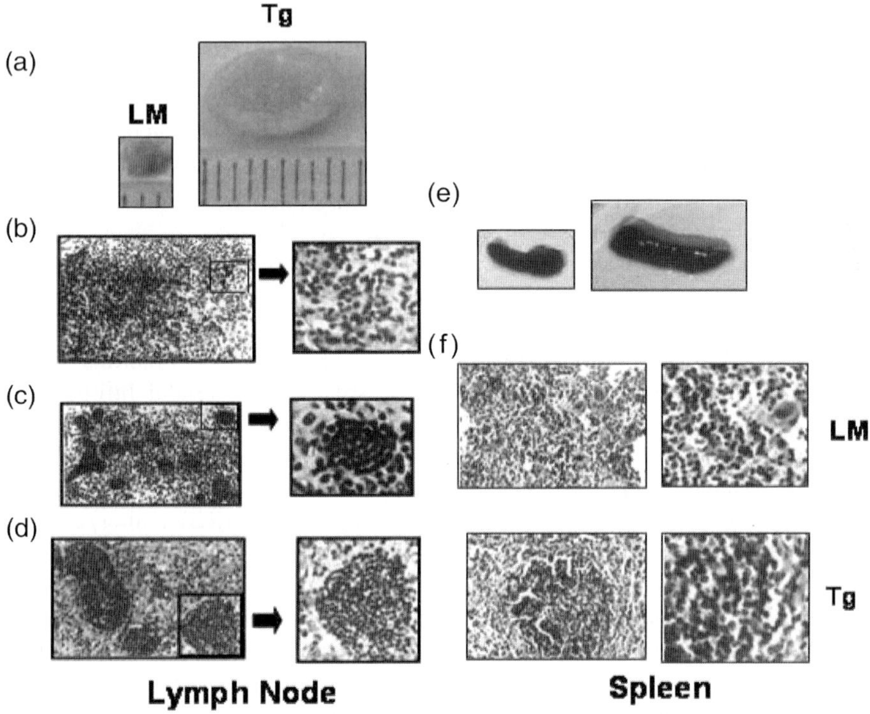

Figure 4. Comparison of lymph node and spleen architecture in SMAR1 transgenic and control littermate mice. (a) Lymph node size of non-transgenic (N) and transgenic (T) mice are shown at the same scale. (b) Histological analysis of lymph node from control at 10X and 40X magnification are displayed. (c and d) Histological sections of Lymph node at 10X and 40X. (e) Enlargement of spleen size shown in transgenic mice compared to control mice. (f) Histological sections of spleen from Littermate normal (LM) and SMAR1 transgenic mice showing strong infiltration of T cells into the lymph node. (See Colour Plate 15.)

chromatin, thereby affecting cellular replication. Since mutations in p53 constitute the most frequent alteration in a single gene in human cancer (Hainaut *et al.*, 1997), the MAR binding potential of the mutant form could form the molecular basis of oncogenic potential documented for mutant p53.

6. MARBPs AS PROGNOSTIC MARKERS AND FUTURE PERSPECTIVES

The aberrant expression of MARBPs upon malignant transformation makes them a reliable marker for diagnosis of advanced diseased conditions. For instance, the NMP 22 levels in urine have now been routinely used to identify bladder and ductal cancer. Similarly, p114 MAR-binding activity has been detected in aggressive tumors, while significantly weaker p114 activity has been observed in less aggressive tumors. Hence, PARP, an interactor of p114 could be used as a marker for identifying invasive breast cancer. The novel isoform of Cux (p75) is a reliable marker for identifying breast cancer. MAR binding protein, p230, is detectable in rat hepatoma cells but not in normal liver and suggests that this protein is a diagnostic and prognostic marker for liver cancer. It is clear that nuclear matrix proteins hold a considerable promise as diagnostic tools for pathologists.

MARBPs with their newly discovered role as linkers between chromatin remodeling, signal transduction and cell cycle regulation form an important part of chromatin biology. Most importantly, direct correlation of disease manifestation and the MARBPs will make them an important tool in understanding disease progression. Present evidence, suggests that nuclear matrix proteins may be useful biomarkers of neoplastic diseases in the serum, body fluids, and tissues. Nuclear matrix proteins are also potential candidates for the use as tumor prognostic factors and targets of anticancer drugs through apoptosis.

ACKNOWLEDGEMENTS

The authors thank Dr. G. C. Mishra, Director, National Center for Cell Science, Pune for his constant support and guidance. Special thanks to Shravanti Rampalli for her ideas, critic and discussion. Thanks to present and past members of our group, notably Surajit Sinha for his valuable inputs. We thank Dr. Jianzhu Chen, MIT, Boston, who helped us carrying out experiments related to V(D)J recombination. Part of the work reported was funded by Department of Biotechnology, Government of India. L. P. is a senior research fellow funded by University Grants Commission, Government of India.

GLOSSARY

SAR/MARs
Scaffold Attachment Region
SAF-A: Scaffold Attachment factor A
Nuclear matrix binding proteins
Matrix Associated Region Binding lproteins(MARBPs)
Tumor suppressor p53
Nucleases
DNase I hypersensitivity assay
Locus control Region (LCR)
SMAR1
Human immunodefficiemcu Virus 1
HTLV
MMTV
TCRβ locus
Cyclin D1
mSin3A complex

V(D)J recombination
SATBI: Special AT-rich sequuence binding protein I
Cux/CDP
CUTL1
Eβ enhancer

REFERENCES

Allen GC, Hall GE, Jr, Childs LC, Weissinger AK, Spiker S, Thompson WF (1993) Scaffold attachment regions increase reporter gene expression in stably transformed plant cells. *Plant Cell* 5(6):603–613

Alvarez JD, Yasui DH, Niida H, Joh T, Loh DY, Kohwi-Shigematsu T (2000) The MAR-binding protein SATB1 orchestrates temporal and spatial expression of multiple genes during T-cell development. *Genes Dev* 14(5):521–535

Amati B, Pick L, Laroche T, Gasser SM (1990) Nuclear scaffold attachment stimulates, but is not essential for ARS activity in Saccharomyces cerevisiae: Analysis of the Drosophila ftz SAR. *EMBO J* 9(12):4007–4016

Ashar HR, Fejzo MS, Tkachenko A, Zhou X, Fletcher JA, Weremowicz S, Morton CC, Chada K (1995) Disruption of the architectural factor HMGI-C: DNA-binding AT hook motifs fused in lipomas to distinct transcriptional regulatory domains. *Cell* 82(1):57–65

Battista S, Fidanza V, Fedele M, Klein-Szanto AJ, Outwater E, Brunner H, Santoro M, Croce CM, Fusco A (1999) The expression of a truncated HMGI-C gene induces gigantism associated with lipomatosis. *Cancer Res* 59(19):4793–4797

Berezney R, Buchholtz LA (1981) Dynamic association of replicating DNA fragments with the nuclear matrix of regenerating liver. *Exp Cell Res* 132(1):1–13

Berezney R, Coffey DS (1974) Identification of a nuclear protein matrix. *Biochem Biophys Res Commun* 60(4):1410–1417

Berezney R, Coffey DS (1977) Nuclear matrix. isolation and characterization of a framework structure from rat liver nuclei. *J Cell Biol* 73(3):616–637

Berrios M, Osheroff N, Fisher PA (1985) In situ localization of DNA topoisomerase II, a major polypeptide component of the Drosophila nuclear matrix fraction. *Proc Natl Acad Sci USA* 82(12):4142–4146

Blasquez VC, Sperry AO, Cockerill PN, Garrard WT (1989) Protein:DNA interactions at chromosomal loop attachment sites. *Genome* 31(2):503–509

Bode J, Kohwi Y, Dickinson L, Joh T, Klehr D, Mielke C, Kohwi-Shigematsu T (1992) Biological significance of unwinding capability of nuclear matrix-associating DNAs. *Science* 255(5041):195–197

Breyne P, Van Montagu M, Depicker N, Gheysen G (1992) Characterization of a plant scaffold attachment region in a DNA fragment that normalizes transgene expression in tobacco. *Plant Cell* 4(4):463–471

Brunagel G, Vietmeier BN, Bauer AJ, Schoen RE, Getzenberg RH (2002) Identification of nuclear matrix protein alterations associated with human colon cancer. *Cancer Res.* 62(8):2437–2442

Bustin M (1999) Regulation of DNA-dependent activities by the functional motifs of the high-mobility-group chromosomal proteins. *Mol Cell Biol* 19(8):5237–5246

Chattopadhyay S, Whitehurst CE, Chen J (1998) A nuclear matrix attachment region upstream of the T cell receptor beta gene enhancer binds Cux/CDP and SATB1 and modulates enhancer-dependent reporter gene expression but not endogenous gene expression. *J Biol Chem* 273(45):29838–29846

Chattopadhyay S, Kaul R, Charest A, Housman D, Chen J (2000) SMAR1, a novel, alternatively spliced gene product, binds the Scaffold/Matrix-associated region at the T cell receptor beta locus. *Genomics* 68(1):93–96

Christian S, Pilch J, Akerman ME, Porkka K, Laakkonen P, Ruoslahti E (2003) Nucleolin expressed at the cell surface is a marker of endothelial cells in angiogenic blood vessels. *J Cell Biol* 163(4):871–878

Chung JH, Whiteley M, Felsenfeld G (1993) A 5' element of the chicken beta-globin domain serves as an insulator in human erythroid cells and protects against position effect in Drosophila. *Cell* 74(3):505–514

Dittmer D, Pati S, Zambetti G, Chu S, AK, Teresky, Moore M, Finlay C, Levine AJ (1993) Gain of function mutations in p53. *Nat Genet* 4(1):42–46

Forrester WC, van Genderen C, Jenuwein T, Grosschedl R (1994) Dependence of enhancer-mediated transcription of the immunoglobulin mu gene on nuclear matrix attachment regions. *Science* 265(5176):1221–1225

Galande S, Kohwi-Shigematsu T (1999) Poly(ADP-ribose) polymerase and Ku autoantigen form a complex and synergistically bind to matrix attachment sequences. *J Biol Chem* 274(29):20521–20528

Galande S, Dickinson LA, Mian IS, Sikorska M, Kohwi-Shigematsu T (2001) SATB1 cleavage by caspase 6 disrupts PDZ domain-mediated dimerization, causing detachment from chromatin early in T-cell apoptosis. *Mol Cell Biol* 21(16):5591–5604

Gohring F, Fackelmayer FO (1997) The scaffold/matrix attachment region binding protein hnRNP-U (SAF-A) is directly bound to chromosomal DNA in vivo: a chemical cross-linking study. *Biochemistry* 36(27):8276–8283

Goulet B, Watson P, Poirier M, Leduy L, Berube G, Meterissian S, Jolicoeur P, (2002) A Nepveu. Characterization of a tissue-specific CDP/Cux isoform, p75, activated in breast tumor cells. *Cancer Res* 62(22):6625–6633

Grosveld F, van Assendelft GB, Greaves DR, Kollias G (1987) Position-independent, high-level expression of the human beta-globin gene in transgenic mice. *Cell* 51(6):975–985

Gu Y, Jin S, Gao Y, Weaver DT, Alt FW (1997) Ku70-deficient embryonic stem cells have increased ionizing radiosensitivity, defective DNA end-binding activity, and inability to support V(D)J recombination. *Proc Natl Acad Sci USA* 94(15):8076–8081

Hainaut P, Soussi T, Shomer B, Hollstein M, Greenblatt M, Hovig E, Harris CC, Montesano R (1997) Database of p53 gene somatic mutations in human tumors and cell lines: updated compilation and future prospects. *Nucleic Acids Res* 25(1):151–157

Harms G, Kraft R, Grelle G, Volz B, Dernedde J, Tauber R (2001) Identification of nucleolin as a new L-selectin ligand. *Biochem. J* 360(Pt 3):531–538

Hart CM, Laemmli UK (1998) Facilitation of chromatin dynamics by SARs. *Cur Opin Genet Dev* 8(5):519–525

Herrscher RF, Kaplan MH, Lelsz DL, Das C, Scheuermann R, Tucker PW (1995) The immunoglobulin heavy-chain matrix-associating regions are bound by Bright: A B cell-specific trans-activator that describes a new DNA-binding protein family. *Genes Dev* 9(24):3067–3082

Hsieh CL, Arlett CF, Lieber MR (1993) V(D)J recombination in ataxia telangiectasia, Bloom's syndrome, and a DNA ligase I-associated immunodeficiency disorder. *J Biol Chem* 268(27):20105–20109

Jackson DA, Cook PR (1985) Transcription occurs at a nucleoskeleton. *EMBO J* 4(4):919–925

Johnson CN, Levy LS (2005) Matrix attachment regions as targets for retroviral integration. *Virol J* 2:68

Kaul R, Mukherjee S, Ahmed F, Bhat MK, Chhipa R, Galande S, Chattopadhyay S (2003) Direct interaction with and activation of p53 by SMAR1 retards cell-cycle progression at G2/M phase and delays tumor growth in mice. *Int J Cancer* 103(5):606–615

Kaul-Ghanekar R, Majumdar S, Jalota A, Gulati N, Dubey N, Saha B, Chattopadhyay S (2005) Abnormal V(D)J recombination of T cell receptor beta locus in SMAR1 transgenic mice. *J Biol Chem* 280(10):9450–9459

Khanuja PS, Lehr JE, Soule HD, Gehani SK, Noto AC, Choudhury S, Chen R, Pienta KJ (1993) Nuclear matrix proteins in normal and breast cancer cells. *Cancer Res* 53(14):3394–3398

Konety BR, Nangia AK, Nguyen T.S, Veitmeier BN, Dhir R, Acierno JS, Becich MJ, Hrebinko RL, Getzenberg RH (1998) Identification of nuclear matrix protein alterations associated with renal cell carcinoma. *J Urol* 159(4):1359–1363

Kulkarni A, Pavithra L, Rampalli S, Mogare D, Babu K, Shiekh G, Ghosh S, Chattopadhyay S (2004) HIV-1 integration sites are flanked by potential MARs that alone can act as promoters. *Biochem Biophys Res Commun* 322(2):672–677

Ledford AW, Brantley JG, Kemeny G, Foreman TL, Quaggin SE, Igarashi P, Oberhaus SM, Rodova M, Calvet JP, Vanden Heuvel GB (2002) Deregulated expression of the homeobox gene Cux-1 in transgenic mice results in downregulation of p27(kip1) expression during nephrogenesis, glomerular abnormalities, and multiorgan hyperplasia. *Dev Biol* 245 (1):157–171

Levine AJ, Wu MC, Chang A, Silver A, Attiyeh EF, Lin J, Epstein CB (1995) The spectrum of mutations at the p53 locus. Evidence for tissue-specific mutagenesis, selection of mutant alleles, and a "gain of function" phenotype. *Ann N Y Acad Sci* 768:111–128

Liebich I, Bode J, Reuter I, Wingender E (2002) Evaluation of sequence motifs found in scaffold/matrix-attached regions (S/MARs). *Nucleic Acids Res* 30(15):3433–3442

Liu J, Barnett A, Neufeld EJ, Dudley JP (1999) Homeoproteins CDP and SATB1 interact: potential for tissue-specific regulation. *Mol Cell Biol* 19(7):4918–4926

Mariman EC, van Eekelen CA, Reinders RJ, Berns AJ, van Venrooij WJ (1982) Adenoviral heterogeneous nuclear RNA is associated with the host nuclear matrix during splicing. *J Mol.Biol* 154(1):103–119

Michl P, Downward J (2006) CUTL1: a key mediator of TGF beta-induced tumor invasion. *Cell Cycle* 5(2):132–134

Mirkovitch J, Mirault ME, Laemmli UK (1984) Organization of the higher-order chromatin loop: Specific DNA attachment sites on nuclear scaffold. *Cell* 39(1):223–232

Mitchell RS, Beitzel BF, Schroder AR, Shinn P, Chen H, Berry CC, Ecker JR, Bushman FD (2004) Retroviral DNA integration: ASLV, HIV, and MLV show distinct target site preferences. *PLoS Biol* 2(8):E234

Moon NS, Zeng W Rong, Premdas P, Santaguida M, Berube G, Nepveu A (2002) Expression of N-terminally truncated isoforms of CDP/CUX is increased in human uterine leiomyomas. *Int J Cancer* 100(4):429–432

Mooslehner K, Karls U, Harbers K (1990) Retroviral integration sites in transgenic Mov mice frequently map in the vicinity of transcribed DNA regions. *J Virol* 64(6):3056–3058

Nepveu A (2001) Role of the multifunctional CDP/Cut/Cux homeodomain transcription factor in regulating differentiation, cell growth and development. *Gene* 270(1–2):1–15

Oesterreich S, Lee AV, Sullivan TM, Samuel SK, Davie JR, Fuqua SA (1997) Novel nuclear matrix protein HET binds to and influences activity of the HSP27 promoter in human breast cancer cells. *J Cell Biochem* 67(2):275–286

Orrick LR, Olson MO, Busch H (1973) Comparison of nucleolar proteins of normal rat liver and Novikoff hepatoma ascites cells by two-dimensional polyacrylamide gel electrophoresis. *Proc Natl Acad Sci USA* 70(5):1316–1320

Rampalli S, Pavithra L, Bhatt A, Kundu TK, Chattopadhyay S (2005) Tumor suppressor SMAR1 mediates cyclin D1 repression by recruitment of the SIN3/histone deacetylase 1 complex. *Mol Cell Biol* 25(19):8415–8429

Razin SV, Petrov P, Hancock R (1991) Precise localization of the alpha-globin gene cluster within one of the 20- to 300-kilobase DNA fragments released by cleavage of chicken chromosomal DNA at topoisomerase II sites *in vivo*: evidence that the fragments are DNA loops or domains. *Proc Natl Acad Sci USA* 88(19):8515–8519

Raziuddin A, Court D, Sarkar FH, Liu YL, Kung H, Raziuddin R (1997) A c-erbB-2 promoter-specific nuclear matrix protein from human breast tumor tissues mediates NF-kappaB DNA binding activity. *J Biol Chem* 272(25):15715–15720

Rogalla P, Drechsler K, Kazmierczak B, Rippe V, Bonk U, Bullerdiek J (1997) Expression of HMGI-C, a member of the high mobility group protein family, in a subset of breast cancers: relationship to histologic grade. *Mol Carcinog* 19(3):153–156

Schanke JT, Marcuzzi A, Podzorski RP, Van Ness B (1994). An AP1 binding site upstream of the kappa immunoglobulin intron enhancer binds inducible factors and contributes to expression. *Nucleic Acids Res* 22(24):5425–5432

Schoenmakers EF, Wanschura S, Mols R, Bullerdiek J, Berghe H. Van den, Van de Ven WJ (1995) Recurrent rearrangements in the high mobility group protein gene, HMGI-C, in benign mesenchymal tumours. *Nat Genet* 10(4):436–444

Schroder AR, Shinn P, Chen H, Berry C, Ecker JR, Bushman F (2002) HIV-1 integration in the human genome favors active genes and local hotspots. *Cell* 110(4):521–529

Schubeler D, Mielke C, Maass K, Bode J (1996) Scaffold/matrix-attached regions act upon transcription in a context-dependent manner. *Biochemistry* 35(34):11160–11169

Shera KA, Shera CA, McDougall JK (2001) Small tumor virus genomes are integrated near nuclear matrix attachment regions in transformed cells. *J Virol* 75(24):12339–12346

Sinclair AM, Lee JA, Goldstein A, Xing D, Liu S, Ju R, Tucker PW, Neufeld EJ, Scheuermann RH (2001) Lymphoid apoptosis and myeloid hyperplasia in CCAAT displacement protein mutant mice. *Blood* 98(13):3658–3667

Smith HC, Berezney R (1980) DNA polymerase alpha is tightly bound to the nuclear matrix of actively replicating liver. *Biochem Biophys Res Commun* 97(4):1541–1547

Smith HC, Puvion E, Buchholtz LA, Berezney R (1984) Spatial distribution of DNA loop attachment and replicational sites in the nuclear matrix. *J Cell Biol* 99(5):1794–1802

Sperry AO, Blasquez VC, Garrard WT (1989) Dysfunction of chromosomal loop attachment sites: Illegitimate recombination linked to matrix association regions and topoisomerase II. *Proc Natl Acad Sci USA* 86(14):5497–5501

Strick R, Laemmli UK (1995) SARs are cis DNA elements of chromosome dynamics: synthesis of a SAR repressor protein. *Cell* 83(7):1137–1148

Stunkel W, Huang Z, Tan SH, O'Connor MJ, Bernard HU (2000) Nuclear matrix attachment regions of human papillomavirus type 16 repress or activate the E6 promoter, depending on the physical state of the viral DNA. *J Virol* 74(6):2489–2501

Tamimi Y, van der Poel HG, Denyn MM, Umbas R, Karthaus HF, Debruyne FM, Schalken JA (1993) Increased expression of high mobility group protein I(Y) in high grade prostatic cancer determined by in situ hybridization. *Cancer Res* 53(22):5512–5516

Townson SM, Sullivan T, Zhang Q, Clark GM, Osborne CK, Lee AV, Oesterreich S (2000) HET/SAF-B overexpression causes growth arrest and multinuclearity and is associated with aneuploidy in human breast cancer. *Clin Cancer Res* 6(9):3788–3796

Tufarelli C, Fujiwara Y, Zappulla DC, Neufeld EJ (1998) Hair defects and pup loss in mice with targeted deletion of the first cut repeat domain of the Cux/CDP homeoprotein gene. *Dev Biol* 200(1):69–81

van Wijnen AJ, Bidwell JP, Fey EG, Penman S, Lian JB, Stein JL, Stein GS (1993) Nuclear matrix association of multiple sequence-specific DNA binding activities related to SP-1, ATF, CCAAT, C/EBP, OCT-1, and AP-1. *Biochemistry* 32(33):8397–8402

von Kries JP, Buhrmester H, Stratling WH (1991) A matrix/scaffold attachment region binding protein: identification, purification, and mode of binding. *Cell* 64(1):123–135

Walker JR, Corpina RA, Goldberg J (2001) Structure of the Ku heterodimer bound to DNA and its implications for double-strand break repair. *Nature* 412(6847):607–614

Will K, Warnecke G, Wiesmuller L, Deppert W (1998) Specific interaction of mutant p53 with regions of matrix attachment region DNA elements (MARs) with a high potential for base-unpairing. *Proc Natl Acad Sci USA* 95(23):13681–13686

Wood LJ, Maher JF, Bunton TE, Resar LM (2000a) The oncogenic properties of the HMG-I gene family. *Cancer Res* 60(15):4256–4261

Wood LJ, Mukherjee M, Dolde CE, Xu Y, Maher JF, Bunton TE, Williams JB, Resar LM (2000b) HMG-I/Y, a new c-Myc target gene and potential oncogene. *Mol Cell Biol* 20(15):5490–5502

Wunderlich V, Bottger M (1997) High-mobility-group proteins and cancer–an emerging link. *J Cancer Res Clin Oncol* 123(3):133–140

Zahn-Zabal M, Kobr M, Girod PA, Imhof M, Chatellard P, de Jesus M, Wurm F, Mermod N (2001) Development of stable cell lines for production or regulated expression using matrix attachment regions. *J Biotechnol* 87(1):29–42

Zeitlin S, Parent A, Silverstein S, Efstratiadis A (1987) Pre-mRNA splicing and the nuclear matrix. *Mol Cell Biol* 7(1):111–120

Zhou X, Benson KF, Ashar HR, Chada K (1995) Mutation responsible for the mouse pygmy phenotype in the developmentally regulated factor HMGI-C. *Nature* 376(6543):771–774

Zhu Q, Gregg K, Lozano M, Liu J, Dudley JP (2000) CDP is a repressor of mouse mammary tumor virus expression in the mammary gland. *J Virol* 74(14):6348–6357

Zweyer M, Riederer BM, Ochs RL, Fackelmayer FO, Kohwi-Shigematsu T, Bareggi R, Narducci P, Martelli AM (1997) Association of nuclear matrix proteins with granular and threaded nuclear bodies in cell lines undergoing apoptosis. *Exp Cell Res* 230(2):325–336

PART III

EPIGENETIC MODIFICATIONS OF CHROMATIN

Implication in disease and therapeutics

CHAPTER 11

ABERRANT FORMS OF HISTONE ACETYLTRANSFERASES IN HUMAN DISEASE

OLIVIER VAN BEEKUM AND ERIC KALKHOVEN*

Department of Metabolic and Endocrine Diseases, UMC Utrecht, Lundlaan 6, 3584 EA, Utrecht, The Netherlands

Abstract: One of the major mechanisms through which eukaryotic cells respond to developmental and environmental signals is by changing their gene expression patterns. This complex and tightly regulated process is largely regulated at the level of RNA polymerase II-mediated transcription. Within this process an important class of transcriptional regulators are the histone acetyltransferases (HATs), proteins that acetylate histones and non-histone substrates. While hyperacetylation of histones is generally associated with active genes, the effect of acetylation of nonhistone proteins varies between substrates resulting in for example alterations in (sub-nuclear) protein localization or protein stability. Given the central role of HATs in transcriptional regulation and other cellular processes, it may not be surprising that genetic alterations in the genes encoding HATs, resulting in aberrant forms of these regulatory proteins, have been linked with various human diseases, including congenital developmental disorders and various forms of cancer, including leukaemia. Here we will review mutations found in genes encoding human HATs and discuss the (putative) functional consequences on the function of these proteins. So far the lessons learned from naturally occurring mutations in humans have proven to be invaluable and recapitulating such genetic alterations in various experimental systems will extend our knowledge even further. This seems particularly relevant given the wide range of diseases in which acetyltransferases have been implicated and may help to open up new therapeutic avenues

Abbreviations: CBP, CREB binding protein; CREB, cAMP response element binding protein; FAT, factor acetyltransferase; HAT, histone acetyltransferase; MLL, mixed lineage leukaemia protein; MORF, MOZ related factor; MOZ, monocytic leukaemia zinc finger protein; PHD, plant homeodomain; RTS, Rubinstein-Taybi syndrome, TIF, transcription intermediary factor

* E-mail: E.Kalkhoven@umcutrecht.nl

T.K. Kundu and D. Dasgupta (eds.), Chromatin and Disease, 233–262.

1. CHROMATIN AND ACETYLATION

Regulation of transcription is a central mechanism by which cells respond to developmental and environmental cues. RNA polymerase II-mediated transcription in eukaryotes is to a large extent regulated at the level of chromatin, which forms a physical barrier for the binding of proteins to the promoter region of a target gene. The basic unit of chromatin is the nucleosome, which consists of an octamer of histone proteins around which the DNA is wrapped (see Fig. 1a).

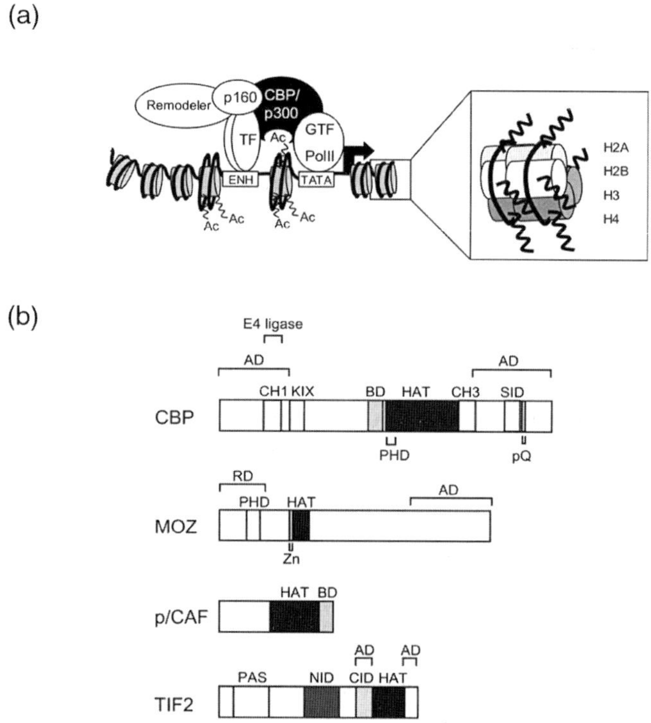

Figure 1. Transcription regulation and HATs. (a) Schematic representation of the promoter region of a gene, with the DNA wrapped around nucleosomes, which consist of octamers the histone proteins H2A, H2B, H3 and H4. A transcription factor dimer (TF), bound to the enhancer region (ENH), recruits remodelling factors (remodeler) and HATs, like CBP or p300 (CBP/p300), sometimes indirectly through other coactivators like the p160 proteins. CBP and p300 can make the DNA more accessible for other regulatory proteins by acetylating the histone tails (Ac). In addition, CBP and p300 can form a physical bridge between transcription factors and the general transcription factors (GTF) and RNA polymerase II (Pol II). See text for details. (b) Domain structure of HATs. Linear representation of proteins representing four different HAT families with regions and functional domains that are highly conserved among species indicated. Shown are CBP (CBP/p300 family), MOZ (MYST family), p/CAF (p/CAF/GCN5 family) and TIF2 (p160 family). AD, activation domain; BD, bromodomain; CH, cyteine-histidine-rich region; CID, CBP/p300 interaction domain; NID, nuclear receptor interaction domain; PAS, PAS/HLH domain; PHD, PHD-type zinc finger; pQ, polyglutamine stretch; RD, repression domain; SID, SRC1 interaction domain; Zn, C2H2 type zinc finger

The nucleosomes form an array that is ordered into higher-order chromatin structures. One of the major enzymatic activities which make the DNA more accessible for the transcription machinery are the histone acetyltransferases (HATs) (Cheung *et al.*, 2000b). The prime targets of the HATs in chromatin are the N-terminal tails of the core histones H2A, H2B, H3 and H4, which protrude away from the DNA. Acetylation of these tails results in neutralization of the positively charged lysines, thereby modifying DNA-histone and histone-histone contacts (Kuo and Allis, 1998; Cheung *et al.*, 2000b). The N-terminal histone tails are also subject to other modifications, like methylation (Kouzarides, 2002), phosphorylation (Cheung *et al.*, 2000a) and sumoylation (Shiio and Eisenman, 2003). Furthermore, the C-terminus of H2A and H2B can be ubiquitinylated (Jason *et al.*, 2002). These modifications not only change accessibility to the DNA and create specific docking sites for proteins (Winston and Allis, 1999), but they also represent the so-called "histone code", a bar code which may be involved in the establishment of epigenetic inheritance (Turner, 2000). The discovery that several known transcriptional coactivator proteins contain HAT activity, while repressors possess histone deacetylase (HDAC) activity, has strongly linked histone acetylation to transcriptional activation, and deacetylation to repression (Kuo and Allis, 1998; Strahl and Allis, 2000).

When a specific gene is activated, a cascade of chromatin modifications mediated by the ATP-dependent remodelling complexes and HATs makes the DNA accessible for transcription factors, including general transcription factors such as TFIID. Following assembly of the correct combination of these proteins RNA polymerase II is recruited and transcription takes place. Interestingly, the combination of the chromatin modifying enzymes required, and their order of recruitment, are promoter-specific, but at least one HAT is involved in promoter activation (Cosma *et al.*, 1999; Agalioti *et al.*, 2000; Shang *et al.*, 2000; Soutoglou and Talianidis, 2002; Martens *et al.*, 2003), indicating the fundamental importance of this protein family in biology.

2. STRUCTURE AND FUNCTION OF HISTONE ACETYLTRANSFERASES

Mammalian HAT enzymes can be divided into subfamilies (Tan, 2001). However, it is currently difficult to classify a protein as a potential HAT enzyme based on its amino acid sequence, since these subfamilies display no obvious similarity in their primary sequence, nor in the size of their HAT domains or the surrounding protein modules (Kuo and Allis, 1998; Marmorstein, 2001). The only region that is partly conserved between HAT subfamilies, either on the amino acid sequence and/or structural level, is a small subdomain first noticed in GCN5-related N-acetyltransferases, which encompasses the coenzyme A (CoA) binding site (Neuwald and Landsman, 1997; Martinez-Balbas *et al.*, 1998; Yan *et al.*, 2000; Marmorstein and Roth, 2001). Four families of mammalian HATs that have been implicated in human disease will be discussed here.

The first family consists of the CREB binding protein (CBP) and the related protein p300. CBP and/or p300 homologues are present in many multicellular organisms, including flies, worms and plants, but not in lower eukaryotes such as yeast (Champagne *et al.*, 1999b; Bordoli *et al.*, 2001b; Yuan and Giordano, 2002). While CBP was originally isolated as a coactivator of the transcription factor CREB (Chrivia *et al.*, 1993), and p300 was cloned as a protein interacting with the transforming adenoviral E1A protein (Eckner *et al.*, 1994), both proteins were subsequently shown to be interchangeable for these functions (Arany *et al.*, 1995; Lee *et al.*, 1996). CBP and p300 are ubiquitously expressed during mouse development (Partanen *et al.*, 1999). Comparison of the amino acid sequences of these multidomain proteins from different species revealed the presence of numerous regions of near identity, including three cysteine-histidine rich regions (CH1, −2 and −3), the binding site for the CREB transcription factor, referred to as the KIX domain, the bromodomain, the HAT domain and the steroid receptor coactivator-1 interaction domain (SID), while other regions are poorly conserved (Arany *et al.*, 1994; see also Fig. 1b). CBP and p300 can interact with the basal transcription factors TATA-binding protein (TBP; (Yuan *et al.*, 1996)) and TFIIB (Yuan *et al.*, 1996; Kwok *et al.*, 1994) and/or form a complex with RNA polymerase II (Nakajima *et al.*, 1997a, b; Cho *et al.*, 1998; Neish *et al.*, 1998). These interactions occur through an N- and C-terminal activation domain (see Fig. 1b). In addition, CBP and p300 can bind to a variety of diverse transcription factors, and other proteins, through their CH1, CH3, KIX and SID domains (Chan and La Thangue, 2001; Vo and Goodman, 2001). By interacting simultaneously with the basal transcription machinery and with one or more upstream transcription factors, CBP and p300 function as physical bridges or scaffolds and thereby stabilize the transcription complex. Interestingly, several of these protein-protein interactions can be regulated by the same post-translational modifications that chromatin is subject to, such as phosphorylation (Janknecht and Nordheim, 1996; Ait-Si-Ali *et al.*, 1998, 1999;), sumoylation (Girdwood *et al.*, 2003) methylation (Xu *et al.*, 2001; Chevillard-Briet *et al.*, 2002), indicating that CBP and p300 are themselves targets of signalling cascades.

The second important aspect of the coactivator function of CBP and p300 is their ability to acetylate promoter proximal nucleosomal histones, resulting in increased accessibility of the DNA for other essential regulators (Bannister and Kouzarides, 1996; Ogryzko *et al.*, 1996; Kundu *et al.*, 2000). Within the HAT domain of CBP and p300 two functionally important regions have been identified, the first of which is the aforementioned CoA binding site (amino acids 1459–1541 in CBP) (Neuwald and Landsman, 1997; Martinez-Balbas *et al.*, 1998; Yan *et al.*, 2000; Marmorstein and Roth, 2001).The second important region in the HAT domain of CBP, which is not found in other HAT proteins, is the PHD type zinc finger (Aasland *et al.*, 1995), also named leukaemia-associated-protein (LAP) finger (Saha *et al.*, 1995) or trithorax consensus (TTC) finger (Koken *et al.*, 1995). This type of zinc-finger is characterised by a cysteine 4-histidinecysteine 3 motif, and is predominantly found in proteins that function at the chromatin level (Aasland *et al.*, 1995). The PHD finger is an integral part of the enzymatic core of the CBP HAT domain (Bordoli *et al.*, 200la; Kalkhoven *et al.*, 2002) but is dispensable for p300 HAT activity (Bordoli *et al.*, 200la). The HAT domain is preceded by the bromodomain, a 110 amino acid domain which is found in many chromatin-associated proteins (Jeanmougin *et al.*, 1997; Winston and Allis, 1999). Bromodomains function as acetyl-lysine binding domains (Dhalluin *et al.*, 1999; Jacobson *et al.*, 2000), and could therefore play a role in tethering CBP and p300 to specific chromosomal sites (Manning *et al.*, 2001). Interestingly, the bromodomain of p300 was recently shown to cooperate with the PHD finger in binding of hyperacetylated nucleosomes (Ragvin *et al.*, 2004). The preferred *in vitro* sites of acetylation on the N-terminal histone tails are lysine 12 (K12) and K15 in histone H2B, K14 and K18 in histone H3 and K5 and K8 in histone H4 (Schiltz *et al.*, 1999). In addition to their ability to acetylate histones, CBP and p300 have also been shown to acetylate other proteins like transcription factors and coactivators, and are therefore also called factor acetyltransferases (FAT; Sterner and Berger, 2000; Yang, 2004). Acetylation of non-histone substrates can result in either positive or negative effects on transcription by affecting for example protein-protein interactions (e.g. the activator of thyroid and retinoid receptors ACTR (Chen *et al.*, 1999b)), protein-DNA interactions (e.g. the high mobility group protein HMGI(Y) (Munshi *et al.*, 1998)), nuclear retention (e.g. the hepatocyte nuclear factor HNF4 (Soutoglou *et al.*, 2000)) or protein half-life (e.g. E2F (Martinez-Balbas *et al.*, 2000)). While CBP and p300 are essential coactivators for a variety of transcription factors, the relative importance of the bridge/scaffold, HAT and FAT functions varies (Korzus *et al.*, 1998; Kurokawa *et al.*, 1998; Kraus *et al.*, 1999).

A second family of histone HATs was named the MYST family, after its founding members *M*OZ, yeast *Y*BF2 (renamed Sas3), yeast *S*as2, and *T*IP60. Proteins in this family share a C2H2 zinc finger and an HAT domain that together comprises the MYST domain (Yang, 2004). Outside this domain, which encompasses approximately 370 amino acids, the overall homology between family members is very limited, and they differ substantially in other structural domains and size (Yang, 2004). Two mammalian members of the family, MOZ (*mo*nocytic leukaemia *z*inc finger protein) and MORF (MOZ related factor) however display significant homology, both in amino acid sequence and in domain architecture. As depicted in Fig. 1b, MOZ and MORF contain 2 PHD-type zinc fingers in the C-terminus and a centrally located MYST domain. In addition, a repression domain was found in the N-terminus, while the C-terminus contains an activation domain (Champagne *et al.*, 1999a, 2001). Despite their high degree of homology MOZ and MORF display substrate specificity: both MOZ and MORF acetylate histones H3 and H4 *in vitro*, but MOZ also acetylates H2B (Champagne *et al.*, 1999a, 2001). Moreover, MORF was reported to acetylate nucleosomes, while MOZ was unable to do so (Champagne *et al.*, 2001). Unlike CBP and p300, no non-histone substrates have been described for either MOZ or MORF up to date. Also in contrast to the situation with CBP and p300, MOZ and MORF seem to function as coactivators for only a limited number of transcription factors, including RUNXl/AMLl/CBFA2 (Kitabayashi *et al.*, 2001; Bristow and Shore, 2003) and RUNX2/AML2/CBFAl (Pelletier *et al.*, 2002). MOZ and MORF reside in multi-subunit protein complexes (Doyon *et al.*, 2006).

A third family of HATs contains, amongst others, the mammalian GCN5 and p/CAF (*p*300/*C*BP *a*ssociated *f*actor) proteins. While both proteins have been shown to be able to interact with CBP and p300 (Yang *et al.*, 1996; Xu *et al.*, 1998), subsequent studies have shown that CBP and p300 are not part of the p/CAF core complex (Ogryzko *et al.*, 1998). GCN5 and p/CAF harbour a C-terminal HAT domain of approximately 160 amino acids, followed by a bromodomain. The N-terminal half of these proteins could be involved in determining substrate specificity. Both GCN5 and p/CAF have been shown to hction as coactivators for many transcription factors, through their ability to acetylate histones or non-histone substrates (Sterner and Berger, 2000; Yang, 2004). GCN5 and p/CAF are part of large multi-protein complexes (Ogryzko *et al.*, 1998; Brand *et al.*, 1999; Martinez *et al.*, 2001), and their enzymatic activity is regulated by other proteins in the complex, since for example p/CAF efficiently acetylates K14 on histone H3 *in vitro*, but only the p/CAF complex and not the recombinant protein can display the same activity towards nucleosomes (Ogryzko *et al.*, 1998).

The fourth family of HATS encompasses the three p160 coactivator proteins SRC1/NCoAl, TIF2/NCoA2/GRIP1 and ACTR/AIBl/RAC3/pCIP/NCoA3 (Leo and Chen, 2000). These three proteins were originally identified through their ligand-dependent interaction with transcription factors of the nuclear receptor family. This interaction requires one or more highly conserved LXXLL motifs (in which L is leucine and X any amino acid), which are grouped together in the nuclear receptor interaction domain (NID) in the p160 proteins (Heery *et al.*, 1997). Outside the NID, the p160 proteins share an N-terminal bHLH-PAS domain, which can mediate interactions with other transcription factors (Belandia and Parker, 2000). C-terminal of the NID, these proteins harbour two activation domains (AD), the first of which corresponds to the CBP/p300 interaction domain while the more C-terminal AD is the binding site for histone methyltransferase CARMl (Chen *et al.*, 1999a). Finally, the C-termini of SRCl and ACTR have been shown to harbour HAT activity (Chen *et al.*, 1997; Spencer *et al.*, 1997). The preferred acetylation substrates of SRCl and ACTR are H3 (K9 and K14) and H4, either as free histones or as mononucleosomes (Chen *et al.*, 1997; Spencer *et al.*, 1997). However, the ability of pl60 proteins to acetylate histones *in vitro* appears to be weak compared to CBP/p300 and GCN5/p/CAF proteins (Voegel *et al.*, 1998). In addition, deletion of the HAT domain of SRCl had no effect in *in vitro* transcription reactions mediated by NRs on chromatin templates (Liu *et al.*, 2001), indicating that the functional significance of this intrinsic HAT activity remains to be established. These findings suggest that histones might not be the primary acetylation targets of the p160 proteins or that the main role of these proteins in transcription is the recruitment of other, more active HATs like CBP/p300 and GCN5/p/CAF.

3. HATs AND HUMAN DISEASE

The importance of HATs in the regulation of normal differentiation, growth control and homeostasis is underscored by the fact that genetic alterations in their genes and functional (in)activation of the proteins are strongly linked to human disease (Giles *et al.*, 1998; Cairns, 2001; Klochendler-Yeivin and Yaniv, 2001; Timmermann *et al.*, 2001). Several different types of genetic aberrations involving HAT genes have been reported. Firstly, chromosomal translocations in which the 5′ end of a gene on one chromosome is juxtaposed to the 3′ end of a gene on another chromosome can result in the generation of gene fusions between a HAT and another gene (see Fig. 2a). Translocations are called balanced when the complementary chromosomal fusion is also detected. Secondly, point mutations can either result in an amino acid change or a stop codon (see Fig. 2b). Examples of such mutations are the S23X mutation, in which a serine (S) at position 23 is changed into a stop codon (X), and the A981T mutation, which changes an alanine (A) at position 981 into a threonine (T) (Table 1). Thirdly, deletions or insertions of a small number

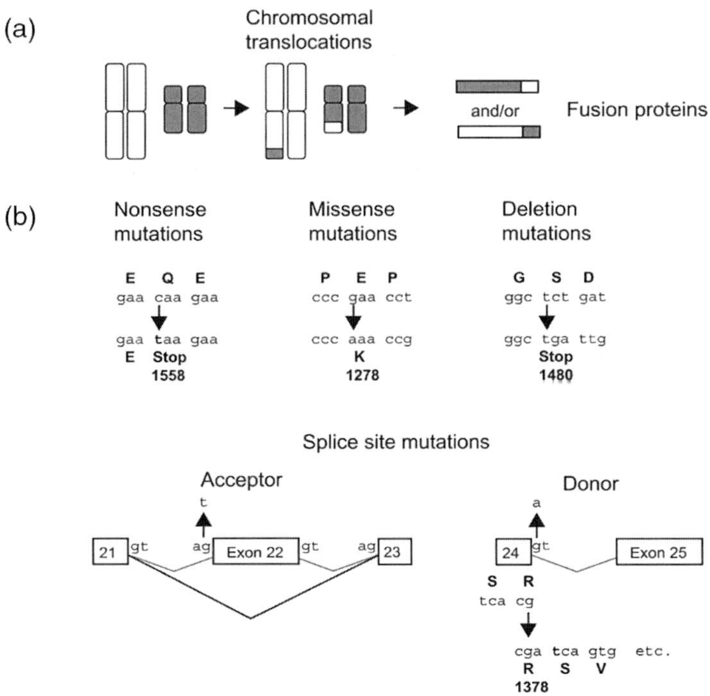

Figure 2. Genetic aberrations observed in HAT genes. (a) Schematic representation of a balanced chromosomal translocation. These translocations result in the formation two new fusion genes, which can give rise to one or two fusion proteins. (b) Examples of nonsense (RTS patient RT163.1), missense (RT209.1), deletion (followed by frame shift; RT231.1) mutations, as well as splice site acceptor (RT211.3) or splice site donor (RT39.1) mutations

Table 1. CBP and p300 mutations found in RTS patients[a]

Mutation type	Amino acid change[b]	Reference
CBP		
Nonsense mutation		
	S23X[c]	(Coupry *et al.*, 2002)
	Q102X	(Roelfsema *et al.*, 2005)
	Q135X	(Petrij *et al.*, 1995)
	Q356X	(Petrij *et al.*, 1995)
	R370X (3x)[d]	(Bartsch *et al.*, 2002), (Coupry *et al.*, 2002)
	R424X	(Bartsch *et al.*, 2005)
	E996X	(Bartsch *et al.*, 2005)
	C1213X	(Bartsch *et al.*, 2005)
	K1267X	(Roelfsema *et al.*, 2005)
	K1269X	(Coupry *et al.*, 2002)
	R1489X	(Coupry *et al.*, 2002)
	Y1466X	(Coupry *et al.*, 2002)
	Ql558X	(Kalkhoven *et al.*, 2003)
	Q1879X	(Bartsch *et al.*, 2005)
	R2004X	(Roelfsema *et al.*, 2005)
	Q2007X	(Bartsch *et al.*, 2005)
	Q2043X	(Coupry *et al.*, 2002)
	Q2045X	(Roelfsema *et al.*, 2005)
	Q2095X	(Roelfsema *et al.*, 2005)
Missense mutation		
	A981T	(Coupry *et al.*, 2002)
	Y1175C	(Bartsch *et al.*, 2002)
	E1278K	(Kalkhoven *et al.*, 2003)
	R1378P	(Murata *et al.*, 2001)
	D1435V	(Bartsch *et al.*, 2005)
	Y1450H	(Roelfsema *et al.*, 2005)
	T14471	(Roelfsema *et al.*, 2005)
	1470R	(Roelfsema *et al.*, 2005)
	1664H (2x)[d]	(Kalkhoven *et al.*, 2003), (Bartsch *et al.*, 2005)
	N1978S (2x)[d]	(Coupry *et al.*, 2002), (Bartsch *et al.*, 2005)
	M2221L	(Coupry *et al.*, 2002)
	A2243L	(Coupry *et al.*, 2002)
Deletion/insertion		
	D29fsX37	(Bartsch *et al.*, 2005)
	G79fsX86	(Roelfsema *et al.*, 2005)
	Q158fsX173	(Bartsch *et al.*, 2005)
	S302fsX348	(Roelfsema *et al.*, 2005)
	G461fsX469	(Roelfsema *et al.*, 2005)
	N494fsX527	(Roelfsema *et al.*, 2005)
	A581fsX586	(Roelfsema *et al.*, 2005)
	A631fsX632	(Bartsch *et al.*, 2005)
	P1132fsX1166	(Roelfsema *et al.*, 2005)
	T1144fsX1168	(Roelfsema *et al.*, 2005)
	Y1167fsX1167	(Murata *et al.*, 2001)

	S1256fsX1256	(Bartsch *et al.*, 2005)
	F1275fsX1282	(Roelfsema *et al.*, 2005)
	R1360fsX1382	(Petrij *et al.*, 2000a)
	V1361fsX1380	(Murata *et al.*, 2001)
	S1419fsX1419	(Kalkhoven *et al.*, 2003)
	R1441fsX1452	(Bartsch *et al.*, 2005)
	Y1466fsX1472	(Murata *et al.*, 2001)
	V1467fsX1467	(Kalkhoven *et al.*, 2003)
	G1469fsX1477	(Murata *et al.*, 2001)
	V1613fsX1634	(Kalkhoven *et al.*, 2003)
	P2017fsX2342	(Bartsch *et al.*, 2005)
Splice site mutation		
	Q405fsX414	(Bartsch *et al.*, 2002)
	T558fsX560	(Bartsch *et al.*, 2002)
	D1124fsX1124	(Coupry *et al.*, 2002)
	R1233fsX1252	(Coupry *et al.*, 2002)
	T1260fsX1260	(Kalkhoven *et al.*, 2003)
	Δ1280–1305	(Kalkhoven *et al.*, 2003)
	S1304fsX1328	(Roelfsema *et al.*, 2005)
	S1378fsX1404	(Kalkhoven *et al.*, 2003)
	T1426fsX1441	(Coupry *et al.*, 2002)
	K1520RfsX1549	(Coupry *et al.*, 2002)
	Δ1521–1576	(Coupry *et al.*, 2002)
p300		
Nonsense mutation		
	R648X	(Roelfsema *et al.*, 2005)
Deletion/insertion		
	S959fsX966	(Roelfsema *et al.*, 2005)

[a] Shown are nonsense, missense, deletion/insertion and splice site mutations.
[b] The amino acid changes are predictions in most cases.
[c] X denotes a stop codon.
[d] Number of independent cases.

of nucleotides have been observed, which often give rise to a frame shift and the addition of a number of new amino acids. An example of such a mutation is the D29fsX37 mutation (Table 1), in which a deletion 3′ of codon 29 (D29) results in a frameshift (fs) and the addition of 8 new amino acids until a stop codon is encountered. Finally, mutations around splice sites can have two different effects. Mutation of the splice acceptor site can give rise to an in-frame deletion of an exon or a frame shift combined with the addition of a number of new amino acids. When a splice donor site is mutated the prediction is that the intron will not be spliced out and will be translated until a stop codon is present (Fig. 2b).

If a genetic aberration is heterozygous, the resultant mutant protein could either (1) inhibit the remaining intact protein in a dominant negative manner, or (2) show a "loss-of-function", resulting in lowering of the functional dosage of the gene product (haploinsufficiency), or (3) display novel characteristics ("gain-of-function"), which can be easily envisaged in the case of chromosomal translocations (see also below). A genetic abberation of both alleles may be required if a certain gene functions as a classical tumour suppressor according to Knudson's two hit model (Knudson, Jr., 1971).

4. CBP, P300 AND RUBINSTEIN–TAYBI SYNDROME

Chromosomal translocations complete or partial deletions and point mutations in one copy of the *CBP* gene result in a congenital developmental disorder named Rubinstein–Taybi syndrome (RTS). The syndrome is characterised by retarded growth and mental function, broad thumbs, broad big toes and typical facial abnormalities (Rubinstein and Taybi, 1963). Furthermore, RTS patients have an increased risk to develop tumours, particularly in the nervous system including oligodendroglioma, medulloblastoma, neuroblastoma, and meningioma (Miller and Rubinstein, 1995). The genetic aberrations found in RTS patients are always *de novo*, meaning that the parents are not affected. Haploinsufficiency of CBP is the likely cause of RTS in humans, since no clear phenotypic differences were observed between patients with a complete deletion of one CBP allele or those with a single heterozygous point mutation were found (Kalkhoven *et al.*, 2003 and references therein). In mice, heterozygous deletion or truncation of CBP also leads to an RTS-like phenotype (Tanaka *et al.*, 1997; Oike *et al.*, 1999; Kung *et al.*, 2000). All nonsense, missense, deletion, insertion and splice site mutations observed in RTS patients so far are indicated in Table 1. When these mutations are indicated on a schematic representation of the CBP protein, it becomes clear that missense and splice site mutations are often predicted to lead to truncated forms of the CBP protein (Fig. 3a). In many cases the resulting protein would lack (part of) the HAT domain, and would therefore not be able to acetylate histones or non-histone substrates. Together with a number of missense mutations which have been shown (R1378P, E1278K) (Murata *et al.*, 2001; Kalkhoven *et al.*, 2003) or are predicted (Y1450H, T1447I, H1470R) (Roelfsema *et al.*, 2005) to abolish the HAT activity, these findings suggest that the full dosage of CBP HAT activity is required to prevent RTS. The fact that a number of genetic aberrations appear to affect the protein C-terminal of the HAT domain (e.g. N1978S; see Fig. 3a) does not necessarily contradict this hypothesis. Firstly, it should be noted that with a small number of exceptions (Petrij *et al.*, 2000b; Kalkhoven *et al.*, 2003), the expression of RTS-associated mutant proteins has not been formally proven. It is therefore possible that the mRNAs containing nonsense or splice site mutations are never translated because they are degraded through nonsense-mediated mRNA decay (NMD; Conti and Izaurralde, 2005), or that the mutant proteins are more prone to degradation by the proteasome. Secondly, if the mutant proteins are being formed, missense mutations that lie clearly outside the enzymatic core domain could still affect their enzymatic function. The first example of this might be a mutation in the bromodomain (Y1175C; Bartsch *et al.*, 2002), since this domain sometimes works in collaboration with the HAT domain (Kraus *et al.*, 1999; Manning *et al.*, 2001;

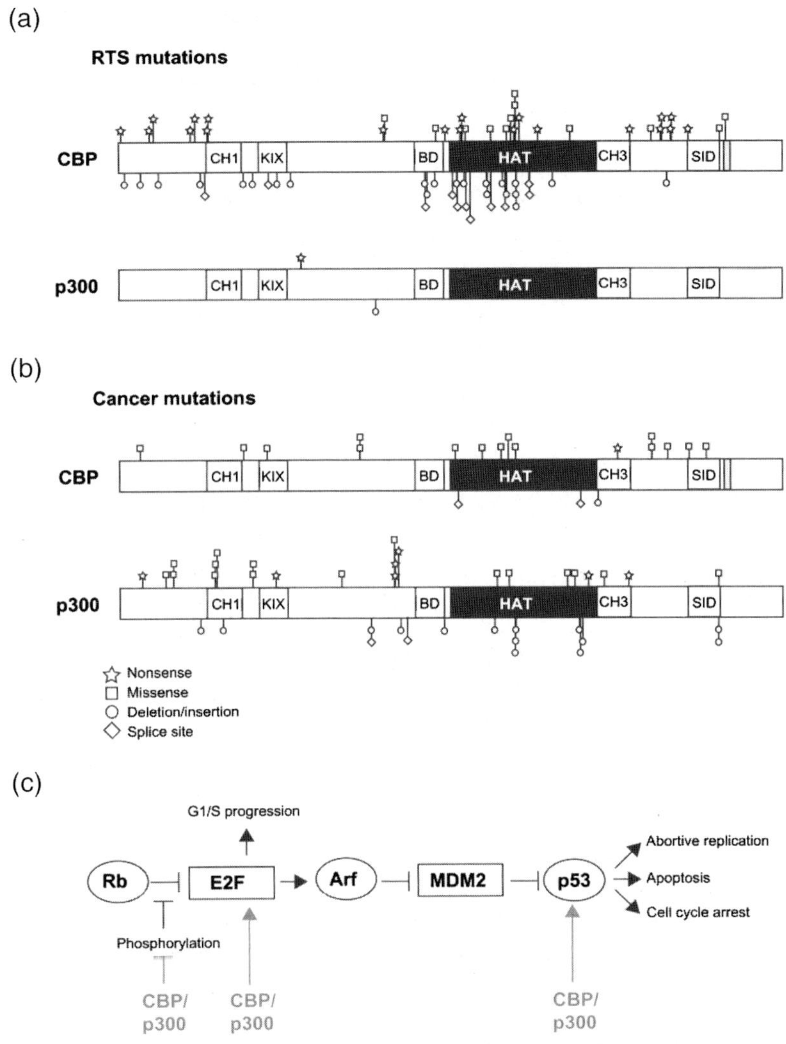

Figure 3. Localisation of mutations in CBP and p300 in RTS patients (panel a) and tumour material (panel b). Indicated are nonsense, missense, deletion/insertion and splice site mutations. (c) Schematic representaion of the Rb-E2F and p53 pathways and the effects of CBP and p300 on these pathways

Mitsiou and Stunnenberg, 2003). Moreover, enzyme-substrate interactions could be disrupted by missense mutations that are located throughout the protein. A clear example of this are the mutations in SRCl, which affect its interaction with CBP, but are located outside the CID region of SRCl (Rowan *et al.*, 2000; Wu *et al.*, 2004) (see also above and Fig. 1b).

Until recently RTS was only linked to the CBP gene, but Roelfsema *et al.* have reported 3 cases of RTS patients in which the p300 gene was affected (Roelfsema *et al.*, 2005). These three p300 mutations, two missense mutations (Table 1 and Fig. 3) and a complete deletion of exon 1 (Roelfsema *et al.*, 2005), are all predicted to result in loss of the HAT function. These interesting findings indicate that CBP and p300 have at least in part overlapping functions during development in humans, as is also the case in mice (Yao *et al.*, 1998; Kung *et al.*, 2000). It is however currently still unclear how the loss of one functional copy of CBP or p300 results in the well-defined Rubinstein-Taybi syndrome, given the plethora of transcriptional events in which CBP and p300 have been implicated (Blobel, 2000; Goodman and Smolik, 2000; Shikama *et al.*, 2000; Vo and Goodman, 2001; Kalkhoven, 2004). Both RTS patients and heterozygous CBP knock-out mice (Tanaka *et al.*, 1997; Yao *et al.*, 1998; Oike *et al.*, 1999; Kung *et al.*, 2000) indicate that the regulation of a subclass of developmentally important genes is extremely sensitive to the total CBP or p300 gene dosage. The identification of RTS-associated HAT mutants would argue that specific regulatory acetylation events on the promoters of these genes are essential in this respect. An important goal for the future will therefore be to identify these CBP/p300 specific target genes. In addition, it should be noted that in only 50% of all RTS patients a defect in either the CBP or the p300 gene was detected (Coupry *et al.*, 2002; Bartsch *et al.*, 2005; Roelfsema *et al.*, 2005). Whether genetic defects in the remainder of RTS patients are located downstream of the HAT, i.e. on the level of acetylation of non-histone substrate protein(s), remains to be established.

5. HATs AND SOLID TUMOURS

Two major pathways involved in cancer development are the Rb-E2F and p53 signalling cascades (see Fig. 3c). Both pathways involve suppression of tumourigenesis and both p53 as well as Rb are mutated in almost 50% of all known human tumours (Hanahan and Weinberg, 2000). In tumours that do not have mutations in these "hotspot" proteins, alterations in or downstream regulatory proteins are often found. One class of regulatory proteins for Rb-E2F and p53 are HATS (Avantaggiati *et al.*, 1997; Gu and Roeder, 1997; Lill *et al.*, 1997; Scolnick *et al.*, 1997; Martinez-Balbas *et al.*, 2000; Marzio *et al.*, 2000; Nguyen *et al.*, 2004). Indeed, many potentially inactivating mutations of CBP and p300, and two mutations in p/CAF, have been found in various cancer cell lines and primary tumours, mainly of epithelial origin (see Table 2 and Fig. 3b). Because only two mutations were described in the pCAF gene, we will focus here on CBP and p300. In some cases were CBP or p300 mutations were detected the second allele was inactivated through deletion (loss of heterozygosity; LOH), silencing (hemizygosity) or a different mutation (compound heterozygosity). At first sight, these findings qualify CBP and p300 as classical tumour suppressor genes according to Knudson's two-hit model (Knudson, Jr., 1971). In agreement with this, heterozygous CBP mice develop haematological malignancies with advancing age, and all the

Table 2. CBP, p300 and p/CAF mutations found solid tumours[a]

Mutation type	Sample	Cancer type	LOH	Reference
CBP				
Nonsense mutation				
E1835X[b,c]	Cell line	Lung	Hetero	(Kishimoto *et al.*, 2005)
Missense mutation				
N83T	Cell line	Lung	Hetero	(Kishimoto *et al.*, 2005)
A467T	Primary	Ovary	Homo	(Ward *et al.*, 2005)
L551I	Primary	Lung	Hetero	(Kishimoto *et al.*, 2005)
S893L	Cell line	Lung	Hetero	(Kishimoto *et al.*, 2005)
S893L	Primary	Breast	?	(Ward *et al.*, 2005)
A1071T	Primary	Ovary	?	(Ward *et al.*, 2005)
A1354V	Primary	Ovary	Homo	(Ward *et al.*, 2005)
G1411E	Primary	Lung	Homo	(Kishimoto *et al.*, 2005)
R1446C	Cell line	Lung	Hetero	(Kishimoto *et al.*, 2005)
W1472C	Cell line	Lung	Allele1	(Kishimoto *et al.*, 2005)
N1275S			Allele 2	
N1978D	Primary	Ovary	Homo	(Ward *et al.*, 2005)
N1978D	Primary	Breast	?	(Ward *et al.*, 2005)
A2044G	Primary	Lung	Hetero	(Kishimoto *et al.*, 2005)
N2111S/N2175S	Cell line	Lung	Hetero	(Kishimoto *et al.*, 2005)
Deletion/insertion				
P1764fsX1769	Cell line	Colon	Hetero	(Ionov *et al.*, 2004)
Splice site mutation				
Δ1280–1305	Cell line	Ovary	Hetero	(Ozdag *et al.*, 2002)
E1724fsX1739[d]	Cell line	Ovary	Hetero	(Ozdag *et al.*, 2002)
p300				
Nonsense mutation				
R86X	Cell line	Colorectal	Hetero	(Bryan *et al.*, 2002)
R580X	Primary	Colorectal	Homo	(Gayther *et al.*, 2000)
E1014X	Cell line	Colon	Hetero	(Gayther *et al.*, 2000)
E1014X	Cell line	Colon	Homo	(Ozdag *et al.*, 2002)

(Continued)

Table 2. (Continued)

Mutation type	Sample	Cancer type	LOH	Reference
E1039X	Cell line	Colon	Homo	(Ionov *et al.*, 2004)
E1039X	Cell line	Colon	Hetero	(Ionov *et al.*, 2004)
S1733X	Cell line	Ovary	Hetero	(Ozdag *et al.*, 2002)
Q1874X	Cell line	Oral	Hemi?	(Suganuma *et al.*, 2002)
Missense mutation				
T169M	Primary	Gastric	Homo	(Koshiishi *et al.*, 2004)
R202L	Primary	Gastric	Hetero	(Koshiishi *et al.*, 2004)
D204N	Primary	Gastric	?	(Koshiishi *et al.*, 2004)
K350E	Primary	Gastric	Hetero	(Koshiishi *et al.*, 2004)
G359E	Primary	Gastric	Homo	(Koshiishi *et al.*, 2004)
K373N[e]	Primary	Gastric	?	(Koshiishi *et al.*, 2004)
S507G[f]	Primary	MDS[g]	Hetero	(Shigeno *et al.*, 2004)
N508S	Primary	Gastric	Homo	(Koshiishi *et al.*, 2004)
L827P	Cell line	Breast	Hemi	(Gayther *et al.*, 2000)
E1013G	Cell line	Breast	Hemi	(Gayther *et al.*, 2000)
D1399Y	Primary	Gastric	Homo	(Muraoka *et al.*, 1996)
Y1446H	Primary	Gastric	Homo	(Koshiishi *et al.*, 2004)
S1650Y	Cell line	Colon	Hetero	(Gayther *et al.*, 2000)
R1680C	Primary	Colon	Homo	(Muraoka *et al.*, 1996)
P1780A	Cell line	Oral	Hemi?	(Suganuma *et al.*, 2002)
P2221Q	Primary	Colon	Homo	(Gayther *et al.*, 2000)
Deletion/insertion				
K291fsX315	Cell line	Colorectal	Homo	(Ionov *et al.*, 2004)
C382fsX404	Primary	Gastric	Hetero	(Koshiishi *et al.*, 2004)
P927fsX947	Cell line	Colorectal	Hetero	(Bryan *et al.*, 2002)

I1048fsX1076	Cell line	Breast	Hetero	(Gayther *et al.*, 2000)
L1409ins6	Primary	Breast	?	(Gayther *et al.*, 2000)
K1469fsX1494	Primary	Gastric	Hetero	(Koshiishi *et al.*, 2004)
K1469fsX1471	Cell line	Colorectal	Hemi?	(Bryan *et al.*, 2002)
K1469fsX1494	Cell line	Colorectal	Hemi?	(Bryan *et al.*, 2002)
E1687fsX1725	Cell line	Breast	Allele 1	(Gayther *et al.*, 2000)
del1198–1243			Allele 2	
C1695fsX1707	Cell line	Colon	Hemi	(Gayther *et al.*, 2000)
K1699fsX1701	Cell line	Colon	Hemi?	(Bryan *et al.*, 2002)
N2209Kde12210–2213	Primary	Breast	Hetero	(Bryan *et al.*, 2002)
		Ovary	Hemi	
Splice site mutation				
de11089–1095	Cell line	Lymphoid	Hetero	(Shigeno *et al.*, 2004)
de1940–1167	Cell line	Cervical	Homo	(Ohshima *et al.*, 2001)
P/CAF				
Missense mutation				
R653W	Cell line	Ovary	?	(Ozdag *et al.*, 2002)
P713T	Primary	Colorectal	Hetero	(Ozdag *et al.*, 2002)

[a] Shown are nonsense, missense, deletion/insertion and splice site mutations.
[b] The amino acid changes are predictions in many cases.
[c] X denotes a stop codon.
[d] Originally annotated as E 1724fsX1795
[e] Originally annotated as K372N.
[f] originally annotated as E507G.
[g] MDS, myelodysplastic syndrome.

malignancies that were checked showed LOH of the second CBP allele with preservation of p300 expression (Kung *et al.*, 2000). Interestingly, these malignancies were not found in p300 heterozygous mice of comparable age, indicating that p300 and CBP have different roles in tumour formation (Kung *et al.*, 2000). However, several lines of evidence indicate that the situation in humans is fundamentally different. Firstly, RTS patients who show heterozygous germline mutations of the CBP gene do not develop the same haematological malignancies found in CBP heterozygous mice, but display an increased disposition to neuronal tumours (Miller and Rubinstein, 1995). It is currently unknown whether the second CBP allele is affected in these tumours. More importantly, with increasing numbers of human tumours being examined it is now becoming clear that in many cases a heterozygous mutation of CBP or p300 is not accompanied by LOH (see also Table 2), while the frequently observed LOH of the p300 and CBP loci is often not found together with mutations in the other allele (Tillinghast *et al.*, 2003; So *et al.*, 2004). CBP and p300 therefore cannot be qualified as classical tumour suppressor genes in humans at this stage. These findings do suggest however that the dosage of their gene products is important in the prevention of uncontrolled cell growth (Fodde and Smits, 2002).

Given the central role of the p53 and Rb-E2F signalling pathways in oncogenesis (see Fig. 3c), it seems possible that tumour-associated CBP and p300 mutations could affect the normal function of these key proteins, thereby contributing to oncogenesis. CBP and p300 function as transcriptional coactivators for the p53 protein (Avantaggiati *et al.*, 1997; Gu and Roeder, 1997; Lill *et al.*, 1997; Scolnick *et al.*, 1997). In addition, the p53 protein was one of the first non-histone acetylation substrates (Gu and Roeder, 1997), and the C-terminus of p53 contains two acetylation sites for CBP/p300 (K373 and K382) and one for p/CAF (K320)(Liu *et al.*, 1999). These modifications were reported to increase DNA binding and stability of p53 and promote coactivator recruitment leading to increased transcriptional activity (Gu and Roeder, 1997; Liu *et al.*, 1999; Barlev *et al.*, 2001; Ito *et al.*, 2001). The significance of p53 acetylation *in vivo* however still remains to be established. Interestingly, the transcriptional activity of p53 is not only regulated through acetylation but also through other post-translational modifications like ubiquitination and phosphorylation and these processes are interrelated (Coutts and La Thangue, 2005). In a more indirect fashion, CBP and p300 affect p53 function by regulating its degradation. Degradation and ubiquitination of p53 is dependent on the murine double minute 2 protein (MDM2), and a ternary complex between these two proteins and CBP/p300 regulates the turnover of p53 in cycling cells (Grossman *et al.*, 1998; Kobet *et al.*, 2000). A more direct role of CBP and p300 in regulating p53 levels could be mediated through their CH1 regions, since this region region exhibits polyubiquitin (E4) ligase activity towards p53 *in vitro* (Grossman *et al.*, 2003).

Tumour-associated CBP and p300 mutations could deregulate p53 function on at least two different levels: transcriptional activity and/or stability. The missense mutations in the HAT domain, some of which have been shown to ablate enzymatic activity (Ito *et al.*, 2001), suggest that the transcriptional activity of p53 might be impaired in cancer cells with CBP or p300 mutations. Mutations affecting the CH3 or SID region could have the same effect, since they potentially disrupt the CBP/p300–p53 interaction. Because the same CH3 and SID mutations, as well as CH1 mutations, might disrupt the CBP/p300–p53-MDM2 complex or disrupt its E4 ligase activity, it is currently very difficult to predict what the net effect on p53 activity and protein levels would be.

If heterozygous mutations in CBP or p300 would be sufficient to dramatically deregulate p53, one might argue that when these enzymes are mutated there would be no need to acquire a mutation in p53 itself. This is however not the case. For example, in a panel of lung cancer sample, mutations in CBP were invariable found together with p53 mutations (Kishimoto *et al.*, 2005). It seems therefore likely that proteins other than p53 are also important HAT targets in oncogenesis. One of these targets could be the Rb-E2F pathway. Both Rb and E2F are acetylation substrates for CBP and p300 (Martinez-Balbas *et al.*, 2000; Marzio *et al.*, 2000; Chan *et al.*, 2001; Nguyen *et al.*, 2004). Acetylation of Rb negatively influences its phosphorylation by the cyclin/CDK complexes leading to enhanced inhibition of the E2F family of transcription factors and therefore sustained G1 arrest (Chan *et al.*, 2001; Nguyen *et al.*, 2004). The mechanism of this inhibition seems quite direct, since the acetylated lysines have been implicated as part of the CDK docking site (Markham *et al.*, 2006). In this model enzymatic mutations of CBP and p300 would increase the level of Rb phosphorylation, thereby stimulating G1/S progression and proliferation. The potential effects of CBP and p300 mutations on E2F function are however more difficult to explain. Acetylation of E2F has been shown to increase its DNA binding, activity and stability (Martinez-Balbas *et al.*, 2000; Marzio *et al.*, 2000). Impaired CBP/p300 function is therefore predicted to inhibit G1/S progression, which would be anti-tumourigenic. One should bear in mind though that since besides the p53 and Rb-E2F pathways discussed here, CBP and p300 have been implicated in a number of other critical pathways in oncogenesis, like the wnt/β-catenin pathway (Barker *et al.*, 2000; Sun *et al.*, 2000), the TGF-β pathway (Arteaga, 2006; Levy and Hill, 2006) and the sonic hedgehog pathway (Villavicencio *et al.*, 2000). The situation is complicated even further by the fact that some experiments suggest that CBP and p300 should not be regarded as tumour suppressor genes but rather as oncogenes. For example, loss of CBP or p300 results in impaired proliferation in mouse fibroblasts (Yao *et al.*, 1998), and high levels of p300 correlate with progression of prostate tumours (Debes *et al.*, 2003). Taken together, these findings indicate that although mutations in the CBP, p300 and p/CAF genes have been found in solid tumours, the exact role of these HATs in human oncogenesis remains to be established and could depend on the cell or tissue type.

6. HATs AND LEUKAEMIA

Current models indicate that (at least) two genetic hits are required for the development of acute leukaemia: one to stimulate self-renewal and a second hit to block differentiation (and/or prevent apoptosis) (Dash and Gilliland, 2001). In these models, which have recently been supported by mouse studies (Dash *et al.*, 2002), activating mutations in tyrosine kinase receptors like flt3 represent the first class, while fusion proteins resulting from chromosomal translocations are supposed to block differentiation (Kelly *et al.*, 2002). The recurrent t(8; 16)(p11;p 13.3) translocation found in AML patients for example fuses the region of the MOZ gene on chromosome 8 to the CBP gene on chromosome 16, and vice versa (Borrow *et al.*, 1996; see Fig. 4a). In other AML patients t(8;22)(p11;q13) fuses the MOZ and p300 genes, t(10; 16)(q22;p 13.3) fuses the MORF and CBP genes and t(10;22)(q22;q13) fuses the MOZ and p300 genes, while the inv(8)(pl lq13.1) inversion on chromosome 8 fuses the MOZ and TIF2 genes. In addition, in therapy-induced AML (t-AML) the CBP and p300 genes are rearranged with the MLL gene in patients with t(11;16)(q23;q23) and (t11;22)(q23;q13), respectively. Although translocations potentially result in the formation of two reciprocal fusion proteins (e.g. MOZ-CBP and CBP-MOZ) the mRNA encoding one of them (CBP-MOZ) is sometimes difficult to detect. Generally speaking, fusion proteins with regions of CBP, p300 or TIF2 at the C-terminus (MOZ-CBP, MORF-CBP, MOZ-p300,

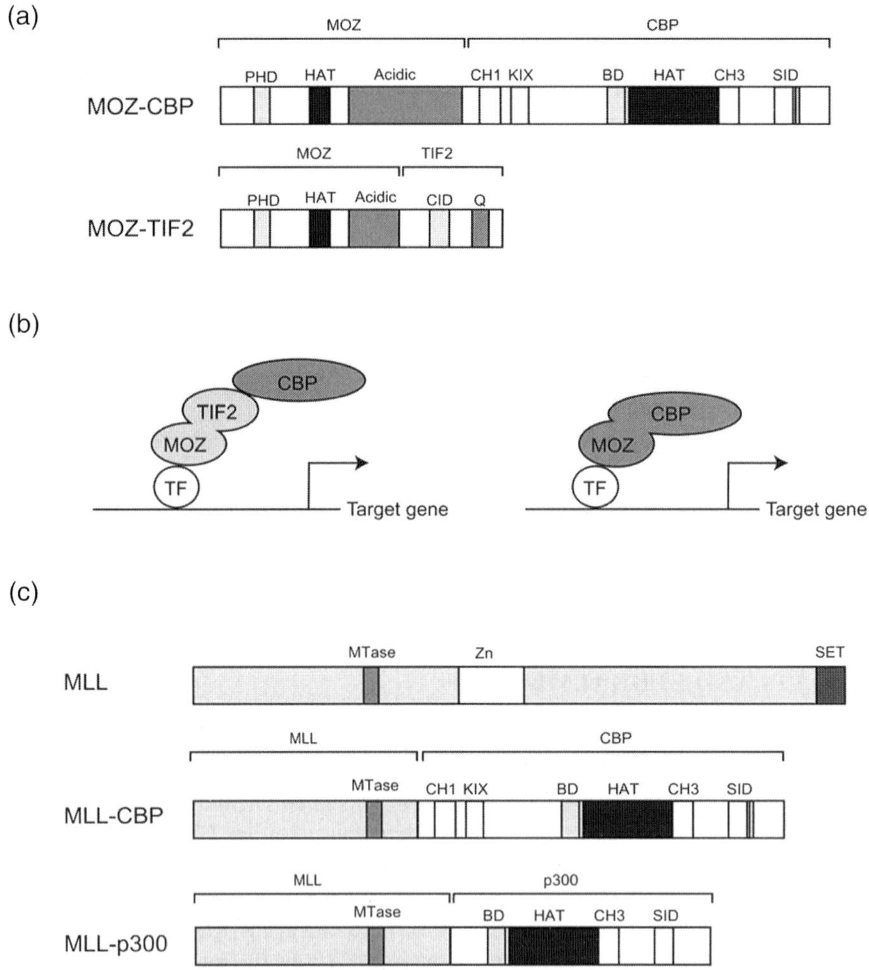

Figure 4. Leukaemic gene fusions involving HATs. (a) The t(8;16) translocation result in the MOZ-CBP fusion protein, while inv(8) gives rise to MOZ-TIF2. (b) Hypothetical model to explain how MOZ-TIF2 and MOZ-CBP fusions result in a similar leukaemia cell phenotype. See text for details. (c) The t(11; 16) and t(11;22) result in the MLL-CBP and MLL-p300 gene fusions, respectively

MOZ-TIF2, MLL-CBP and MLL-p300) appear to play the critical role in leukaemogenesis, since mRNAs encoding the reciprocal proteins (e.g. CBP-MOZ) could not be detected in all patients (Sobulo *et al.*, 1997; Taki *et al.*, 1997; Carapeti *et al.*, 1998; Liang *et al.*, 1998; Rozman *et al.*, 2004) or are out-of-frame (Borrow *et al.*, 1996).

Detailed inspection of the chromosomal fusion point has shown that in the leukaemic fusions the N-terminus of MOZ or MORF, which includes the PHD fingers and the HAT domain, is fused to CBP or p300. Although several variants have been described, the region of CBP and p300 in these fusions consists of almost the full-length protein in all cases, including an intact HAT domain (Borrow *et al.*, 1996; Chaffanet *et al.*, 2000; Panagopoulos *et al.*, 2000, 2001, 2002, 2003; Kojima *et al.*, 2003; Vizmanos *et al.*, 2003; Murati *et al.*, 2004; Rozman *et al.*, 2004; Schmidt *et al.*, 2004; Crowley *et al.*, 2005). The region of TIF2 that is fused to MOZ encompasses the CBP/p300 interaction domain, the putative HAT domain and the second activation domain, but lacks the LXXLL motifs required for interaction with nuclear hormone receptors (Carapeti *et al.*, 1998; Liang *et al.*, 1998; Billio *et al.*, 2002). The fusion proteins are likely to affect cellular function in multiple ways, including (i) reduction of the normal amount of the parental proteins, and/or (ii) aberrant function of the resultant proteins including both dominant negative and gain-offunction effects. Interestingly, the MOZ-CBP/p300 and MOZ-TIF2 gene fusions occur in a specific subset of AML patients (AML-M4/5), and these fusion proteins have therefore been proposed to give rise to the same phenotype. The CBP interaction domain (CID) is retained in the MOZ-TIF2 fusion (Fig. 4a), and this domain is required for myeloid cell proliferation and transformation (Deguchi *et al.*, 2003; Kindle *et al.*, 2005). These findings suggest that the role of the MOZ-TIF2 protein in leukaemia is in fact indirect and occurs through its effects on CBP (Fig. 4b). In agreement with this, both MOZ-TIF2 and MOZ-CBP act as dominant negative inhibitors of a number of CBP-dependent activators including nuclear receptors and p53, while activating the MOZ-dependent activator RUNX1/AML1 (Kindle *et al.*, 2005; Collins *et al.*, 2006). In addition, MOZ-TIF2 lowers cellular CBP levels, leading to depletion of CBP from subnuclear structures named PML bodies (Kindle *et al.*, 2005). These data together with the functional domains involved (Fig. 4a) lead to the hypothesis that disruption of the normal function of CBP, and in particular its HAT function, is a key event in AML, resulting in aberrant acetylation patterns. This idea is supported by the fact that the HAT domain of CBP is required for the MOZ-CBP mediated block of differentiation of a myeloid cell line (Kitabayashi *et al.*, 2001) and also more recently MOZ-TIF2 was indeed shown to alter histone modifications *in vivo* on the RARβ promoter (Collins *et al.*, 2006). It is currently however unknown whether histones and/or non-histone proteins are the most relevant substrates in this respect. In addition, the target genes of the leukaemic MOZ fisions, i.e. the set of genes on which these aberrant acetylation events have a positive or negative effect, await identification.

The t(11; 16) and t(11;22) translocations involve the MLL gene, which encodes a histone methyltransferase (Milne *et al.*, 2002; Nakamura *et al.*, 2002). The MLL can methylate histone H3 on K4, a modification that is associated with transcriptional activation. Recurrent chromosomal translocations resulting in fusion of the MLL gene on chromosome 1 lq23 to >30 different partner genes are found in 80% of children under 1 year with leukaemia (Dimartino and Cleary, 1999; Ayton and Cleary, 2001). In addition, MLL translocations are probably induced by treatment of AML patients with topoisomerase II inhibitors (Rowley *et al.*, 1997). Recent studies on large numbers of leukaemia patients have shown that patients with MLL rearrangements form a distinct subgroup based on their gene expression profile (Yokota *et al.*, 1997; Armstrong *et al.*, 2002). In the MLL fusions the methyltransferase domain of MLL is lost, which might suggest that in this case a reduction in histone methyltransfersase activity is involved (Ida *et al.*, 1997; Satake *et al.*, 1997; Sobulo *et al.*, 1997; Hayashi, 2000). Although such loss-of-function perhaps contributes to the development of leukaemia, MLL clearly requires the bromodomain and HAT domain of CBP for full *in vitro* transformation and for the induction of the leukaemic phenotype *in vivo* (Lavau *et al.*, 2000). A gain-of-function model for the MLL fusion proteins, in which CBP or p300 donate specific "leukaemic" functions to the MLL protein, is therefore more appropriate. In contrast to the situation with leukaemic MOZ and MORF fusion proteins, at least one critical target gene of MLL-CBP has been identified. Using a conditional knockin strategy, Wang *et al.* have shown that MLL-CBP requires the Hox a9 gene for its effects on granulocyte/macrophage progenitor self renewal/proliferation (Wang *et al.*, 2005). It will be interesting to see whether MLL-CBP alters the acetylation pattern of histone or nonhistone proteins at the Hox a9 promoter.

Finally, it is worth noticing that additional reciprocal fusions involving HATs could exist, based on less well-characterized translocations in leukaemia. For example, the t(2;8)(p23;p11) translocation potentially generates a fusion between MOZ and SRC1, a close relative of TIF2 (Imamura *et al.*, 2003). A fusion between MLL and SRC1 might be the result of the t(2;11)(p23;q23) translocation (Mathew *et al.*, 2000). If these translocations indeed result in the aforementioned gene fusions, it will be interesting to see whether for example the MOZ-SRC1 fusion protein harbours the same leukaemogenic properties as MOZ-TIF2 (Deguchi *et al.*, 2003; Kindle *et al.*, 2005).

7. CONCLUDING REMARKS

Genetic aberrations in HATs, and in particular CBP and p300, have been implicated in several human diseases: heterozygous mutations in CBP and more recently also p300 result in the congenital developmental disorder RTS, heterozygous mutations, with or without LOH, have been reported in solid tumours of different origin, and chromosomal translocations involving HATs are found in AML. Inactivation of HATs on the protein level has been proposed to be important in a number of other human diseases. For example, inactivation of CBP has been implicated in several neurological disorders (Saha and Pahan, 2006), such as polyglutamine-related pathologies (Huntington's disease) (Zoghbi and Orr, 2000), Alzheimer's disease and amyotrophic lateral sclerosis (Rouaux *et al.*, 2004). Furthermore, the range of human diseases in which HATs play a role could be much more extensive, based on studies in mice. For example, mice heterozygous for p300 display defective development of the heart (Yao *et al.*, 1998), while CBP+/− mice exhibited growth retardation and craniofacial abnormalities (Tanaka *et al.*, 1997; Kung *et al.*, 2000), reminiscent of RTS in humans (see above), an increased incidence of haematological malignancies (Kung *et al.*, 2000), and lipodystrophy (Yamauchi *et al.*, 2002). While it is unknown if human patients suffering from these diseases express genetic variants of one or more HATs, it should be noted that the activity of CBP and p300 could also be disrupted indirectly, e.g. through alterations in regulatory proteins. This is of course also true in RTS patients and tumours in which no CBP or p300 mutations have been detected. Of particular interest in this respect are the ING proteins, a family of tumour suppressor proteins that regulate the activity of several HATs. While both ING4 and ING5 interact with p300 and enhance the acetylation and activity of p53 (Shiseki *et al.*, 2003), ING5 is also part of the MOZ-MORF complex and enhances the activity of these MYST proteins (Doyon *et al.*, 2006). While mutations in ING proteins are rare, their nuclear expression, and thereby their ability to regulate HATs, is often lost in tumours (Russell *et al.*, 2006).

As described above, evidence suggests that both in RTS and cancer the enzymatic function of HAT proteins could play a central role. However, our knowledge on the critical target proteins and/or genes is currently limited and future studies should be aimed at identifying these key acetylation events. Given the wide range of human diseases in which the HAT function of these proteins have been implicated, understanding the mechanisms that regulate this enzymatic activity *in vivo* could help to develop novel approaches for therapeutic strategies.

ACKNOWLEDGEMENTS

We would like to thank Wendy van Veelen and Paul Coffer (Molecular Immunology, UMC Utrecht) for critically reading the manuscript. E.K. is supported by a fellowship from the Royal Netherlands Academy of Arts and Sciences.

REFERENCES

Aasland R, Gibson TJ, Stewart AF (1995) The PHD finger: implications for chromatin-mediated transcriptional regulation. Trends Biochem Sci 20:56–59

Agalioti T, Lomvardas S, Parekh B, Yie J, Maniatis T, Thanos D (2000) Ordered recruitment of chromatin modifying and general transcription factors to the IFN-beta promoter. Cell 103:667–678

Ait-Si-Ali S, Ramirez S, Barre FX, Dkhissi F, Magnaghi-Jaulin L, Girault JA, Robin P, Knibiehler M, Pritchard LL, Ducommun B, Trouche D, Harel-Bellan A (1998) Histone acetyltransferase activity of CBP is controlled by cycle-dependent kinases and oncoprotein E 1 A. Nature 396:184–186

Ait-Si-Ali S, Carlisi D, Ramirez S, Upeguí-Gonzalez LC, Duquet A, Robin P, Rudkin B, Harel-Bellan A, Trouche D (1999) Phosphorylation by p44 MAP KinaseJERKl stimulates CBP histone acetyl-transferase activity in vitro. Biochem Biophys Res Commun 262:157–162

Arany Z, Sellers WR, Livingston DM, Eckner R (1994) E1A-associated p300 and CREB-associated CBP belong to a conserved family of coactivators. Cell 77:799–800

Arany Z, Newsome D, Oldread E, Livingston DM, Eckner R (1995) A family of transcriptional adaptor proteins targeted by the ElA oncoprotein. Nature 374:81–84

Armstrong SA, Staunton JE, Silverrnan LB, Pieters R, den Boer ML, Minden MD, Sallan SE, Lander ES, Golub TR, Korsmeyer SJ (2002) MLL translocations specify a distinct gene expression profile that distinguishes a unique leukemia. Nat Genet 30:41–47

Arteaga CL (2006) Inhibition of TGFbeta signaling in cancer therapy. Curr Opin Genet Dev 16:30–37

Avantaggiati ML, Ogryzko V, Gardner K, Giordano A, Levine AS, Kelly K (1997) Recruitment of p300lCBP in p53-dependent signal pathways. Cell 89:1175–1184

Ayton PM, Cleary ML (2001) Molecular mechanisms of leukemogenesis mediated by MLL fusion proteins. Oncogene 20:5695–5707

Bannister AJ, Kouzarides T (1996) The CBP co-activator is a histone acetyltransferase. Nature 384:641–643

Barker N, Morin PJ, Clevers H (2000) The Yin-Yang of TCFlbeta-catenin signaling. Adv Cancer Res 77:1–24

Barlev NA, Liu L, Chehab NH, Mansfield K, Harris KG, Halazonetis TD, Berger SL (2001) Acetylation of p53 activates transcription through recruitment of coactivators/histone acetyltransferases. Mol Cell 8:1243–1254

Bartsch O, Locher K, Meinecke P, Kress W, Seemanova E, Wagner A, Ostermann K, Rodel G (2002) Molecular studies in 10 cases of Rubinstein-Taybi syndrome, including a mild variant showing a missense mutation in codon 1175 of CREBBP. J Med Genet 39:496–501

Bartsch O, Schmidt S, Richter M, Morlot S, Seemanova E, Wiebe G, Rasi S (2005) DNA sequencing of CREBBP demonstrates mutations in 56% of patients with Rubinstein-Taybi syndrome (RSTS) and in another patient with incomplete RSTS. Hum Genet 117:485–493

Belandia B, Parker MG (2000) Functional interaction between the p 160 coactivator proteins and the transcriptional enhancer factor family of transcription factors. J Biol Chem 275:30801–30805

Billio A, Steer EJ, Pianezze G, Svaldi M, Casin M, Amato B, Coser P, Cross NC (2002) A further case of acute myeloid leukaemia with inv(8)(p1 lq13) and MOZ-TIF2 fusion. Haematologica 87, ECRI 5

Blobel GA (2000) CREB-binding protein and p300: molecular integrators of hematopoietic transcription. Blood 95:745–755

Bordoli L, Husser S, Luthi U, Netsch M, Osmani H, Eckner R (200la) Functional analysis of the p300 acetyltransferase domain: the PHD finger of p300 but not of CBP is dispensable for enzymatic activity. Nucleic Acids Res 29:4462–4471

Bordoli L, Netsch M, Luthi U, Lutz W, Eckner R (2001b) Plant orthologs of p300lCBP: conservation of a core domain in metazoan p300lCBP acetyltransferaserelated proteins. Nucleic Acids Res 29:589–597

Borrow J, Stanton VP, Jr., Andresen JM, Becher R, Behm FG, Chaganti RS, Civin CI, Disteche C, Dube I, Frischauf AM, Horsman D, Mitelrnan F, Volinia S, Watmore AE, Housman DE (1 996) The translocation t(8; 16)(p 1 1;p 13) of acute myeloid leukaemia fuses a putative acetyltransferase to the CREB-binding protein [see comments]. Nat Genet 14:33–41

Brand M, Yamamoto K, Staub A, Tora L (1999) Identification of TATA-binding protein-free TAFII-containing complex subunits suggests a role in nucleosome acetylation and signal transduction. J Biol Chem 274:18285–18289

Bristow CA, Shore P (2003) Transcriptional regulation of the human MIP-lalpha promoter by RUNXl and MOZ. Nucleic Acids Res 31:2735–2744

Bryan EJ, Jokubaitis VJ, Chamberlain NL, Baxter SW, Dawson E, Choong DY, Campbell IG (2002) Mutation analysis of EP300 in colon, breast and ovarian carcinomas. Int J Cancer 102:137–141

Cairns BR (2001) Emerging roles for chromatin remodeling in cancer biology. Trends Cell Biol 11:S15–S21

Carapeti M, Aguiar RC, Goldman JM, Cross NC (1998) A novel fusion between MOZ and the nuclear receptor coactivator TIF2 in acute myeloid leukemia. Blood 91:3127–3133

Chaffanet M, Gressin L, Preudhomme C, Soenen-Cornu V, Birnbaum D, Pebusque MJ (2000) MOZ is fused to p300 in an acute monocytic leukemia with t(8;22). Genes Chromosomes. Cancer 28:138–144

Champagne N, Bertos NR, Pelletier N, Wang AH, Vezmar M, Yang Y, Heng HH, Yang XJ (1999a) Identification of a human histone acetyltransferase related to monocytic leukemia zinc finger protein. J Biol Chem 274:28528–28536
Champagne N, Bertos NR, Pelletier N, Wang AH, Vezmar M, Yang Y, Heng HH, Yang XJ (1999b) Identification of a human histone acetyltransferase related to monocytic leukemia zinc finger protein. J Biol Chem 274:28528–28536
Champagne N, Pelletier N, Yang XJ (2001) The monocytic leukemia zinc finger protein MOZ is a histone acetyltransferase. Oncogene 20:404–409
Chan HM, La Thangue NB (2001) p300lCBP proteins: HATS for transcriptional bridges and scaffolds. J Cell Sci 114:2363–2373
Chan HM, Krstic-Demonacos M, Smith L, Demonacos C, La Thangue NB (2001) Acetylation control of the retinoblastoma tumour-suppressor protein. Nat Cell Biol 3:667–674
Chen H, Lin RJ, Schiltz RL, Chakravarti D, Nash A, Nagy L, Privalsky ML, Nakatani Y, Evans RM (1997) Nuclear receptor coactivator ACTR is a novel histone acetyltransferase and forms a multimeric activation complex with P/CAF and CBP/p300. Cell 90:569–580
Chen D, Ma H, Hong H, Koh SS, Huang SM, Schurter BT, Aswad DW, Stallcup MR (1999a) Regulation of transcription by a protein methyltransferase. Science 284:2174–2177
Chen H, Lin RJ, Xie W, Wilpitz D, Evans RM (1999b) Regulation of hormone induced histone hyperacetylation and gene activation via acetylation of an acetylase. Cell 986:75–686
Cheung P, Allis CD, Sassone-Corsi P (2000a) Signaling to chromatin through histone modifications. Cell 103:263–271
Cheung WL, Briggs SD, Allis CD (2000b) Acetylation and chromosomal functions. Curr Opin Cell Biol 12:326–333
Chevillard-Briet M, Trouche D, Vandel L (2002) Control of CBP co-activating activity by arginine methylation. EMBO J 21:5457–5466
Cho H, Orphanides G, Sun X, Yang XJ, Ogryzko V, Lees E, Nakatani Y, Reinberg D (1998) A human RNA polymerase II complex containing factors that modify chromatin structure. Mol Cell Biol 18:5355–5363
Chrivia JC, Kwok RP, Lamb N, Hagiwara M, Montminy MR, Goodman RH (1993) Phosphorylated CREB binds specifically to the nuclear protein CBP. Nature 365:855–859
Collins HM, Kindle KB, Matsuda S, Ryan C, Troke PJ, Kalkhoven E, Heery DM (2006) MOZ-TIF2 alters cofactor recruitment and histone modification at the RARbeta 2 promoter: Differential effects of MOZ fusion proteins on CBP- and MOZ dependent activators. J Biol Chem 281:17124–17133
Conti E, Izaurralde E (2005) Nonsense-mediated mRNA decay: molecular insights and mechanistic variations across species. Curr Opin Cell Biol 17:316–325
Cosma MP, Tanaka T, Nasmyth K (1999) Ordered recruitment of transcription and chromatin remodeling factors to a cell cycle- and developmentally regulated promoter. Cell 97:299–311
Coupry I, Roudaut C, Stef M, Delrue MA, Marche M, Burgelin J, Taine L, Cruaud C, Lacombe D, Arveiler B (2002) Molecular analysis of the CBP gene in 60 patients with Rubinstein-Taybi syndrome. J Med Genet 39:415–421
Coutts AS, La Thangue NB (2005) The p53 response: emerging levels of co-factor complexity. Biochem Biophys Res Commun 331:778–785
Crowley JA, Wang Y, Rapoport AP, Ning Y (2005) Detection of MOZ-CBP fusion in acute myeloid leukemia with 8;16 translocation. Leukemia 19:2344–2345
Dash A, Gilliland DG (2001) Molecular genetics of acute myeloid leukaemia. Best Pract Res Clin Haematol 14:49–64
Dash AB, Williams IR, Kutok, JL, Tomasson MH, Anastasiadou E, Lindahl K, Li S, Van Etten RA, Borrow J, Housman D, Druker B, Gilliland DG (2002) A murine model of CML blast crisis induced by cooperation between BCRIABL and NUP98lHOXA9. Proc Natl Acad Sci U S A 99:7622–7627
Debes JD, Sebo TJ, Lohse CM, Murphy LM, Haugen de AL, Tindall,D J (2003) p300 in prostate cancer proliferation and progression. Cancer Res 63:7638–7640
Deguchi K, Ayton PM, Carapeti M, Kutok JL, Snyder CS, Williams IR, Cross NC, Glass CK, Cleary ML, Gilliland DG (2003) MOZ-TIF2-induced acute myeloid leukemia requires the MOZ nucleosome

binding motif and TIF2-mediated recruitment of CBP. Cancer Cell 3:259–271

Dhalluin C, Carlson JE, Zeng L, He C, Aggarwal AK, Zhou MM (1999) Structure and ligand of a histone acetyltransferase brornodomain. Nature 399:491–496

Dimartino JF, Cleary ML (1999) M11 rearrangements in haematological malignancies: lessons from clinical and biological studies. Br J Haematol 106:614–626

Doyon Y, Cayrou C, Ullah M, Landry AJ, Cote V, Selleck W, Lane WS, Tan S, Yang XJ, Cote J (2006) ING tumor suppressor proteins are critical regulators of chromatin acetylation required for genome expression and perpetuation. Mol Cell 21:51–64

Eckner R, Ewen ME, Newsome D, Gerdes M, DeCaprio JA, Lawrence JB, Livingston DM (1994) Molecular cloning and functional analysis of the adenovirus E1A-associated 300-kD protein (p300) reveals a protein with properties of a transcriptional adaptor. Genes Dev 8:869–884

Fodde R, Smits R (2002) Cancer biology. A matter of dosage. Science 298:761–763

Gayther SA, Batley SJ, Linger L, Bannister A, Thorpe K, Chin SF, Daigo Y, Russell P, Wilson A, Sowter HM, Delhanty JD, Ponder BA, Kouzarides T, Caldas C (2000) Mutations truncating the EP300 acetylase in human cancers. Nat Genet 24:300–303

Giles RH, Peters DJ, Breuning MH (1998) Conjunction dysfunction: CBP/p300 in human disease. Trends Genet 14:178–183

Girdwood D, Bumpass D, Vaughan OA, Thain A, Anderson LA, Snowden A W, Garcia-Wilson E, Perkins ND, Hay RT (2003) P300 transcriptional repression is mediated by SUMO modification. Mol Cell 11:1043–1054

Goodman RH, Smolik S (2000) CBP/p300 in cell growth, transformation, and development. Genes Dev 14:1553–1577

Grossman SR, Perez M, Kung AL, Joseph M, Mansur C, Xiao ZX, Kurnar, S, Howley PM, Livingston DM (1998) p300lMDM2 complexes participate in MDM2-mediated p53 degradation. Mol Cell 2:405–415

Grossman SR, Deato ME, Brignone C, Chan HM, Kung AL, Tagami H, Nakatani Y, Livingston DM (2003) Polyubiquitination of p53 by a ubiquitin ligase activity of p300. Science 300:342–344

Gu W, Roeder RG (1997) Activation of p53 sequence-specific DNA binding by acetylation of the p53 C-terminal domain. Cell 90:595–606

Hanahan D, Weinberg RA (2000) The hallmarks of cancer. Cell 100:57–70

Hayashi Y (2000) The molecular genetics of recurring chromosome abnormalities in acute myeloid leukemia [In Process Citation]. Semin Hematol 37:368–380

Heery DM, Kalkhoven E, Hoare S, Parker MG (1997) A signature motif in transcriptional co-activators mediates binding to nuclear receptors [see comments]. Nature 387:733–736

Ida K, Kitabayashi I, Taki T, Taniwaki M, Noro K, Yamamoto M, Ohki M, Hayashi Y (1997) Adenoviral E 1 A-associated protein p300 is involved in acute myeloid leukemia with t(ll;22)(q23;q13). Blood 90:4699–4704

Imamura T, Kakam N, Hibi S, Morimoto A, Fukushima Y, Ijuin I, Hada S, Kitabayashi I, Abe T, Imashuku S (2003) Rearrangement of the MOZ gene in pediatric therapy-related myelodysplastic syndrome with a novel chromosomal translocation t(2;8)(p23;pll). Genes Chromosomes. Cancer 36:413–419

Ionov,Y, Nowak N, Perucho M, Markowitz S, Cowell JK (2004) Manipulation of nonsense mediated decay identifies gene mutations in colon cancer Cells with microsatellite instability. Oncogene 23:639–645

Ito A, Lai CH, Zhao X, Saito S, Hamilton MH, Appella E, Yao TP (2001) p300/CBP-mediated p53 acetylation is commonly induced by p53-activating agents and inhibited by MDM2. EMBO J 20:1331–1340

Jacobson RH, Ladurner AG, King DS, Tjian R (2000) Structure and function of a human TAFII250 double bromodomain module. Science 288:1422–1425.

Janknecht R, Nordheim A (1996) MAP kinase-dependent transcriptional coactivation by Elk-1 and its cofactor CBP. Biochem Biophys Res Commun 228:831–837

Jason LJ, Moore SC, Lewis JD, Lindsey G, Ausio J (2002) Histone ubiquitination: a tagging tail unfolds? Bioessays 24, 166–174

Jeanmougin F, Wurtz JM, Le Douarin B, Chambon P, Losson R (1997) The bromodomain revisited. Trends Biochem Sci 22:151–153

Kalkhoven E (2004) CBP and p300: HATS for different occasions. Biochem Pharmacol 68:1145–1155

Kalkhoven E, Teunissen H, Houweling A, Verrijzer CP, Zantema A (2002) The PHD type zinc finger is an integral part of the CBP acetyltransferase domain. Mol Cell Biol 22:1961–1970

Kalkhoven E, Roelfsema JH, Teunissen H, den Boer A, Ariyurek Y, Zantema A, Breuning MH, Hennekam RC, Peters DJ (2003) Loss of CBP acetyltransferase activity by PHD finger mutations in Rubinstein-Taybi syndrome. Hum Mol Genet 12:441–450

Kelly LM, Kutok, JL, Williams IR, Boulton CL, Amaral SM, Curley DP, Ley TJ, Gilliland DG (2002) PMLJRARalpha and FLT3-ITD induce an APL-like disease in a mouse model. Proc Natl Acad Sci U S A 99:8283–8288

Kindle KB, Troke PJ, Collins HM, Matsuda S, Bossi D, Bellodi C, Kalkhoven E, Salomoni P, Pelicci PG, Minucci S, Heery DM (2005) MOZ-TIF2 inhibits transcription by nuclear receptors and p53 by impairment of CBP fimction. Mol Cell Biol 25:988–1002

Kishimoto M, Kohno T, Okudela K, Otsuka A, Sasaki H, Tanabe C, Sakiyama T, Hirama C, Kitabayashi I, Minna JD, Takenoshita S, Yokota J (2005) Mutations and deletions of the CBP gene in human lung cancer. Clin Cancer Res 11:512–519

Kitabayashi I, Aikawa Y, Nguyen LA, Yokoyama A, Ohki M (2001) Activation of AMLl -mediated transcription by MOZ and inhibition by the MOZ-CBP hsion protein. EMBO J 20:7184–7196

Klochendler-Yeivin A, Yaniv M (2001) Chromatin modifiers and tumor suppression. Biochim Biophys Acta 1551:MI-10

Knudson AG, Jr. (1971) Mutation and cancer: statistical study of retinoblastoma. Proc Natl Acad Sci U S A 68:820–823

Kobet E, Zeng X, Zhu Y, Keller D, Lu H (2000) MDM2 inhibits p300-mediated p53 acetylation and activation by forming a ternary complex with the two proteins. Proc Natl Acad Sci U S A 97:12547–12552

Kojima K, Kaneda K, Yoshida C, Dansako H, Fujii N, Yano T, Shinagawa K, Yasukawa M, Fujita S, Tanimoto M (2003) A novel fusion variant of the MORF and CBP genes detected in therapy-related myelodysplastic syndrome with t(10; 16)(q22;p 13). Br J Haematol 120:271–273

Koken MH, Saib A, de The H (1995) A C4HC3 zinc finger motif. C R Acad Sci III 318:733–739

Korzus E, Torchia J, Rose DW, Xu L, Kurokawa R, McInerney EM, Mullen TM, Glass CK, Rosenfeld MG(1.9 98) Transcription factor-specific requirements for coactivators and their acetyltransferase functions. Science 279:703–707

Koshiishi N, Chong JM, Fukasawa T, Ikeno R, Hayashi Y, Funata N, Nagai H, Miyaki M, Matsumoto Y, Fukayama M (2004) p300 gene alterations in intestinal and diffuse types of gastric carcinoma. Gastric Cancer 7:85–90

Kouzarides T (2002) Histone methylation in transcriptional control. Curr Opin Genet Dev 12:198–209

Kraus WL, Manning ET, Kadonaga JT (1999) Biochemical analysis of distinct activation functions in p300 that enhance transcription initiation with chromatin templates. Mol Cell Biol 19:8123–8135

Kundu TK, Palhan VB, Wang Z, An, W, Cole PA, Roeder RG (2000) Activator-dependent transcription from chromatin in vitro involving targeted histone acetylation by p300. Mol Cell 6:551–561

Kung AL, Rebel VI, Bronson RT, Ch'ng LE, Sieff CA, Livingston DM, Yao TP (2000) Gene dose-dependent control of hematopoiesis and hematologic tumor suppression by CBP. Genes Dev 14:272–277

Kuo MH, Allis CD (1998) Roles of histone acetyltransferases and deacetylases in gene regulation. Bioessays 20:615–626

Kurokawa R, Kalafus D, Ogliastro MH, Kioussi C, Xu L, Torchia J, Rosenfeld MG, Glass CK (1998) Differential use of CREB binding protein-coactivator complexes. Science 279:700–703

Kwok RP, Lundblad JR, Chrivia,JC, Richards JP, Bachinger HP, Brennan RG, Roberts SG, Green MR, Goodman RH (1994) Nuclear protein CBP is a coactivator for the transcription factor CREB. Nature 370:223–226

Lavau C, Du C, Thirman M, Zeleznik-Le N (2000) Chromatin-related properties of CBP fused to MLL generate a myelodysplastic-like syndrome that evolves into myeloid leukemia. EMBO J 19:4655–4664

Lee JS, Zhang X, Shi Y (1996) Differential interactions of the CREBIATF family of transcription factors with p300 and adenovirus El A. J Biol Chem 271:17666–17674

Leo C, Chen JD (2000) The SRC family of nuclear receptor coactivators. Gene 245:1–11

Levy L, Hill CS (2006) Alterations in components of the TGF-beta superfamily signaling pathways in human cancer. Cytokine Growth Factor Rev 17:41–58

Liang J, Prouty L, Williams BJ, Dayton MA, Blanchard KL (1998) Acute mixed lineage leukemia with an inv(8)(pl lq13) resulting in fbsion of the genes for MOZ and TIF2. Blood 92:2118–2122

Lill NL, Grossman SR, Ginsberg D, DeCaprio J, Livingston DM (1 997) Binding and modulation of p53 by p300lCBP coactivators. Nature 387:823–827

Liu L, Scolnick DM, Trievel RC, Zhang HB, Marmorstein R, Halazonetis TD, Berger SL (1999) p53 sites acetylated in vitro by PCAF and p300 are acetylated in vivo in response to DNA damage. Mol Cell Biol 19:1202–1209

Liu Z, Wong J, Tsai SY, Tsai MJ, O'Malley BW (2001) Sequential recruitment of steroid receptor coactivator-1 (SRC-1) and p300 enhances progesterone receptordependent initiation and reinitiation of transcription from chromatin. Proc Natl Acad Sci U S A 98:12426–12431

Manning ET, Ikehara T, Ito T, Kadonaga JT, Kraus WL (2001) p300 forms a stable, template-committed complex with chromatin: role for the bromodomain. Mol Cell Biol 21:3876–3887

Markham D, Munro S, Soloway J, OlConnor DP, La Thangue NB (2006) DNAdamage-responsive acetylation of pRb regulates binding to E2F-1. EMBO Rep 7:192–198

Marmorstein R (2001) Protein modules that manipulate histone tails for chromatin regulation. Nat Rev Mol Cell Biol 2:422–432

Marmorstein R, Roth SY (2001) Histone acetyltransferases: function, structure, and catalysis. Curr Opin Genet Dev 11:155–161

Martens JH, Verlaan M, Kalkhoven E, Zantema A (2003) Cascade of distinct histone modifications during collagenase gene activation. Mol Cell Biol 23:1808–1816

Martinez E, Palhan VB, Tjernberg A, Lymar ES, Gamper AM, Kundu TK, Chait BT, Roeder RG (2001) Human STAGA complex is a chromatin-acetylating transcription coactivator that interacts with pre-rnRNA splicing and DNA damagebinding factors in vivo. Mol Cell Biol 21:6782–6795

Martinez-Balbas MA, Bannister AJ, Martin K, Haus-Seuffert P, Meisterernst M, Kouzarides T (1998) The acetyltransferase activity of CBP stimulates transcription. EMBO J 17:2886–2893

Martinez-Balbas MA, Bauer UM, Nielsen SJ, Brehm A, Kouzarides T (2000) Regulation of E2F 1 activity by acetylation. EMBO J 19:662–671

Marzio G, Wagener C, Gutierrez MI, Cartwright P, Helin K, Giacca M (2000) E2F family members are differentially regulated by reversible acetylation. J Biol. Chem. 275:10887–10892

Mathew S, Head D, Rubnitz JE, Raimondi SC (2000) Concurrent translocations of MLL and CBFA2 (AML1) genes with new partner breakpoints in a child with secondary myelodysplastic syndrome after treatment of acute lymphoblastic leukemia. Genes Chromosomes. Cancer 28:227–232

Miller RW, Rubinstein JH (1995) Tumors in Rubinstein-Taybi syndrome. Am J Med Genet 56:112–115

Milne TA, Briggs SD, Brock HW, Martin ME, Gibbs D, Allis CD, Hess JL (2002) MLL targets SET domain methyltransferase activity to Hox gene promoters. Mol. Cell 10:1107–1117

Mitsiou DJ, Stunnenberg HG (2003) p300 is involved in formation of the TBPTFIIA-containing basal transcription complex, TAC. EMBO J 22:4501–4511

Munshi N, Merika M, Yie J, Senger K, Chen sG, Thanos D (1998) Acetylation of HMG I(Y) by CBP turns off IFN beta expression by disrupting the enhanceosome. Mol Cell 2:457–467

Muraoka M, Konishi M, Kikuchi-Yanoshita R, Tanaka K, Shitara N, Chong JM, Iwama T, Miyaki M (1996) p300 gene alterations in colorectal and gastric carcinomas. Oncogene 12:1565–1569

Murata T, Kurokawa R, Krones A, Tatsumi K, Ishii M, Taki T, Masuno M, Ohashi H, Yanagisawa M, Rosenfeld MG, Glass CK, Hayashi Y (2001) Defect of histone acetyltransferase activity of the nuclear transcriptional coactivator CBP in Rubinstein-Taybi syndrome. Hum Mol Genet 10: 1071–1076

Murati A, Adelaide J, Mozziconacci MJ, Popovici C, Carbuccia N, Letessier A, Birg F, Birnbaum D, Chaffanet M (2004) Variant MYST4-CBP gene fusion in a t(10; 16) acute myeloid leukaemia. Br J Haematol 125:601–604

Nakajima T, Uchida C, Anderson SF, Lee CG, Hurwitz J, Parvin JD, Montminy M (1997a) RNA helicase A mediates association of CBP with RNA polymerase 11. Cell 90:1107–1112.

Nakajima T, Uchida C, Anderson SF, Parvin JD, Montminy M (1997b) Analysis of a CAMP-responsive activator reveals a two-component mechanism for transcriptional induction via signal-dependent factors. Genes Dev 11:738–747

Nakamura T, Mori T, Tada S, Krajewski W, Rozovskaia T, Wassell R, Dubois G, Mazo A, Croce CM, Canaani E (2002) ALL-1 is a histone methyltransferase that assembles a supercomplex of proteins involved in transcriptional regulation. Mol Cell 10:1119–1128

Neish AS, Anderson SF, Schlegel BP, Wei W, Parvin JD (1998) Factors associated with the mammalian RNA polymerase II holoenzyrne. Nucleic Acids Res. 26:847–853

Neuwald AF, Landsman D (1997) GCN5-related histone N-acetyltransferases belong to a diverse superfamily that includes the yeast SPTIO protein. Trends Biochem Sci 22:154–155

Nguyen DX, Baglia LA, Huang SM, Baker CM, McCance DJ (2004) Acetylation regulates the differentiation-specific functions of the retinoblastoma protein. EMBO J 23:1609–1618

Ogryzko VV, Kotani T, Zhang X, Schlitz RL, Howard T, Yang XJ, Howard BH, Qin J, Nakatani Y (1998) Histone-like TAFs within the PCAF histone acetylase complex [see comments]. Cell 94:35–44

Ogryzko VV, Schiltz RL, Russanova V, Howard BH, Nakatani Y (1996) The transcriptional coactivators p300 and CBP are histone acetyltransferases. Cell 87:953–959

Ohshima T, Suganuma T, Ikeda M (2001) A novel mutation lacking the bromodomain of the transcriptional coactivator p300 in the SiHa cervical carcinoma cell line. Biochem Biophys Res Commun 281:569–575

Oike Y, Hata A, Mamiya T, Kaname T, Noda Y, Suzuki M, Yasue H, Nabeshima T, Araki K, Yamamura K (1999) Truncated CBP protein leads to classical Rubinstein-Taybi syndrome phenotypes in mice: implications for a dominant-negative mechanism. Hum Mol Genet 8:387–396

Ozdag H, Batley SJ, Forsti A, Iyer NG, Daigo Y, Boutell J, Arends MJ, Ponder BA, Kouzarides T, Caldas C (2002) Mutation analysis of CBP and PCAF reveals rare inactivating mutations in cancer cell lines but not in primary tumours. Br J Cancer 87:1162–1165

Panagopoulos I, Isaksson M, Lindvall C, Bjorkholm M, Ahlgren T, Fioretos T, Heim S, Mitelman F, Johansson B (2000) RT-PCR analysis of the MOZ-CBP and CBP-MOZ chimeric transcripts in acute myeloid leukemias with t(8; 16)(pll;p13). Genes Chromosomes. Cancer 28:415–424

Panagopoulos I, Fioretos T, Isaksson M, Samuelsson U, Billstrom R, Strombeck B, Mitelman F, Johansson B (2001) Fusion of the MORF and CBP genes in acute myeloid leukemia with the t(10;16)(q22;p13). Hum Mol Genet 10:395–404

Panagopoulos I, Fioretos T, Isaksson M, Mitelman F, Johansson B, Theorin N, Juliusson G (2002) RT-PCR analysis of acute myeloid leukemia with t(8; 16)(p 1 1;p 13): identification of a novel MOZICBP transcript and absence of CBP/MOZ expression. Genes Chromosomes. Cancer 35:372–374

Panagopoulos I, Isaksson M, Lindvall C, Hagemeijer A, Mitelman F, Johansson B (2003) Genomic characterization of MOZICBP and CBP/MOZ chimeras in acute myeloid leukemia suggests the involvement of a damage-repair mechanism in the origin of the t(8; 16)(p11;p13). Genes Chromosomes. Cancer 36:90–98

Partanen A, Motoyama J, Hui CC (1 999) Developmentally regulated expression of the transcriptional cofactors/histone acetyltransferases CBP and p300 during mouse embryogenesis. Int J Dev Biol 43:487–494

Pelletier N, Champagne N, Stifani S, Yang XJ (2002) MOZ and MORF histone acetyltransferases interact with the Runt-domain transcription factor Runx2. Oncogene 21:2729–2740

Petrij F, Giles RH, Dauwerse HG, Saris JJ, Hennekam RC, Masuno M, Tommerup N, van Ommen GJ, Goodman RH, Peters DJ, Breuning MH (1995) Rubinstein-Taybi syndrome caused by mutations in the transcriptional co-activator CBP. Nature 376:348–351

Petrij F, Dauwerse HG, Blough RI, Giles RH, van der Smagt JJ, Wallerstein R, Maaswinkel-Mooy PD, van Karnebeek CD, van Ommen GJ, van Haeringen A, Rubinstein JH, Saal HM, Hennekam RC, Peters DJ, Breuning MH (2000a) Diagnostic analysis of the Rubinstein-Taybi syndrome: five cosrnids should be used for microdeletion detection and low number of protein truncating mutations. J Med Genet 37:168–176

Petrij F, Dorsman JC, Dauwerse HG, Giles RH, Peeters T, Hennekam RC, Breuning MH, Peters DJ (2000b) Rubinstein-Taybi syndrome caused by a De Novo reciprocal translocation t(2;16)(q36.3;p13.3). Am J Med Genet 92:47–52

Ragvin A, Valvatne H, Erdal S, Arskog V, Tufteland KR, Breen K, OYan AM, Eberharter A, Gibson TJ, Becker PB, Aasland R (2004) Nucleosome Binding by the Bromodomain and PHD Finger of the Transcriptional Cofactor p300. J Mol Biol 337:773–788

Roelfsema JH, White SJ, Ariyurek Y, Bartholdi D, Niedrist D, Papadia F, Bacino CA, den Dunnen JT, van Ommen GJ, Breuning MH, Hennekam RC, Peters DJ (2005) Genetic heterogeneity in Rubinstein-Taybi syndrome: mutations in both the CBP and EP300 genes cause disease. Am J Hum Genet 76:572–580

Rouaux C, Loeffler JP, Boutillier AL (2004) Targeting CREB-binding protein (CBP) loss of function as a therapeutic strategy in neurological disorders. Biochem Pharmacol 68:1157–1164

Rowan BG, Garrison N, Weigel NL, OIMalley BW (2000) 8-Bromo-cyclic AMP induces phosphorylation of two sites in SRC-1 that facilitate ligand-independent activation of the chicken progesterone receptor and are critical for functional cooperation between SRC-1 and CREB binding protein. Mol. Cell Biol. 20:8720–8730

Rowley JD, Reshmi S, Sobulo O, Musvee T, Anastasi J, Raimondi S, Schneider NR, Barredo JC, Cantu ES, Schlegelberger B, Behm F, Doggett NA, Borrow,J, Zeleznik-Le N (1997) All patients with the T(11; 16)(q23;p13.3) that involves MLL and CBP have treatment-related hematologic disorders. Blood 90:535–541

Rozman M, Camos M, Colomer D, Villamor N, Esteve J, Costa D, Carrio A, Aymerich M, Aguilar JL, Domingo A, Sole F, Gomis F, Florensa L, Montserrat E, Campo E (2004) Type I MOZICBP (MYST3lCREBBP) is the most common chimeric transcript in acute myeloid leukemia with t(8; 16)(p11;p13) translocation. Genes Chromosomes. Cancer 40:140–145

Rubinstein JH, Taybi H (1963) Broad thumbs and toes and facial abnormalities. Am J Dis Child 105:588–608

Russell M, Berardi P, Gong W, Riabowol K (2006) Grow-ING, Age-ING and Die-ING: ING proteins link cancer, senescence and apoptosis. Exp Cell Res 312:951–961

Saha RN, Pahan K (2006) HATS and HDACs in neurodegeneration: a tale of disconcerted acetylation homeostasis. Cell Death Differ 13:539–550

Saha V, Chaplin T, Gregorini A, Ayton P, Young BD (1995) The leukemiaassociated-protein (LAP) domain, a cysteine-rich motif, is present in a wide range of proteins, including MLL, AFIO, and MLLT6 proteins. Proc Natl Acad Sci U S A 92:9737–9741

Satake N, Ishida Y, Otoh Y, Hinohara S, Kobayashi H, Sakashita A, Maseki N, Kaneko Y (1997) Novel MLL-CBP fusion transcript in therapy-related chronic myelomonocytic leukemia with a t(l1; 16)(q23;p13) chromosome translocation. Genes Chromosomes. Cancer 20:60–63

Schiltz RL, Mizzen CA, Vassilev A, Cook RG, Allis CD, Nakatani Y (1999) Overlapping but distinct patterns of histone acetylation by the human coactivators p300 and PCAF within nucleosomal substrates. J Biol Chem 274:1189–1192

Schmidt HH, Strehl S, Thaler D, Strunk D, Sill H, Linkesch W, Jager U, Sperr W, Greinix HT, Konig M, Emberger W, Haas O A (2004) RT-PCR and FISH analysis of acute myeloid leukemia with t(8; 16)(p 1 1;p 13) and chimeric MOZ and CBP transcripts: breakpoint cluster region and clinical implications. Leukemia 18:1115–1121

Scolnick DM, Chehab NH, Stavridi ES, Lien MC, Caruso L, Moran E, Berger SL, Halazonetis TD (1 997) CREB-binding protein and p300lCBP-associated factor are transcriptional coactivators of the p53 tumor suppressor protein. Cancer Res. 57:3693–3696

Shang Y, Hu X, DiRenzo J, Lazar MA, Brown M (2000) Cofactor dynamics and sufficiency in estrogen receptor-regulated transcription. Cell 103:843–852

Shigeno K, Yoshida H, Pan L, Luo JM, Fujisawa S, Naito K, Nakamura S, Shinjo K, Takeshita A, Ohno R, Ohnishi K (2004) Disease-related potential of mutations in transcriptional cofactors CREB-binding protein and p300 in leukemias. Cancer Lett. 213:11–20

Shiio Y, Eisenman RN (2003) Histone sumoylation is associated with transcriptional repression. Proc Natl Acad Sci U S A 100:13225–13230

Shikama N, Lyon J, La Thangue NB (2000) The p300lCBP family: Integrating signals with transcription factors and chromatin. Trends Cell Biol 7:230–236

Shiseki M, Nagashima M, Pedeux RM, Kitahama-Shiseki M, Miura K, Okamura S, Onogi H, Higashimoto Y, Appella E, Yokota J, Harris CC (2003) p29ING4 and p28ING5 bind to p53 and p300, and enhance p53 activity. Cancer Res 63:2373–2378

So CK, Nie Y, Song Y, Yang GY, Chen S, Wei C, Wang LD, Doggett NA, Yang CS (2004) Loss of heterozygosity and internal tandem duplication mutations of the CBP gene are frequent events in human esophageal squamous cell carcinoma. Clin Cancer Res 10:19–27

Sobulo OM, Borrow J, Tomek R, Reshmi S, Harden A, Schlegelberger B, Housman D, Doggett NA, Rowley JD, Zeleznik L (1997) MLL is hsed to CBP, a histone acetyltransferase, in therapy-related acute myeloid leukemia with a t(l1;16)(q23;p13.3). Proc Natl Acad Sci U S A 94:8732–8737

Soutoglou E, Talianidis I (2002) Coordination of PIC assembly and chromatin remodeling during differentiation-induced gene activation. Science 295:1901–1904

Soutoglou E, Katrakili N, Talianidis I (2000) Acetylation regulates transcription factor activity at multiple levels. Mol Cell 5:745–751

Spencer TE, Jenster G, Burcin MM, Allis CD, Zhou J, Mizzen CA, McKenna NJ, Onate SA, Tsai SY, Tsai MJ, O'Malley BW (1 997) Steroid receptor coactivator-1 is a histone acetyltransferase. Nature 389:194–198

Sterner DE, Berger SL (2000) Acetylation of histones and transcription-related factors. Microbiol. Mol. Biol. Rev. 64:435–459

Strahl BD, Allis CD (2000) The language of covalent histone modifications. Nature 403:41–45

Suganuma T, Kawabata M, Ohshima T, Ikeda MA (2002) Growth suppression of human carcinoma cells by reintroduction of the p300 coactivator. Proc Natl Acad Sci U S A 99:13073–13078

Sun Y, Kolligs FT, Hottiger MO, Mosavin R, Fearon ER, Nabel GJ(2000) Regulation of beta-catenin transformation by the p300 transcriptional coactivator. Proc Natl Acad Sci U S A 97:12613–12618

Taki T, Sako M, Tsuchida M, Hayashi Y (1997) The t(11; 16)(q23;p13) translocation in myelodysplastic syndrome fhses the MLL gene to the CBP gene. Blood 89:3945–3950

Tan S (2001) One HAT size fits all? Nat Struct Biol 8:8–10

Tanaka Y, Naruse I, Maekawa T, Masuya H, Shiroishi T, Ishii S (1997) Abnormal skeletal patterning in embryos lacking a single Cbp allele: a partial similarity with Rubinstein-Taybi syndrome. Proc Natl Acad Sci U S A 94:10215–10220

Tillinghast GW, Partee J, Albert P, Kelley JM, Burtow KH, Kelly K (2003) Analysis of genetic stability at the EP300 and CREBBP loci in a panel of cancer cell lines. Genes Chromosomes. Cancer 37:121–131

Timmermann S, Lehrrnann H, Polesskaya A, Harel-Bellan A (2001) Histone acetylation and disease. Cell Mol Life Sci 58:728–736

Turner BM (2000) Histone acetylation and an epigenetic code. Bioessays 22:836–845

Villavicencio EH, Walterhouse DO, Iannaccone PM (2000) The sonic hedgehogpatched-gli pathway in human development and disease. Am J Hum Genet 67:1047–1054

Vizmanos, JL., Larrayoz MJ, Lahortiga I, Floristan F, Alvarez C, Odero MD, Novo FJ, Calasanz MJ (2003) t(10; 16)(q22;p13) and MORF-CREBBP fusion is a recurrent event in acute myeloid leukemia. Genes Chromosomes. Cancer 36:402–405

Vo N, Goodman RH (2001) CREB-binding protein and p300 in transcriptional regulation. J Biol Chem 276:13505–13508

Voegel JJ, Heine MJ, Tini M, Vivat V, Chambon P, Gronemeyer H (1998) The coactivator TIF2 contains three nuclear receptor-binding motifs and mediates transactivation through CBP binding-dependent and -independent pathways. EMBO J 17:507–519

Wang J, Iwasaki H, Krivtsov A, Febbo PG, Thorner AR, Ernst P, Anastasiadou E, Kutok JL, Kogan SC, Zinkel SS., Fisher, JK, Hess, JL, Golub TR, Armstrong SA, Akashi K, Korsmeyer SJ (2005) Conditional MLL-CBP targets GMP and models therapy-related myeloproliferative disease. EMBO J 24:368–381

Ward R, Johnson M, Shridhar V, van Deursen J, Couch FJ (2005) CBP truncating mutations in ovarian cancer. J Med Genet 42:514–518

Winston F, Allis CD (1999) The bromodomain: a chromatin-targeting module? Nat Struct Biol 6:601–604

Wu RC, Qin J, Yi P, Wong J, Tsai SY, Tsai M J, O'Malley BW (2004) Selective phosphorylations of the SRC-3lAIB 1 coactivator integrate genomic reponses to multiple cellular signaling pathways. Mol Cell 15:937–949

Xu W, Edmondson DG, Roth SY (1998) Mammalian GCN5 and PICAF acetyltransferases have homologous amino-terminal domains important for recognition of nucleosomal substrates. Mol Cell Biol 18:5659–5669

Xu W, Chen H, Du K, Asahara H, Tini M, Emerson BM, Montrniny M, Evans RM (2001) A transcriptional switch mediated by cofactor methylation. Science 294:2507–2511

Yamauchi T, Oike Y, Kamon J, Waki H, Komeda K, Tsuchida A, Date Y, Li MX, Miki H, Akanuma Y, Nagai R, Kimura S, Saheki T, Nakazato M, Naitoh T, Yamamura K, Kadowaki T (2002) Increased insulin sensitivity despite lipodystrophy in Crebbp heterozygous mice. Nat Genet 30:221–226

Yan Y, Barlev NA, Haley RH, Berger SL, Marmorstein R (2000) Crystal structure of yeast esal suggests a unified mechanism for catalysis and substrate binding by histone acetyltransferases. Mol. Cell 6:1195–1205

Yang XJ (2004) The diverse superfamily of lysine acetyltransferases and their roles in leukemia and other diseases. Nucleic Acids Res 32:959–976

Yang XJ, Ogryzko VV, Nishikawa J, Howard BH, Nakatani Y (1996) A p300/CBP-associated factor that competes with the adenoviral oncoprotein El A. Nature 382:319–324

Yao TP, Oh SP, Fuchs M, Zhou ND, Ch'ng LE, Newsome D, Bronson RT, Li E, Livingston DM, Eckner R (1998) Gene dosage-dependent embryonic development and proliferation defects in mice lacking the transcriptional integrator p300. Cell 93:361–372

Yokota S, Kiyoi H, Nakao M, Iwai T, Misawa S, Okuda T, Sonoda Y, Abe T, Kahsima K, Matsuo Y, Naoe T (1997) Internal tandem duplication of the FLT3 gene is preferentially seen in acute myeloid leukemia and myelodysplastic syndrome among various hematological malignancies. A study on a large series of patients and cell lines. Leukemia 11:1605–1609

Yuan LW, Giordano A (2002) Acetyltransferase machinery conserved in p300lCBP-family proteins. Oncogene 21:2253–2260

Yuan W, Condorelli G, Caruso M, Felsani A, Giordano A (1996) Human p300 protein is a coactivator for the transcription factor MyoD. J Biol Chem 271:9009–9013

Zoghbi HY, Orr HT (2000) Glutamine repeats and neurodegeneration. Annu Rev Neurosci 23:217–247

CHAPTER 12

CHROMATIN ACETYLATION STATUS IN THE MANIFESTATION OF NEURODEGENERATIVE DISEASES

HDAC inhibitors as therapeutic tools

ANNE-LAURENCE BOUTILLIER*, CAROLINE ROUAUX, IRINA PANTELEEVA, AND JEAN-PHILIPPE LOEFFLER

Laboratoire de Signalisation Moléculaire et Neurodégénérescence – INSERM U692 11, rue Humann, 67085 Strasbourg Cedex – France

Abstract: During the development and maintenance of the central nervous system, neurons receive specific instructions to differentiate, survive or die, the correct choice being crucial for the maturation of a functional brain and to face pathological conditions. At the transcriptional level, chromatin remodeling enzymes participates in such processes. In this paper, we will see that disruption of the Histone acetyl transferase (HAT)/Deacetylase (HDAC) balance is often observed in different contexts of neurological disorders and more particularly during neuronal apoptosis. During the last 5 years, it has been evidenced that the chromatin acetylation status was greatly impaired in different neurodegenerative diseases, a common mechanism being the loss of function of a specific HAT: the CREB-binding protein (CBP). We will review the last attempts of the use of small molecules antagonizing HDAC activity (HDAC inhibitors) to restore proper levels of acetylation and enhance neuronal survival, both in *in vitro* and *in vivo* models of neurodegenerative diseases such as polyglutamine-related diseases and amyotrophic lateral sclerosis. Although this strategy lacks specificity towards CBP, certain of these molecules display promising therapeutic properties

1. INTRODUCTION

Histones are master-piece of DNA compaction, able to form a highly ordered and condensed nucleoproteic complex termed chromatin. Thereby, they play a major role in genomic function regulations. They are the target of multiple post-translational

* E-mail: laurette@neurochem.u-strasbg.fr

T.K. Kundu and D. Dasgupta (eds.), Chromatin and Disease, 263–293.

modifications (acetylation; methylation, phosphorylation,...) bringing the epigenetic information. The term 'epigenetic' represents inheritable parameters through cellular divisions, by which cells regulate some of their functions, but without directly affecting the DNA sequence. Thus, the combinatorial nature of distinct posttranslational modifications on histone amino termini, so-called the 'histone code', has a critical regulatory role in most biological functions including cell fate decisions and both normal and pathological development (Lehrmann *et al.,* 2002; Fischle *et al.,* 2003).

One of the post-translational modifications utilized in the 'histone code' is acetylation. Acetylation of histones neutralizes their positive charges, weakening their interaction with DNA, thus, resulting in an opened chromatin conformation. Such conformation facilitates access for transcriptional regulators and hyperacetylation is thus generally associated with transcriptional activation. This phenomenon is reversible and deacetylation of histones is thought to promote the return to a repressive, higher order chromatin structure. Apart from acetylation of core histones, several other proteins are targets of acetylation as well. These include non-histone chromatin components, transcription factors, nuclear receptor coactivators, and proteins not involved in gene expression. Acetylation of transcription related factors, such as p53, Myo D or E2F-1, usually leads to their activation, and in turn, increased transcription of their target genes but in some cases, acetylation can also mediate deactivation (reviewed in Glozak *et al.,* 2005). Thus, particular HAT enzymes may influence the transcription of a much higher number of genes than they regulate directly by acetylating nucleosomes at their promoters.

These acetylation regulations are finely tuned by the opposing activities of two enzyme families: histone acetyltransferases (HATs) and histone deacetylases (HDACs). The balance of these two activities regulates the proper transcriptional competence of target genes, ultimately triggering important control on the cell fate. In neurons, a subset of specific genes, such as those from the cell cycle for example, not to mention proapoptotic genes, have to stay silent, whereas a battery of neuroprotective genes should remain transcriptionally active. Consequently, in adult neurons, if normally silenced regions become hyperacetylated, or normally transcriptionally active regions, deacetylated, this can lead to major disorders, potentially triggering neuronal death as observed in neurodegenerative diseases (Fig. 1).

2. THE DIFFERENT PLAYERS OF THE HAT/HDAC BALANCE

2.1. HATs and Their Functions

So far, six families of HAT have been described. The different groups have been made according to their sequence similarities between homolog regions and between their different HAT domains (reviewed in Torchia *et al.,* 1998; Sterner and Berger 2000; see references therein).

The GNAT family is related to GCN5 acetyltransferases that are represented in mammals by two closely related proteins, GCN5 and p300/CREB-binding

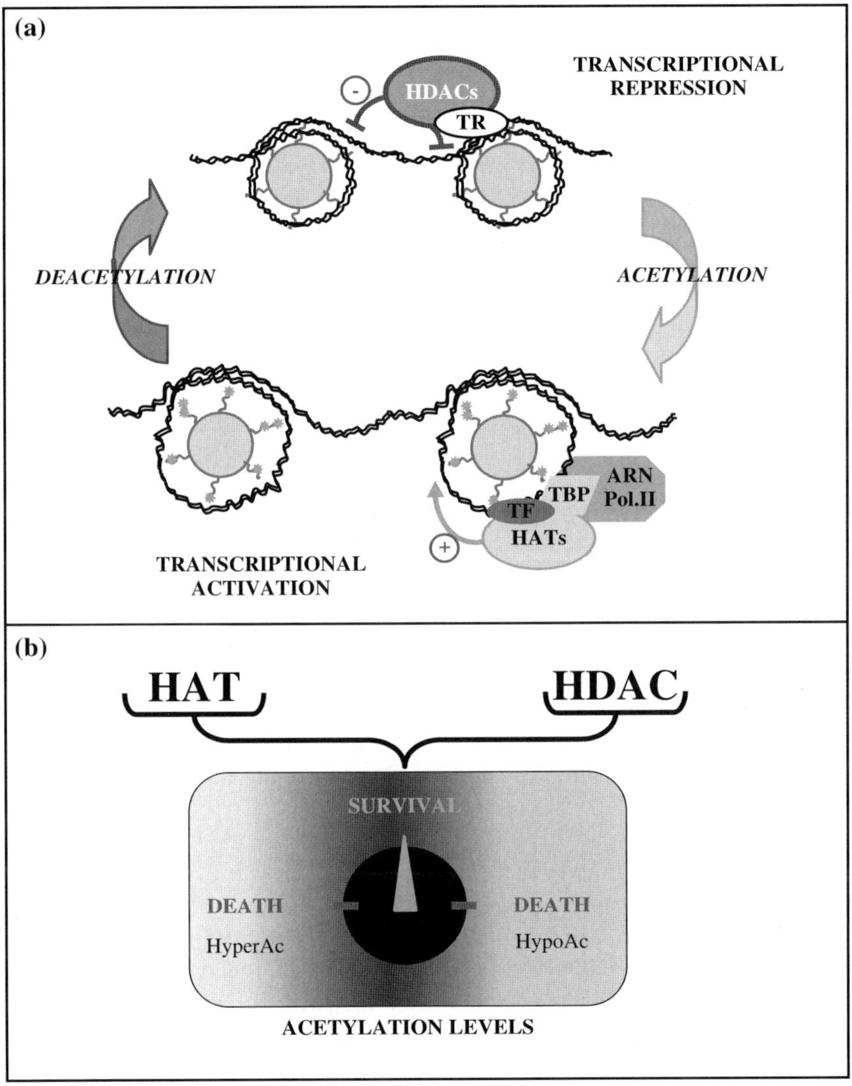

Figure 1. Chromatin acetylation status, transcription and survival: a balance between HAT and HDAC activities. (a) Transcriptional activationlrepression relies on the chromatin acetylation status of histones. TBP: TATA-Binding Protein, TF: Transcription Factor, TR: Transcriptional Repressor. (b) A fine-tuning of HAT/HDAC activities orchestrates neuronal death and survival. On one hand, acetylation levels can be decreased (HypoAc) because of CBP loss of function, as observed during apoptosis and neurodegeneration. On the other hand, when the threshold of acetylation is exceeded (HyperAc), this ultimately leads to neuronal death. (See Colour Plate 16.)

protein-associated factor (P/CAF). P/CAF is required as a HAT and coactivator for myogenesis, nuclear receptor-mediated and growth factor-signaled activation, among other processes. P/CAF is able to acetylate histones and also various non histone transcription-related proteins.

The MYST family, named for its founding members: MOZ, Ybf2/Sas3, Sas2, and Tip60, is involved in a wide range of regulatory functions in various organisms. Sas-type proteins are involved in gene silencing in yeast. The first MYST protein to be discovered was Tip60 (Tatinteracting protein, 60 kDa), involved in DNA repair processes and apoptosis. Finally, the MOZ (monocytic leukemia zinc finger) protein is a MYST protein involved in oncogenic transformation leading to leukemia, as it has been found to be able to recombinate with other HATs to give chimeric proteins (CBP or TIF2, transcriptional intermediary factor 2).

The p300/CBP family. The CREB binding protein (CBP) is one of most extensively characterized coactivator proteins. CBP was first identified through its ability to link the phosphorylated form of CREB by the cyclic AMP protein kinase (PKA) to components of the basal transcriptional machinery, including TFIIB, TATA-binding protein, and the RNA polymerase II holoenzyme complex. CBP is highly related to p300, and CBP and p300 are considered to be functional homologues, although a few differences in their activities have been reported (Kawasaki *et al.,* 1998; Kalkhoven 2004). CBP and p300 associate with a wide variety of transcriptional activators in addition to CREB, suggesting that each may serve as a transcriptional integrator of different signaling cascades (reviewed in references (reviewed in Vo and Goodman 2001). CBP and p300 have also been proposed to mediate transcriptional activation via intrinsic and associated HAT activity. CBP's HAT activity has been identified notably as a critical component of memory consolidation. Inducible expression of a CBP mutant, which acts as a dominant-negative for CBP HAT activity, impairs the stabilization of short-term memory into long-term memory in adult mice (Alarcon *et al.,* 2004).

The nuclear co-activator family. SRC-1 (steroid co-activator-1) stimulates ligand-dependent activation by hormone nuclear receptors. ACTR was firstly shown to bind the human retinoic acid receptor RARß and then shown to bind and activate multiple nuclear hormone receptor. TIF2 displays similarities with SRC-1 and ACTR, but its particularity is its recombination potential with MOZ found in leukemia. The three nuclear receptor coactivators discussed above are part of an evolutionarily and functionally related HAT family. All three interact with p300/CBP, and at least two interact with PCAF. Moreover, CBP can acetylate the nuclear co-activator family, likely activating SRC-1 function, while preventing ACTR receptor binding hence its activation potential. However, p300/CBP and PCAF can also directly interact with nuclear receptors, as does Tip60.

$TAF_{II}250$: TBP-Associated Factor II 250. Another direct connection between acetylation and transcriptional activation was demonstrated with the discovery that one of the TAFII (TATA-binding protein [TBP]-associated factor) subunits of the general transcription factor TFIID is itself a HAT. TFIID is one of the general factors required for the assembly of the RNA polymerase II transcription preinitiation

complex. The HAT activity of TAFII250 would allow the binding of TBPs to the TATA Box, as well as formation of the preinitiation complex (reviewed in Nogales 2000).

TFIIIC subunits: TFIIIC90 and TFIIICllO. The known function of TFIIIC is to initiate transcription complex formation by binding to promoter DNA and recruiting TBP-containing TFIIIB and RNA polymerase III, that directs synthesis of tRNA precursors. TFIIIC have an intrinsic HAT activity.

2.2. HDACs and Their Functions

In humans, 18 proteins HDACs have been identified and categorized in three classes based on homology to yeast HDACs (reviewed in de Ruijter *et al.,* 2003). The first class of human deacetylases (Class I) have high sequence similarity to yeast Rpd3. Since then, two new classes of yeast HDACs have been isolated, the HDAl family (Class II) and the Sir2 family (Class III). Human homologs of each of these two new classes have been isolated (for review see Gray and Ekstrom 2001). The sequences outside the catalytic domain are highly divergent, indicating that these enzymes might have different biological functions and a broader substrate repertoire beyond histones. However, the active sites of Class I and Class II have a high degree of sequence similarity. Both of these classes function by a Zinc-dependent mechanism. The more recently discovered Sir2 family (Class III) differs in that they are NAD-dependent.

• *Class I (RPD3-like)* comprises HDAC 1,2,3 and 8, that are homologs of the transcriptional regulator Rpd3 (Reduced Potassium Dependency) in yeast (Taunton *et al.,* 1996; Yang *et al.,* 1996). This class of HDACs is ubiquitous and localized in the nucleus, where their predominant substrate is found. A general mode of regulation for Class I HDACS involves association with proteins that modulate their deacetylase activity and recruitment to genomic area. A large body of evidence exists to demonstrate that these HDACs associate with corepressor complexes to direct gene specific transcriptional repression, such as the Sin3 corepressor complexes and the NuRD complex (Knoepfler and Eisenman 1999; Ng and Bird 2000). Co-repressors recruitment will allow local histone-DNA interaction modifications, using the free energy liberated by ATP hydrolysis. HDAC activity of these isoforms is modulated by post-translational modifications, such as phosphorylation (Galasinski *et al.,* 2002), a modification that can also affect their ability to form complexes with co-repressors. This class of HDACs also associates with other factors, probably allowing their recruitment at specific DNA sequences, among which the glucocorticoid receptor, Rb, the DNA methyltransferase 1 (dnmtl) or the SP1 transcription factor, to mention a few.

HDAC8 lies close to the phylogenetic boundary between the class I and class II HDACs (Buggy *et al.,* 2000; Hu *et al.,* 2000). It is a PKA substrate and its phosphorylation, contrary to the phosphorylation-mediated gain of function of all HDACs, is associated with decreased activity (Lee *et al.,* 2004).

All four members of this class of HDAC have been demonstrated to be sensitive to HDAC-specific inhibitors (HDACi) such as trichostatin A (TSA). It is noteworthy that the HDACi Valproic acid (VPA), in addition to selectively inhibiting the catalytic activity of class I HDACs, induces proteasomal degradation of HDAC2, in contrast to other inhibitors such as TSA (Kramer *et al.*, 2003).

• *Class II (HDA-1 like)* includes HDAC 4,5,6,7,9 and 10, homologs of the yeast Hda1 (Grozinger *et al.*, 1999; Zhou *et al.*, 2001; Guardiola and Yao 2002). Like the Class I enzymes, members of the Class II HDAC have been demonstrated to form large multiprotein complexes (Fischle *et al.*, 1999; Grozinger *et al.*, 1999). Class II HDACs demonstrate tissue-specific expression and are primarily expressed in striated muscle and brain (Marks *et al.*, 2001; Johnstone 2002; Vigushin and Coombes 2002). They are regulated by subcellular compartimentalization, where active nucleocytoplasmic trafficking delimits the availability of these enzyme for epigenetic regulation. Members of the class II HDAC have been shown to be actively maintained in the cytoplasm through binding with the chaperone protein 14-3-3. Phosphorylation favors their binding to 14-3-3 protein in the nucleus, thus leading to their nuclear expulsion (Grozinger and Schreiber 2000; Verdel *et al.*, 2000; Wang *et al.*, 2000). Evidence point to a role of Class II HDAC in cellular proliferationldifferentiation (reviewed in Lehrmann *et al.*, 2002). It is well described that HDAC 4, 5 and 7 associate with the MEF2 family of transcription factor, repress specific gene expression in a calcium-dependent manner, thus playing a critical role during myogenesis (Miska *et al.*, 1999; Wang *et al.*, 1999; Lemercier *et al.*, 2000; Lu *et al.*, 2000) and neuronal survival (Chawla *et al.*, 2003; Bolger and Yao 2005).

HDAC6 is localized exclusively in the cytoplasm, to punctate structures concentrated perinuclearly, as well as the leading edge of the cell (Verdel *et al.*, 2000; Barlow *et al.*, 2001). Recent evidence implicate HDAC6 in the stability of microtubules by adjusting the acetylation levle of alpha-tubulin (Hubbert *et al.*, 2002), thereby suggesting its critical role in neuronal functions. HDAC10 was classified as a class II subfamily member based upon similarity to HDAC6 (Fischer *et al.*, 2002). HDAC10 is able to associate with many other HDACs indicating that it might function as a recruiter rather than as a deacetylase (Guardiola and Yao 2002; Tong *et al.*, 2002).

HDAC9 is the predominant member of the class II HDAC family expressed in heart (Zhang *et al.*, 2002). Its major product was shown to encode the splice variant MEF2-interacting transcription repressor/histone deacetylase-related protein (MITRIHDRP), which lacks the enzymatic domain but forms complexes with both HDAC1 and HDAC3 (Zhou *et al.*, 2000; Zhou *et al.*, 2001) and has been recently implicated in skeletal muscle chromatin acetylation and gene expression under motor innervation control (Mejat *et al.*, 2005).

Like Class I HDACs, all Class II HDACS are inhibited by trichostatin A (TSA). However, unlike other family members, HDAC6 is uniquely resistant to the potent HDAC inhibitors trapoxin-B (Furumai *et al.*, 2001) and sodium butyrate as a result of an internal duplication of two catalytic domains, which appear to function independently of each other (Grozinger *et al.*, 1999; Guardiola and Yao 2002).

• *HDAC11* lies at the boundary between these two classes. HDAC11 contains a catalytic domain situated at the N-terminus with proven HDAC activity that can be inhibited by trapoxin (a TSA analog). HDAC11 was found not to reside in any of the known HDAC complexes (Sin3, N-CoR/SMRT), possibly indicating a biochemically distinct function of HDAC11 (Gao *et al.*, 2002).

• *Class III (Sir2-like)* have catalytic domains similar to that of the yeast NAD+-dependent deacetylase Sir2 (Silent Information Regulator 2) (reviewed in Buck *et al.*, 2004). For this reason, this class of HDAC is insensitive to TSA (Yoshida *et al.*, 1990). While class I deacetylases function as transcriptional co-repressors, Sir2-related proteins appear to be involved in gene silencing. In mammalian cells, SIRTs (named sirtuins SIRTl to 7) appear not to have histones as their primary substrates, being involved in the deacetylation of other proteins, and hence are not considered classical HDACs (Guarente 2000).

3. NEURODEGENERATIVE DISEASES AND APOPTOSIS

Neuronal apoptosis is a process that naturally occurs during development of the brain, as more neurons than needed are originally produced and those which are not correctly connected will die in order to set the proper number of neurons to form a functional network. Neuronal death underlies the symptoms of many human neurological disorders, including Alzheimer's, Parkinson's and Huntington's diseases, and amyotrophic lateral sclerosis. Neurological symptoms will depend of which neuronal population is targeted. The identification of specific genetic and environmental factors responsible for these diseases has bolstered evidence for a shared pathway of neuronal death or apoptosis (reviewed in Mattson 2000; Fadeel and Orrenius 2005; Krantic *et al.*, 2005), which main mechanisms and signalings are depicted in Fig. 2.

3.1. Neuronal Death Mechanisms and Signalings

Classical apoptosis, the best-known phenotypic expression of neuronal death, consists of at least two phases: initiation and execution. These ultimately lead to a series of stereotypic morphological and biochemical events resulting from the activation of a set of specific apoptotic cysteine proteases (reviewed in Hengartner 2000). These death proteases are homologous to each other, and are part of a large protein family known as the caspases (Alnemri *et al.*, 1996). Caspases are highly conserved through evolution. All known caspases possess an active-site cysteine, and cleave substrates at Asp-Xxx bonds; a caspase's distinct substrate specificity is determined by the four residues amino-terminal to the cleavage site (Thornberry *et al.*, 1997).

Caspase-mediated cleavage of specific substrates can explain several of the characteristic features of apoptosis. For example, cleavage of the nuclear lamins and cytoplasmic proteins such as fodrin and gelsolin are required for nuclear and cellular

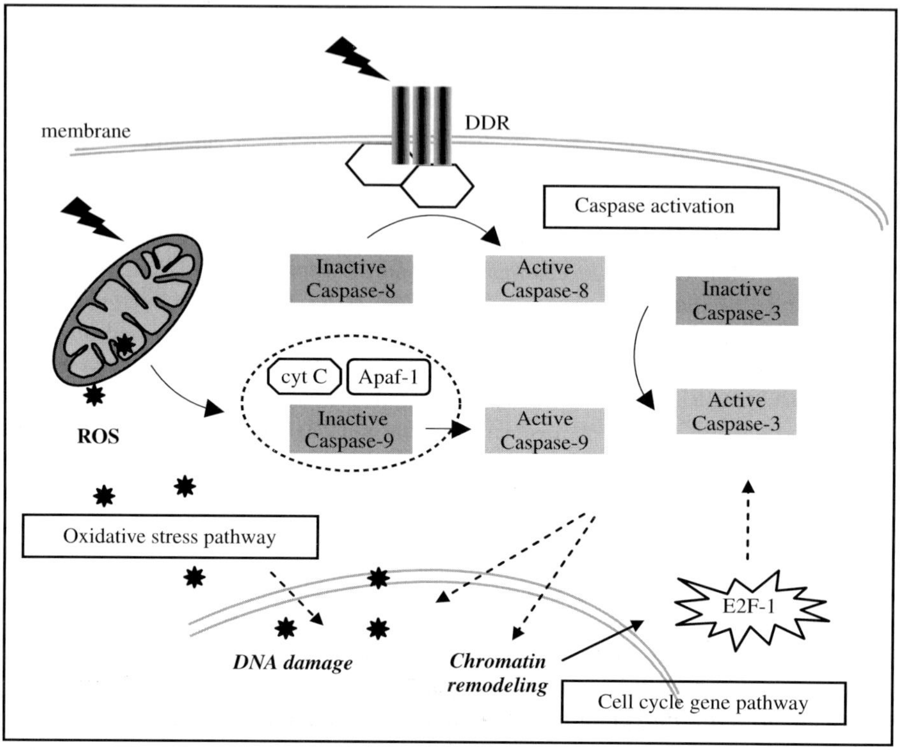

Figure 2. Mechanisms and signalings of neuronal death. Death can be initiated at the membrane by activation of death domain receptors (DDR), or by intracellular signalings through oxidative stress (and the production of reactive oxigen species, ROS), perturbed calcium homeostasis, mitochondrial dysfunction (release of cytochrome c, cytC), activation of caspases, as well as reactivation of cell cycle genes such as the transcription factor E2F (see text). Interconnections have been demonstrated (dotted lines) depending on the apoptotic context

shrinking. Inactivation of the inhibitory subunit ICAD by caspase cleavage leads to the release and activation of the caspaseactivated DNase (CAD) that triggers the 'typical' inter-nucleosomal DNA fragmentation (reviewed in Nagata 2000; Fadeel and Orrenius 2005). The canonical pathways of caspase activation during initiation include the 'death-receptor-mediated' recruitment of procaspase-2, procaspase-8 and procaspase-10, and a 'mitochondrial' pathway through which caspase-9 is activated via release of cytochrome *c*. The two pathways converge, leading to activation of procaspase-3 and, further downstream, to activation of caspase-6 and caspase-7. Of note, the apoptotic program is much more than just caspases, and in many cell types, activation of the apoptotic program inevitably leads to death, with or without caspases (Borner and Monney 1999).

In terms of neuronal death signalings, oxidative stress has been implicated as playing a role in degenerative disorders, such as Alzheimer's disease, Huntington's disease, Parkinson's disease, amyotrophic lateral sclerosis (reviewed in Langley and Ratan 2004). The generation of deleterious oxidants is mostly attributed to metabolism, a major source of reactive oxygen species (ROS) being the mitochondrial electron transport chain. For numerous reasons, the central nervous system appears particularly vulnerable to oxidative stress. Ultimately, the increased and unopposed ROS can lead to death via a number of pathways, including apoptosis or necrosis.

Finally, neuronal death is distinguished from general apoptotic machinery of dividing cells by the expression and involvement of cell-cycle regulators. Normally, differentiated adult neurons do not undergo cellular division and remain in a quiescent state. During neurodegeneration, uncoordinated ectopic re-entry of neurons into the cell-cycle proves abortive subsequently leading to apoptosis (Greene *et al.,* 2004; Herrup *et al.,* 2004). In neurons, the proapoptotic functions of the E2F-1 transcription factor, which is a retinoblastoma-controlled key factor in the progression of the cell cycle toward the S phase, have also largely been documented (Liu and Greene 2001; Trinh *et al.,* 2001; Greene *et al.,* 2004).

3.2. Neuronal Death in Pathologies

In this section, we will more particularly detail two neurodegenerative diseases for which CBP loss of function has clearly been reported by several studies and in which a therapeutic strategy based on resetting acetylation levels has been tested *in vivo*: Polyglutamine disorders and Amyotrophic lateral sclerosis (ALS).

3.2.1. Polyglutamine diseases

Polyglutamine diseases are hereditary neurodegenerative disorders caused by the expansion of a trinucleotide cytosine adenine guanine (CAG) repeat in the disease gene. The resulting elongated polyglutamine (polyQ) repeats are thought to cause structural changes in the affected proteins, leading to aberrant interactions, such as those that allow formation of extra- and intra- nuclear aggregates.

Huntington disease (HD) is the prototypic disease caused by expansion of unstable CAG repeat. It primarily affects striatal neurons. It is a mid-life onset disorder characterized by unvoluntary movements (chorea), personality changes and dementia that progress to death within 10–20 years of onset. There are currently no treatment to delay or prevent appearance of the symptoms in the patients. Other diseases in this class include spinocerebellar ataxias (SCA) 1, 2, 3 (also known as Machado-Joseph disease, MJD), 6, and 7, DRPLA, and spinobulbar muscular atrophy (SMA, also known as Kennedy's disease) (Zoghbi and Orr 2000).

HD is caused by polyQ expansion in the amino-terminal region of a protein huntingtin (Htt). Disruption of the HD gene in mice causes increased neuronal apoptosis and progressive behavioural and motor dysfunction reminiscent of the disease in humans (Sanchez *et al.,* 1999) but the exact molecular mechanisms by which mutant Htt induces cell death as well as the function of Htt are not totally understood. Several studies indicate that huntingtin is an antiapoptotic protein that could regulate intracellular dynamic. It was recently demonstrated, that huntingtin specifically enhances vesicular transport of brain-derived neurotrophic factor (BDNF) along microtubules (Gauthier *et al.,* 2004). xperimental evidence also indicates that Ca^{2+} homeostasis is abnormal in mitochondria isolated from lymphoblasts of HD patients. The mutated Htt protein would induce cytosolic and mitochondrial Ca^{2+} overload of striatal neurons, thus playing an important role in the pathogenesis of HD (Bezprozvanny and Hayden 2004).

Htt is a substrate of caspase-3, or a closely related protease, and its cleavage is modulated by the length of the polyglutamine tract (Goldberg *et al.,* 1996). Neuronal death in HD is associated with the accumulation of intracellular inclusions of Htt fragments in the HD brain, as well as in the transgenic mouse brain. Cytosolic aggregates of polyglutamine repeat proteins may recruit procaspase-8, resulting in the activation of caspase-8 (Sanchez *et al.,* 1999). Recent findings reveal that Htt under normal circumstances forms a complex with Htt-interacting protein-1 (Hip-1), while Mutant Htt rather complexes with Hip-1, caspase-8 and the DEDcontaining protein Hippi (Hip-1 protein interactor), thus leading to activation of caspase-8 as well as caspase-3, independently of death receptors (Gervais *et al.,* 2002).

3.2.2. Amyotrophic lateral sclerosis (ALS)

ALS is a progressive neurological disorder that is associated with the death of motor neurons in the spinal cord and brain, leading to paralysis. ALS occurs in both sporadic and familial forms, and about 20% of the familial forms are associated with mutations in the gene encoding the cytosolic CuIZn superoxide dismutase-1 (SOD 1) (Rosen *et al.,* 1993). Transgenic mice, which express human mutant SOD1, develop a motor neuron degenerative syndrome that closely resembles ALS (Gurney *et al.,* 1994). The mutations do not decrease antioxidant activity of the enzyme, but result in the gain of an adverse pro-apoptotic activity that may involve increased peroxidase activity. Through interactions with hydrogen peroxide or superoxide anion, the mutant enzyme may induce oxidative damage to membranes and disturbances in mitochondrial function that make neurons vulnerable to excitotoxic apoptosis (Kruman *et al.,* 1999).

The selective degeneration of motor neurons involves increased oxidative stress, overactivation of glutamate receptors and cellular calcium overload (Cookson and Shaw 1999). In people with ALS, DNA starts to fragment as a ladder, typical of apoptosis, in neurons within the spinal-cord anterior horn and motor cortex. The DNA damage is associated with increased mitochondrial localization of Bax and decreased association of Bcl-2 (Martin 1999). The involvement of apoptosis in ALS is further suggested by the fact that overexpression of Bcl-2 and administration of caspase inhibitors delays degeneration (Kostic *et al.,* 1997) and and a potential role for caspases in ALS was demonstrated by a significant decrease in the progression of symptoms following inhibition of caspase-1 in mice (Friedlander *et al.,* 1997).

Supportive evidence comes from a study showing that intraventricular administration of z-VADfluoromethylketone (fmk), a pan-caspase inhibitor increases the lifespan of SOD1 transgenic mice by approximately 25% (Li *et al.,* 2000). Furthermore, overexpression of XIAP, a mammalian inhibitor of caspases 3,7 and 9, in spinal motor neurons of mutant SOD1 mice attenuated disease progression without delaying onset, whilst expression of p35, a baculoviral caspase inhibitor that does not inhibit caspase-9, delayed onset without decreasing disease progression (Inoue *et al.,* 2003). Moreover, caspase-9 was activated in spinal motor neurons of ALS patients.

Cell cycle proteins were evidenced in port-mortem tissues of ALS patients. Recent results from Ranganathan and Bowser (2003) indicate hyperphosphorylation of the retinoblastoma protein in motor neurons during ALS, concurrent with increased levels of cyclin D, and redistribution of E2F-1 into the cytoplasm of motor neurons and glia, while Nguyen *et al.,* (2003) found increased Cdk4 activity in SODl mutant mice, associated with an increase in nuclear Cdk4, cyclin Dl and abnormal Rb phosphorylation. These results suggest that G(1) to S phase activation occurs during ALS and may participate in molecular mechanisms regulating motorneuron death.

4. ALTERATION OF THE HAT/HDAC BALANCE IN THE MANIFESTATION OF NEUROLOGICAL DISEASES

4.1. Chromatin Acetylation Status During Neuronal Apoptosis

During normal conditions, protein levels and enzymatic activity of HATS and HDACs regulate chromatin and transcription factor acetylation in a controlled fashion. Such equilibrium manifests neuronal homeostasis and is responsible for regulated gene expression leading to normal neuro-physiological outputs like long-term potentiation, learning and memory. This equilibrium is maintained very stringently and any damagelinsult that disrupts the HAT/HDAC balance will ultimately alter cellular homeostasis.

For example, increasing hyperacetylation by treating neurons with a general HDAC inhibitors (HDACi) such as trichostatin A (TSA) in normal conditions induces neuronal apoptosis (Salminen *et al.,* 1998; Boutillier *et al.,* 2003). This neuronal death is accompanied by typical features of neuronal apoptosis, such as caspase-3 activation (Salminen *et al.,* 1998; Boutillier *et al.,* 2003), internucleosomal DNA cleavage and increased levels of E2F1 transcription factor (Boutillier *et al.,* 2003). Similarly, elevating acetylation levels by overexpression of CBP in resting neurons, we also observed chromatin condensation and neuronal death (Rouaux *et al.,* 2003). Overexpression of a HAT domain-deleted CBP mutant did not generate similar results, thus supporting the idea that hyperacetylation is fatal to the neuron.

By contrast, apoptosis studied in the model of potassium deprivation of cerebellar granule neurons (CGN) revealed an early drop in histone acetylation levels

(Rouaux *et al.*, 2003), that preceded caspase 3 activation, internucleosomal DNA cleavage, and E2F1 activation (Trinh *et al.*, 2001). It is norteworthy that in both situations – histone hyper- and hypo-acetylation – increased levels of E2F1 transcription are measured, suggesting a fine tuning operated by the HAT/HDAC balance, independently of the net global acetylation status of the cell. This regulation would be possible if recruitment of specific HAT/HDAC isoforms or other epigenetic enzymes are intimately associated with certain promoters. Although the acetylation status of the e2f-1 promoter is currently not known in response to potassium deprivation, it displayed increased acetylated-histone H3 in response to TSA treatment as measured bu Chromatin Immunoprecipitation (ChIP) (Boutillier *et al.*, 2003). In this line, hyperacetylated histone H3 have been evidenced on the cJun promoter in the same apoptotic model (Morrison *et al.*, 2006), thus proving that discrete and local hyperacetylation can indeed be induced, despite a global hypoacetylation status. In this case, an explanation is that the class II HDAC-like HDRP (see section 2.2) exerts neuroprotective functions, at least in part by repressing c-Jun gene expression through HDAC1 interaction with the *c-Jun* promoter and that HDRP levels are repressed during neuronal apoptosis, hence increased promoter acetylation (Morrison *et al.*, 2006). Histone deacetylation has also been observed in the context of oxidative stress-induced neuronal death. Indeed, upon transient oxigen peroxide treatment, the motoneuronal cell line NSC34 displays hypoacetylated histone H3, as also do CGN undertaking the same stress (our unpublished data).

Taken together, these results show that while hypoacetylation is a feature of neuronal apoptosis, hyperacetylation can also induce death, through common mechanisms.

4.2. CBP, a HAT of Particular Interest in Neurological Disorders

An explanation for this net deacetylation throughout neurodegeneration is that neuronal cells progressively lose a HAT, and more particularly, the CREB binding protein (CBP) (reviewed below and in (Rouaux *et al.*, 2004; Saha and Pahan 2006)

Genetic studies indicate an essential role for CBP in cellular function and development. Gene knockouts in mice indicated that CBP and p300 are required for normal embryonic development and viability (Oike *et al.*, 1999). In humans, *CBP* loss of function is associated with Rubinstein-Taybi syndrome, a haploinsufficiency disorder (Petrij *et al.*, 1995; Kalkhoven *et al.*, 2003) characterized by mental retardation, developmental defects, and an increased predisposition to cancer (Miller and Rubinstein 1995). Recent studies suggest that mutations affecting *cbp* gene in RTS patients would mainly target the HAT domain, leading to abolition of CBP's HAT activity as well as its ability to transactivate CREB (Murata *et al.*, 2001; Kalkhoven *et al.*, 2003).

One mechanism by which CBP and p300 could contribute to neuronal survival is through the transcription factor CREB (cAMP response element binding protein). Normally, activated CBP binds to CREB and acetylates the latter at three putative acetylation sites within its activation domain (Lu *et al.*, 2003). It has been proposed that during resting conditions, these nonacetylated lysine residues restrain CREB-mediated transactivation by maintaining the molecule in an inactive conformation. However, once acetylated by the bound HAT and consequential charge neutralization at the lysine residues, CREB attains the active state and dimerizes to bind CREB responsive element (CRE) found in promoters of many survival genes, such as Bcl-2 (Freeland *et al.*, 2001), brain-derived neurotrophic factor (BDNF) (West *et al.*, 2001), and arginase I (Morris 2002). CREB-dependent gene expression is heavily contingent on the presence and enzymatic activity of CBP and may thus fall short of the threshold level during neurodegenerative conditions when the quantity and functionality of HATs are severely compromised.

Different mechanisms can account for CBP disappearance during apoptosis, depending on the death context (Fig. 3). In a classical model of apoptosis, we recently defined CBP as a new caspase-6 substrate (Rouaux *et al.*, 2003). In the context of death induced in polyQ diseases, it was reported that CBP is sequestred in nuclear inclusions formed by polyglutamine-containing proteins, and that CBP ubiquitylation and degradation could be selectively enhanced by the mutated polyQ-expanded huntingtin protein (reviewed in McCampbell and Fischbeck 2001; Langley *et al.*, 2005). Oxidative stress-induced death of the motorneuronal NSC34 cell line results in low levels of CBP protein, that in this case, rely on decreased *cbp* gene expression (our unpublished data). Overexpression of CBP was neuroprotective and the neuroprotective capacity, dependent on its HAT activity (Rouaux *et al.*, 2003).These recent findings obtained on cellular models define CBP as a potential neuroprotective protein.

4.2.1. Link CBP – polyglutamine diseases

The sequestration model of polyQ diseases hypothesizes that aberrant protein interactions lead to the disturbance of cellular functions and ultimately neuronal cell death by decreasing the concentration and activity of critical cellular factors. CBP has been found to co-localize with the nuclear aggregates found in polyQ-repeat disorders (Steffan *et al.*, 2000), such as a wide variety of other cellular proteins including p53 (Steffan *et al.*, 2000), mSin3A, N-CoR (Boutell *et al.*, 1999), TAFII-130 (Shimohata *et al.*, 2000), SP1 (Dunah *et al.*, 2002) and CAI50 (Holbert *et al.*, 2001). For some of these proteins, direct binding with expanded polyQ bearing proteins have been shown. In particular, strong physical interactions were found between the Htt exon 1 protein and the acetyltransferase neighbouring CH3 domains of recombinant CBP protein (Steffan *et al.*, 2001). In the same study, *in vitro* experiments demonstrated that the Htt exon 1 protein interacted with p300/CBP Associated Factor (P/CAF) as well as p300, thereby reducing the enzymatic activity of the affected proteins (Steffan *et al.*, 2001). Similarly, physical interactions have been reported between CBP, p300 and P/CAF and both wild type and mutated forms of Ataxin-3 (Li *et al.*, 2002).

Several cell culture experiments and *in vivo* observations support these *in vitro* findings. Redistribution of CBP in nuclear or cytoplasmic inclusions upon overexpression of expanded HD constructs or ataxin-3 has been described in different cellular

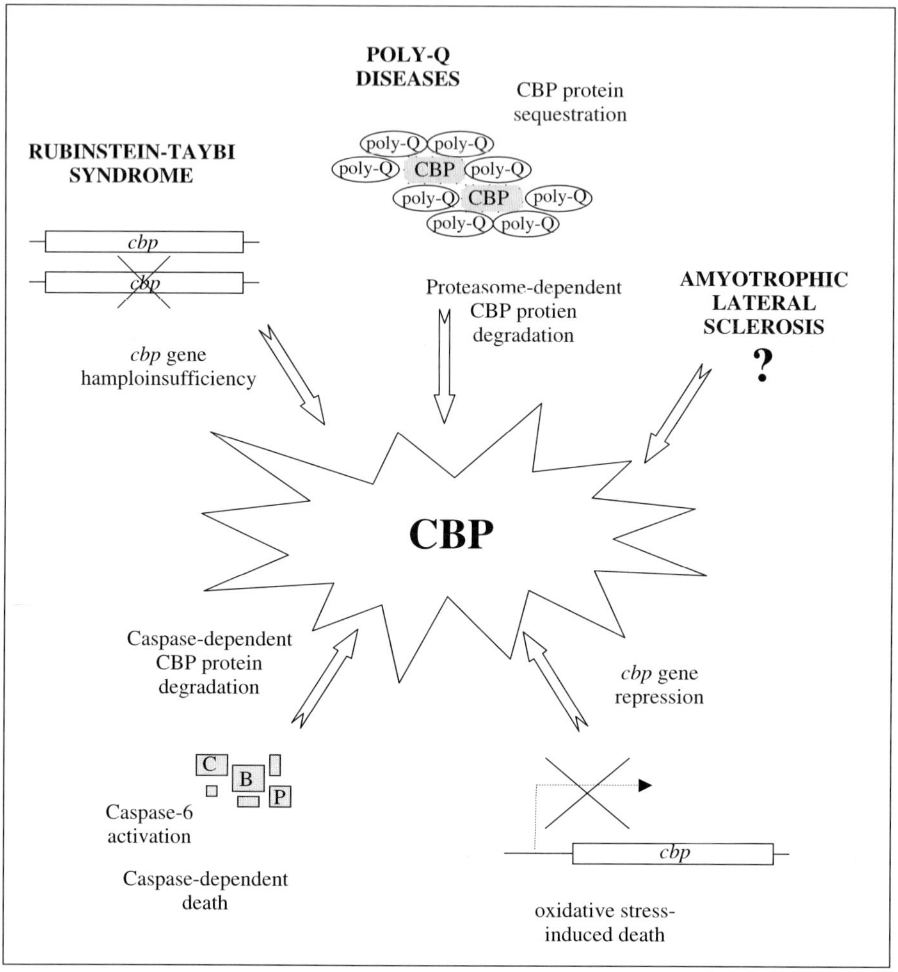

Figure 3. The many ways to lose a HAT. Decreased amounts of functional CBP protein and subsequent CBP's loss of function has been observed in different contexts of neurological disorders and neuronal apoptosis. RTS (Rubinstein-Taybi Syndrome) results from a mutation on one *cbp* gene allele. In several cases of polyQ diseases, CBP can be sequestred by the mutated polyQ proteins, forming aggregates in the cytoplasm or the nucleus. CBP proteasomal degradation was also shown to be favored by polyQ proteins. CBP is a caspase-6 substrate in cerebellar granule neurons (CGN) deprived of potassium modeling caspase-dependent apoptosis. Finally, *cbp* gene repression has been observed in oxidative stress-induced death of a motorneuronal cell line. The mechanisms by which CBP levels are reduced in motorneurons of ALS mice is still unknown

models (Kazantsev *et al.,* 1999; McCampbell *et al.,* 2000; Chai *et al.,* 2001), a phenomenon accompanied with inhibition of its HAT activity and further leading to global histone deacetylation and cell death (Igarashi *et al.,* 2003; Taylor *et al.,* 2003). Alternatively, recruitment of CBP by the mutated form of htt has been shown to enhance its processing by the ubiquitin-proteasome pathway (Jiang *et al.*, 2003; Cong *et al.*, 2005). Nevertheless, one study performed on transgenic animals bearing mutant proteins with long polyQ repeats (up to 150) could not reproduce transcription factors depletion by intranuclear huntingtin inclusions, and further suggested that soluble mutant huntingtin binds to CBP (and other transcription factors) and affects gene expression (Yu *et al.*, 2002). The study from Cong and colleagues further confirmed the binding of CBP to soluble mutant huntingtin in a specific way as p300 was unaffected by mutant huntingtin (Cong *et al.*, 2005). In addition to the aberrant redistribution of the protein, soluble CBP protein levels were reduced in postmortem HD brain samples, further emphasizing the relevance of reduced CBP levels associated with polyQ diseases (Nucifora *et al.*, 2001). If we should keep in mind that the disease protein context can influence the interactions mediated by the polyQ domain, it is postulated that CBP sequestration leads to a decrease in CBP enzymatic activity responsible for neuronal function alterations.

Gene expression screens have shown that genes that are regulated by CBP, such as enkephalin and Jun, are down-regulated in HD transgenic mice and HD postmortem brain tissue (Richfield *et al.*, 1995; Luthi-Carter *et al.*, 2000). For example, BDNF, a cell survival protein whose expression is regulated by CREB, has recently been shown to be down-regulated in HD patient tissue (Ferrer *et al.*, 2000). Mutant huntingtin-mediated sequestration of CBP away from transcription pockets may be recalled in this respect.

4.2.2. Link CBP – ALS

We recently evidenced that CBP levels were decreased in the lumbar spinal cord of symptomatic ALS mice model (SOD1G86R) when compared to wild type animals. Moreover, immunohistochemistries showed that CBP loss was specifically observed in motorneurons nuclei, that also displayed decreased histone acetylation levels (Rouaux *et al.*, 2003). Interestingly, we found that *cbp gene* transcription could be downregulated by oxidative stress-induced death in a motorneuronal-derived cell line (our unpublished data). Moreover, we observed that overexpression of the SOD1G86R mutant could reduce luciferase activity driven by the mouse *cbp promoter* (our unpublished data). Taken together, these results suggest that the chronic dysregulation of the SOD1 enzyme observed in some cases of ALS could be the direct cause of *cbp* gene downregulation and this needs further investigations to be confirmed. Supporting this, the *cbp* gene was downregulated in a gene expression profile study performed on laser microdissected degenerating motorneurons from autopsied patients with sporadic ALS (Jiang *et al.*, 2005).

5. IMPLICATION FOR THERAPEUTIC STRATEGIES

Despite differences in disease etiology and cellular properties of affected neurons in various forms of neurodegenerative disorders, loss of acetylation homeostasis appears to represent a critical and decisive mechanism commonly underlying neuronal dysfunction and degeneration. Once this was realized, research in this field focused on experimental therapeutics and drug development that would allow proper acetylation levels resetting. The first attempts were performed by HD researchers, who tried to fine-tune the altered HAT-HDAC balance by using general HDAC inhibitors to re-establish the acetylation threshold in dying neurons. On the opposite, targeting HAT loss reversal have also been proposed. These therapeutic strategies are represented Fig. 4. In the following section, we have evaluated the efficacy of both modes of resetting acetylation homeostasis.

5.1. Hat Functionality Resetting: Hat Overexpression, Hat Activators, Caspase/proteasome Inhibitors

Resetting HAT/CBP functionality is possible by three means: either overexpressing the protein, enhance the activity of the remaining protein with HAT activators or blocking its degradation in the first place.

Overexpression studies of CBP have been conducted by different laboratories and they all demonstrated increased neuronal survival in different context of apoptosis. For example, the androgen receptor with an expanded polyglutamine repeat, which is the cause of spinal and bulbar muscular atrophy (SBMA), interacts with CBP, leading to cell toxicity. This cell death can be mitigate in part by increased expression of CBP, which retains histone acetylase activity (McCampbell *et al.,* 2001). Overexpression of CBP rescued polyglutamine-induced (huntingtin or atrophin-1 with expanded polyglutamine repeats) neuronal toxicity of N2a cells (Nucifora *et al.,* 2001). Using a *Drosophila* model of polyglutamine toxicity, Taylor et al., demonstrated complete functional and morphological rescue by up-regulation of endogenous Drosophila CBP. Rescue of the degenerative phenotype was associated with eradication of polyglutarnine aggregates, recovery of histone acetylation, and normalization of the transcription profile (Taylor *et al.,* 2003). Thus, the success of using CBP overexpression to rescue polyQ toxicity in cell models suggests that a gene therapy approach could be of benefit. Our attempt to prevent cell death in CGN deprived of potassium by overexpression of CBP also proved succesfull, but only with a CBP protein that still beared an intact HAT domain (Rouaux *et al.,* 2003). However, high levels of CBP expressed in neuroprotective conditions ultimately leads to cell death, likely because of unwanted hyperacetylation as again, a HAT deletion mutant of CBP was without effect (Rouaux *et al.,* 2003). This therapeutic strategy thus implies a restricted gene targeting to damaged neurons.

Several small molecule modulators (SMM) of p300 and PCAF have been developed (Varier *et al.,* 2004). Recently, the first naturally occurring HAT inhibitor anacardic acid was isolated from cashew nut shell liquid, which inhibits the HAT activity of both p300 and PCAF very effectively (Balasubramanyam *et al.,* 2003). By using anacardic acid as a synthon, an amide derivative of anacardic acid, CTPB, has been synthesized, which is the only known small molecule activator of any histone acetyltransferase, in this case, p300. However, cells are impermeable or

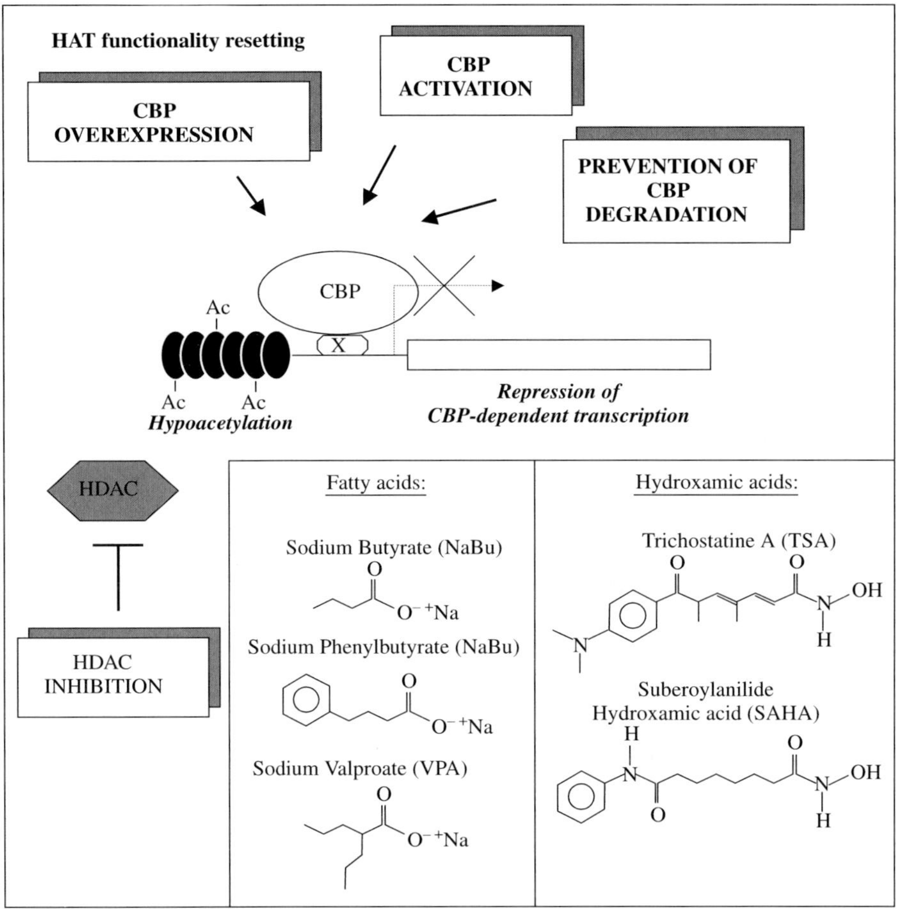

Figure 4. Therapeutic strategies to counteract CBP loss of function. CBP loss of function leads to a decrease in histone acetylation levels as well as a decrease in CBP-dependent transcription. Two main approaches can be tested to reverse this process: either resetting HAT functionality or resetting global acetylation levels with the use of HDAC inhibitors. Whereas both strategies would increase histone acetylation levels, HDAC inhibition would act on a broad range of genes, while CBP activation (overexpression or by a pharmacological approach) would specifically target both CBP-dependent histone acetylation and transcription. The structure of some of the HDACi that have been tested in different models, such as small fatty acids and hydroxamic acids, are represented in the boxes

poorly permeable to both anacardic acid and CTPB, so progress must still be made in that direction. Here again, the same restrictions as above apply, concerning diseased-cell targeting in case of therapeutic use.

One could turn to the use of a pharmacological approach with small molecule drugs that prevent CBP degradation. This can be achieved by designing molecules to block caspases-6 activity on CBP or blocking attachment of CBP-to ubiquitin so as to subsequently prevent its proteosomal degradation. So far, we are not aware of the use of such molecules *in vivo*. Finally, one can speculate that a molecule reducing or preventing aggregation could represent a way to favor CBP protein availability within cells in HD pathology. Different recent studies have reported beneficial effect of such molecules tested in *in vitro* and *in vivo* models (Heiser *et al.,* 2002; Kazantsev *et al.,* 2002; Pollitt *et al.,* 2003; Sanchez *et al.,* 2003; Zhang *et al.,* 2005).

5.2. Acetylation Levels Resetting: Use of HDAC Inhibitors

Since a few years, HDAC inhibitors have been tested as potential neuroprotective agents in several cellular and animal models of neurodegenerative diseases. The logic behind using these inhibitors is based on the argument that transcriptional repression is the predominant reason of transcriptional dysregulation by global deacetylation. Thus, inhibiting HDAC activity could remove repressive blocks from promoters of essential genes - including those that were under CBP's control (Fig. 4).

5.2.1. HDAC inhibitors (HDACi)

One of the first HDAC inhibitors to be identified and characterized was sodium butyrate, where it was found to alter the histone acetylation state (Riggs *et al.,* 1977), and further determined to inhibit HDAC activity both *in vitro* and *in vivo* (Candido *et al.,* 1978). Almost a decade later trichostatin A (TSA), a fungistatic antibiotic, was found to induce murine erythroleukemia cell differentiation (Yoshida *et al.,* 1987). To date, a wide range of molecules have been described that inhibit the activity of Class I and Class II HDAC enzymes, and with a few exceptions, can be divided into structural classes including: (1) small-molecule hydroxamates, such as TSA, suberoylanilide hydroxamic acid (SAHA), scriptaid and oxamflatin; (2) short-chain fatty-acids, such as sodium butyrate, sodium phenylbutyrate and valproic acid (VPA); (3) cyclic tetrapeptides, such as apicidin, trapoxin and the depsipeptide FK-228; and (4) benzamides, such as MS-275 and C1-994 (for reviews see Remiszewski *et al.,* 2002; Miller *et al.,* 2003). Some of these molecules are represented in Fig. 4.

Many of the small molecule hydroxamate HDAC inhibitors are thought to function by interacting with the HDAC active site Zn2+ ion, thereby inhibiting the catalytic activity of the enzyme. This is supported by crystal structure studies of TSA or SAHA (Finnin *et al.,* 1999), and more recently, eukaryotic HDAC8 (Vannini *et al.,* 2004). TSA, with its hydroxamic acid group and its five-carbon atom linker to the phenyl group, has the optimal conformation to fit into the active site (Finnin *et al.,* 1999). TSA is the most potent reversible HDACi currently known, with an IC50 in low nanomolar range (Vannini *et al.,* 2004). All Class I and II HDACs are thought to be approximately equally sensitive to inhibition by TSA (Marks *et al.,* 2001; Marmorstein 2001; Johnstone 2002). Whether all the structural classes of HDAC inhibitors inhibit HDACs by interfering with the active site remains to be determined.

Class 2) containing the short-chain fatty-acids is therapeutically interesting in neurodegenerative disorders as they possess the natural characteristic of penetrating rapidly through the Blood Brain Barrier because of their small molecular weight (Vigushin and Coombes 2002). For example, Valproic acid (VPA), a drug that has been used for decades in the treatment of epilepsy and as a mood stabilizer, has been recently shown to modulate the epigenome by inhibiting histone deacetylases (HDACs) (Gottlicher *et al.,* 2001; Phiel *et al.,* 2001). VPA is also able to exert epigenetic functions by enhancing intracellular demethylase activity through its effects on histone acetylation (Detich *et al.,* 2003). By contrast, SAHA does not appreciably penetrate the blood brain barrier by itself, and when administred *in vivo*, needs to be complexed with a cyclodextrin derivative in solution, thus raising some specificity and cost issues.

5.2.2. HDACi treatment in vitro

The neuroprotective effects of HDACis have been examined in different cell culture models of neuronal death.

Treatment with TSA, SAHA or sodium butyrate increased histone acetylation and neuron survival in a model of expanded polyglutamine repeat disease, in which CBP was suggested to be repressed by the overexpression of expanded polyQ proteins (McCampbell *et al.,* 2001). Nevertheless, we found that neither TSA nor NaBu were able to reverse low potassium-induced CGN death, despite they were able, at each dose tested, to increase histone acetylation (Rouaux *et al.,* 2004). Interestingly, TSA was able to reverse oxidative stress-induced CGN death (our unpublished data), while Ryu *et al.,* (2005) demonstrated that HDACi could also prevent oxidative neuronal death. Using a motorneuronal cell line (NSC34), we thouroughly measured the ability of 3 HDAC is (TSA, NaBu and VPA) to reverse oxidative stress-induced death, by comparing histone acetylation levels to survival, and found that each HDACi dose that was able to reverse acetylation, at least back to the original control levels, had a significant effect on survival (our unpublished data). As mentioned above, high doses producing too much acetylation levels where toxic. Nevertheless, TSA was the most neuroprotective molecule, while NaBu was the less effective, despite both drugs ultimately induced comparable global acetylation levels. These results suggest that reversing acetylation is 'not good enough' for neuroprotection, most likely because they do not have the same affinity for each HDAC in that exact situation (see conclusion).

It is likely that the ability of HDACi to prevent neuronal death will depend on the underlying death signaling mechanisms. For example, neuronal death in the potassium-deprivation model is largely dependent on CBP loss by caspase-6 activation (Rouaux *et al.,* 2003), as well as E2F1-induced mechanisms (Trinh *et al.,* 2001), reviewed in Greene *et al.,* (2004), so even if HDACi can increase specific (yet unidentified) neuroprotective target genes, it may not be sufficient to improve survival, as (1) CBP will still be cleaved off by activated caspase-6, and (2) E2F1 expression still induced. By contrast, oxidative stress-induced death or perturbed redox homeostasis relies on other mechanisms, that appear to be context dependent and determined by the level and duration of oxidative stress (reviewed in Langley and Ratan 2004). It is thus possible that in such case, HDACi could be more effective in neuroprotection by acting on other signaling involving for example SP1 acetylation (Ryu *et al.,* 2003), NfKB regulation (Ryu *et al.,* 2005), or by resetting levels of the *cbp* gene. Indeed, we recently found that *cbp* levels could be transcriptionally up-regulated by HDACi treatment during oxidative stress-induced motorneuronal death (our unpublished data), thus providing a molecular basis to neuroprotection in these conditions.

In summary, studies of HDACi effects in *in vitro* models have proven that HDACi could be potent neuroprotectors, and much effort should be taken to thoroughly define which genes are transcriptionally altered in response to these molecules.

Of note, a molecule screening gave the HDACi scriptaid as a potent molecule able to prevent aggresome formation in cultured cells, in the ALS context (Corcoran *et al.,* Curr Biol, 2004), emphasizing here, a novel mechanism of HDACi action in neurodegenerative disease.

5.2.3. HDACi treatment in vivo

HDAC inhibitors are also effective in animal models of neurodegeneration demonstrating that these *in vitro* findings are translatable *in vivo.* As mentioned above, butyrates are considered to be most effective *in vivo* in terms of their ability in crossing the blood brain barrier.

HD: In *Drosophila* models of Huntington's disease, the HDAC inhibitors SAHA and sodium butyrate arrest the progressive neuronal degeneration and lethality (Steffan *et al.,* 2001). SAHA and sodium butyrate have also been demonstrated to extend survival, ameliorate motor deficits and delay characteristic neuropathology in the mouse Huntington's disease model, R6/2 (Ferrante *et al.,* 2003; Hockly *et al.,* 2003). In NaBu-treated animals, animals displayed enhanced acetylation status of histones and pro-survival transcription factors like Spl and reduction in several neuropathological hallmarks like striatal neuronal atrophy (Ferrante *et al.,* 2003). Consistent with the idea that HDAC inhibition relieves transcriptional repression and that protection is downstream of mutant htt, neither SAHA nor sodium butyrate decreased mutant htt expression or aggregates (Ferrante *et al.,* 2003; Hockly *et al.,* 2003).

In a different study, phenylbutyrate was adrninistred after onset of symptoms in the N171- 82Q transgenic mouse model of HD and significantly extended survival, attenuating both gross brain and neuronal atrophy (Gardian *et al.,* 2005). Phenylbutyrate increased histone acetylation and decreased histone methylation levels. Moreover, phenylbutyrate increased expression of ubiquitin-proteosomal component and down-regulated caspases implicated in apoptotic cell death, such as caspase-3 (Gardian *et al.,* 2005).

SBMA: In addition to HD, NaBu has also been reported to be effective in treating spinal and bulbar muscular atrophy (SBMA) (Minamiyama *et al.,* 2004), an inheritable motorneuron disease caused by an expanded polyglutamine-repeat within the androgen receptor. SBMA is associated with motorneuron loss in the spinal cord and brain stem as well as a sub-clinical loss of sensory neurons in the dorsal root ganglia resulting in skeletal muscle atrophy and weakness (Sobue *et al.,* 1989;

La Spada *et al.,* 1991). Treatment of SBMA transgenic mice with sodium butyrate, by oral administration, increased histone acetylation in spinal cord tissue and ameliorated the functional and histopathological defects associated with SBMA (Minamiyama *et al.,* 2004).

ALS: Surprisingly, the first report of an *in vivo* assay on ALS with VPA did not mention its HDACi potency (Sugai *et al.,* 2004), but the authors utilized VPA based on the fact that it linked with neuroprotection - together with Lithium ion - in the context of mood bipolar disorders (Coyle and Duman 2003). Indeed, this molecule has neuroprotective effects occurring through changes in a variety of intracellular signalling pathways (including upregulation of Bcl-2 protein and inhibiting glycogen synthase kinase 3-beta). VPA was also shown to induce persistent activation of extracellular signal-regulated kinase (ERK), and its downstream effectors RSK and CREB, thus promoting neurite growth (Yuan *et al.,* 2001). When tested in the ALS mice model (G93A), a pre-onset VPA treatment induced prolongation of the disease duration (Sugai *et al.,* 2004) while a post-onset treatement was less effective. VPA was able to protect spinal motorneuron degeneration induced by glutamate toxicity in organotypic slice cultures *in vitro*. When tested in the G86R strain of ALS mice model, VPA also displayed a strong neuroprotective response *in vivo*: the number of large motorneurons before death was higher in VPA-treated than in non-treated animals, and motorneurons nuclei from treated animals displayed a strong immunoreactivity to CBP compared to non-treated (our unpublished data). Taken together, these results support a beneficial effect of VPA on spinal motor neurons protection from death. Nevertheless, mean survival of VPA-treated SODG86R mice was not significantly increased, while we found a significant delay in the onset of the disease. Although both studies showed a neuroprotective effect, our results are in contrast with those of Sugai *et al.,* (2004) described above, which reports an increase in the length of the disease and no onset retardation. Several reasons can account for this discrepancy, such as the dose, the duration of VPA treatment, as well as the mice strain, which does not bear the same SOD1 mutation and thus, does not display the same pathological time course. However, it should be pointed out that now some studies question the origin of the ALS disease, suggesting that biochemical and molecular events occurring in the muscle fibers at the neuromuscular junction – such as hypermetabolism – could participate to the increasing spinal motorneuron vulnerability observed in ALS (Dupuis *et al.,* 2004). It is thus conceivable that protecting motorneurons from death might not be sufficient to ameliorate the whole ALS condition as also emphasized in Gould *et al.* (2006). Using the G86R mice strain, we found that muscle denervation was retarded in VPA-treated animals as measured by EMG studies (our unpublished data). This is in line with a recent report, showing that VPA favors reinervation of muscle fibers in axotomized rats (Cui *et al.,* 2003). It has been shown recently that another butyrate derivative, Phenylbutyrate, resulted in significant enhancement of survival and improvement in clinical and neuropathological phenotypes of the ALS G93A strain mice when administered at the optimal dose of 400 mg/kg/day, pre-onset (Ryu *et al.,* 2005) or at disease onset (Petri *et al.,* 2006). At the molecular level, the authors provided evidence that the neuroprotection observed resulted from regulation of the NfKB-dependent pathway. In these conditions, bc12 was upregulated, cytochrome c release blocked, as well as subsequent caspase activation (Ryu *et al.,* 2005). Additive neuroprotective effect was seen when Phenylbutyrate was administered together with an antioxidant (Petri *et al.,* 2006). Altogether, the authors observed transcriptional regulations that may contribute to buffering of free radicals, increase in proteasome activity and decrease in apoptotic cell death in the treated mice.

Of note, a very recent report shows that SIRTl (sirtuin class of HDAC), that was shown to participate in axonal protection (Araki *et al.,* 2004), could be acting in a neuroprotective way in ALS (Fischer *et al.,* 2005). As noted earlier, this class of HDAC is insensitive to classical HDAC inhibitors, but it should be kept in mind reinstating acetylation homeostasis with classical HDACi might indirectly biased sirtuin's regulations as well.

6. CONCLUSIONS

Taken together, HDACis are promising candidates for harnessing neuronal loss, both in apoptotic models *in vitro*, and in several models of neurodegenerescence *in vivo*. However, they are not the 'wonder' drug and improvement in different ways need to be performed.

6.1. Lack of Specificity Towards CBP

First of all, the mode of action for HDAC inhibitors is highly non-specific and nontargeted: if the context of neurodegeneration were CBP's loss of function, clearly, HDACi do not specifically target CBP's regulations only, but rather unlock repressive conformations at promoters of essential genes. This implies that more genes 'than needed' will be activated, and this might reveal toxic for the cell. In this respect, while HDACi have been proved efficient in several *in vitro* and *in vivo* models of apoptosis, it is worth to consider the death signaling pathway involved in each pathological situation.

It is noteworthy that the CBP/p300 proteins are present in limited quantities thus creating a competition between transcription factors. CBP sequestration or CBP level drop, while inducing specific gene downregulation, could also represent a way to allow specific re-activation of a given gene. HDACi treatment may worsen this problem.

Determining the genes that promote survival, especially those under CBP's control, could facilitate the development of novel drugs and specific therapeutic strategies with lower adverse side effects than those currently available.

6.2. HDACi Doses

As largely detailed in this review, neuroprotection depends on a balance between the enzymes HATS and HDACs and pushing the balance towards hyperacetylation in healthy cells will lead to cell death. For example, at 100 nM, TSA displays prominent toxic effects in normal neuronal cells by activating the caspase-dependent pathway and upregulating the E2F1 transcription factor (Boutillier *et al.,* 2003), while it is not at 10 nM, a concentration that displays neuroprotective functions against oxidative stress-induced death (our unpublished data). Clearly, target genes will differ upon a given HDACi concentration used. For example, gene expression profiles have been performed in response to high and low concentration of VPA and the authors described different panels of up- and down-regulated genes (Massa *et al.,* 2005). Thus, *in vivo,* it will be of prime importance to monitor the dose of HDACi treatment, as general administration of HDAC inhibitors may enhance survival potency of previously challenged neurons at the very high cost of endangering other normal brain cells in vicinity.

6.3. HDACi Targets

A major challenge will be to identify the genes and transcription factors targeted by these inhibitors. HDACi do not have the same affinity for each HDAC enzyme and, depending on the role this latter will have in neuroprotection/death, the expected cellular response can be very different. HDACis affect expression of 2% genes in mammals (Davie 2003). Comparing gene expression profiles in response to each HDACi in a specific degenerative context could be interesting for the choice of HDACi treatment in a given situation. In this respect, one need also to develop new HDACis, more specific to the different HDACs within the three classes.

In conclusion, some of these compounds seems promising therapeutic tools in neurodegenerative contexts, especially as neuroprotecting agents. Even if all these compounds do not have all the hoped beneficial effects in neurodegenerative diseases, they may prove to be powerful probes for understanding the molecular signaling involved in these pathologies.

ACKNOWLEDGEMENTS

C.R. is a recipient of a fellowship from the French Research Ministry. IP obtained a post-doctoral fellowship from 'Association Frangaise pour les Myopathies' (AFM). The laboratory is supported by grants from 'Association pour la Recherche sur la SLA' (ARS), 'Association pour la Recherche et le développement de moyens de lutte contre les Maladies Neurodégénératives' (AREMANE), 'Association pour la Recherche contre le Cancer' (ARC n°4306 and 7653). We apologize for all the authors we could not cite in this review due to space limitation.

REFERENCES

Alarcon JM, Malleret G, Touzani K, Vronskaya S, Ishii S, Kandel ER, Barco A (2004) Chromatin acetylation, memory, and LTP are impaired in CBP+/- mice: a model for the cognitive deficit in Rubinstein-Taybi syndrome and its amelioration. Neuron 42(6):947–959

Alnernri ES, Livingston DJ, Nicholson DW, Salvesen G, Thornberry NA, Wong WW, Yuan J (1996) Human ICEICED-3 protease nomenclature. Cell 87(2):171

Araki T, Sasaki Y, Milbrandt J (2004) Increased nuclear NAD biosynthesis and SIRT1 activation prevent axonal degeneration. Science 305(5686):1010–1013

Balasubramanyam K, Swaminathan V, Ranganathan A, Kundu TK (2003) Small molecule modulators of histone acetyltransferase p300. J Biol Chem 278(21):19134–19140

Barlow AL, van Drunen CM, Johnson CA, Tweedie S, Bird A, Turner BM (2001) dSIR2 and dHDAC6: two novel, inhibitor-resistant deacetylases in Drosophila melanogaster. Exp Cell Res 265(1):90–103

Bezprozvanny I, Hayden MR (2004) Deranged neuronal calcium signaling and Huntington disease. Biochem Biophys Res Commun 322(4):1310–1317

Bolger TA, Yao TP (2005) Intracellular trafficking of histone deacetylase 4 regulates neuronal cell death. J Neurosci 25(41):9544–9553

Borner C, Monney L (1999) Apoptosis without caspases: an inefficient molecular guillotine? Cell Death Differ 6(6):497–507

Boutell JM, Thomas P, Neal JW, Weston VJ, Duce J, Harper PS, Jones AL (1999) Aberrant interactions of transcriptional repressor proteins with the Huntington's disease gene product, huntingtin. Hum Mol Genet 8(9):1647–1655

Boutillier AL, Trinh E, Loeffler JP (2003) Selective E2F-dependent gene transcription is controlled by histone deacetylase activity during neuronal apoptosis. J Neurochem 84(4):814–828

Buck SW, Gallo CM, Smith JS (2004) Diversity in the Sir2 family of protein deacetylases. J Leukoc Biol 75(6):939–950

Buggy JJ, Sideris ML, Mak P, Lorimer DD, McIntosh B, Clark JM (2000) Cloning and characterization of a novel human histone deacetylase, HDAC8. Biochem J 350(Pt 1):199–205

Candido EP, Reeves R, Davie JR (1978) Sodium butyrate inhibits histone deacetylation in cultured cells. Cell 14(1):105–113

Chai Y, Wu L, Griffin JD, Paulson HL (2001) The role of protein composition in specifying nuclear inclusion formation in polyglutamine disease. J Biol Chem 276(48):44889–44897

Chawla S, Vanhoutte P, Arnold FJ, Huang CL, Bading H (2003) Neuronal activity-dependent nucleocytoplasmic shuttling of HDAC4 and HDAC5. J Neurochem 85(1):151–159

Cong SY, Pepers BA, Evert BO, Rubinsztein DC, Roos RA, van Ommen GJ, Dorsman JC (2005) Mutant huntingtin represses CBP, but not p300, by binding and protein degradation. Mol Cell Neurosci 30(4):560–571

Cookson MR, Shaw PJ (1999) Oxidative stress and motor neurone disease. Brain Pathol 9(1):165–186

Corcoran LJ, Mitchison TJ, Liu Q (2004) A novel action of histone deacetylase inhibitors in a protein aggresome disease model. Curr Biol 14(6):488–492

Coyle JT, Duman RS (2003) Finding the intracellular signaling pathways affected by mood disorder treatments. Neuron 38(2):157–160

Cui SS, Yang CP, Bowen RC, Bai O, Li XM, Jiang W, Zhang X (2003) Valproic acid enhances axonal regeneration and recovery of motor function after sciatic nerve axotomy in adult rats. Brain Res 975(1–2):229–236

Davie JR (2003) Inhibition of histone deacetylase activity by butyrate. J Nutr 133(7 Suppl):2485S–2493S

Detich N, Bovenzi V, Szyf M (2003) Valproate induces replication-independent active DNA demethylation. J Biol Chem 278(30):27586–27592

de Ruijter AJ, van Gennip AH, Caron HN, Kemp S, van Kuilenburg AB (2003) Histone deacetylases (HDACs): characterization of the classical HDAC family. Biochem J 370(Pt 3):737–749

Dunah AW, Jeong H, Griffin A, Kim YM, Standaert DG, Hersch SM, Mouradian MM, Young AB, Tanese N, Krainc D (2002) Spl and TAFII130 transcriptional activity disrupted in early Huntington's disease. Science 296(5576):2238–2243

Dupuis L, Oudart H, Rene F, Gonzalez de Aguilar JL, Loeffler JP (2004) Evidence for defective energy homeostasis in amyotrophic lateral sclerosis: benefit of a high-energy diet in a transgenic mouse model. Proc Natl Acad Sci U S A 101(30):11159–11164

Fadeel B, Orrenius S (2005) Apoptosis: a basic biological phenomenon with wide-ranging implications in human disease. J Intern Med 258(6):479–517

Ferrante RJ, Kubilus JK, Lee J, Ryu H, Beesen A, Zucker B, Smith K, Kowall NW, Ratan RR, Luthi-Carter R, Hersch SM (2003) Histone deacetylase inhibition by sodium butyrate chemotherapy ameliorates the neurodegenerative phenotype in Huntington's disease mice. J Neurosci 23(28):9418–9427

Ferrer I, Goutan E, Marin C, Rey MJ, Ribalta T (2000) Brain-derived neurotrophic factor in Huntington disease. Brain Res 866(1–2):257–261

Finnin MS, Donigian JR, Cohen A, Richon VM, Rifkind RA, Marks PA, Breslow R, Pavletich NP (1999) Structures of a histone deacetylase homologue bound to the TSA and SAHA inhibitors. Nature 401(6749):188–193

Fischer DD, Cai R, Bhatia U, Asselberg FA, Song C, Terry R, Trogani N, Widmer R, Atadja P, Cohen D (2002) Isolation and charaterization of a novel class II histone deacetylase, HDAC 10. J Biol Chem 277(8):6656–6666

Fischer LR, Culver DG, Davis AA, Tennant P, Wang M, Coleman M, Asress S, Adalbert R, Alexander GM, Glass JD (2005) The WldS gene modestly prolongs survival in the SOD1G93A fALS mouse. Neurobiol Dis 19(1–2):293–300

Fischle W, Emiliani S, Hendzel MJ, Nagase T, Nomura N, Voelter W, Verdin E (1999) A new family of human histone deacetylases related to Saccharomyces cerevisiae HDAlp. J Biol Chem 274(17):11713–11720

Fischle W, Wang Y, Allis CD (2003) Binary switches and modification cassettes in histone biology and beyond. Nature 425(6957):475–479

Freeland K, Boxer LM, Latchman DS (2001) The cyclic AMP response element in the Bcl-2 promoter confers inducibility by hypoxia in neuronal cells. Brain Res Mol Brain Res 92(1–2):98–106

Friedlander RM, Brown RH, Gagliardini V, Wang J, Yuan J (1997) Inhibition of ICE slows ALS in mice. Nature 388(6637):31

Furumai R, Komatsu Y, Nishino N, Khochbin S, Yoshida M, Horinouchi S (2001) Potent histone deacetylase inhibitors built from trichostatin A and cyclic tetrapeptide antibiotics including trapoxin. Proc Natl Acad Sci U S A 98(1):87–92

Galasinski SC, Resing KA, Goodrich JA, Ahn NG (2002) Phosphatase inhibition leads to histone deacetylases 1 and 2 phosphorylation and disruption of corepressor interactions. J Biol Chem 277(22):19618–19626

Gao L, Cueto MA, Asselbergs F, Atadja P (2002) Cloning and functional characterization of HDAC11s, a novel member of the human histone deacetylase family. J Biol Chem 277(28):25748–25755

Gardian G, Browne SE, Choi DK, Klivenyi P, Gregorio J, Kubilus JK, Ryu H, Langley B, Ratan RR, Ferrante RJ, Beal MF (2005) Neuroprotective effects of phenylbutyrate in the N171-824 transgenic mouse model of Huntington's disease. J Biol Chem 280(1):556–563

Gauthier LR, Charrin BC, Borrell-Pages M, Dompierre JP, Rangone H, Cordelieres FP, De Mey J, MacDonald ME, Lessmann V, Humbert S, Saudou F (2004) Huntingtin controls neurotrophic support and survival of neurons by enhancing BDNF vesicular transport along microtubules. Cell 118(1):127–138

Gervais FG, Singaraja R, Xanthoudakis S, Gutekunst CA, Leavitt BR, Metzler M, Hackam AS, Tam J, Vaillancourt JP, Houtzager V, Rasper DM, Roy S, Hayden MR, Nicholson DW (2002) Recruitment and activation of caspase-8 by the Huntingtin-interacting protein Hip-1 and a novel partner Hippi. Nat Cell Biol 4(2):95–105

Glozak MA, Sengupta N, Zhang X, Seto E (2005) Acetylation and deacetylation of nonhistone proteins. Gene 363:15–23

Goldberg YP, Nicholson DW, Rasper DM, Kalchman MA, Koide HB, Graham RK, Bromm M, Kazemi-Esfarjani P, Thornberry NA, Vaillancourt JP, Hayden MR (1996) Cleavage of huntingtin by apopain, a proapoptotic cysteine protease, is modulated by the polyglutamine tract. Nat Genet 13(4):442–449

Gottlicher M, Minucci S, Zhu P, Kramer OH, Schimpf A, Giavara S, Sleeman JP, Lo Coco F, Nervi C, Pelicci PG, Heinzel T (2001) Valproic acid defines a novel class of HDAC inhibitors inducing differentiation of transformed cells. Embo J 20(24):6969–6978

Gray SG, Ekstrom TJ (2001) The human histone deacetylase family. Exp Cell Res 262(2):75–83

Greene LA, Biswas SC, Liu DX (2004) Cell cycle molecules and vertebrate neuron death: E2F at the hub. Cell Death Differ 11(1):49–60

Gould TW, Buss RR, Vinsant S, Prevette D, Sun W, Knudson CM, Millican CE, Oppenheim RW (2006) Complete dissociation of motor neuron death from motor dysfunction by Bax deletion in a mouse model of ALB. J Neurosc. 26(34):8774–8786

Grozinger CM, Schreiber SL (2000) Regulation of histone deacetylase 4 and 5 and transcriptional activity by 14-3-3-dependent cellular localization. Proc Natl Acad Sci U S A 97(14):7835–7840

Grozinger CM, Hassig CA, Schreiber SL (1999) Three proteins define a class of human histone deacetylases related to yeast Hdalp. Proc Natl Acad Sci U S A 96(9):4868–4873

Guardiola AR, Yao TP (2002) Molecular cloning and characterization of a novel histone deacetylase HDAC10. J Biol Chem 277(5):3350–3356

Guarente L (2000) Sir2 links chromatin silencing, metabolism, and aging. Genes Dev 14(9):1021–1026

Gurney ME, Pu H, Chiu AY, Dal Canto MC, Polchow CY, Alexander DD, Caliendo J, Hentati A, Kwon YW, Deng HX *et al* (1994) Motor neuron degeneration in mice that express a human Cu,Zn superoxide dismutase mutation. Science 264(5166):1772–1775

Heiser V, Engemann S, Brocker W, Dunkel I, Boeddrich A, Waelter S, Nordhoff E, Lurz R, Schugardt N, Rautenberg S, Herhaus C, Barnickel G, Bottcher H, Lehrach H, Wanker EE (2002) Identification of benzothiazoles as potential polyglutamine aggregation inhibitors of Huntington's disease by using an automated filter retardation assay. Proc Natl Acad Sci U S A 99(Suppl 4): 16400–16406

Hengartner MO (2000) The biochemistry of apoptosis. Nature 407(6805):770–776

Herrup K, Neve R, Ackerrnan SL, Copani, A (2004) Divide and die: cell cycle events as triggers of nerve cell death. J Neurosci 24(42):9232–9239

Hockly E, Richon VM, Woodman B, Smith DL, Zhou X, Rosa E, Sathasivam K, Ghazi-Noori S, Mahal A, Lowden PA, Steffan JS, Marsh JL, Thompson LM, Lewis CM, Marks PA, Bates GP (2003) Suberoylanilide hydroxamic acid, a histone deacetylase inhibitor, ameliorates motor deficits in a mouse model of Huntington's disease. Proc Natl Acad Sci U S A 100(4):2041–2046

Holbert S, Denghien I, Kiechle T, Rosenblatt A, Wellington C, Hayden MR, Margolis RL, Ross CA, Dausset J, Ferrante RJ, Neri C (2001) The Gln-Ala repeat transcriptional activator CAI50 interacts with huntingtin: neuropathologic and genetic evidence for a role in Huntington's disease pathogenesis. Proc Natl Acad Sci U S A 98(4):1811–1816

Hu E, Chen Z, Fredrickson T, Zhu Y, Kirkpatrick R, Zhang GF, Johanson K, Sung CM, Liu R, Winkler J (2000) Cloning and characterization of a novel human class I histone deacetylase that functions as a transcription repressor. J Biol Chem 275(20):15254–15264

Hubbert C, Guardiola A, Shao R, Kawaguchi Y, Ito A, Nixon A, Yoshida M, Wang XF, Yao TP (2002) HDAC6 is a microtubule-associated deacetylase. Nature 417(6887):455–458

Igarashi S, Morita H, Bennett KM, Tanaka Y, Engelender S, Peters MF, Cooper JK, Wood JD, Sawa A, Ross CA (2003) Inducible PC12 cell model of Huntington's disease shows toxicity and decreased histone acetylation. Neuroreport 14(4):565–568

Inoue H, Tsukita K, Iwasato T, Suzuki Y, Tomioka M, Tateno M, Nagao M, Kawata A, Saido TC, Miura M, Misawa H, Itohara S, Takahashi R (2003) The crucial role of caspase-9 in the disease progression of a transgenic ALS mouse model. Embo J 22(24):6665–6674

Jiang H, Nucifora FC, Jr., Ross CA, DeFranco DB (2003) Cell death triggered by polyglutamine-expanded huntingtin in a neuronal cell line is associated with degradation of CREB-binding protein. Hum Mol Genet 12(1):1–12

Jiang YM, Yamamoto M, Kobayashi Y, Yoshihara T, Liang Y, Terao S, Takeuchi H, Ishigaki S, Katsuno M, Adachi H, Niwa J, Tanaka F, Doyu M, Yoshida M, Hashizume Y, Sobue G (2005) Gene expression profile of spinal motor neurons in sporadic amyotrophic lateral sclerosis. Ann Neurol 57(2):236–251

Johnstone RW (2002) Histone-deacetylase inhibitors: novel drugs for the treatment of cancer. Nat Rev Drug Discov 1(4):287–299

Kalkhoven E (2004) CBP and p300: HATS for different occasions. Biochem Pharmacol 68(6):1145–1155

Kalkhoven E, Roelfsema JH, Teunissen H, den Boer A, Ariyurek Y, Zantema A, Breuning MH, Hennekam RC, Peters DJ (2003) Loss of CBP acetyltransferase activity by PHD finger mutations in Rubinstein-Taybi syndrome. Hum Mol Genet 12(4):441–450

Kawasaki H, Eckner R, Yao TP, Taira K, Chiu R, Livingston DM, Yokoyama KK (1998) Distinct roles of the co-activators p300 and CBP in retinoic-acid-induced F9-cell differentiation. Nature 393(6682):284–289

Kazantsev A, Preisinger E, Dranovsky A, Goldgaber D, Housman D (1999) Insoluble detergent-resistant aggregates form between pathological and nonpathological lengths of polyglutamine in mammalian cells. Proc Natl Acad Sci U S A 96(20):11404–11409

Kazantsev A, Walker HA, Slepko N, Bear JE, Preisinger E, Steffan JS, Zhu YZ, Gertler FB, Housman DE, Marsh JL, Thompson LM (2002) A bivalent Huntingtin binding peptide suppresses polyglutamine aggregation and pathogenesis in Drosophila. Nut Genet 30(4):367–376

Knoepfler PS, Eisenman RN (1999) Sin meets NuRD and other tails of repression. Cell 99(5):447–450

Kostic V, Jackson-Lewis V, de Bilbao F, Dubois-Dauphin M, Przedborski S (1997) Bcl-2: prolonging life in a transgenic mouse model of familial amyotrophic lateral sclerosis. Science 277(5325):559–562

Kramer OH, Zhu P, Ostendorff HP, Golebiewski M, Tiefenbach J, Peters MA, Brill B, Groner B, Bach I, Heinzel T, Gottlicher M (2003) The histone deacetylase inhibitor valproic acid selectively induces proteasomal degradation of HDAC2. Embo J 22(13):3411–3420

Krantic S, Mechawar N, Reix S, Quirion R (2005) Molecular basis of programmed cell death involved in neurodegeneration. Trends Neurosci 28(12):670–676

Kruman II, Pedersen WA, Springer JE, Mattson MP (1999) ALS-linked Cu/Zn-SOD mutation increases vulnerability of motor neurons to excitotoxicity by a mechanism involving increased oxidative stress and perturbed calcium homeostasis. Exp Neurol 160(1):28–39

La Spada AR, Wilson EM, Lubahn DB, Harding AE, Fischbeck KH (1991) Androgen receptor gene mutations in X-linked spinal and bulbar muscular atrophy. Nature 352(6330):77–79

Langley B, Ratan RR (2004) Oxidative stress-induced death in the nervous system: cell cycle dependent or independent? J Neurosci Res 77(5):621–629

Langley B, Gensert JM, Beal MF, Ratan RR (2005) Remodeling chromatin and stress resistance in the central nervous system: histone deacetylase inhibitors as novel and broadly effective neuroprotective agents. Curr Drug Targets CNS Neurol Disord 4(1):41–50

Lee H, Rezai-Zadeh N, Seto E (2004) Negative regulation of histone deacetylase 8 activity by cyclic AMP-dependent protein kinase A. Mol Cell Biol 24(2):765–773

Lehrmann H, Pritchard LL, Harel-Bellan A (2002) Histone acetyltransferases and deacetylases in the control of cell proliferation and differentiation. Adv Cancer Res 86:41–65

Lemercier C, Verdel A, Galloo B, Curtet S, Brocard MP, Khochbin S (2000) mHDA1/HDAC5 histone deacetylase interacts with and represses MEF2A kanscriptional activity. J Biol Chem 275(20):15594–15599

Li F, Macfarlan T, Pittman RN, Chakravarti D (2002) Ataxin-3 is a histone-binding protein with two independent transcriptional corepressor activities, J Biol Chem 277(47):45004–45012

Li M, Ona VO, Guegan C, Chen M, Jackson-Lewis V, Andrews LJ, Olszewski AJ, Stieg PE, Lee JP, Przedborski S, Friedlander RM (2000) Functional role of caspase-1 and caspase-3 in an ALS transgenic mouse model. Science 288(5464):335–339

Liu DX, Greene LA (2001) Regulation of neuronal survival and death by E2F-dependent gene repression and derepression. Neuron 32(3):425–438

Lu J, McKinsey TA, Nicol RL, Olson EN (2000) Signal-dependent activation of the MEF2 transcription factor by dissociation from histone deacetylases. Proc Natl Acad Sci U S A 97(8):4070–4075

Lu Q, Hutchins AE, Doyle CM, Lundblad JR, Kwok RP (2003) Acetylation of CAMP-responsive element-binding protein (CREB) by CREB-binding protein enhances CREB-dependent transcription. J Biol Chem 278(18):15727–15734

Luthi-Carter R, Strand A, Peters NL, Solano SM, Hollingsworth ZR, Menon AS, Frey AS, Spektor BS, Penney EB, Schilling G, Ross CA, Borchelt DR, Tapscott SJ, Young AB, Cha JH, Olson JM (2000) Decreased expression of striatal signaling genes in a mouse model of Huntington's disease. Hum Mol Genet 9(9):1259–1271

Marks PA, Richon VM, Breslow R, Rifkind RA (2001) Histone deacetylase inhibitors as new cancer drugs. Curr Opin Oncol 13(6):477–483

Marmorstein R (2001) Structure of histone deacetylases: insights into substrate recognition and catalysis. Structure 9(12):1127–1133
Martin LJ (1999) Neuronal death in amyotrophic lateral sclerosis is apoptosis: possible contribution of a programmed cell death mechanism. J Neuropathol Exp Neurol 58(5):459–471
Massa V, Cabrera RM, Menegola E, Giavini E, Finnell RH (2005) Valproic acidinduced skeletal malformations: associated gene expression cascades. Pharmacogenet Genomics 15(11):787–800
Mattson MP (2000) Apoptosis in neurodegenerative disorders. Nat Rev Mol Cell Biol 1(2):120–129
McCampbell A, Fischbeck KH (2001) Polyglutamine and CBP: fatal attraction? Nat Med 7(5):528–530
McCampbell A, Taylor JP, Taye AA, Robitschek J, Li M, Walcott J, Merry D, Chai Y, Paulson H, Sobue G, Fischbeck KH (2000) CREB-binding protein sequestration by expanded polyglutamine. Hum Mol Genet 9(14):2197–2202
McCampbell A, Taye AA, Whitty L, Penney E, Steffan JS, Fischbeck KH (2001) Histone deacetylase inhibitors reduce polyglutamine toxicity. Proc Natl Acad Sci U S A 98(26):15179–15184
Mejat A, Ramond F, Bassel-Duby R, Khochbin S, Olson EN, Schaeffer L (2005) Histone deacetylase 9 couples neuronal activity to muscle chromatin acetylation and gene expression. Nat Neurosci 8(3):313–321
Miller RW, Rubinstein JH (1995) Tumors in Rubinstein-Taybi syndrome. Am J Med Genet 56(1):112–115
Miller TA, Witter DJ, Belvedere S (2003) Histone deacetylase inhibitors. J Med Chem 46(24):5097–5116
Minamiyama M, Katsuno M, Adachi H, Waza M, Sang C, Kobayashi Y, Tanaka F, Doyu M, Inukai A, Sobue G (2004) Sodium butyrate ameliorates phenotypic expression in a transgenic mouse model of spinal and bulbar muscular atrophy. Hum Mol Genet 13(11):1183–1192
Miska EA, Karlsson C, Langley E, Nielsen SJ, Pines J, Kouzarides T (1999) HDAC4 deacetylase associates with and represses the MEF2 transcription factor. Embo J 18(18):5099–5107
Morris SM, Jr (2002) Regulation of enzymes of the urea cycle and arginine metabolism. Annu Rev Nutr 22:87–105
Morrison BE, Majdzadeh N, Zhang X, Lyles A, Bassel-Duby R, Olson EN, D'Mello, SR (2006) Neuroprotection by histone deacetylase-related protein. Mol Cell Biol 26(9):3550–3564
Murata T, Kurokawa R, Krones A, Tatsurni K, Ishii M, Taki T, Masuno M, Ohashi H, Yanagisawa M, Rosenfeld MG, Glass CK, Hayashi Y (2001) Defect of histone acetyltransferase activity of the nuclear transcriptional coactivator CBP in Rubinstein-Taybi syndrome. Hum Mol Genet 10(10):1071–1076
Nagata S (2000) Apoptotic DNA fragmentation. Exp Cell Res 256(1):12–18
Ng HH, Bird A (2000) Histone deacetylases: silencers for hire. Trends Biochem Sci 25(3):121–126
Nguyen MD, Boudreau M, Kriz J, Couillard-Despres S, Kaplan DR, Julien JP (2003) Cell cycle regulators in the neuronal death pathway of amyotrophic lateral sclerosis caused by mutant superoxide disrnutase 1. J Neurosci 23(6):2131–2140
Nogales E (2000) Recent structural insights into transcription preinitiation complexes. J Cell Sci 113 Pt 24:4391–4397
Nucifora FC, Jr., Sasaki M, Peters MF, Huang H, Cooper JK, Yamada M, Takahashi H, Tsuji S, Troncoso J, Dawson VL, Dawson TM, Ross CA (2001) Interference by huntingtin and atrophin-1 with cbp-mediated transcription leading to cellular toxicity. Science 291(5512):2423–2428
Oike Y, Takakura N, Hata A, Kaname T, Akizuki M, Yarnaguchi Y, Yasue H, Araki K, Yamamura K, Suda T (1999) Mice homozygous for a truncated form of CREBbinding protein exhibit defects in hematopoiesis and vasculo-angiogenesis. Blood 93(9):2771–2779
Petri S, Kiaei M, Kipiani K, Chen J, Calingasan NY, Crow JP, Beal MF (2006) Additive neuroprotective effects of a histone deacetylase inhibitor and a catalytic antioxidant in a transgenic mouse model of amyotrophic lateral sclerosis. Neurobiol Dis 22(1):40–49
Petrij F, Giles RH, Dauwerse HG, Saris JJ, Hennekam RC, Masuno M, Tommerup N, van Ommen GJ, Goodman RH, Peters DJ *et al* (1995) Rubinstein-Taybi syndrome caused by mutations in the transcriptional co-activator CBP. Nature 376(6538):348–351
Phiel C J, Zhang F, Huang EY, Guenther MG, Lazar MA, Klein PS (2001) Histone deacetylase is a direct target of valproic acid, a potent anticonvulsant, mood stabilizer, and teratogen. J Biol Chem 276(39):36734–36741

Pollitt SK, Pallos J, Shao J, Desai UA, Ma AA, Thompson LM, Marsh JL, Diamond MI (2003) A rapid cellular FRET assay of polyglutamine aggregation identifies a novel inhibitor. Neuron 40(4):685–694

Ranganathan S, Bowser, R (2003) Alterations in G(1) to S phase cell-cycle regulators during amyotrophic lateral sclerosis. Am J Pathol 162(3):823–835

Remiszewski SW, Sambucetti LC, Atadja P, Bair KW, Cornell WD, Green MA, Howell KL, Jung M, Kwon P, Trogani N, Walker H (2002) Inhibitors of human histone deacetylase: synthesis and enzyme and cellular activity of straight chain hydroxamates. J Med Chem 45(4):753–757

Richfield EK, Maguire-Zeiss KA, Cox C, Gilmore J, Voorn P (1995) Reduced expression of preproenkephalin in striatal neurons from Huntington's disease patients. Ann Neurol 37(3):335–343

Riggs MG, Whittaker RG, Neumann JR, Ingram VM (1977) n-Butyrate causes histone modification in HeLa and Friend erythroleukaemia cells. Nature 268(5619):462–464

Rosen DR, Siddique T, Patterson D, Figlewicz DA, Sapp P, Hentati A, Donaldson D, Goto J, O'Regan, JP, Deng HX *et al* (1993) Mutations in CuIZn superoxide dismutase gene are associated with familial amyotrophic lateral sclerosis. Nature 362(6415):59–62

Rouaux C, Jokic N, Mbebi C, Boutillier S, Loeffler JP, Boutillier AL (2003) Critical loss of CBP/p300 histone acetylase activity by caspase-6 during neurodegeneration. Embo J 22(24):6537–6549

Rouaux C, Loeffler JP, Boutillier AL (2004) Targeting CREB-binding protein (CBP) loss of function as a therapeutic strategy in neurological disorders. Biochem Pharmacol 68(6):1157–1164

Ryu H, Lee J, Zaman K, Kubilis J, Ferrante RJ, Ross BD, Neve R, Ratan RR (2003) Spl and Sp3 are oxidative stress-inducible, antideath transcription factors in cortical neurons. J Neurosci 23(9):3597–3606

Ryu H, Smith K, Camelo SI, Carreras I, Lee J, Iglesias AH, Dangond F, Cormier KA, Cudkowicz ME, Brown RH, Jr, Ferrante RJ (2005) Sodium phenylbutyrate prolongs survival and regulates expression of anti-apoptotic genes in transgenic amyotrophic lateral sclerosis mice. J Neurochem 93(5):1087–1098

Saha RN, Pahan K (2006) HATS and HDACs in neurodegeneration: a tale of disconcerted acetylation homeostasis. Cell Death Differ 13(4):539–550

Salminen A, Tapiola T, Korhonen P, Suuronen T (1998) Neuronal apoptosis induced by histone deacetylase inhibitors. Brain Res Mol Brain Res 61(1–2):203–206

Sanchez I, Xu CJ, Juo P, Kakizaka A, Blenis J, Yuan J (1999) Caspase-8 is required for cell death induced by expanded polyglutamine repeats. Neuron 22(3):623–633

Sanchez I, Mahlke C, Yuan J (2003) Pivotal role of oligomerization in expanded polyglutamine neurodegenerative disorders. Nature 421(6921):373–379

Shimohata T, Onodera O, Tsuji S (2000) Interaction of expanded polyglutamine stretches with nuclear transcription factors leads to aberrant transcriptional regulation in polyglutamine diseases. Neuropathology 20(4):326–333

Sobue G, Hashizume Y, Mukai E, Hirayama M, Mitsuma T, Takahashi A (1989) Xlinked recessive bulbospinal neuronopathy. A clinicopathological study. Brain 112(Pt 1):209–232

Steffan JS, Kazantsev A, Spasic-Boskovic O, Greenwald M, Zhu YZ, Gohler H, Wanker EE, Bates GP, Housman DE, Thompson LM (2000) The Huntington's disease protein interacts with p53 and CREB-binding protein and represses transcription. Proc Natl Acad Sci U S A 97(12):6763–6768

Steffan JS, Bodai L, Pallos J, Poelman M, McCampbell A, Apostol BL, Kazantsev A, Schmidt E, Zhu YZ, Greenwald M, Kurokawa R, Housman DE, Jackson GR, Marsh JL, Thompson LM (2001) Histone deacetylase inhibitors arrest polyglutamine-dependent neurodegeneration in Drosophila. Nature 413(6857):739–743

Sterner DE and Berger SL (2000) Acetylation of histones and transcription-related factors. Microbiol Mol Biol Rev 64(2):435–459

Sugai F, Yamamoto Y, Miyaguchi K, Zhou Z, Sumi H, Hamasaki T, Goto M, Sakoda S (2004) Benefit of valproic acid in suppressing disease progression of ALS model mice. Eur J Neurosci 20(11):3179–3183

Taunton J, Hassig CA, Schreiber SL (1996) A mammalian histone deacetylase related to the yeast transcriptional regulator Rpd3p. Science 272(5260):408–411

Taylor JP, Taye AA, Campbell C, Kazemi-Esfarjani P, Fischbeck KH, Min KT (2003) Aberrant histone acetylation, altered transcription, and retinal degeneration in a Drosophila model of polyglutamine disease are rescued by CREB-binding protein. Genes Dev 17(12):1463–1468

Thornberry NA, Rano TA, Peterson EP, Rasper DM, Timkey T, Garcia-Calvo M, Houtzager VM, Nordstrom PA, Roy S, Vaillancourt JP, Chapman KT, Nicholson DW (1997) A combinatorial approach defines specificities of members of the caspase family and granzyme B. Functional relationships established for key mediators of apoptosis. J Biol Chem 272(29):17907–17911

Tong JJ, Liu J, Bertos NR, Yang XJ (2002) Identification of HDAC10, a novel class II human histone deacetylase containing a leucine-rich domain. Nucleic Acids Res 30(5):1114–1123

Torchia J, Glass C, Rosenfeld MG (1998) Co-activators and co-repressors in the integration of transcriptional responses. Curr Opin Cell Biol 10(3):373–383

Trinh E, Boutillier AL, Loeffler JP (2001) Regulation of the retinoblastoma-dependent Mdm2 and E2F-1 signaling pathways during neuronal apoptosis. Mol Cell Neurosci 17(2):342–353

Vannini A, Volpari C, Filocamo G, Casavola EC, Brunetti M, Renzoni D, Chakravarty P, Paolini C, De Francesco R, Gallinari P, Steinkuhler C, Di Marco S (2004) Crystal structure of a eukaryotic zinc-dependent histone deacetylase, human HDAC8, complexed with a hydroxamic acid inhibitor. Proc Natl Acad Sci U S A 101(42):15064–15069

Varier RA, Swaminathan V, Balasubramanyam K, Kundu TK (2004) Implications of small molecule activators and inhibitors of histone acetyltransferases in chromatin therapy. Biochem Pharmacol 68(6):1215–1220

Verdel A, Curtet S, Brocard MP, Rousseaux S, Lemercier C, Yoshida M, Khochbin S (2000) Active maintenance of mHDA2/mHDAC6 histone-deacetylase in the cytoplasm. Curr Biol 10(12): 747–749

Vigushin DM, Coombes RC (2002) Histone deacetylase inhibitors in cancer treatment. Anticancer Drugs 13(1):1–13

Vo N, Goodman RH (2001) CREB-binding protein and p300 in transcriptional regulation. J Biol Chem 276(17):13505–13508

Wang AH, Bertos NR, Vezmar M, Pelletier N, Crosato M, Heng HH, Th'ng J, Han J, Yang XJ (1999) HDAC4, a human histone deacetylase related to yeast HDA1, is a transcriptional corepressor. Mol Cell Biol 19(11):7816–7827

Wang AH, Kruhlak MJ, Wu J, Bertos NR, Vezmar M, Posner BI, Bazett-Jones DP, Yang XJ (2000) Regulation of histone deacetylase 4 by binding of 14-3-3 proteins. Mol Cell Biol 20(18):6904–6912

West AE, Chen WG, Dalva MB, Dolmetsch RE, Kornhauser JM, Shaywitz AJ, Takasu MA, Tao X, Greenberg ME (2001) Calcium regulation of neuronal gene expression. Proc Natl Acad Sci U S A 98(20):11024–11031

Yang WM, Inouye C, Zeng Y, Bearss D, Seto E (1996) Transcriptional repression by YY1 is mediated by interaction with a mammalian homolog of the yeast global regulator RPD3. Proc Natl Acad Sci U S A 93(23):12845–12850

Yoshida M, Nomura S, Beppu T (1987) Effects of trichostatins on differentiation of murine erythroleukemia cells. Cancer Res 47(14):3688–3691

Yoshida M, Kijima M, Akita M, Beppu T (1990) Potent and specific inhibition of mammalian histone deacetylase both in vivo and in vitro by trichostatin A. J Biol Chem 265(28):17174–17179

Yu ZX, Li SH, Nguyen HP, Li XJ (2002) Huntingtin inclusions do not deplete polyglutamine-containing transcription factors in HD mice. Hum Mol Genet 11(8):905–914.

Yuan PX, Huang LD, Jiang YM, Gutkind JS, Manji HK, Chen G (2001) The mood stabilizer valproic acid activates mitogen-activated protein kinases and promotes neurite growth. J Biol Chem 276(34):31674–31683

Zhang CL, McKinsey TA, Chang S, Antos CL, Hill JA, Olson EN (2002) Class II histone deacetylases act as signal-responsive repressors of cardiac hypertrophy. Cell 110(4):479–488

Zhang X, Smith DL, Meriin AB, Engemann S, Russel DE, Roark M, Washington SL, Maxwell MM, Marsh JL, Thompson LM, Wanker EE, Young AB, Housman DE, Bates GP, Sherman MY, Kazantsev AG (2005) A potent small molecule inhibits polyglutamine aggregation in Huntington's disease neurons and suppresses neurodegeneration in vivo. Proc Natl Acad Sci U S A 102(3):892–897

Zhou X, Richon VM, Rifkind RA, Marks PA (2000) Identification of a transcriptional repressor related to the noncatalytic domain of histone deacetylases 4 and 5. Proc Natl Acad Sci U S A 97(3):1056–1061
Zhou X, Marks PA, Rifkind RA, Richon VM (2001) Cloning and characterization of a histone deacetylase, HDAC9. Proc Natl Acad Sci U S A 98(19):10572–10577
Zoghbi HY, Orr HT (2000) Glutamine repeats and neurodegeneration. Annu Rev Neurosci 23:217–247

CHAPTER 13

FUNCTIONS OF MYST FAMILY HISTONE ACETYLTRANSFERASES AND THEIR LINK TO DISEASE

NIKITA AVVAKUMOV AND JACQUES CÔTÉ*

Laval University Cancer Research Center, Hôtel-Dieu de Québec (CHUQ), 9 McMahon Street, Quebec City Qc, Canada G1R 2J6

Abstract: The MYST family of histone acetyltransferases is highly conserved in eukaryotes and is responsible for the majority of acetylation events. These enzymes are exclusively found in multisubunit protein complexes, which structure is also very well conserved. Recent studies have shed light on the precise functions of these HAT complexes. They play critical roles in gene-specific transcription regulation, DNA damage response and repair, as well as DNA replication. Such roles in basic nuclear functions suggest that alteration of these MYST HAT complexes could lead to malfunctioning cells, leading to cell death, uncontrolled growth and/or disease. Indeed, many of these enzymes and their associated factors have been implicated in several forms of cancers. This chapter summarizes the current knowledge on MYST HAT complexes, their functions and link to human diseases.

GLOSSASRY

ATM Ataxia telangiectasia mutated. Kinase, plays a key role in repair of double-stranded DNA breaks.

DSB DNA double-stranded break.

HBO1 Histone acetyltransferase bound to ORC1 (origin-recognition complex subunit 1). A MYST family HAT; involved in DNA replication.

ING Inhibitor of growth. A family of tumour suppressors (ING1-5). ING3, 4 and 5 are found in multisubunit complexes with the Tip60, MOZ/MORF and HBO1 MYST HATs. Function in a number of key cellular processes and are frequently inactivated in human cancer.

MOF Males-absent on the first. A MYST family HAT; subunit of the *Drosophila* MSL dosage-compensation complex and a highly homologous human complex.

* E-mail Jacques.Cote@crhdq.ulaval.ca

T.K. Kundu and D. Dasgupta (eds.), Chromatin and Disease, 295–317.

MOZ/MORF Monocytic leukemia zinc-finger protein and MOZ-related factor. MYST family HATs; frequently fused to other proteins as a result of chromosomal translocation in acute myeloid leukemia.

MYST HATs A family of acetyltransferases named after its founding members MOZ, Ybf2/Sas3, Sas2 and Tip60. Human MYST HATs include MOZ, Tip60, MOZ/MORF, HBOl and MOF.

NuA4 Nucleosome acetyltransferase of histone H4. A multisubunit complex which includes, notably, TRRAP, p400 and Tip60 proteins.

Tip60 Tat-interacting protein of 60kDa. A MYST family HAT, catalytic subunit of the NuA4 complex. Performs key functions in repair of double-stranded DNA breaks.

TRRAP transformation/transcription domain-associated protein. Subunit of the Tip60/NuA4 HAT complex. Required for cell transformation and transcriptional activation by c-Myc, E2F1 and adenovirus E1A.

EPC1 enhancer of polycomb 1. Subunit of the Tip60/NuA4 HAT complex. Physically links the Tip60-EPC1-ING3 trimer to the rest of the complex. Fused to PHF1/polycomb-like human protein in uterine sarcomas.

1. INTRODUCTION

The eukaryotic genome is packaged into chromatin – a complex, multi-tiered structure consisting of DNA and proteins. Formation of chromatin is necessary to contain the sizable genome in the relatively tiny confines of the nucleus. The basic unit of chromatin – the nucleosome – is made up of 146 bp of DNA wrapped around a histone octamer containing two molecules of each core histone: H2A, H2B, H3 and H4. While the main "body" of each histone is tightly packed within the octamer, their N-terminal tails protrude outward from the nucleosome. These form additional histone–DNA and histone–histone contacts with other nucleosomes, facilitating formation of higher-order chromatin structure. However, the highly condensed nature of chromatin poses a barrier to nuclear processes that require access to DNA, such as gene transcription and DNA replication and repair. In response, the eukaryotic cell has developed intricate mechanisms that make the chromatin structure more accessible to "third parties". Such instances of chromatin opening are tightly regulated and occur in concert with specific nuclear events.

Most eukaryotic cells employ three principal mechanisms capable of inducing changes in chromatin structure. Histone chaperones are involved in assembly/disassembly of nucleosomes, as well as in controlling their dynamics during DNA transcription and replication. Chromatin remodeling complexes, such as SWI/SNF, utilize the energy of ATP hydrolysis to remove or reposition nucleosomes, thereby exposing or masking DNA. The third method involves covalent modification of histones, predominantly on the exposed tails, including phosphorylaton, acetylation, methylation, ubiquitination and others. According to the "histone code" hypothesis (Jenuwein and Allis 2001), the various modifications within the histone tails combine to produce very specific effects on the chromatin structure and send exquisitely-tuned signals to histone- and DNA-binding proteins. The entire "alphabet" of histone modifications is currently the subject of intense investigation.

Acetylation of lysine residues was one of the first histone modifications observed. As acetylation neutralizes a lysine's positive charge, it was proposed to weaken histone–DNA and histone–histone interactions, leading to relaxation of the higher-order chromatin structure and allowing access to DNA. Indeed, transcriptionally active regions of chromatin were observed to be enriched with acetylated histones, while their paucity was correlated with transcriptional silence (Howe *et al.*, 1999). Identification of the first transcription-related histone acetyltransferase (HAT) p55 in the protozoan *Tetrahymena thermophila* (Brownell and Allis, 1995), and the observation that it was homologous to the known yeast transcriptional co-activator Gcn5, further solidified the link between histone acetylation and transcription. Today, the many known HATs are grouped based on structural similarity, with the GCN5/pCAF (GNAT), p300/CBP and the MYST families being the most prominent (Sterner and Berger, 2000).

This chapter focuses on the MYST family HATs, describes their functions in the many diverse nuclear processes and discusses the established and putative links between their activities and human disease.

2. THE MYST FAMILY

The acronym MYST derives from the four founding members of this HAT family: mammalian *M*OZ, yeast *Y*bf2/Sas3 and *S*as2, and mammalian *T*ip60. A number of other MYST HATs were later discovered in various species, including mammalian MORF and its mouse orthologue Querkopf, MOF in mammals and *Drosophila*, and mammalian HBO1 and its *Drosophila* orthologue Chameau. Several other uncharacterized or less characterized MYST HATs are known to exist in other species (Utley and Côté, 2003). Thus, our current understanding of their functions combines observations made not only in mammalian cells but also in the lower eukaryotic model organisms.

The yeast Sas (*s*omething *a*bout *s*ilencing) genes are involved in transcriptional silencing in *Saccharomyces cerevisiae* (Reifsnyder *et al.*, 1996), while Esa1 (*e*ssential *S*as2-related *a*cetyltransferase 1, homologue of the human Tip60 gene) has been linked to transcriptional activation (Allard *et al* 1999; Clarke *et al.*, 1999; Smith *et al.*, 1998) and DNA repair (Bird *et al.*, 2002). The *Drosophila* MYST HAT Chameau ("camel" in French) has been linked to both JNK-mediated activation of transcription (Miotto *et al.*, 2006), and to gene silencing (Grienenberger *et al.*, 2002). Another *Drosophila* MYST HAT related to MOZ/MORF, Enok, has been shown to play a role in the development of olfactory centers of the brain (Scott *et al.*, 2001). The mouse gene Querkopf ("squarehead" in German) plays a significant role in development of the mouse cerebral cortex (Thomas *et al.*, 2000). The studies of this family of histone acetyltransferases in model organisms such as yeast and fruit flies have laid important foundations for the subsequent research into the functions of their human counterparts. Even today, such work is providing significant insight into this highly conserved protein family. In keeping with the theme of this book, however, this chapter will focus mainly on the human MYST HATs and their roles in human diseases.

The defining feature of the MYST family members is the presence of the highly conserved MYST domain which comprises an acetyl-CoA binding motif and a C_2HC zinc finger. The acetyl-CoA binding motif facilitates the acetyltransferase activity of the MYST HATs, while the zinc finger is also essential and is likely involved in interaction with substrates (Yan *et al.*, 2000). The MYST family members are further subdivided into subgroups based on the presence or absence of additional structural and functional features. These variously include the chromodomain, plant homeodomain (PHD), and other regions of homology unique only to subsets of MYST HATs (Fig. 1a). For more detailed descriptions of MYST domains, the reader is referred to the reviews by Utley and Côté (2003) and Yang (2004).

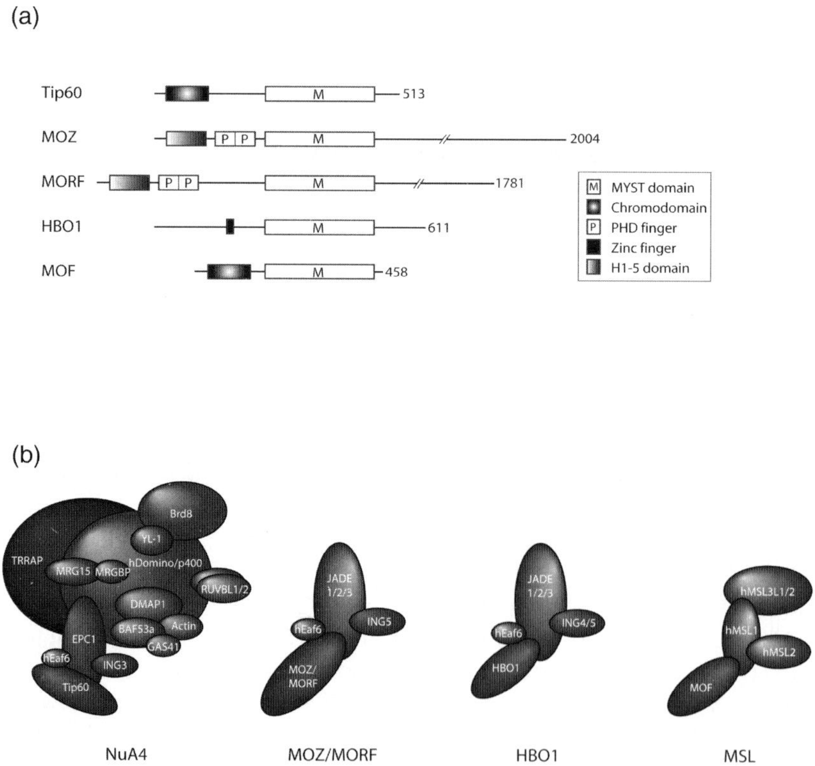

Figure 1. MYST histone acetyltransferases and their associated complexes. (a) domain structure of human MYST HAT proteins. (b) Protein complexes containing MYST HAT proteins

While all MYST family members possess intrinsic HAT activity, they do not function in isolation *in vivo* but, rather, are found in multisubunit protein complexes. Thus, to fully understand or to plausibly speculate about the potential roles of MYST HATs in disease, it is necessary to incorporate the known facts about other complex members. To this end, the sections of this chapter that discuss individual MYST HATs will also describe the complexes within which they function and summarize the available disease-related information about other complex components.

3. TIP60

The *T*at-*i*nteracting *p*rotein of *60* kDa (Tip60) was originally identified through a yeast two-hybrid screen as a binding partner of the HIV1 Tat protein (Kamine *et al.*, 1996). It is the catalytic subunit of the evolutionarily conserved NuA4 (*Nu*cleosome *A*cetyltransferase of histone H*4*) complex (Fig. 1b), which in humans also notably contains the p400 ATP-dependent chromatin remodeler and the catalytically inactive TRRAP PI3-kinase-like kinase (Doyon and Cote, 2004; Doyon *et al.*, 2006). The NuA4 complex acetylates histones H4 and H2A in chromatin (Ikura *et al.*, 2000).

3.1. Tip60 Functions in Transcriptional Regulation

The same study that identified Tip60 also demonstrated that its overexpression augmented the transcriptional activation ability of HIV1 Tat (Kamine *et al.*, 1996). While a follow-up investigation did not confirm the positive influence of Tip60 on Tat-mediated trans-activation, it did observe that Tat had a negative effect on the ability of Tip60 to activate the expression of the cellular Mn-dependent superoxide dismutase gene (Mn-SOD) (Creaven *et al.*, 1999). Thus, despite the disagreement, both studies linked Tip60 to regulation of transcription. Numerous subsequent studies have conclusively shown that Tip60 is indeed involved in transcription control, linking it to trans-activation. Importantly, Tip60 acts as a co-activator for a number of key cellular proteins such as p53 (Legube *et al.*, 2004), nuclear hormone receptors (Brady *et al.*, 1999; Gaughan *et al.*, 2001 2002), beta-catenin (Sierra *et al.*, 2006), NF-κB (Baek *et al.*, 2002), c-Myc oncoprotein (Frank *et al.*, 2003) and E2F (Taubert *et al.*, 2004). Notably, a large-scale RNAi-based screen demonstrated that Tip60 is required for induction of apoptosis by the tumour suppressor p53 (Berns *et al.*, 2004). A further intriguing connection between these two proteins comes from the observation that Tip60 can be targeted for degradation through ubiquitination by Mdm2, a p53 regulator (Legube *et al.*, 2002).

3.2. Roles of Tip60 in DNA Damage Repair

In recent years, additional transcription-unrelated roles of Tip60 have emerged. Perhaps most notably, it has been demonstrated to play key roles in a number of nuclear processes involved in DNA damage repair. Efficient and accurate repair of

DNA lesions is critical for cell survival, and errors in this process can lead not only to cellular death but to oncogenesis.

One of the seminal studies in this field reported not only a defect in repair of *d*ouble-*s*trand *b*reaks (DSBs) in DNA, but also inefficient induction of apoptosis in response to such DNA damage in cells expressing a HAT-deficient mutant of Tip60 (Ikura *et al.*, 2000). Ensuing research has revealed two major functions of Tip60 in DNA repair. First, it is involved in DNA damage checkpoint activation by acetylating the *a*taxia *t*elangiectasia *m*utated (ATM) kinase (Sun *et al.*, 2005). The inactive form of this kinase is recruited to sites of DNA breaks by damage-sensing protein complexes where it is activated. ATM then initiates a series of phosphorylation events that lead to activation of cell cycle checkpoints and initiation of DSB repair (Sancar *et al.*, 2004; Shiloh 2003). One of the key substrates of activated ATM is the variant histone H2AX (Shroff *et al.*, 2004). Phosphorylated H2AX (γ-H2AX) marks a wide area around the break site as it can be found in nucleosomes megabases away from the DSB (Rogakou *et al.*, 1999). As Tip60 interacts with both ATM and TRRAP via their FATC domains, it is suggested to activate ATM in isolation from TRRAP and, possibly, the rest of the NuA4 complex (Jiang *et al.*, 2006).

In addition to helping set off the alarm, Tip60 is also involved in the repair process itself. *Ch*romatin *i*mmuno*p*recipitation (ChIP) assays have detected Tip60 and TRRAP at the site of DSB in mammalian cells (Murr *et al.*, 2006). Several yeast NuA4 (yNuA4) components have also been found at DSBs in *S. cerevisiae*, their recruitment assisted by the Arp4 subunit which recognizes phosphorylated histone H2A (γ-H2A, the yeast equivalent of γ-H2AX) (Downs *et al.*, 2004). In mammalian cells, depletion of TRRAP or introduction of a dominant-negative mutant of Tip60 results in impairment of DSB repair due to inefficient recruitment of repair machinery to the site of the break. This is apparently caused by the failure of the surrounding chromatin to relax – an event that is normally facilitated by hyperacetylation of histone H4 by the human NuA4 complex (Murr *et al.*, 2006).

Once repair is complete, the phosphorylation mark on histone H2AX needs to be removed so as not to needlessly tie up DNA repair complexes. Again, the NuA4 complex appears to play a role. A study carried out in *Drosophila* has shown that both the dTip60 and p400/Domino components of dNuA4 catalyze the exchange of phosphorylated H2Av (γ-H2AX equivalent) with unmodified H2Av (Kusch *et al.*, 2004). The free γ-H2Av would then be dephosphorylated as signal for checkpoint release (Chowdhury *et al.*, 2005; Keogh *et al.*, 2006). Thus, the "red flag" is removed from the now-fixed chromatin and the repair machinery is released to function elsewhere.

3.3. Tip60 and Links to Human Disease

Given the deep involvement of Tip60 in some of the key nuclear processes, it is hardly surprising that (mal)functions of this MYST HAT have been implicated in a number of human diseases. Indeed, its original identification as an HIV1 Tat-binding protein signified that Tip60 was likely to have a profound impact on human health.

3.3.1. Tip60 and the human immunodeficiency virus

The HIV1 Tat protein has been reported to inhibit the HAT activity of Tip60, preventing the latter from activating transcription of the Mn-SOD gene (Creaven *et al.*, 1999). Downregulation of Mn-SOD by Tat had been observed previously and presumed to be responsible, at least in part, for the oxidative stress leading to T-cell depletion in AIDS (Westendorp *et al.*, 1995). More recently, Tat was demonstrated to target Tip60 for degradation via ubiquitination (Col *et al.*, 2005). Linking this observation to the functions of Tip60 in DNA repair (Ikura *et al.*, 2000), the same study observed that such depletion of Tip60 impaired induction of apoptosis in response to DNA damage. Thus, Tat-mediated destruction of Tip60 may inure the cell to genotoxic stress and keep it alive, allowing HIV to replicate. It remains to be discovered whether the Tat/Tip60 interaction similarly affects expression of other cellular genes in order to facilitate viral survival and propagation.

3.3.2. Tip60 and other human diseases

As with HIV, most other links between Tip60 and disease are related to its transcriptional effects. For instance, a series of studies has strongly implicated Tip60 in the onset of Alzheimer's disease. The β-amyloid precursor protein (APP) is a cell membrane protein that, when cleaved by the action of β- and γ-secretases, produces the extracellular amyloid β-peptide which is deposited into amyloid plaques in the brains of Alzheimer's patients (Steiner and Haass, 2001). The resulting intracellular C-terminal portion of APP, variously referred to as AICD or APP-CT, in turn has been shown to associate with the APP-binding protein Fe65 and with Tip60 (Cao and Sudhof, 2001 2004). This tri-molecular complex is able to reverse transcriptional repression of the KAI1 metastasis suppressor mediated by the *n*uclear receptor *co-r*epressor (N-CoR) (Baek *et al.*, 2002; Kim *et al.*, 2005). Significantly, this suggests a functional link between Tip60 and N-CoR, whose role in normal development and in disease is rapidly emerging (Jepsen and Rosenfeld, 2002). Interestingly, another study has reported that association of AICD/APP-CT with Tip60 could lead to induction of apoptosis in a manner dependent on Tip60's HAT function (Kinoshita *et al.*, 2002). This activity of Tip60 has been proposed to lead to Alzheimer's-related neurodegeneration in parallel with release of the amyloid β-peptide.

Several cancer links have been suggested for Tip60. It has been shown to acetylate and thereby activate the *a*ndrogen *r*eceptor (AR) – a hormone-dependent transcription factor strongly implicated in development of prostate cancer (Brady *et al.*, 1999; Gaughan *et al.*, 2001 2002). More recently, Tip60 has also been implicated in progression of prostate cancer toward hormone-independence and resistance to therapy which leads to a fatal outcome. A study conducted in androgen-dependent and -independent cell lines demonstrated that cellular Tip60 levels increase with progression to hormone independence. Tip60 then accumulates in the nucleus where it co-activates AR-responsive genes in a ligand-independent fashion. Thus, its activity may be responsible for hormone-independent survival of advanced prostate cancer cells (Halkidou *et al.*, 2003).

The observed involvement of Tip60 in expression of NF-κB-regulated genes (Baek *et al.*, 2002) suggests that the HAT may play a general role in the wide variety of NF-κB activities. This family of transcription factors is involved in regulation of a large number of genes that participate in diverse cellular functions (Radhakrishnan and Kamalakaran, 2006). Notably, these include signaling molecules like cytokines and chemokines, and affectors of apoptosis and cell proliferation. NF-kB is considered an oncogene due to its ability to activate growth promoting and anti-apoptotic genes, and can also promote metastasis. Of similar potential importance is the recruitment of Tip60 and other NuA4 components by the c-Myc transcription factor (Frank *et al.*, 2003). c-Myc is a potent promoter of cellular growth and proliferation, and is often found to be deregulated in a variety of human cancers (Nesbit *et al.*, 1999). Thus, the initial observations that Tip60 is associated with the functions of NF-kB and c-Myc may be indicative of its significant involvement in a variety of normal cellular processes and in development of a number of human maladies.

Two recent studies have linked Tip60 to the tumour suppressor p14ARF. This protein has a well-established role in controlling p53-mediated cell cycle arrest, but is also known to function independently of p53. In support of this, one study demonstrated that, in response to genotoxic stress, a direct physical interaction between p14ARF and Tip60 results in activation of the ATM/CHK signaling cascade and arrest of the cell cycle in the G2 phase (Eymin *et al.*, 2006). The second study observed that p14ARF can prevent degradation of the *r*etino*b*lastoma (Rb) tumour suppressor by blocking its acetylation by Tip60 (Leduc *et al.*, 2006). These observations suggest that Tip60 may be involved in regulation of Rb homeostasis in the cell, and further highlight its role in cellular response to DNA damage.

An interaction between Tip60 and *i*nter*l*eukin-9 (IL-9) receptor has also been reported (Sliva *et al.*, 1999). While the functional consequences of this association are unknown, it may prove to have a significant impact on human health as IL-9 signaling has been linked to Hodgkin's lymphoma and asthma (Knoops and Renauld 2004; Renauld, 2001). An additional connection between Tip60 and lymphoma comes from a study that shows that a protein expressed by the human T-cell lymphotropic virus type-1 stabilizes the interaction between this MYST HAT and the Myc oncoprotein, leading to enhanced cellular transformation (Awasthi *et al.*, 2005).

The discovery of the Tip60 splice variant PLIP (cytosolic *p*hospho*l*ipase A_2 – *i*nteracting *p*rotein) is also of potential interest as its overexpression was shown to lead to apoptosis (Sheridan *et al.*, 2001). As the phospholipase itself is involved in a number of cellular processes and dysfunctions, further investigation of this interaction is likely to reveal additional key roles of PLIP.

3.3.3. Other NuA4 subunits and human disease

As mentioned previously, an additional consideration needs to be given to the known Tip60-interacting proteins and their connection with human maladies. Several components of the NuA4 complex (Fig. 1b), in fact, have been linked to disease and thus Tip60 may be "guilty by association". For instance, allelic loss of the NuA4 subunit ING3, a member of the ING (*in*hibitor of *g*rowth) family of tumour suppressors, has been observed in human head and neck cancers (Gunduz *et al.*, 2002). ING3 has been shown to function as a co-activator of p53-dependent genes, thereby regulating cell cycle progression and apoptosis (Nagashima *et al.*, 2003). Interestingly, recently it was also observed to facilitate UV-induces apoptosis through the Fas/caspase-8 pathway, independently of p53 (Wang and Li, 2006).

Additionally, several isolated studies have linked other NuA4 components to disease, though we must wait for further research to provide detailed mechanistic evidence. For instance, Brd8/p120 has been demonstrated to function as a co-activator of the thyroid hormone receptor (Monden *et al.*, 1997) whose malfunction can lead to a number of human diseases such as RTH (*r*esistance to *t*hyroid *h*ormone), characterized by a number of clinical phenotypes (Cheng, 2005). Amplified copy numbers of the GAS41 (*g*lioma-*a*mplified *s*equence 41) gene have been observed in glioblastoma multiforme and astrocytoma (Fischer *et al.*, 1997). More recently, GAS41 has been implicated in suppression of the p53 tumour suppressor pathway during normal cell growth (Park and Roeder, 2006). Like Tip60 itself, the NuA4 subunits BAF53, TRRAP and p400 have been implicated in Myc oncogene-mediated cellular transformation (Fuchs *et al.*, 2001; Nikiforov *et al.*, 2002; Park *et al.*, 2002). Both TRRAP and p400 are also targeted by the adenoviral E1A oncoprotein and are essential for E1A-mediated cellular transformation (Fuchs *et al.*, 2001; Lang and Hearing, 2003). Finally, MRG15, another human NuA4 subunit (Cai *et al.*, 2003; Doyon *et al.*, 2004), belongs to a family of proteins that have been linked to cellular senescence (Bertram and Pereira-Smith, 2001; Bertram *et al.*, 1999).

The NuA4 subunit EPC1 was recently shown to fuse to the coding sequence of *PH*D *f*inger protein 1 (PHF1) through a chromosomal rearrangement observed in a subset of uterine sarcomas (Micci *et al.*, 2006). The resulting chimeric protein has an open reading frame consisting of the first 581 residues from EPC1 and the entire coding region of PHF1 (Fig. 2a). Within NuA4, the N-terminus of EPC1 binds Tip60 and ING3 forming the trimeric piccolo NuA4 (picNuA4), while the C-terminus of EPC1 bridges it to the rest of the NuA4 complex (Boudreault *et al.*, 2003; Doyon *et al.*, 2004) (Fig. 1b). Thus, picNuA4 contains HAT activity and, in fact, is the minimal module within NuA4 capable of acetylating chromatin substrates (Doyon *et al.*, 2004; Selleck *et al.*, 2005). On the other hand, EPC1's fusion partner PHF1 is homologous to the *Drosophila* polycomblike protein (PCL), which is associated with transcriptional repression/heterochromatin formation. Polycomblike interacts with enhancer of zeste, an H3 lysine 27 histone methyltransferase, both in fruit fly and human cells (O'Connell *et al.*, 2001). Thus, the detection of the EPC1/PCL

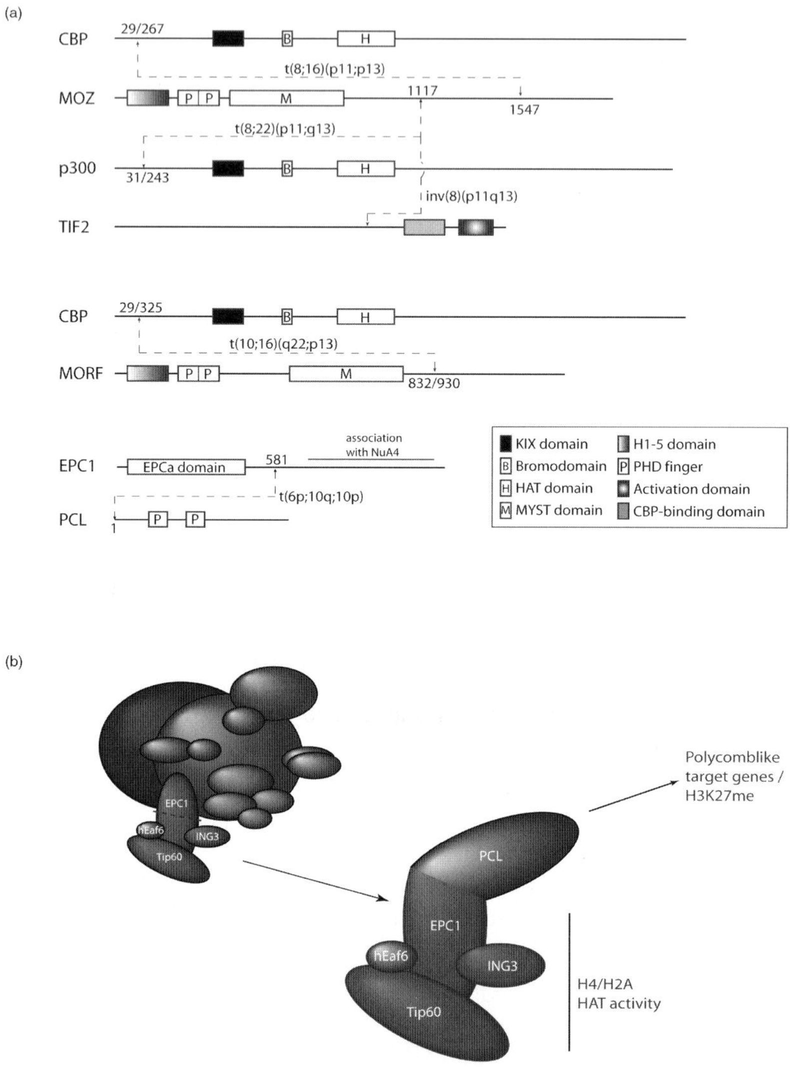

Figure 2. Chromosomal translocations leading the production of a MYST HAT fusion protein and cancer. (a) structure of different MYST HAT fusion proteins leading to cancer. Numbers indicate amino acid positions at break points. (b) Model for the consequences of EPC1-PCL fusion on chromatin function

fusion raises a very intriguing possibility that the chimeric protein diverts the HAT activity of picNuA4 to the polycomblike's normal targets – genomic regions that are normally maintained in the repressed state (Fig. 2b). Mis-targeted acetylation of histone H4 by picNuA4 would lead to unraveling of the heterochromatin, resulting in aberrant gene expression which could easily account for appearance of malignancy.

While the precise roles played by these NuA4 components in various maladies remain to be elucidated, in many cases it seems plausible that the entire complex is affected or involved. Thus, the function of the NuA4 complex may be subverted through one of its subunits, linking the catalytic component Tip60 to human disease.

4. MOZ AND MORF

The *m*onocytic leukemia *z*inc-finger protein (MOZ) was identified in a study that examined the t(8;16)(p11;p13) chromosomal translocation common in *a*cute *m*yeloid *l*eukemia (AML) (Borrow *et al.*, 1996). The involved gene on chromosome 16 was found to be the *C*REB-*b*inding *p*rotein (CBP) – a well-known transcriptional co-activator. Its fusion counterpart on chromosome 8 was a novel protein that possessed a C_2HC zinc finger and a putative acetyltransferase signature. MOZ's ability to function as a HAT was confirmed in a later *in vitro* study (Champagne *et al.*, 2001). The *M*OZ-*r*elated *f*actor (MORF) was identified based on its homology to MOZ and similarly demonstrated to possess an *in vitro* acetyltransferase activity (Champagne *et al.*, 1999).

Both MOZ and MORF have also been linked to control of transcription. In initial analyses, these MYST HATs were found to contain potent trans-activation domains in their C-termini (Champagne *et al.*, 1999 2001). Follow-up studies demonstrated that the two proteins can indeed function as transcriptional co-activators for two Runt-domain transcription factors – Runx1 and 2 (also referred to as AML1 & 3) (Kitabayashi *et al.*, 2001a; Pelletier *et al.*, 2002). Furthermore, MORF has been found in the co-activator complex associated with the peroxisome proliferator-activated receptor alpha in rat liver (Surapureddi *et al.*, 2002).

More recently, a MOZ/MORF-containing human complex was purified and found to contain the ING5 tumour suppressor as well as *br*omodomain- and *P*HD *f*inger-containing (BRPF) proteins 1, 2 and 3 (Fig. 1b) (Doyon *et al.*, 2006). This complex can also function as a co-activator for Runx2 and acetylates nucleosomal histone H3. Additionally, the ING5-MOZ/MORF-BRPF complex appears to play a role in DNA replication, as knockdown of ING5 led to inability of the cells to complete the S phase.

4.1. MOZ/MORF and Human Disease

As described above, MOZ was originally identified in an investigation of a recurrent chromosomal translocation in AML that resulted in its fusion to CBP (Borrow *et al.*, 1996). Numerous subsequent studies have described the in-frame fusions of MOZ to CBP and its homologue p300 (Kitabayashi *et al.*, 2001b; Panagopoulos *et al.*, 2002 2002; Rozman *et al.*, 2004), and to the TIF2 nuclear receptor co-activator (Carapeti *et al.*, 1998; Kindle *et al.*, 2005). MORF has also been found fused in-frame to CBP following a chromosomal translocation in AML (Fig. 2a) (Kojima *et al.*, 2003; Panagopoulos *et al.*, 2001). Additional evidence for the key role of MOZ translocation in induction of leukemia came from its identification as a common retroviral integration site (Lund *et al.*, 2002). Intriguingly, a sequencing analysis of the chromosomal breakpoints suggested involvement of a DNA repair complex in the appearance of these translocations (Panagopoulos *et al.*, 2003).

The functions of these fusion proteins and their precise roles in the induction of cancer are currently the subject of intense investigation (Collins *et al.*, 2006; Huntly *et al.*, 2004; Kindle *et al.*, 2005). In general, it appears that since the proteins are fused in-frame, their catalytic domains continue to function, but these activities are mis-directed by the targeting domains of their fusion partners. The result is aberrant acetylation of chromatin and non-chromatin proteins, as well as sequestration of key nuclear regulators. These mechanisms are discussed in-depth in Chapter 8 of this book. Furthermore, two recent studies have demonstrated that wild-type MOZ is essential for development of hematopoietic stem cells (Katsumoto *et al.*, 2006; Thomas *et al.*, 2006). This suggests that AML may arise not only due to the aberrant activity of the fusion proteins, but also due to the loss of their original function.

Several additional lines of evidence link these MYST HATs to human disease. Disruption of the MORF coding region due to a chromosomal rearrangement distinct from AML has been observed in benign uterine smooth muscle tumours – a condition that afflicts up to 75% of women in the latter reproductive years (Moore *et al.*, 2004). It remains to be determined how the loss of MORF is related to the onset of this disease. One possibility is that it affects the composition and/or stability of the ING5-MOZ/MORF-BPRF complex (Doyon *et al.*, 2006). This, in turn, could disrupt the reported ability of the ING5 tumour suppressor to modulate the activity of p53 (Shiseki *et al.*, 2003), promoting oncogenesis. Additionally, disruption of the MORF homologue Querkofp in mice leads to abnormal development of the nervous system and craniofacial features (Thomas *et al.*, 2000). Intriguingly, this suggests that despite their structural and functional similarities, MOZ and MORF may play distinct roles in various cellular processes.

5. HBO1

The HBO1 MYST HAT derives its name from its original identification as the *h*istone acetyltransferase *b*ound to *O*RC1 – the largest subunit of the *o*rigin *r*ecognition *c*omplex (ORC) (Iizuka and Stillman, 1999). ORC is a key nuclear protein complex that recognizes and binds the origins of replication in the eukaryotic genome. This, in turn, facilitates binding and assembly of the *pre-r*eplication *c*omplexes (pre-RC), leading to bi-directional DNA replication (DePamphilis, 2003). HBO1 also interacts with *mini*chromosome *m*aintenance protein 2 (MCM2), a subunit of the MCM complex which is involved in DNA replication along with ORC (Burke *et al.*, 2001). Given the presence of a putative MYST domain in the HBO1 cDNA sequence, the original study examined HBO1 for HAT activity and found that it was capable of acetylating free histones H3 and H4 (Iizuka and Stillman, 1999). The presence of a HAT at replication origins suggested that histone acetylation may be involved in control of pre-RC assembly and, thus, DNA replication.

More recently, two native HBO1-containing complexes have been purified from human cells and their functions analyzed (Fig. 1b) (Doyon *et al.*, 2006). In addition to HBO1 itself, one was found to contain the ING4 tumour suppressor, hEaf6 (also a component of the human NuA4 complex), and *g*ene for *a*poptosis and *d*ifferentiation in *e*pithelia (Jade)-1/2/3. The second complex was very similar, but included ING5 instead of ING4. Both complexes were found to acetylate nucleosomal histone H4 and, to a lesser extent, H3. Remarkably, RNA-interference analysis revealed that HBO1 was responsible for the majority of H4 acetylation in the cell. Peptides corresponding to MCM proteins were co-purified with ING5- but not ING4-HBO1, suggesting that the ING5-HBO1 complex is involved in DNA replication. Subsequent knock-down of HBO1 demonstrated that it plays an essential role in this process, as cells lacking HBO1 were severely impaired in their ability to complete the S phase.

Two other recent studies have also confirmed HBO1's involvement in control of DNA replication. Using human cells and *Xenopus* egg extracts, it was demonstrated that depletion of HBO1 abolishes loading of the MCM complex onto chromatin (Iizuka *et al.*, 2006). In *Drosophila*, the HBO1 homologue Chameau appears to be involved in hyperacetylation of chromatin that results in increased activity of DNA replication origins (Aggarwal and Calvi, 2004). Thus, HBO1 and its homologues appear to play an essential role in DNA replication by remodeling the chromatin at origins and facilitating assembly of the pre-replication complex.

In addition to DNA replication, HBO1 has also been implicated in control of transcription. In one assay, it was found to possess a trans-repression ability and down-regulate androgen receptor-mediated transcription in transient transfection assays (Sharma *et al.*, 2000). Conversely, a recent investigation has demonstrated that HBO1 can enhance transcription by a number of steroid receptors, in a fashion that appears to be quite distinct from its HAT activity (Georgiakaki *et al.*, 2006). Two studies have observed that *Drosophila* Chameau can play dual roles in transcriptional regulation as well. In one instance, it was found to be required for maintenance of transcriptional silencing of Hox genes in *Drosophila*, and could substitute for the yeast HAT Sas2 in mediating silencing at telomeres (Grienenberger *et al.*, 2002). On the other hand, the same group has recently demonstrated that Chameau can act to promote trans-activation ability of the *Drosophila* Fos cofactor (Miotto *et al.*, 2006). Intriguingly, Chameau's HAT activity was required in both cases.

5.1. HBO1 and Links to Human Disease

As HBO1 appears to be essential for DNA replication, it is reasonable to speculate that defects in its function will have profound negative consequences on cellular viability. However, no such evidence exists in the current literature, and thus the discussion of how the role of HBO1 in replication of genomic DNA may affect human health must await further research. However, other potential connections between HBO1 and human disease have been revealed. One such link involves the function of HBO1 in replication of genomes of invading pathogens. Numerous viruses that infect humans do not encode a full complement of DNA replication proteins, but rely instead on the replication machinery of the host cell. Indeed, ORC and MCM complexes are involved in replication of genomes of the Epstein-Barr virus (Chaudhuri *et al.*, 2001; Dhar *et al.*, 2001) and the *K*aposi's *s*arcoma-associated *h*erpes*v*irus (KSHV) (Stedman *et al.*, 2004), and HBO1 has been co-purified with KSHV's origins of replication (Stedman *et al.*, 2004). Other viruses may also rely on the same components of the human DNA replication machinery.

Like MOZ, the HBO1 coding region has been identified as a common retroviral integration site leading to myeloid leukemia and B/T cell lymphoma in mice (Suzuki *et al.*, 2002). This observation suggests that its loss may lead to development of human cancer.

Other putative links between HBO1 and disease involve its binding partners. Its interaction with the androgen receptor (Sharma *et al.*, 2000) putatively connects it to development of prostate cancer. The interaction between the HBO1 complex subunit Jade-1 and the *v*on *H*ippel-*L*indau (VHL) tumour suppressor correlates with protection from renal cancer (Zhou *et al.*, 2005), suggesting that HBO1 may be connected to that malignancy and to the von Hippel-Lindau disease. The presence of HBO1 in complexes with the tumour suppressors ING4 and 5 (Doyon *et al.*, 2006) further implies that it may play a role in protection from malignancy. ING4 has been shown to suppress loss of contact inhibition in cultured cells and found to be inactivated in various human cancer cell lines (Kim *et al.*, 2004). It is also a potent inhibitor of glioma and angiogenesis in mouse models (Garkavtsev *et al.*, 2004; Kim *et al.*, 2004). Both ING4 and ING5 have been reported to enhance transcriptional activity of p53 (Shiseki *et al.*, 2003), and HBO1's histone acetylation activity may facilitate this.

6. ING TUMOUR SUPPRESSORS

Of particular significance with respect to human health is the fact that the Tip60, MOZ/MORF and HBO1 MYST HATs have been each found in a stable complex with a specific member of the ING family of tumour suppressors (Fig. 1b) (Doyon *et al.*, 2006). Among several distinguishing structural features of the ING proteins is the highly conserved PHD zinc finger, often found in proteins involved in chromatin remodeling. The ING proteins impinge on a number of key cellular processes, such as DNA damage repair, cell cycle progression, senescence, apoptosis and tumourigenesis, and are frequently found to be inactivated in various human cancers (Russell *et al.*, 2006). In human, each of the five ING tumour suppressors is exclusively found as a native subunit of a specific HAT or *h*istone *deac*etylase (HDAC) complex, linking chromatin modification to cellular regulation (Doyon *et al.*, 2006; Shi and Gozani, 2005). Furthermore, ING subunits are essential factors that enable their associated HAT/HDAC enzymes to modify chromatin substrates (Boudreault *et al.*, 2003; Doyon *et al.*, 2006; Nourani *et al.*, 2003; Selleck *et al.*, 2005).

Of note, the ING PHD finger was recently shown to recognize histone 3 lysine 4 trimethylation (H3K4me3) – a mark deposited by the *m*ixed *l*ineage *l*eukemia (MLL) methyltransferase and associated with high transcription levels. The particular observation that ING2 binds H3K4me3 and recruits its associated repressor complex suggests that this mechanism is capable of rapid shut-off of actively transcribed genes (Shi *et al.*, 2006). Such a role would be invaluable to the cell in times of genotoxic stress as it would allow for a prompt silencing of proliferation genes, arresting the cell cycle until the DNA damage can be repaired.

Significantly, MLL is a frequent subject of chromosomal rearrangement in several types of leukemia, and its methyltransferase activity becomes deregulated as a result (Linggi *et al.*, 2005). By recognizing the H3K4me3 mark, the ING proteins may be able to recruit their associated histone modification complexes, preventing aberrant transcription. Thus, the ING family members may play important roles in preventing the onset of leukemia.

7. MOF

The *mof* (*m*ales-absent *o*n the *f*irst) gene was originally identified in *Drosophila* and found to encode a putative histone acetyltransferase. This HAT is an essential catalytic subunit of the *Drosophila* dosage-compensation MSL complex which acetylates lysine 16 on histone H4 (H4K16). This modification leads to upregulation of gene expression from the single X chromosome in the male fruit flies (Akhtar and Becker 2000; Hilfiker *et al.*, 1997). In addition to modifying chromatin, the *Drosophila* mof was also shown to functionally acetylate another MSL component, MSL-3 (Buscaino *et al.*, 2003). The human MOF (hMOF) HAT was identified based on its homology to the *Drosophila* protein and found to be responsible for the majority of H4K16 acetylation *in vivo* (Smith *et al.*, 2005; Taipale *et al.*, 2005). Furthermore, hMOF has been demonstrated to associate with human MSL proteins, forming an MSL complex very similar to the one found in *Drosophila* (Fig. 1b) (Smith *et al.*, 2005; Taipale *et al.*, 2005). In contrast to the observations in fruit fly, the human MSL complex is not restricted to the X chromosome, but instead is found throughout the genome.

Additionally, hMOF has been co-purified with the MLL1 methyltransferase whose activity results in methylation of histone H3 on lysine 4 (H3K4) – also a mark of actively transcribed genes. Co-ordinated activities of MLL1 and hMOF were shown to enhance p53-dependent trans-activation, and to synergistically upregulate expression of the Hoxa9 gene (Dou *et al.*, 2005). As mentioned previously, MLL1 is frequently found to be deregulated in several types of leukemia. hMOF has also been shown to interact with the ATM kinase involved in DNA repair (see section on Tip60, above). Depletion of hMOF resulted in decreased activation of ATM in response to DNA damage, decreased DNA repair and failure of the damage-induced cell cycle checkpoint (Gupta *et al.*, 2005).

7.1. Putative Connections of MOF to Disease

The recent observation that hMOF is involved in activation of the DNA repair process (Gupta *et al.*, 2005) opens the possibility that malfunctions of this MYST HAT may result in inefficient or erroneous DNA repair, which in turn can lead to malignancy. Additionally, hMOF's observed involvement in control of Hoxa9 expression (Dou *et al.*, 2005) putatively links it to regulation of the HOX gene family. HOX genes play essential roles in embryonic development and have been linked to a number of human diseases including cancer (Grier *et al.*, 2005). Hoxa9, specifically, is known to function in hematopoiesis and may be involved in development of leukemia (Tedeschi *et al.*, 2006).

The finding that hMOF is responsible for the bulk of H4K16 acetylation in the cell (Smith *et al.*, 2005) also suggests that loss or malfunction of this HAT can have a significant negative impact on normal cell function and viability. This histone modification is known to play critical roles in various chromatin-related processes and has been linked to a number of diseases (Shia *et al.*, 2006). Perhaps most notably, loss of H4K16 acetylation is a common hallmark of human cancers (Fraga *et al.*, 2005). The molecular mechanisms underlying the importance of H4K16 acetylation are now becoming clearer. This histone modification has a profound effect on chromatin structure, as it both inhibits the formation of the higher-order 30 nm fibre and interferes with the activity of chromatin remodeling enzymes (Shogren-Knaak *et al.*, 2006). Thus, the observation that downregulation of hMOF results in significant loss of H4K16 acetylation and arrest of the cell cycle (Smith *et al.*, 2005) has provided a very solid link between this HAT and human disease.

8. CONCLUDING REMARKS

The HATs belonging to the MYST family play essential roles in a wide variety of critical cellular processes. Further research will undoubtedly reveal additional functions of these enzymes that profoundly impact cellular growth, differentiation and replication. It is not surprising, therefore, that their malfunctions or subversion of their regular functions by cellular or viral machinery have been linked to a number of human diseases. A comprehensive understanding of these mechanisms is crucial to design of successful therapies, such as small-molecule inhibitors, that will be able to specifically target the undesirable activities of MYST HATs.

ACKNOWLEDGEMENTS

Work in our laboratory is supported by operating grants from the Canadian Institutes of Health Research (CIHR). N.A. holds a CIHR/Institute of Aging Fellowship and J.C. is a CIHR Investigator.

REFERENCES

Aggarwal BD, Calvi BR (2004) Chromatin regulates origin activity in Drosophila follicle cells. Nature 430:372–376

Akhtar A, Becker PB (2000) Activation of transcription through histone H4 acetylation by MOF, an acetyltransferase essential for dosage compensation in Drosophila. Mol Cell 5:367–375

Allard S, Utley RT, Savard J, Clarke A, Grant P, Brandl CJ, Pillus L, Workman JL, Cote J (1999) NuA4, an essential transcription adaptor/histone H4 acetyltransferase complex containing Esa1p and the ATM-related cofactor Tra1p. Embo J 18:5108–5119

Awasthi S, Sharma A, Wong K, Zhang J, Matlock EF, Rogers L, Motloch P, Takemoto S, Taguchi H, Cole MD et al (2005) A human T-cell lymphotropic virus type 1 enhancer of Myc transforming potential stabilizes Myc-TIP60 transcriptional interactions. Mol Cell Biol 25:6178–6198

Baek SH, Ohgi KA, Rose DW, Koo EH, Glass CK, Rosenfeld MG (2002) Exchange of N-CoR corepressor and Tip60 coactivator complexes links gene expression by NF-kappaB and beta-amyloid precursor protein. Cell 110:55–67

Berns K, Hijmans EM, Mullenders J, Brummelkamp TR, Velds A, Heimerikx M, Kerkhoven RM, Madiredjo M, Nijkamp W, Weigelt B et al (2004) A large-scale RNAi screen in human cells identifies new components of the p53 pathway. Nature 428:431–437

Bertram MJ, Berube NG, Hang-Swanson X, Ran Q, Leung JK, Bryce S, Spurgers K, Bick RJ, Baldini A, Ning Y et al (1999) Identification of a gene that reverses the immortal phenotype of a subset of cells and is a member of a novel family of transcription factor-like genes. Mol Cell Biol 19:1479–1485

Bertram MJ, Pereira-Smith OM (2001) Conservation of the MORF4 related gene family: identification of a new chromo domain subfamily and novel protein motif. Gene 266:111–121

Bird AW, Yu DY, Pray-Grant MG, Qiu Q, Harmon KE, Megee PC, Grant PA, Smith MM, Christman MF (2002) Acetylation of histone H4 by Esa1 is required for DNA double-strand break repair. Nature 419:411–415

Borrow J, Stanton VP Jr, Andresen JM, Becher R, Behm FG, Chaganti RS, Civin CI, Disteche C, Dube I, Frischauf AM et al (1996) The translocation t(8;16)(p11;p13) of acute myeloid leukaemia fuses a putative acetyltransferase to the CREB-binding protein. Nat Genet 14:33–41

Boudreault AA, Cronier D, Selleck W, Lacoste N, Utley RT, Allard S, Savard J, Lane WS, Tan S, Cote J (2003) Yeast enhancer of polycomb defines global Esa1-dependent acetylation of chromatin. Genes Dev 17:1415–1428

Brady ME, Ozanne DM, Gaughan L, Waite I, Cook S, Neal DE, Robson CN (1999) Tip60 is a nuclear hormone receptor coactivator. J Biol Chem 274:17599–17604

Brownell JE, Allis CD (1995) An activity gel assay detects a single, catalytically active histone acetyltransferase subunit in Tetrahymena macronuclei. Proc Natl Acad Sci USA 92:6364–6368

Burke TW, Cook JG, Asano M, Nevins JR (2001) Replication factors MCM2 and ORC1 interact with the histone acetyltransferase HBO1. J Biol Chem 276:15397–15408

Buscaino A, Kocher T, Kind JH, Holz H, Taipale M, Wagner K, Wilm M, Akhtar A (2003) MOF-regulated acetylation of MSL-3 in the Drosophila dosage compensation complex. Mol Cell 11:1265–1277

Cai Y, Jin J, Tomomori-Sato C, Sato S, Sorokina I, Parmely TJ, Conaway RC, Conaway JW (2003) Identification of new subunits of the multiprotein mammalian TRRAP/TIP60-containing histone acetyltransferase complex. J Biol Chem 278:42733–42736

Cao X, Sudhof TC (2001) A transcriptionally [correction of transcriptively] active complex of APP with Fe65 and histone acetyltransferase Tip60. Science 293:115–120

Cao X, Sudhof TC (2004) Dissection of amyloid-beta precursor protein-dependent transcriptional transactivation. J Biol Chem 279:24601–24611

Carapeti M, Aguiar RC, Goldman JM, Cross NC (1998) A novel fusion between MOZ and the nuclear receptor coactivator TIF2 in acute myeloid leukemia. Blood 91:3127–3133

Champagne N, Bertos NR, Pelletier N, Wang AH, Vezmar M, Yang Y, Heng HH, Yang XJ (1999) Identification of a human histone acetyltransferase related to monocytic leukemia zinc finger protein. J Biol Chem 274:28528–28536

Champagne N, Pelletier N, Yang XJ (2001) The monocytic leukemia zinc finger protein MOZ is a histone acetyltransferase. Oncogene 20:404–409

Chaudhuri B, Xu H, Todorov I, Dutta A, Yates JL (2001) Human DNA replication initiation factors, ORC and MCM, associate with oriP of Epstein-Barr virus. Proc Natl Acad Sci USA 98:10085–10089

Cheng SY (2005) Thyroid hormone receptor mutations and disease: beyond thyroid hormone resistance. Trends Endocrinol Metab 16:176–182

Chowdhury D, Keogh MC, Ishii H, Peterson CL, Buratowski S, Lieberman J (2005) gamma-H2AX dephosphorylation by protein phosphatase 2A facilitates DNA double-strand break repair. Mol Cell 20:801–809.

Clarke AS, Lowell JE, Jacobson SJ, Pillus L (1999) Esa1p is an essential histone acetyltransferase required for cell cycle progression. Mol Cell Biol 19:2515–2526

Col E, Caron C, Chable-Bessia C, Legube G, Gazzeri S, Komatsu Y, Yoshida M, Benkirane M, Trouche D, Khochbin S (2005) HIV-1 Tat targets Tip60 to impair the apoptotic cell response to genotoxic stresses. Embo J 24:2634–2645

Collins HM, Kindle KB, Matsuda S, Ryan C, Troke PJ, Kalkhoven E, Heery DM (2006) MOZ-TIF2 alters cofactor recruitment and histone modification at the RARbeta 2 promoter: Differential effects of MOZ fusion proteins on CBP-and MOZ-dependent activators. J Biol Chem 281:17124–17133

Creaven M, Hans F, Mutskov V, Col E, Caron C, Dimitrov S, Khochbin S (1999) Control of the histone-acetyltransferase activity of Tip60 by the HIV-1 transactivator protein, Tat. Biochemistry 38:8826–8830

DePamphilis ML (2003) The 'ORC cycle': a novel pathway for regulating eukaryotic DNA replication. Gene 310:1–15

Dhar SK, Yoshida K, Machida Y, Khaira P, Chaudhuri B, Wohlschlegel JA, Leffak M, Yates J, Dutta A (2001) Replication from oriP of Epstein-Barr virus requires human ORC and is inhibited by geminin. Cell 106:287–296.

Dou Y, Milne TA, Tackett AJ, Smith ER, Fukuda A, Wysocka J, Allis CD, Chait BT, Hess JL, Roeder RG (2005) Physical association and coordinate function of the H3 K4 methyltransferase MLL1 and the H4 K16 acetyltransferase MOF. Cell 121:873–885

Downs JA, Allard S, Jobin-Robitaille O, Javaheri A, Auger A, Bouchard N, Kron SJ, Jackson SP, Cote J (2004) Binding of chromatin-modifying activities to phosphorylated histone H2A at DNA damage sites. Mol Cell 16:979–990

Doyon Y, Cayrou C, Ullah M, Landry AJ, Cote V, Selleck W, Lane WS, Tan S, Yang XJ, Cote J (2006) ING tumor suppressor proteins are critical regulators of chromatin acetylation required for genome expression and perpetuation. Mol Cell 21:51–64

Doyon Y, Cote J (2004) The highly conserved and multifunctional NuA4 HAT complex. Curr Opin Genet Dev 14:147–154

Doyon Y, Selleck W, Lane WS, Tan S, Cote J (2004) Structural and functional conservation of the NuA4 histone acetyltransferase complex from yeast to humans. Mol Cell Biol 24:1884–1896

Eymin B, Claverie P, Salon C, Leduc C, Col E, Brambilla E, Khochbin S, Gazzeri S (2006) p14ARF Activates a Tip60-Dependent and p53-Independent ATM/ATR/CHK Pathway in Response to Genotoxic Stress. Mol Cell Biol 26:4339–4350

Fischer U, Heckel D, Michel A, Janka M, Hulsebos T, Meese E (1997) Cloning of a novel transcription factor-like gene amplified in human glioma including astrocytoma grade I. Hum Mol Genet 6:1817–1822

Fraga MF, Ballestar E, Villar-Garea A, Boix-Chornet M, Espada J, Schotta G, Bonaldi T, Haydon C, Ropero S, Petrie K et al (2005) Loss of acetylation at Lys16 and trimethylation at Lys20 of histone H4 is a common hallmark of human cancer. Nat Genet 37:391–400

Frank SR, Parisi T, Taubert S, Fernandez P, Fuchs M, Chan HM, Livingston DM, Amati B (2003) MYC recruits the TIP60 histone acetyltransferase complex to chromatin. EMBO Rep 4:575–580

Fuchs M, Gerber J, Drapkin R, Sif S, Ikura T, Ogryzko V, Lane WS, Nakatani Y, Livingston DM (2001) The p400 complex is an essential E1A transformation target. Cell 106:297–307

Garkavtsev I, Kozin SV, Chernova O, Xu L, Winkler F, Brown E, Barnett GH, Jain RK (2004) The candidate tumour suppressor protein ING4 regulates brain tumour growth and angiogenesis. Nature 428:328–332

Gaughan L, Brady ME, Cook S, Neal DE, Robson CN (2001) Tip60 is a co-activator specific for class I nuclear hormone receptors. J Biol Chem 276:46841–46848

Gaughan L, Logan IR, Cook S, Neal DE, Robson CN (2002) Tip60 and histone deacetylase 1 regulate androgen receptor activity through changes to the acetylation status of the receptor. J Biol Chem 277:25904–25913

Georgiakaki M, Chabbert-Buffet N, Dasen B, Meduri G, Wenk S, Rajhi L, Amazit L, Chauchereau A, Burger CW, Blok LJ *et al* (2006) Ligand-controlled Interaction of HBO1 with the N-terminal Transactivating Domain of Progesterone Receptor Induces SRC-1-dependent Co-activation of Transcription. Mol Endocrinol 20:2122–2140

Grienenberger A, Miotto B, Sagnier T, Cavalli G, Schramke V, Geli V, Mariol MC, Berenger H, Graba Y, Pradel J (2002) The MYST domain acetyltransferase Chameau functions in epigenetic mechanisms of transcriptional repression. Curr Biol 12:762–766

Grier DG, Thompson A, Kwasniewska A, McGonigle GJ, Halliday HL, Lappin TR (2005) The pathophysiology of HOX genes and their role in cancer. J Pathol 205:154–171

Gunduz M, Ouchida M, Fukushima K, Ito S, Jitsumori Y, Nakashima T, Nagai N, Nishizaki K, Shimizu K (2002) Allelic loss and reduced expression of the ING3, a candidate tumor suppressor gene at 7q31, in human head and neck cancers. Oncogene 21:4462–4470

Gupta A, Sharma GG, Young CS, Agarwal M, Smith ER, Paull TT, Lucchesi JC, Khanna KK, Ludwig T, Pandita TK (2005) Involvement of human MOF in ATM function. Mol Cell Biol 25:5292–5305

Halkidou K, Gnanapragasam VJ, Mehta PB, Logan IR, Brady ME, Cook S, Leung HY, Neal DE, Robson CN (2003) Expression of Tip60, an androgen receptor coactivator, and its role in prostate cancer development. Oncogene 22:2466–2477

Hilfiker A, Hilfiker-Kleiner D, Pannuti A, Lucchesi JC (1997) mof, a putative acetyl transferase gene related to the Tip60 and MOZ human genes and to the SAS genes of yeast, is required for dosage compensation in Drosophila. Embo J 16:2054–2060

Howe L, Brown CE, Lechner T, Workman JL (1999) Histone acetyltransferase complexes and their link to transcription. Crit Rev Eukaryot Gene Expr 9:231–243

Huntly BJ, Shigematsu H, Deguchi K, Lee BH, Mizuno S, Duclos N, Rowan R, Amaral S, Curley D, Williams IR et al (2004) MOZ-TIF2, but not BCR-ABL, confers properties of leukemic stem cells to committed murine hematopoietic progenitors. Cancer Cell 6:587–596

Iizuka M, Matsui T, Takisawa H, Smith MM (2006) Regulation of replication licensing by acetyltransferase Hbo1. Mol Cell Biol 26:1098–1108

Iizuka M, Stillman B (1999) Histone acetyltransferase HBO1 interacts with the ORC1 subunit of the human initiator protein. J Biol Chem 274:23027–23034

Ikura T, Ogryzko VV, Grigoriev M, Groisman R, Wang J, Horikoshi M, Scully R, Qin J, Nakatani Y (2000) Involvement of the TIP60 histone acetylase complex in DNA repair and apoptosis. Cell 102:463–473

Jenuwein T, Allis CD (2001) Translating the histone code. Science 293:1074–1080

Jepsen K, Rosenfeld MG (2002) Biological roles and mechanistic actions of co-repressor complexes. J Cell Sci 115:689–698

Jiang X, Sun Y, Chen S, Roy K, Price BD (2006) The FATC domains of PIKK proteins are functionally equivalent and participate in the Tip60-dependent activation of DNA-PKcs and ATM. J Biol Chem 281:15741–15746

Kamine J, Elangovan B, Subramanian T, Coleman D, Chinnadurai G (1996) Identification of a cellular protein that specifically interacts with the essential cysteine region of the HIV-1 Tat transactivator. Virology 216:357–366

Katsumoto T, Aikawa Y, Iwama A, Ueda S, Ichikawa H, Ochiya T, Kitabayashi I (2006) MOZ is essential for maintenance of hematopoietic stem cells. Genes Dev 20:1321–1330

Keogh MC, Kim JA, Downey M, Fillingham J, Chowdhury D, Harrison JC, Onishi M, Datta N, Galicia S, Emili A et al (2006) A phosphatase complex that dephosphorylates gammaH2AX regulates DNA damage checkpoint recovery. Nature 439:497–501

Kim JH, Kim B, Cai L, Choi HJ, Ohgi KA, Tran C, Chen C, Chung CH, Huber O, Rose DW et al (2005) Transcriptional regulation of a metastasis suppressor gene by Tip60 and beta-catenin complexes. Nature 434:921–926

Kim S, Chin K, Gray JW, Bishop JM (2004) A screen for genes that suppress loss of contact inhibition: identification of ING4 as a candidate tumor suppressor gene in human cancer. Proc Natl Acad Sci USA 101:16251–16256

Kindle KB, Troke PJ, Collins HM, Matsuda S, Bossi D, Bellodi C, Kalkhoven E, Salomoni P, Pelicci PG, MinucciS, Heery DM (2005) MOZ-TIF2 inhibits transcription by nuclear receptors and p53 by impairment of CBP function. Mol Cell Biol 25:988–1002

Kinoshita A, Whelan CM, Berezovska O, Hyman BT (2002) The gamma secretase-generated carboxyl-terminal domain of the amyloid precursor protein induces apoptosis via Tip60 in H4 cells. J Biol Chem 277:28530–28536

Kitabayashi I, Aikawa Y, Nguyen LA, Yokoyama A, Ohki M (2001a) Activation of AML1-mediated transcription by MOZ and inhibition by the MOZ-CBP fusion protein. Embo J 20:7184–7196

Kitabayashi I, Aikawa Y, Yokoyama A, Hosoda F, Nagai M, Kakazu N, Abe T, Ohki M (2001b) Fusion of MOZ and p300 histone acetyltransferases in acute monocytic leukemia with a t(8;22)(p11;q13) chromosome translocation. Leukemia 15:89–94

Knoops L, Renauld JC (2004) IL-9 and its receptor: from signal transduction to tumorigenesis. Growth Factors 22:207–215

Kojima K, Kaneda K, Yoshida C, Dansako H, Fujii N, Yano T, Shinagawa K, Yasukawa M, Fujita S, Tanimoto M (2003) A novel fusion variant of the MORF and CBP genes detected in therapy-related myelodysplastic syndrome with t(10;16)(q22;p13). Br J Haematol 120:271–273

Kusch T, Florens L, Macdonald WH, Swanson SK, Glaser RL, Yates JR III, Abmayr SM, Washburn MP, Workman JL (2004) Acetylation by Tip60 is required for selective histone variant exchange at DNA lesions. Science 306:2084–2087

Lang SE, Hearing P (2003) The adenovirus E1A oncoprotein recruits the cellular TRRAP/GCN5 histone acetyltransferase complex. Oncogene 22:2836–2841

Leduc C, Claverie P, Eymin B, Col E, Khochbin S, Brambilla E, Gazzeri S (2006) p14(ARF) promotes RB accumulation through inhibition of its Tip60-dependent acetylation. Oncogene 25:4147–4154

Legube G, Linares LK, Lemercier C, Scheffner M, Khochbin S, Trouche D (2002) Tip60 is targeted to proteasome-mediated degradation by Mdm2 and accumulates after UV irradiation. Embo J 21:1704–1712

Legube G, Linares LK, Tyteca S, Caron C, Scheffner M, Chevillard-Briet M, Trouche D (2004) Role of the histone acetyl transferase Tip60 in the p53 pathway. J Biol Chem 279:44825–44833

Linggi BE, Brandt SJ, Sun ZW, Hiebert SW (2005) Translating the histone code into leukemia. J Cell Biochem 96:938–950

Lund AH, Turner G, Trubetskoy A, Verhoeven E, Wientjens E, Hulsman D, Russell R, DePinho RA, Lenz J, van Lohuizen M (2002) Genome-wide retroviral insertional tagging of genes involved in cancer in Cdkn2a-deficient mice. Nat Genet 32:160–165

Micci F, Panagopoulos I, Bjerkehagen B, Heim S (2006) Consistent rearrangement of chromosomal band 6p21 with generation of fusion genes JAZF1/PHF1 and EPC1/PHF1 in endometrial stromal sarcoma. Cancer Res 66:107–112

Miotto B, Sagnier T, Berenger H, Bohmann D, Pradel J, Graba Y (2006) Chameau HAT and DRpd3 HDAC function as antagonistic cofactors of JNK/AP-1-dependent transcription during Drosophila metamorphosis. Genes Dev 20:101–112

Monden T, Wondisford FE, Hollenberg AN (1997) Isolation and characterization of a novel ligand-dependent thyroid hormone receptor-coactivating protein. J Biol Chem 272:29834–29841

Moore SD, Herrick SR, Ince TA, Kleinman MS, Cin PD, Morton CC, Quade BJ (2004) Uterine leiomyomata with t(10;17) disrupt the histone acetyltransferase MORF. Cancer Res 64:5570–5577

Murr R, Loizou JI, Yang YG, Cuenin C, Li H, Wang ZQ, Herceg Z (2006) Histone acetylation by Trrap-Tip60 modulates loading of repair proteins and repair of DNA double-strand breaks. Nat Cell Biol 8:91–99

Nagashima M, Shiseki M, Pedeux RM, Okamura S, Kitahama-Shiseki M, Miura K, Yokota J, Harris CC (2003) A novel PHD-finger motif protein, p47ING3, modulates p53-mediated transcription, cell cycle control, and apoptosis. Oncogene 22:343–350

Nesbit CE, Tersak JM, Prochownik EV (1999) MYC oncogenes and human neoplastic disease. Oncogene 18:3004–3016

Nikiforov MA, Chandriani S, Park J, Kotenko I, Matheos D, Johnsson A, McMahon SB, Cole MD (2002) TRRAP-dependent and TRRAP-independent transcriptional activation by Myc family oncoproteins. Mol Cell Biol 22:5054–5063

Nourani A, Howe L, Pray-Grant MG, Workman JL, Grant PA, Cote J (2003) Opposite role of yeast ING family members in p53-dependent transcriptional activation. J Biol Chem 278:19171–19175

O'Connell S, Wang L, Robert S, Jones CA, Saint R, Jones RS (2001) Polycomblike PHD fingers mediate conserved interaction with enhancer of zeste protein. J Biol Chem 276:43065–43073

Panagopoulos I, Fioretos T, Isaksson M, Mitelman F, Johansson B, Theorin N, Juliusson G (2002) RT-PCR analysis of acute myeloid leukemia with t(8;16)(p11;p13): identification of a novel MOZ/CBP transcript and absence of CBP/MOZ expression. Genes Chromosomes Cancer 35:372–374

Panagopoulos I, Fioretos T, Isaksson M, Samuelsson U, Billstrom R, Strombeck B, Mitelman F, Johansson B (2001) Fusion of the MORF and CBP genes in acute myeloid leukemia with the t(10;16)(q22;p13). Hum Mol Genet 10:395–404

Panagopoulos I, Isaksson M, Lindvall C, Hagemeijer A, Mitelman F, Johansson B (2003) Genomic characterization of MOZ/CBP and CBP/MOZ chimeras in acute myeloid leukemia suggests the involvement of a damage-repair mechanism in the origin of the t(8;16)(p11;p13). Genes Chromosomes Cancer 36:90–98

Park J, Wood M A, Cole MD (2002) BAF53 forms distinct nuclear complexes and functions as a critical c-Myc-interacting nuclear cofactor for oncogenic transformation. Mol Cell Biol 22:1307–1316

Park JH, Roeder RG (2006) GAS41 Is Required for Repression of the p53 Tumor Suppressor Pathway during Normal Cellular Proliferation. Mol Cell Biol 26:4006–4016

Pelletier N, Champagne N, Stifani S, Yang XJ (2002) MOZ and MORF histone acetyltransferases interact with the Runt-domain transcription factor Runx2. Oncogene 21:2729–2740

Radhakrishnan SK, Kamalakaran S (2006) Pro-apoptotic role of NF-kappaB: Implications for cancer therapy. Biochim Biophys Acta 1766:53–62

Reifsnyder C, Lowell J, Clarke A, Pillus L (1996) Yeast SAS silencing genes and human genes associated with AML and HIV-1 Tat interactions are homologous with acetyltransferases. Nat Genet 14:42–49

Renauld JC (2001) New insights into the role of cytokines in asthma. J Clin Pathol 54:577–589

Rogakou EP, Boon C, Redon C, Bonner WM (1999) Megabase chromatin domains involved in DNA double-strand breaks in vivo. J Cell Biol 146:905–916

Rozman M, Camos M, Colomer D, Villamor N, Esteve J, Costa D, Carrio A, Aymerich M, Aguilar JL, Domingo A et al (2004) Type I MOZ/CBP (MYST3/CREBBP) is the most common chimeric transcript in acute myeloid leukemia with t(8;16)(p11;p13) translocation. Genes Chromosomes Cancer 40:140–145

Russell M, Berardi P, Gong W, Riabowol K (2006) Grow-ING, Age-ING and Die-ING: ING proteins link cancer, senescence and apoptosis. Exp Cell Res 312:951–961

Sancar A, Lindsey-Boltz LA, Unsal-Kacmaz K, Linn S (2004) Molecular mechanisms of mammalian DNA repair and the DNA damage checkpoints. Annu Rev Biochem 73:39–85

Scott EK, Lee T, Luo L (2001) enok encodes a Drosophila putative histone acetyltransferase required for mushroom body neuroblast proliferation. Curr Biol 11:99–104

Selleck W, Fortin I, Sermwittayawong D, Cote J, Tan S (2005) The Saccharomyces cerevisiae Piccolo NuA4 histone acetyltransferase complex requires the Enhancer of Polycomb A domain and chromodomain to acetylate nucleosomes. Mol Cell Biol 25:5535–5542

Sharma M, Zarnegar M, Li X, Lim B, Sun Z (2000) Androgen receptor interacts with a novel MYST protein, HBO1. J Biol Chem 275:35200–35208

Sheridan AM, Force T, Yoon HJ, O'Leary E, Choukroun G, Taheri MR, Bonventre JV (2001) PLIP, a novel splice variant of Tip60, interacts with group IV cytosolic phospholipase A(2), induces apoptosis, and potentiates prostaglandin production. Mol Cell Biol 21:4470–4481

Shi X, Gozani O (2005) The fellowships of the INGs. J Cell Biochem 96:1127–1136

Shi X, Hong T, Walter KL, Ewalt M, Michishita E, Hung T, Carney D, Pena P, Lan F, Kaadige MR, Lacoste N, Cayrou C, Davrazou F, Saha A, Cairns BR, Ayer DE, Kutateladze TG, Shi Y, Cote J, Chua KF, Gozani O (2006) ING2 PHD domain links histone H3 lysine 4 methylation to active gene repression. Nature. 442:96–99

Shia WJ, Pattenden SG, Workman JL (2006) Histone H4 lysine 16 acetylation breaks the genome's silence. Genome Biol 7:217

Shiloh Y (2003) ATM and related protein kinases: safeguarding genome integrity. Nat Rev Cancer 3:155–168

Shiseki M, Nagashima M, Pedeux RM, Kitahama-Shiseki M, Miura K, Okamura S, Onogi H, Higashimoto Y, Appella E, Yokota J, Harris CC (2003) p29ING4 and p28ING5 bind to p53 and p300, and enhance p53 activity. Cancer Res 63:2373–2378

Shogren-Knaak M, Ishii H, Sun JM, Pazin MJ, Davie JR, Peterson CL (2006) Histone H4-K16 acetylation controls chromatin structure and protein interactions. Science 311:844–847

Shroff R, Arbel-Eden A, Pilch D, Ira G, Bonner WM, Petrini JH, Haber JE, Lichten M (2004) Distribution and dynamics of chromatin modification induced by a defined DNA double-strand break. Curr Biol 14:1703–1711

Sierra J, Yoshida T, Joazeiro CA, Jones KA (2006) The APC tumor suppressor counteracts beta-catenin activation and H3K4 methylation at Wnt target genes. Genes Dev 20:586–600

Sliva D, Zhu YX, Tsai S, Kamine J, Yang YC (1999) Tip60 interacts with human interleukin-9 receptor alpha-chain. Biochem Biophys Res Commun 263:149–155

Smith ER, Cayrou C, Huang R, Lane WS, Cote J, Lucchesi JC (2005) A human protein complex homologous to the Drosophila MSL complex is responsible for the majority of histone H4 acetylation at lysine 16. Mol Cell Biol 25:9175–9188

Smith ER, Eisen A, Gu W, Sattah M, Pannuti A, Zhou J, Cook RG, Lucchesi JC, Allis CD (1998) ESA1 is a histone acetyltransferase that is essential for growth in yeast. Proc Natl Acad Sci USA 95:3561–3565

Stedman W, Deng Z, Lu F, Lieberman PM (2004) ORC, MCM, and histone hyperacetylation at the Kaposi's sarcoma-associated herpesvirus latent replication origin. J Virol 78:12566–12575

Steiner H, Haass C (2001) Nuclear signaling: a common function of presenilin substrates? J Mol Neurosci 17:193–198

Sterner DE, Berger SL (2000) Acetylation of histones and transcription-related factors. Microbiol Mol Biol Rev 64:435–459

Sun Y, Jiang X, Chen S, Fernandes N, Price BD (2005) A role for the Tip60 histone acetyltransferase in the acetylation and activation of ATM. Proc Natl Acad Sci USA 102:13182–13187

Surapureddi S, Yu S, Bu H, Hashimoto T, Yeldandi AV, Kashireddy P, Cherkaoui-Malki M, Qi C, Zhu YJ, Rao MS, Reddy JK (2002) Identification of a transcriptionally active peroxisome proliferator-activated receptor alpha -interacting cofactor complex in rat liver and characterization of PRIC285 as a coactivator. Proc Natl Acad Sci USA 99:11836–11841

Suzuki T, Shen H, Akagi K, Morse HC, Malley JD, Naiman DQ, Jenkins NA, Copeland NG (2002) New genes involved in cancer identified by retroviral tagging. Nat Genet 32:166–174

Taipale M, Rea S, Richter K, Vilar A, Lichter P, Imhof A, Akhtar A (2005) hMOF histone acetyltransferase is required for histone H4 lysine 16 acetylation in mammalian cells. Mol Cell Biol 25:6798–6810

Taubert S, Gorrini C, Frank SR, Parisi T, Fuchs M, Chan HM, Livingston DM, Amati B (2004) E2F-dependent histone acetylation and recruitment of the Tip60 acetyltransferase complex to chromatin in late G1. Mol Cell Biol 24:4546–4556

Tedeschi FA, Zalazar FE (2006) HOXA9 gene expression in the chronic myeloid leukemia progression. Leuk Res 30:1453–1456

Thomas T, Corcoran LM, Gugasyan R, Dixon MP, Brodnicki T, Nutt SL, Metcalf D, Voss AK (2006) Monocytic leukemia zinc finger protein is essential for the development of long-term reconstituting hematopoietic stem cells. Genes Dev 20:1175–1186

Thomas T, Voss AK, Chowdhury K, Gruss P (2000) Querkopf, a MYST family histone acetyltransferase, is required for normal cerebral cortex development. Development 127:2537–2548

Utley RT, Côte J (2003) The MYST family of histone acetyltransferases. Curr Top Microbiol Immunol 274:203–236

Wang Y, Li G (2006) ING3 Promotes UV-induced Apoptosis via Fas/Caspase-8 Pathway in Melanoma Cells. J Biol Chem 281:11887–11893

Westendorp MO, Shatrov VA, Schulze-Osthoff K, Frank R, Kraft M, Los M, Krammer PH, Droge W, Lehmann V (1995) HIV-1 Tat potentiates TNF-induced NF-kappa B activation and cytotoxicity by altering the cellular redox state. Embo J 14:546–554

Yan Y, Barlev NA, Haley RH, Berger SL, Marmorstein R (2000) Crystal structure of yeast Esa1 suggests a unified mechanism for catalysis and substrate binding by histone acetyltransferases. Mol Cell 6:1195–1205

Yang XJ (2004) The diverse superfamily of lysine acetyltransferases and their roles in leukemia and other diseases. Nucleic Acids Res 32:959–976

Zhou MI, Foy RL, Chitalia VC, Zhao J, Panchenko MV, Wang H, Cohen HT (2005) Jade-1, a candidate renal tumor suppressor that promotes apoptosis. Proc Natl Acad Sci USA 102:11035–11040

CHAPTER 14

ROLE OF HISTONE PHOSPHORYLATION IN CHROMATIN DYNAMICS AND ITS IMPLICATIONS IN DISEASES

MASAYA OKI, HITOSHI AIHARA AND TAKASHI ITO*

Department of Biochemistry, Nagasaki University School of Medicine, 1-12-4 Sakamoto, Nagasaki, Nagasaki 852-8523, Japan

Abstract: In eukaryotic cells, relaxed interphase chromatin undergoes pronounced changes resulting in formation of highly condensed mitotic chromosomes. Moreover, chromatin condensation is particularly evident during mitosis and apoptotic cell death, whereas chromatin relaxation is necessary for replication, repair, recombination and transcription. The post-translational modifications of histone tails such as reversible acetylation, phosphorylation and methylation play a critical role in dynamic condensation/relaxation that occurs during the cell cycle. Histone phosphorylation is believed to play a direct role in mitosis, cell death, repair, replication and recombination. However, definitive roles for this modification in these processes have not yet been elucidated. In this review, we discuss recent progress in studies of histone phosphorylation

1. INTRODUCTION

DNA in eukaryotic nuclei is highly compacted by association with two molecules each of the core histone proteins H2A, H2B, H3 and H4 and one molecule of linker histone H1 in a chromatin complex (for review see (Ito *et al.*, 1997; Ito, 2003). The linker histone H1 is believed to play important roles in the stabilization of higher-order chromatin structure. The phosphorylation of histone H1 increases during the cell cycle (Ord and Stocken, 1968), however, this function is poorly understood. CDKs (cyclin dependent kinases) are involved in cell cycle dependent phosphorylation of histone H1 (Swank *et al.*, 1997) and dysregulation of cell cycle control such as in cancer, resulting in changes to histone phosphorylation. Posphorylation of Histone H3 at Ser 10 during interphase correlates with chromatin relaxation and gene expression, whereas in mitosis it correlates with chromosome condensation. Thus, one of the roles of H3 phosphorylation at Ser 10 is to stimulate

*E-mail: tito@net.nagasaki-u.ac.jp

T.K. Kundu and D. Dasgupta (eds.), Chromatin and Disease, 319–336.

transcription and multiple examples of its biological importance are shown by the finding that H3 at ser 10 is phosphorylated by several signaling activated kinases, such as RSK2, MSK1/2, PKA, Aurora kinase, Nima kinase and IKKα (Prigent and Dimitrov, 2003). Some of these kinases are associated with human diseases. For example, RSK2 is associated with Coffin-Lowry syndrome (Young, 1988) and Aurora B kinase is associated with colorectal cancer (Ota *et al.*, 2002). In addition to H3 phosphorylation, it has been found that phosphorylation of H2A correlates with mitotic chromosome condensation, and mammalian histone variant H2AX is rapidly phosphorylated at ser 139 upon exposure to DNA damaging agents (Thiriet and Hayes, 2005). These data suggest that phosphorylation mediates an alteration of chromatin structure, facilitates DNA repair and maintains genome integrity. It is also suggested that H2AX functions as a tumor suppressor gene, playing an important role in cancer. A defective DNA damage response causes a variety of diseases, such as Seckel syndrome, ataxia telangiectasia (AT), Nijmegen breakage syndrome, Bloom's syndrome, ataxia-telangiectasia-like disease (ATLD), radiosensitive sever combined immunodeficiency (RS-SCID), LIG4 syndrome, Xeroderma pigmentosum (XP), Trichothiodystrophy (TTD) and Cockayne syndrome (CS) (O'Driscoll *et al.*, 2004). In Drosophila embryos, it is found that NHK-1 (Nucleosomal Histone Kinase) phosphorylates Thr 119 of H2A (Aihara *et al.*, 2004). Its phosphorylation site is related to cell cycle progression and chromosome dynamics. In this review, we focus on the importance of histone phosphorylation in biological events.

2. HISTONE H1 PHOSPHORYLATION

CDK1 is the major kinase that phosphorylates histone H1, while numerous data suggest that CDK2 is another histone H1 kinase (Langan *et al.*, 1989; Roth *et al.*, 1991; Bhattacharjee *et al.*, 2001). CDK2 complex cannot phosphorylate a mutant histone H1, in which the CDK phosphorylation sites have been changed to alanine (Threonine 18, Threonine 146, Threonine 154, Serine 172, Serine 187) (Contreras *et al.*, 2003) (Fig. 1). In FRAP (Fluorescence recovery after photobleaching) experiments (Lever *et al.*, 2000) (Misteli *et al.*, 2000), the mobility of a GFP fused mutant histone H1 is decreased compared to that of a GFP fused wild type histone H1 and this recovery correlated with CDK2 activity (Contreras *et al.*, 2003). Furthermore histone H1 phosphorylation regulates ATP-dependent chromatin remodeling enzyme (Dou *et al.*, 2002; Horn *et al.*, 2002) and transcription of some specific genes, including oncogenes (Dou *et al.*, 1999; Herrera *et al.*, 1996; Chadee *et al.*, 1995; Taylor *et al.*, 1995; Chadee *et al.*, 1997, 2002). These data

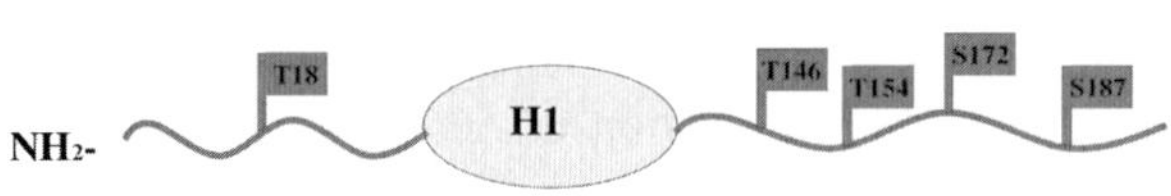

Figure 1. Linker Histone H1 CDK phosphorylation site in mammalian cells. (See Colour plate 17.)

suggest that the consequence of increased histone H1 phosphorylation appears to be relaxation of chromatin structure, and regulation of cell cycle progression and transcription.

3. HISTONE H2A PHOSPHORYLATION

In Drosophila embryos, Nucleosomal Histone Kinase-1 (NHK-1) is a histone H2A kinase (Aihara *et al.*, 2004). NHK-1 has a high affinity for chromatin and phosphorylates Thr119, at the C terminus of H2A (Fig. 2) (Table 1). NHK-1 specifically phosphorylates nucleosomal H2A, but not free H2A in solution. Immunostaining of NHK-1 revealed that it travels to chromatin during mitosis and is excluded from chromatin during S phase (Aihara *et al.*, 2004). Comparison of the predicted amino acid sequence of this unique NHK-1 with known proteins revealed similarities to other known kinases, such as human Vaccinia related kinase 1 (VRK1) (Nezu *et al.*, 1997), mouse VRK1, *Xenopus* VRK1 and *Caenorhabditis elegans* VRK, with conservation of 44%, 43%, 41% and 37% of this kinase domain, respectively. In addition to a conserved kinase domain, there is a common structure, namely an acidic amino acid region between the basic amino acid regions. This basic-acidic-basic amino acid motif (BAB motif) is conserved among species.

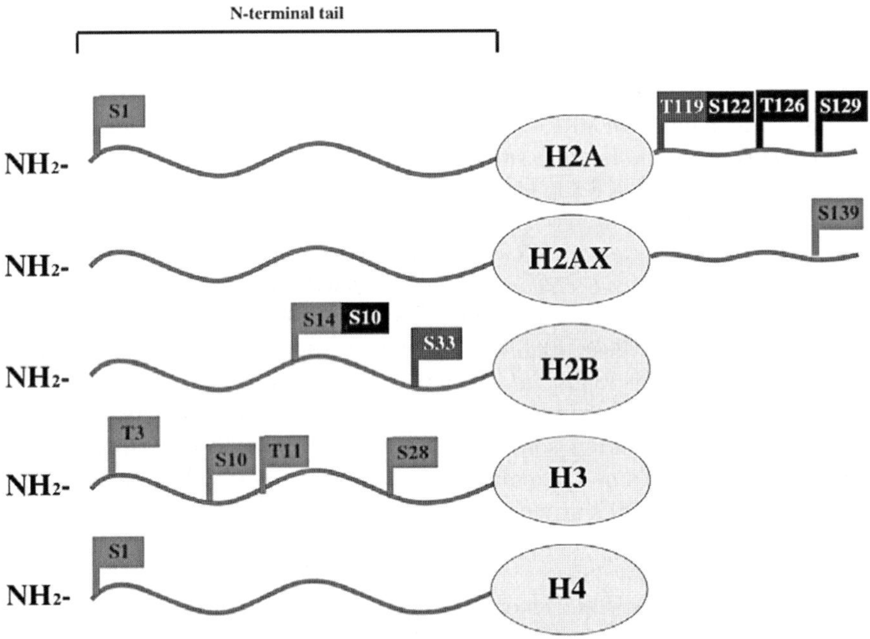

Figure 2. Current view of post-translational histone phosphorylation. Red flag : mammalian specific or common, Blue flag : *Drosophila melanogaster* specific, Black flag : *Saccharomyces cerevisiae* specific. (See Colour Plate 18.)

Table 1. Current view of histone phosphorylation sites and specific kinases

Histone	Phosphorylation site	Kinase
H2A	S1	MSK1
	T119 (D.melanogaster)	NHK-1
	S122 (S.cerevisiae)	?
	T126 (S.cerevisiae)	?
	S129 (S.cerevisiae)	Mec1p,Tel1p
H2AX	S139	ATM,ATR,DNA-PK
H2B	S10 (S.cerevisiae)	?
	S14	MST1
	S33 (D.melanogaster)	TAF1
H3	T3	?
	S10	MSK1/MSK2
		PKA
		NIMA
		Aurora B
		RSK2
		IKKα
	T11	Dlk/ZIP
	S28	Aurora B
CENP-A	S7	Aurora A/B
H3.3	S31	?
H4	S1	CK2

Concerning the functions of VRK1, it has been reported that human VRK1 phosphorylates Thr18 of p53, the binding site of mdm2 (Vega *et al.*, 2003). However, the biological role of human VRK1, like that of NHK-1, is not known (Lopez-Borges and Lazo, 2000; Barcia *et al.*, 2002; Vega *et al.*, 2004; Sevilla *et al.*, 2004a, 2004b; Santos *et al.*, 2006). *C.elegans* VRK1 was characterized in a systematic analysis using RNAi. Inactivation of *C.elegans* VRK1 by RNAi revealed embryonic lethality, with large cytoplasmic granules and failure to form a pronucleus (Piano *et al.*, 2002; Kamath *et al.*, 2003). These studies suggest that VRK has important functions for viability. In a recent report, *nhk-1* mutation leads to female sterility due to defects in the formation of the meiotic chromosomal structure, including failure to assemble a karyosome (chromosomal structure of oocyte nucleus in prophase I), a metaphase I spindle and a normal polar body (Ivanovska *et al.*, 2005). Histone H2A is phosphorylated at Thr119 in meiosis and NHK-1 is phosphorylated by itself in mitosis and meiosis (Cullen *et al.*, 2005). Phosphorylation of NHK-1 itself may be regulated by other mitotic kinases and therefore may play a part in coordinating mitotic or meiotic progression. Histone H3(K14) and histone H4(K5) are not acetylated in the *nhk-1* mutant, implying that histone H2A(T119) phosphorylation is a prerequisite for acetylation of these residues in meiosis. Histone H1 phosphorylation and histone H4 (K12) acetylation have no effect in this mutant (Ivanovska *et al.*, 2005) (Fig. 3). Concerning another phosphorylation site of histone H2A,

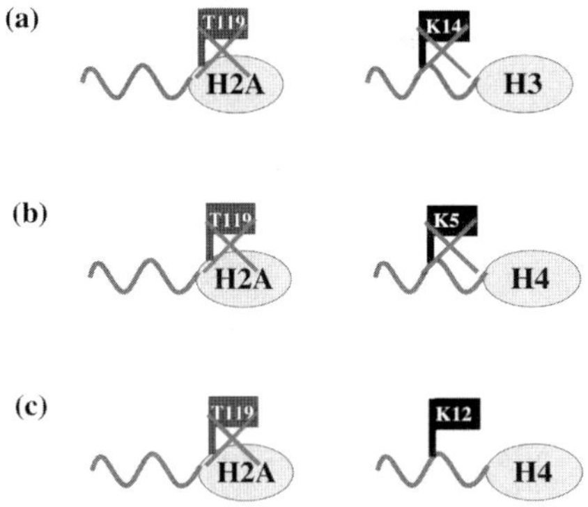

Figure 3. Histone modification cross-talk between histone H2A phosphorylation on T119 and another histones methylation or acetylation in nhk-1 mutants (a) Histone H3 (K14) and (b) Histone H4 (K5) are not acetylated in the nhk-1 mutant. (c) Histone H4 (K12) is acetylated in the nhk-1 mutant (Ivanovska *et al.* 2005). Phosphorylation is represented by the blue flag, and acetylation is represented by the black flag

Barber CM *et al.*, report that histone H2A is highly phosphorylated at Serine 1 residues during mitosis in the worm, fly, and mammalian cells (Barber *et al.*, 2004). This phosphorylation by MSK1 negatively regulated transcription on chromatin templates (Zhang *et al.*, 2004).

In *Saccharomyces cerevisiae*, multiple histone H2A phosphorylation sites have been characterized (Serine 122, Serine 129, Threonine 126) (Wyatt *et al.*, 2003; Harvey *et al.*, 2005; Redon *et al.*, 2006). Histone H2A (S129) is essential for DNA double-strand-break responses (see Section 4) and histone H2A (S122) is important for survival in the presence of DNA damage (Harvey *et al.*, 2005) (Fig. 2).

These studies demonstrate that H2A (T119) phosphorylation by NHK-1 regulates mitotic and meiotic progression. However, it remains unclear what is the biological function of other histone H2A posphorylation sites and how they impact the many other histone modifications.

4. HISTONE H2AX PHOSPHORYLATION

Histone H2AX, one of several variants of H2A is phosphorylated on Ser 139 (Rogakou *et al.*, 1998, 1999; Kurose *et al.*, 2005; Ichijima *et al.*, 2005; McManus and Hendzel, 2005) (Fig. 2). The Serine phosphorylated form of H2AX, termed γ-H2AX, occurs in response to DNA double-strand breaks for recruitment of DNA-damage-response proteins, including DNA damage checkpoint proteins (Fernandez-Capetillo *et al.*, 2002; Nakamura *et al.*, 2004) , cohesin complex (Strom *et al.*, 2004; Unal *et al.*, 2004) and chromatin remodeling protein (Morrison *et al.*, 2004; van Attikum *et al.*, 2004) to regions of damaged chromatin. The phosphatidylinositol-3-OH kinase related kinase (PIKK) family includes ataxia-telangiectasia mutated (ATM), ATM and Rad3-related (ATR) and DNA-dependent protein kinase (DNA-PK) (Smith and Jackson, 1999; Shiloh, 2001) (Table 1). These kinases are important for DNA-damage signaling (Burma *et al.*, 2001), however there is functional redundancy among the pathways such that γ-H2AX is detectable in individual kinase dead mutants (Burma *et al.*, 2001; Ward and Chen, 2001; Fernandez-Capetillo *et al.*, 2002; Brown and Baltimore, 2003; Fernandez-Capetillo *et al.*, 2003). In mice lacking functional H2AX, the efficiency of DSB repair is impeded, the DNA damage-induced checkpoint fails and numerous morphological abnormalities are observed (Bassing *et al.*, 2002; Celeste *et al.*, 2002). Recent studies have demonstrated that mammalian MDC1 directly binds to γ-H2AX and this direct interaction plays a central role in the mammalian response to DNA damage (Stucki *et al.*, 2005). Bassing *et al.*, 2003 reported that H2AX had been shown to function as a dosage-dependent suppressor of oncogenic translocations and tumors in mice, whereas in humans, H2AX maps to a cytogenetic region frequently altered in human cancers. These observations are consistent with a role for altered or mutated H2AX expression in the onset and progression of at least some human tumors. Current evidence shows that H2AX is also phosphorylated during apoptotic DNA fragmentation (Talasz *et al.*, 2002; Huang *et al.*, 2005; Kurose *et al.*, 2005; Mukherjee *et al.*, 2006).

In *S.cerevisiae*, histone H2A is more similar to mammalian H2AX than to mammalian H2A. The counterparts of ATM and ATR, Tel1 and Mec1, similarly phosphorylate histone H2A (S129) in response to DNA damage (Downs *et al.*, 2000). Furthermore, recent work has shown that the phosphorylated histone H2A (S129) interacts with the NuA4 histone acetylferase and INO80 ATP-dependent chromatin remodeling complexes (Downs *et al.*, 2004; Morrison *et al.*, 2004; van Attikum *et al.*, 2004).

We have discussed recent studies about phosphorylation of histone H2AX in mammals or H2A in yeast, and it is one of the key events in the response to DNA damage. This phosphorylation recruits cohesin complex, chromatin remodeling complex and histone acetyltranseferase to DSBs, and promotes the accumulation of checkpoint and repair proteins at these sites. However, some groups report that another complex, for example SWR1 complexes including H2AZ (another histone H2A variant), also recruit to double strand break regions. There still remains some questions about how other histone modification or other histone variants respond to DNA damage.

5. HISTONE H2B PHOSPHORYLATION

Histone H2B amino-terminal tail is essential for chromatin condensation (de la Barre *et al.*, 2001). In Xenopus, chicken, and human cells phosphorylation of H2B at Serine 14 by Mst1 (Mammalian Sterile Twenty) kinase has been linked to chromatin compaction during apoptosis (Ajiro, 2000; Cheung *et al.*, 2003), and DNA double-strand breaks (Fernandez-Capetillo *et al.*, 2004) (Fig. 2) (Table 1). At late time points after irradiation, phosphorylated H2B (S14) accumulates into irradiation-induced foci and is dependent on phosphorylated H2AX (Fernandez-Capetillo *et al.*, 2004). Recent studies show that H2B is specifically phosphorylated at Serine 10 in a hydrogen peroxide-induced cell death pathway in *S. cerevisiae.* S10A point mutants in H2B exhibit increased cell survival accompanied by a loss of both DNA fragmentation and chromatin condensation (Ahn *et al.*, 2005a). Another H2B phosphorylation at Serine 10 is a biological event that is associated with dramatic alterations in higher-order chromatin structure during meiosis (Ahn *et al.*, 2005b). Regarding another phosphorylation site of H2B, Maile T *et al.*, 2004 reported that histone H2B was phosphorylated at Serine 33 by the carboxyl-terminal kinase domain (CTK) of the Drosophila TFIID subunit TAF1 (Fig. 2) (Table 1). Histone H2B (S33) phosphorylation is essential for transcriptional activation events that promote cell cycle progression and development (Maile *et al.*, 2004).

Recent studies demonstrate that histone H2B phosphorylation also plays a key role in the response to DNA double-strand breaks, apoptosis, meiosis and transcription activation events. However, the details of this mechanism are poorly understood.

6. HISTONE H3 PHOSPHORYLATION

Phosphorylation of histone H3 at Ser 10 has been observed in different processes, including the activation of transcription and chromosome condensation during mitosis (Gurley *et al.*, 1978; Paulson and Taylor, 1982; Hendzel *et al.*, 1997; Van Hooser *et al.*, 1998, Wei *et al.*, 1998, de la Barre *et al.*, 2000, De Souza *et al.*, 2000, Hsu *et al.*, 2000) and meiosis (Kaszas and Cande, 2000) (Fig. 2). Mutation of this phosphorylation site impairs chromosome condensation and segregation *in vivo* (Wei *et al.*, 1999). Histone H3 at S10 is phosphorylated by multiple kinases, such as mitogen and stress-activated protein kinases 1 and 2 (MSK1 and MSK2) (Thomson *et al.*, 1999, Soloaga *et al.*, 2003, Lim *et al.*, 2004), cAMP-dependent protein kinase A (PKA) (Taylor, 1982, Kogel *et al.*, 1998, DeManno *et al.*, 1999, Schmitt *et al.*, 2002), NIMA kinase (De Souza *et al.*, 2000), Aurora B kinase (Hsu *et al.*, 2000, Adams *et al.*, 2001, Giet and Glover 2001, Murnion *et al.*, 2001, Petersen *et al.*, 2001, Scrittori *et al.*, 2001, Crosio *et al.*, 2002, MacCallum *et al.*, 2002, Sugiyama *et al.*, 2002, Hirota *et al.*, 2005, Fischle *et al.*, 2005), ribosomal S6 kinase 2 (RSK2) (Erikson and Maller 1985, Sassone-Corsi *et al.*, 1999) and IκB kinase α (IKKα) (Yamamoto *et al.*, 2003, Duncan *et al.*, 2006) (Table 1). MSK1/2 are downstream targets of the Ras mitogen-activated protein kinase (MAPK) signal transduction pathway. Histone H3 phosphorylation at S10 by MSK1/2 has been linked to transcriptional activation of mitogen-stimulated immediate-early response genes, such as c-fos and c-jun (Mahadevan *et al.*, 1991, Chadee *et al.*, 1999, Sassone-Corsi *et al.*, 1999, Thomson *et al.*, 1999, Dyson *et al.*, 2005) and, more recently, has been reported in conjunction with many other inducible genes and oncogenes (Clayton and Mahadevan 2003, Ge *et al.*, 2006). Recent studies have demonstrated that there exists a cross-talk between phosphorylation of histone H3 Ser 10 and acetylation or methylation of histone H3. Histone

H3 phoshorylation at S10 can enhance acetylation of histone H3 at K14 (Cheung *et al.*, 2000, Lo *et al.*, 2000), abolish acetylation of histone H3 at K9 (Edmondson *et al.*, 2002) and inhibit methylation of histone H3 at K9 (Rea *et al.*, 2000). Furthermore, methylation of histone H3 at K9 interferes with phosphorylation of histone H3 at S10 (Rea *et al.*, 2000) (Fig. 4). Further evidence about phosphorylation of histone H3 Ser 10, is that overexpression of Aurora kinase, which has been observed in many cancer cell lines, causes increased phosphorylation (Ota *et al.*, 2002). Increased phosphorylation was found during the heat shock response (Nowak and Corces, 2000) and in oncogene-transformed mouse fibroblasts (Chadee *et al.*, 1999). Phosphorylation of histone H3 at Ser 10 by AuroraB disrupts the HP1 (Heterochromatin protein 1) and methylation of histone H3 (K9) interaction, and release of HP1 from chromosomes during mitosis (Fischle *et al.*, 2005, Hirota *et al.*, 2005). These data suggest that switching of methylation and phosphorylation roles permit the progression of mitosis.

Recent studies have demonstrated other phosphorylation sites of histone H3 at Threonine 3 (Polioudaki *et al.*, 2004) and Threonine 11 (Preuss *et al.*, 2003) (Fig. 2).

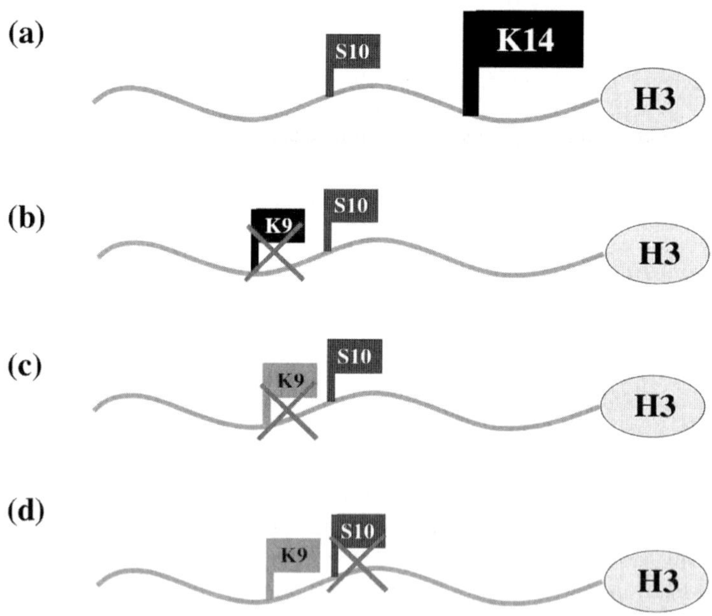

Figure 4. Histone modification cross-talk between phosphorylation of histone H3 (S10) and acetylation of histone H3 (K14) or methylation of histone H3 (K9). Histone H3 phoshorylation at S10 can enhance acetylation of histone H3 at K14 (Cheung *et al.*, 2000), (Lo *et al.*, 2000). (b) Histone H3 phoshorylation at S10 abolish acetylation of histone H3 at K9 (Edmondson *et al.*, 2002). (c) Histone H3 phoshorylation at S10 inhibits methylation of histone H3 at K9 (Rea *et al.*, 2000). (d) methylation of histone H3 at K9 interferes with phosphorylation of histone H3 at S10. Phosphorylation is represented by the blue flag, acetylation is represented by the black flag, and methylation represented by the green flag

Histone H3 (T11) phosphorylation occurs during mitosis by Dlk/ZIP kinase (Dlk: Death-associated protein (DAP)-like kinase, ZIP: Zipper interacting protein kinase) (Preuss *et al.*, 2003) (Table 1). Histone H3 at Serine 28 is phosphorylated by Aurora B kinase at mitosis and this phosphorylation coincides with chromosome condensation (Goto *et al.*, 1999, Goto *et al.*, 2002) (Fig. 2), (Table 1). Histone H3 (S28) phosphorylation initiates at prophase, whereas histone H3 (S10) phosphorylation initiates during the late G2 phase (Hendzel *et al.*, 1997).

Centromere protein A (CENP-A), one of several variants of histone H3, is phosphorylated on Ser 7 by Aurora B kinase which is equivalent to Ser 10 of histone H3 (Zeitlin *et al.*, 2001). Recent studies demonstrate that Aurora A kinase also phosphorylates CENP-A (S7) (Kunitoku *et al.*, 2003) (Table 1). The presence of CENP-A in centromeric nucleosomes is required for kinetochore organization and function (Choo 2001). Loss of CENP-A phosphorylation function at Ser 7 caused a mislocalisation of Aurora B, a putative partner phosphatase (PP1γ1) and inner centromere protein (INCENP). H3.3, another variant of histone H3 is phosphorylated on Ser 31 *in vivo* (Table 1). H3.3 (S31) is a mitosis-specific modification that is present only in late prometaphase and metaphase. Furthermore, H3.3 (S31) is excluded from centromeres. However it is enriched in distinct chromosomal areas immediately adjacent to centromeres (Hake *et al.*, 2005).

We have discussed phosphorylation of histone H3, which has been studied in many organisms. Phosphorylation of histone H3 (S10) has two opposite main functions. One is necessary to initiate chromosome condensation during mitosis and meiosis, while the other is transcriptional activation. Current evidence shows that a combination of phosphorylation of H3 (S10) and methylation of H3 (K9) or acetylation H3 (K9, K14) play important roles in these phenomena including cell cycle related chromosome dynamics and transcriptional activation. These results suggest that a combination of different histone modifications excute different biological outcomes.

7. HISTONE H4 PHOSPHORYLATION

Phosphorylation of histone H4 (S1) increases during the cell-cycle, similar to histone H2A (S1) (Ruiz-Carrillo *et al.*, 1975; Barber *et al.*, 2004) (Fig. 2) (Table 1). Directly comparing the timing of histone H3 (S10 or S28) and H2A/H4 (S1) phosphorylation during the different stages of mitosis with individual antibodies, Barber *et al.*, (Chromosoma, 2004) found that the overall timing of these modifications was similar. They also detected a lower level of histone H2A/H4 phosphorylation in early S-phase cells, possibly on newly deposited histones on replicating DNA. The phosphorylation of H2A/H4 (S1) is an evolutionarily conserved modification that may have separate roles during mitosis and S-phase (Barber *et al.*, 2004). In *S.cerevisiae*, histone H4 (S1) is also associated with DNA double-strand breaks. Casein kinase II (CK2) can phosphorylate histone H4 (S1) *in vitro* and that null or temperature-sensitive mutants are defective for induction of H4 (S1) phosphorylation upon DNA damage *in vivo* (Cheung *et al.*, 2005).

Another function of H4 (S1) regulates chromatin acetylation by the NuA4 complex and this process is important for normal gene expression and DNA repair (Utley *et al.*, 2005).

Recent studies have demonstrated that histone H4 (S1) phosphorylation is also a key role in the response to DNA double-strand breaks, cell-cycle progression and gene expression. In particular, this modification may have important roles during mitosis and S-phase-associated events in the cell-cycle and its phosphorylation found on newly synthesized histones during S-phase. However this phosphorylated residue is a novel histone modification site, and the details of this mechanism will be made evident by future experimentation.

8. HISTONE PHOSPHORYLATION AND HUMAN DISEASES

In mammalian cells, DNA double-strand breaks (DSBs) are highly toxic lesions that, if not repaired or repaired incorrectly, can cause cell death, mutations, and chromosomal translocations and can lead to cancer. Phosphorylation of H2AX (γ-H2AX) is induced in response to DSBs. A defective DNA damage response has caused some diseases, such as ataxia telangiectasia (AT) (Boder and Sedgwick, 1958), Seckel syndrome (O'Driscoll *et al.*, 2003), Nijmegen breakage syndrome (Tauchi *et al.*, 2002), Bloom's syndrome (Chester *et al.*, 1998), ataxia telangiectasia-like disease (ATLD), Radiosensitive sever combined immunodeficiency (RS-SCID), LIG4 syndrome (O'Driscoll *et al.*, 2004), Xeroderma pigmentosum (XP), Trichothiodystrophy (TTD) and Cockayne syndrome (CS)(Theron *et al.*, 2005). In humans, mutations of the ATM gene cause AT, mutations of the ATR gene cause Seckel syndrome, mutations of the NBS1 gene cause Nijmegen breakage syndrome, mutations of the BLM gene cause Bloom's syndrome, mutations of the MRE11 gene cause ATLD, mutations of the Artemis gene, specifying a protein that plays a subsidiary role in non-homologous end-joining (NHEJ) although it is not an essential component, cause RS-SCID, mutations of DNA ligase IV, an essential component of DNA NHEJ, cause LIG4 syndrome, and mutations of XPD gene cause XP, TTD and CS (Table 2). For example, Seckel syndrome patients have dramatic microcephaly and marked growth and development delay, Nijmegen breakage syndrome is a recessive genetic disorder characterized by immunodeficiency with a high frequency of malignancies, and LIG4 syndrome is associated with pancytopenia, developmental and growth delay and dysmorphic facial features.

Coffin-Lowry syndrome (CLS) is a syndromic form of X-linked mental retardation, with some diagnostic features, such as skeletal abnormalities (Hanauer and Young, 2002; Young, 1988). Trivier *et al.*, (1996) had data suggesting that CLS is caused by loss of function mutations in RSK2 (Trivier *et al.*, 1996). RSK2 protein is a member of a family of serine/threonine protein kinases that act at the distal end of the mitogen induced Ras mitogen-activated protein kinase (MAPK) signaling pathway. It can phosphorylate histone H3 (S10) (see section 6).

Table 2. Syndromes and associated mutated genes

Syndrome	Mutated gene
Ataxia telangiectasia (AT)	ATM
Seckel syndrome	ATR
Nijmegen breakage syndrome	Nbs1
Blooms syndrome	BLM
Ataxia telangiectasia like disease (ATLD)	Mre11
Radiosensitive sever combined immunodeficiency (RS-SCID)	Artemis
LIG4 syndrome	DNA ligase IV
Xeroderma pigmentosum (XP)	XPD
Trichothiodystrophy (TTD)	XPD
Cockayne syndrome	XPD

9. CONCLUSIONS AND PERSPECTIVES

Here we discuss recent progress in understanding what occur as a consequence of histone phosphorylation. Given the number of different histone tails and the number of different histone modifications, the potential complexity of regulation is immense. Strahl and Allis have proposed the histone code hypothesis (Strahl and Allis, 2000), according to which each combination of post-translational modifications on a histone tail has a specific function (Jenuwein and Allis, 2001). For example, mitotic chromatin condensation is associated with histone H3 (S10) phosphorylation, but this modification has not been observed during apoptosis-induced chromatin condensation. Both mitosis and apoptosis have chromatin condensation phenomena, however histone marks that regulate these phenomena are different. There is insufficient knowledge to address how these different histone modifications regulate chromatin condensation in the different contexts. Histone phosphorylation function is still poorly understood and there are many questions. Future experimentation will provide answers for many of these questions.

REFERENCES

Adams RR, Maiato H, Earnshaw WC, Carmena M (2001) Essential roles of Drosophila inner centromere protein (INCENP) and aurora B in histone H3 phosphorylation, metaphase chromosome alignment, kinetochore disjunction, chromosome segregation. J Cell Biol 153(4):865–880

Ahn SH, Cheung WL, Hsu JY, Diaz RL, Smith MM, Allis CD (2005a) Sterile 20 kinase phosphorylates histone H2B at serine 10 during hydrogen peroxide-induced apoptosis in S. cerevisiae. Cell 120(1):25–36

Ahn SH, Henderson KA, Keeney S, Allis CD (2005b) H2B (Ser10) phosphorylation is induced during apoptosis and meiosis in S. cerevisiae. Cell Cycle 4(6):780–783

Aihara H, Nakagawa T, Yasui K, Ohta T, Hirose S, Dhomae N, Takio K, Kaneko M, Takeshima Y, Muramatsu M, Ito T (2004) Nucleosomal histone kinase-1 phosphorylates H2A Thr 119 during mitosis in the early Drosophila embryo. Genes Dev 18(8):877–888

Ajiro K (2000) Histone H2B phosphorylation in mammalian apoptotic cells. An association with DNA fragmentation. J Biol Chem 275(1):439–443

Barber CM, Turner FB, Wang Y, Hagstrom K, Taverna SD, Mollah S, Ueberheide B, Meyer BJ, Hunt DF, Cheung P, Allis CD (2004) The enhancement of histone H4 and H2A serine 1 phosphorylation during mitosis and S-phase is evolutionarily conserved. Chromosoma 112(7):360–371

Barcia R, Lopez-Borges S, Vega, FM, Lazo PA (2002) Kinetic properties of p53 phosphorylation by the human vaccinia-related kinase 1. Arch Biochem Biophys 399(1):1–5

Bassing CH, Chua KF, Sekiguchi J, Suh H, Whitlow SR, Fleming JC, Monroe BC, Ciccone DN, Yan C, Vlasakova K, Livingston DM, Ferguson DO, Scully R, Alt FW (2002) Increased ionizing radiation sensitivity and genomic instability in the absence of histone H2AX. Proc Natl Acad Sci USA 99(12):8173–8178

Bassing CH, Suh H, Ferguson DO, Chua KF, Manis J, Eckersdorff M, Gleason M, Bronson R, Lee C, Alt FW (2003) Histone H2AX: a dosage-dependent suppressor of oncogenic translocations and tumors. Cell 114(3):359–370

Bhattacharjee RN, Banks GC, Trotter KW, Lee HL, Archer TK (2001) Histone H1 phosphorylation by Cdk2 selectively modulates mouse mammary tumor virus transcription through chromatin remodeling. Mol Cell Biol 21(16):5417–5425

Boder E, Sedgwick RP (1958) Ataxia-telangiectasia; a familial syndrome of progressive cerebellar ataxia, oculocutaneous telangiectasia and frequent pulmonary infection. Pediatrics 21(4):526–554

Brown EJ, Baltimore D (2003) Essential and dispensable roles of ATR in cell cycle arrest and genome maintenance. Genes Dev 17(5):615–628

Burma S, Chen BP, Murphy M, Kurimasa A, Chen DJ (2001) ATM phosphorylates histone H2AX in response to DNA double-strand breaks. J Biol Chem 276(45):42462–42467

Celeste A, Petersen S, Romanienko PJ, Fernandez-Capetillo O, Chen HT, Sedelnikova OA, Reina-San-Martin B, Coppola V, Meffre E, Difilippantonio MJ, Redon C, Pilch DR, Olaru A, Eckhaus M, Camerini-Otero RD, Tessarollo L, Livak F, Manova K, Bonner WM, Nussenzweig MC, Nussenzweig A (2002) Genomic instability in mice lacking histone H2AX. Science 296(5569):922–927

Chadee DN, Allis CD, Wright JA, Davie JR (1997) Histone H1b phosphorylation is dependent upon ongoing transcription and replication in normal and ras-transformed mouse fibroblasts. J Biol Chem 272(13):8113–8116

Chadee DN, Hendzel MJ, Tylipski CP, Allis CD, Bazett-Jones DP, Wright JA, Davie JR (1999) Increased Ser-10 phosphorylation of histone H3 in mitogen-stimulated and oncogene-transformed mouse fibroblasts. J Biol Chem 274(35):24914–24920

Chadee DN, Peltier CP, Davie JR (2002) Histone H1(S)-3 phosphorylation in Ha-ras oncogene-transformed mouse fibroblasts. Oncogene 21(55):8397–8403

Chadee DN, Taylor WR, Hurta RA, Allis CD, Wright JA, Davie JR (1995) Increased phosphorylation of histone H1 in mouse fibroblasts transformed with oncogenes or constitutively active mitogen-activated protein kinase kinase. J Biol Chem 270(34):20098–20105

Chester N, Kuo F, Kozak C, O'Hara CD, Leder P (1998) Stage-specific apoptosis, developmental delay, and embryonic lethality in mice homozygous for a targeted disruption in the murine Bloom's syndrome gene. Genes Dev 12(21):3382–3393

Cheung P, Tanner KG, Cheung WL, Sassone-Corsi P, Denu JM, Allis CD (2000) Synergistic coupling of histone H3 phosphorylation and acetylation in response to epidermal growth factor stimulation. Mol Cell 5(6):905–915

Cheung WL, Ajiro K, Samejima K, Kloc M, Cheung P, Mizzen CA, Beeser A, Etkin LD, Chernoff J, Earnshaw WC, Allis CD (2003) Apoptotic phosphorylation of histone H2B is mediated by mammalian sterile twenty kinase. Cell 113(4):507–517

Cheung WL, Turner FB, Krishnamoorthy T, Wolner B, Ahn SH, Foley M, Dorsey JA, Peterson CL, Berger SL, Allis CD (2005) Phosphorylation of histone H4 serine 1 during DNA damage requires casein kinase II in S. cerevisiae. Curr Biol 15(7):656–660

Choo KH (2001) Domain organization at the centromere and neocentromere. *Dev* Cell 1(2):165–177

Clayton AL, Mahadevan LC (2003) MAP kinase-mediated phosphoacetylation of histone H3 and inducible gene regulation. FEBS Lett 546(1):51–58

Contreras A, Hale TK, Stenoien DL, Rosen JM, Mancini MA, Herrera RE (2003) The dynamic mobility of histone H1 is regulated by cyclin/CDK phosphorylation. Mol Cell Biol 23(23):8626–8636

Crosio C, Fimia GM, Loury R, Kimura M, Okano Y, Zhou H, Sen S, Allis CD, Sassone-Corsi P (2002) Mitotic phosphorylation of histone H3: spatio-temporal regulation by mammalian Aurora kinases. Mol Cell Biol 22(3):874–885

Cullen CF, Brittle AL, Ito T, Ohkura H (2005) The conserved kinase NHK-1 is essential for mitotic progression and unifying acentrosomal meiotic spindles in Drosophila melanogaster. J Cell Biol 171(4):593–602

de la Barre AE, Angelov D, Molla A, Dimitrov S (2001) The N-terminus of histone H2B, but not that of histone H3 or its phosphorylation, is essential for chromosome condensation. Embo J 20(22):6383–6393

de la Barre AE, Gerson V, Gout S, Creaven M, Allis CD, Dimitrov S (2000) Core histone N-termini play an essential role in mitotic chromosome condensation. Embo J 19(3):379–391

De Souza CP, Osmani AH, Wu LP, Spotts JL, Osmani SA (2000) Mitotic histone H3 phosphorylation by the NIMA kinase in Aspergillus nidulans. Cell 102(3):293–302

DeManno DA, Cottom JE, Kline MP, Peters CA, Maizels ET, Hunzicker-Dunn M (1999) Follicle-stimulating hormone promotes histone H3 phosphorylation on serine-10. Mol Endocrinol 13(1):91–105

Dou Y, Bowen J, Liu Y, Gorovsky, MA (2002) Phosphorylation and an ATP-dependent process increase the dynamic exchange of H1 in chromatin. J Cell Biol 158(7):1161–1170

Dou Y, Mizzen, CA, Abrams M, Allis CD, Gorovsky MA (1999) Phosphorylation of linker histone H1 regulates gene expression in vivo by mimicking H1 removal. Mol Cell 4(4):641–647

Downs JA, Allard S, Jobin-Robitaille O, Javaheri A, Auger A, Bouchard N, Kron SJ, Jackson SP, Cote J (2004) Binding of chromatin-modifying activities to phosphorylated histone H2A at DNA damage sites. Mol Cell 16(6):979–990

Downs JA, Lowndes NF, Jackson SP (2000) A role for Saccharomyces cerevisiae histone H2A in DNA repair. Nature 408(6815):1001–1004

Duncan EA, Anest V, Cogswell P, Baldwin AS (2006) The Kinases MSK1 and MSK2 Are Required for Epidermal Growth Factor-induced, but Not Tumor Necrosis Factor-induced, Histone H3 Ser10 Phosphorylation. J Biol Chem 281(18):12521–12525

Dyson MH, Thomson S, Inagaki M, Goto H, Arthur SJ, Nightingale K, Iborra FJ, Mahadevan LC (2005) MAP kinase-mediated phosphorylation of distinct pools of histone H3 at S10 or S28 via mitogen- and stress-activated kinase 1/2. J Cell Sci 118(Pt 10):2247–2259

Edmondson DG, Davie JK, Zhou J, Mirnikjoo B, Tatchell K, Dent SY (2002) Site-specific loss of acetylation upon phosphorylation of histone H3. *J Biol Chem* 277(33):29496–29502

Erikson E, Maller JL (1985) A protein kinase from Xenopus eggs specific for ribosomal protein S6. Proc Natl Acad Sci USA 82(3):742–746

Fernandez-Capetillo O, Allis CD, Nussenzweig A (2004) Phosphorylation of histone H2B at DNA double-strand breaks. J Exp Med 199(12):1671–1677

Fernandez-Capetillo O, Chen HT, Celeste A, Ward I, Romanienko PJ, Morales JC, Naka K, Xia Z, Camerini-Otero RD, Motoyama N, Carpenter PB, Bonner WM, Chen J, Nussenzweig A (2002) DNA damage-induced G2-M checkpoint activation by histone H2AX and 53BP1. Nat Cell Biol 4(12):993–997

Fernandez-Capetillo O, Liebe B, Scherthan H, Nussenzweig A (2003) H2AX regulates meiotic telomere clustering. J Cell Biol 163(1):15–20

Fischle W, Tseng BS, Dormann HL, Ueberheide BM, Garcia BA, Shabanowitz J, Hunt DF, Funabiki H, Allis CD (2005) Regulation of HP1-chromatin binding by histone H3 methylation and phosphorylation. Nature 438(7071):1116–1122

Ge Z, Liu C, Bjorkholm M, Gruber A, Xu D (2006) Mitogen-activated protein kinase cascade-mediated histone H3 phosphorylation is critical for telomerase reverse transcriptase expression/telomerase activation induced by proliferation. Mol Cell Biol 26(1):230–237

Giet R, Glover DM (2001) Drosophila aurora B kinase is required for histone H3 phosphorylation and condensin recruitment during chromosome condensation and to organize the central spindle during cytokinesis. J Cell Biol 152(4):669–682

Goto H, Tomono Y, Ajiro K, Kosako H, Fujita M, Sakurai M, Okawa K, Iwamatsu A, Okigaki T, Takahashi T, Inagaki M (1999) Identification of a novel phosphorylation site on histone H3 coupled with mitotic chromosome condensation. J Biol Chem 274(36):25543–25549

Goto H, Yasui Y, Nigg EA, Inagaki M (2002) Aurora-B phosphorylates histone H3 at serine28 with regard to the mitotic chromosome condensation. Genes Cells 7(1):11–17

Gurley LR, D'Anna JA, Barham SS, Deaven LL, Tobey RA (1978) Histone phosphorylation and chromatin structure during mitosis in Chinese hamster cells. Eur J Biochem 84(1):1–15

Hake SB, Garcia BA, Kauer M, Baker SP, Shabanowitz J, Hunt DF, Allis CD (2005) Serine 31 phosphorylation of histone variant H3.3 is specific to regions bordering centromeres in metaphase chromosomes. Proc Natl Acad Sci USA 102(18):6344–6349

Hanauer A, Young ID (2002) Coffin-Lowry syndrome: clinical and molecular features. J Med Genet 39(10):705–713

Harvey AC, Jackson SP, Downs JA (2005) Saccharomyces cerevisiae histone H2A Ser122 facilitates DNA repair. Genetics 170(2):543–553

Hendzel MJ, Wei Y, Mancini MA, Van Hooser A, Ranalli T, Brinkley BR, Bazett-Jones DP, Allis CD (1997) Mitosis-specific phosphorylation of histone H3 initiates primarily within pericentromeric heterochromatin during G2 and spreads in an ordered fashion coincident with mitotic chromosome condensation. Chromosoma 106(6):348–360

Herrera RE, Chen F, Weinberg RA (1996) Increased histone H1 phosphorylation and relaxed chromatin structure in Rb-deficient fibroblasts. Proc Natl Acad Sci USA 93(21):11510–11515

Hirota T, Lipp JJ, Toh BH, Peters JM (2005) Histone H3 serine 10 phosphorylation by Aurora B causes HP1 dissociation from heterochromatin. Nature 438(7071):1176–1180

Horn PJ, Carruthers LM, Logie C, Hill DA, Solomon MJ, Wade PA, Imbalzano AN, Hansen JC, Peterson CL (2002) Phosphorylation of linker histones regulates ATP-dependent chromatin remodeling enzymes. Nat Struct Biol 9(4):263–267

Hsu JY, Sun ZW, Li X, Reuben M, Tatchell K, Bishop DK, Grushcow JM, Brame CJ, Caldwell JA, Hunt DF, Lin R, Smith MM, Allis CD (2000) Mitotic phosphorylation of histone H3 is governed by Ipl1/aurora kinase and Glc7/PP1 phosphatase in budding yeast and nematodes. Cell 102(3):279–291

Huang X, Halicka HD, Traganos F, Tanaka T, Kurose A, Darzynkiewicz Z (2005) Cytometric assessment of DNA damage in relation to cell cycle phase and apoptosis. Cell Prolif 38(4):223–243

Ichijima Y, Sakasai R, Okita N, Asahina K, Mizutani S, Teraoka H (2005) Phosphorylation of histone H2AX at M phase in human cells without DNA damage response. Biochem Biophys Res Commun 336(3):807–812

Ito T (2003) Nucleosome assembly and remodeling. Curr Top Microbiol Immunol 274:1–22

Ito T, Tyler JK, Kadonaga JT (1997) Chromatin assembly factors: a dual function in nucleosome formation and mobilization? Genes Cells 2(10):593–600

Ivanovska I, Khandan T, Ito T, Orr-Weaver TL (2005) A histone code in meiosis: the histone kinase, NHK-1, is required for proper chromosomal architecture in Drosophila oocytes. Genes Dev 19(21):2571–2582

Jenuwein T, Allis CD (2001) Translating the histone code. Science 293(5532):1074–1080

Kamath RS, Fraser AG, Dong Y, Poulin G, Durbin R, Gotta M, Kanapin A, Le Bot N, Moreno S, Sohrmann M, Welchman DP, Zipperlen P, Ahringer J (2003) Systematic functional analysis of the Caenorhabditis elegans genome using RNAi. Nature 421(6920):231–237

Kaszas E, Cande WZ (2000) Phosphorylation of histone H3 is correlated with changes in the maintenance of sister chromatid cohesion during meiosis in maize, rather than the condensation of the chromatin. J Cell Sci 113 (Pt 18):3217–3226

Kogel D, Plottner O, Landsberg G, Christian S, Scheidtmann KH (1998) Cloning and characterization of Dlk, a novel serine/threonine kinase that is tightly associated with chromatin and phosphorylates core histones. Oncogene 17(20):2645–2654

Kunitoku N, Sasayama T, Marumoto T, Zhang D, Honda S, Kobayashi O, Hatakeyama K, Ushio Y, Saya H, Hirota T (2003) CENP-A phosphorylation by Aurora-A in prophase is required for enrichment of Aurora-B at inner centromeres and for kinetochore function. Dev Cell 5(6):853–864

Kurose A, Tanaka T, Huang X, Halicka HD, Traganos F, Dai W, Darzynkiewicz Z (2005) Assessment of ATM phosphorylation on Ser-1981 induced by DNA topoisomerase I and II inhibitors in relation to Ser-139-histone H2AX phosphorylation, cell cycle phase, apoptosis. Cytometry A 68(1):1–9

Langan TA, Gautier J, Lohka M, Hollingsworth R, Moreno S, Nurse P, Maller J, Sclafani RA (1989) Mammalian growth-associated H1 histone kinase: a homolog of cdc2+/CDC28 protein kinases controlling mitotic entry in yeast and frog cells. Mol Cell Biol 9(9):3860–3868

Lever MA, Th'ng JP, Sun X, Hendzel MJ (2000) Rapid exchange of histone H1.1 on chromatin in living human cells. Nature 408(6814):873–876

Lim JH, Catez F, Birger Y, West KL, Prymakowska-Bosak M, Postnikov YV, Bustin M (2004) Chromosomal protein HMGN1 modulates histone H3 phosphorylation. Mol Cell 15(4):573–584

Lo WS, Trievel RC, Rojas JR, Duggan L, Hsu JY, Allis CD, Marmorstein R, Berger SL (2000) Phosphorylation of serine 10 in histone H3 is functionally linked *in vitro* and in vivo to Gcn5-mediated acetylation at lysine 14. Mol Cell 5(6):917–926

Lopez-Borges S, Lazo PA (2000) The human vaccinia-related kinase 1 (VRK1) phosphorylates threonine-18 within the mdm-2 binding site of the p53 tumour suppressor protein. Oncogene 19(32): 3656–3664

MacCallum DE, Losada A, Kobayashi R, Hirano T (2002) ISWI remodeling complexes in Xenopus egg extracts: identification as major chromosomal components that are regulated by INCENP-aurora B. Mol Biol Cell 13(1):25–39

Mahadevan LC, Willis AC, Barratt MJ (1991) Rapid histone H3 phosphorylation in response to growth factors, phorbol esters, okadaic acid, protein synthesis inhibitors. Cell 65(5):775–783

Maile T, Kwoczynski S, Katzenberger RJ, Wassarman DA, Sauer F (2004) TAF1 activates transcription by phosphorylation of serine 33 in histone H2B. Science 304(5673):1010–1014

McManus KJ, Hendzel MJ (2005) ATM-dependent DNA damage-independent mitotic phosphorylation of H2AX in normally growing mammalian cells. Mol Biol Cell 16(10):5013–5025

Misteli T, Gunjan A, Hock R, Bustin M, Brown DT (2000) Dynamic binding of histone H1 to chromatin in living cells. Nature 408(6814):877–881

Morrison AJ, Highland J, Krogan NJ, Arbel-Eden A, Greenblatt JF, Haber JE, Shen X (2004) INO80 and gamma-H2AX interaction links ATP-dependent chromatin remodeling to DNA damage repair. Cell 119(6):767–775

Mukherjee B, Kessinger C, Kobayashi J, Chen BP, Chen DJ, Chatterjee A, Burma S (2006) DNA-PK phosphorylates histone H2AX during apoptotic DNA fragmentation in mammalian cells. DNA Repair (Amst) 5(5):575–590

Murnion ME, Adams RR, Callister DM, Allis CD, Earnshaw WC, Swedlow JR (2001) Chromatin-associated protein phosphatase 1 regulates aurora-B and histone H3 phosphorylation. J Biol Chem 276(28):26656–26665

Nakamura TM, Du LL, Redon C, Russell P (2004) Histone H2A phosphorylation controls Crb2 recruitment at DNA breaks, maintains checkpoint arrest, influences DNA repair in fission yeast. Mol Cell Biol 24(14):6215–6230

Nezu J, Oku A, Jones MH, Shimane M (1997) Identification of two novel human putative serine/threonine kinases, VRK1 and VRK2, with structural similarity to vaccinia virus B1R kinase. Genomics 45(2):327–331

Nowak SJ, Corces VG (2000) Phosphorylation of histone H3 correlates with transcriptionally active loci. Genes Dev 14(23):3003–3013

O'Driscoll M, Gennery AR, Seidel J, Concannon P, Jeggo PA (2004) An overview of three new disorders associated with genetic instability: LIG4 syndrome, RS-SCID and ATR-Seckel syndrome. DNA Repair (Amst) 3(8–9):1227–1235

O'Driscoll M, Ruiz-Perez VL, Woods CG, Jeggo PA, Goodship JA (2003) A splicing mutation affecting expression of ataxia-telangiectasia and Rad3-related protein (ATR) results in Seckel syndrome. Nat Genet 33(4):497–501

Ord MG, Stocken LA (1968) Variations in the phosphate content and thioldisulphide ratio of histones during the cell cycle. Studies with regenerating rat liver and sea urchins. Biochem J 107(3):403–410

Ota T, Suto S, Katayama H, Han ZB, Suzuki F, Maeda M, Tanino M, Terada Y, Tatsuka M (2002) Increased mitotic phosphorylation of histone H3 attributable to AIM-1/Aurora-B overexpression contributes to chromosome number instability. Cancer Res 62(18):5168–5177

Paulson JR, Taylor SS (1982) Phosphorylation of histones 1 and 3 and nonhistone high mobility group 14 by an endogenous kinase in HeLa metaphase chromosomes. J Biol Chem 257(11):6064–6072

Petersen J, Paris J, Willer M, Philippe M, Hagan IM (2001) The S. pombe aurora-related kinase Ark1 associates with mitotic structures in a stage dependent manner and is required for chromosome segregation. J Cell Sci 114(Pt 24):4371–4384

Piano F, Schetter AJ, Morton DG, Gunsalus KC, Reinke V, Kim SK, Kemphues KJ (2002) Gene clustering based on RNAi phenotypes of ovary-enriched genes in C. elegans. Curr Biol 12(22):1959–1964

Polioudaki H, Markaki Y, Kourmouli N, Dialynas G, Theodoropoulos PA, Singh PB, Georgatos SD (2004) Mitotic phosphorylation of histone H3 at threonine 3. FEBS Lett 560(1–3):39–44

Preuss U, Landsberg G, Scheidtmann KH (2003) Novel mitosis-specific phosphorylation of histone H3 at Thr11 mediated by Dlk/ZIP kinase. Nucleic Acids Res 31(3):878–885

Prigent C, Dimitrov S (2003) Phosphorylation of serine 10 in histone H3, what for? J Cell Sci 116(Pt 18):3677–3685

Rea S, Eisenhaber F, O'Carroll D, Strahl BD, Sun ZW, Schmid M, Opravil S, Mechtler K, Ponting CP, Allis CD, Jenuwein T (2000) Regulation of chromatin structure by site-specific histone H3 methyltransferases. Nature 406(6796):593–599

Redon C, Pilch DR, Bonner WM (2006) Genetic analysis of Saccharomyces cerevisiae H2A serine 129 mutant suggests a functional relationship between H2A and the sister-chromatid cohesion partners Csm3-Tof1 for the repair of topoisomerase I-induced DNA damage. Genetics 172(1):67–76

Rogakou EP, Boon C, Redon C, Bonner WM (1999) Megabase chromatin domains involved in DNA double-strand breaks in vivo. J Cell Biol 146(5):905–916

Rogakou EP, Pilch DR, Orr AH, Ivanova VS, Bonner WM (1998) DNA double-stranded breaks induce histone H2AX phosphorylation on serine 139. J Biol Chem 273(10):5858–5868

Roth SY, Collini MP, Draetta G, Beach D, Allis CD (1991) A cdc2-like kinase phosphorylates histone H1 in the amitotic macronucleus of Tetrahymena. Embo J 10(8):2069–2075

Ruiz-Carrillo A, Wangh LJ, Allfrey VG (1975) Processing of newly synthesized histone molecules. Science 190(4210):117–128

Santos CR, Rodriguez-Pinilla M, Vega FM, Rodriguez-Peralto JL, Blanco S, Sevilla A, Valbuena A, Hernandez T, van Wijnen AJ, Li F, de Alava E, Sanchez-Cespedes M, Lazo PA (2006) VRK1 signaling pathway in the context of the proliferation phenotype in head and neck squamous cell carcinoma. Mol Cancer Res 4(3):177–185

Sassone-Corsi P, Mizzen CA, Cheung P, Crosio C, Monaco L, Jacquot S, Hanauer A, Allis CD (1999) Requirement of Rsk-2 for epidermal growth factor-activated phosphorylation of histone H3. Science 285(5429):886–891

Schmitt A, Gutierrez GJ, Lenart P, Ellenberg J, Nebreda AR (2002) Histone H3 phosphorylation during Xenopus oocyte maturation: regulation by the MAP kinase/p90Rsk pathway and uncoupling from DNA condensation. FEBS Lett 518(1–3):23–28

Scrittori L, Hans F, Angelov D, Charra M, Prigent C, Dimitrov S (2001) pEg2 aurora-A kinase, histone H3 phosphorylation, and chromosome assembly in Xenopus egg extract. J Biol Chem 276(32):30002–30010

Sevilla A, Santos CR, Barcia R, Vega FM, Lazo PA (2004a) c-Jun phosphorylation by the human vaccinia-related kinase 1 (VRK1) and its cooperation with the N-terminal kinase of c-Jun (JNK). Oncogene 23(55):8950–8958

Sevilla A, Santos CR, Vega FM, Lazo PA (2004b) Human vaccinia-related kinase 1 (VRK1) activates the ATF2 transcriptional activity by novel phosphorylation on Thr-73 and Ser-62 and cooperates with JNK. J Biol Chem 279(26):27458–27465

Shiloh Y (2001) ATM and ATR: networking cellular responses to DNA damage. Curr Opin Genet Dev 11(1):71–77

Smith GC, Jackson SP (1999) The DNA-dependent protein kinase. Genes Dev 13(8):916–934

Soloaga A, Thomson S, Wiggin GR, Rampersaud N, Dyson MH, Hazzalin CA, Mahadevan LC, Arthur sJS (2003) MSK2 and MSK1 mediate the mitogen- and stress-induced phosphorylation of histone H3 and HMG-14. Embo J 22(11):2788–2797

Strahl BD, Allis CD (2000) The language of covalent histone modifications. Nature 403(6765):41–45

Strom L, Lindroos HB, Shirahige K, Sjogren C (2004) Postreplicative recruitment of cohesin to double-strand breaks is required for DNA repair. Mol Cell 16(6):1003–1015

Stucki M, Clapperton JA, Mohammad D, Yaffe MB, Smerdon SJ, Jackson SP (2005) MDC1 directly binds phosphorylated histone H2AX to regulate cellular responses to DNA double-strand breaks. Cell 123(7):1213–1226

Sugiyama K, Sugiura K, Hara T, Sugimoto K, Shima H, Honda K, Furukawa K, Yamashita S, Urano T (2002) Aurora-B associated protein phosphatases as negative regulators of kinase activation. Oncogene 21(20):3103–3111

Swank RA, Th'ng JP, Guo XW, Valdez J, Bradbury EM, Gurley LR (1997) Four distinct cyclin-dependent kinases phosphorylate histone H1 at all of its growth-related phosphorylation sites. Biochemistry 36(45):13761–13768

Talasz H, Helliger W, Sarg B, Debbage PL, Puschendorf B, Lindner H (2002) Hyperphosphorylation of histone H2A.X and dephosphorylation of histone H1 subtypes in the course of apoptosis. Cell Death Differ 9(1):27–39

Tauchi H, Matsuura S, Kobayashi J, Sakamoto S, Komatsu K (2002) Nijmegen breakage syndrome gene, NBS1, molecular links to factors for genome stability. Oncogene 21(58):8967–8980

Taylor SS (1982) The *in vitro* phosphorylation of chromatin by the catalytic subunit of cAMP-dependent protein kinase. J Biol Chem 257(11):6056–6063

Taylor WR, Chadee DN, Allis CD, Wright JA, Davie JR (1995) Fibroblasts transformed by combinations of ras, myc and mutant p53 exhibit increased phosphorylation of histone H1 that is independent of metastatic potential. FEBS Lett 377(1):51–53

Theron T, Fousteri MI, Volker M, Harries LW, Botta E, Stefanini M, Fujimoto M, Andressoo JO, Mitchell J, Jaspers NG, McDaniel LD, Mullenders LH, Lehmann AR (2005) Transcription-associated breaks in xeroderma pigmentosum group D cells from patients with combined features of xeroderma pigmentosum and Cockayne syndrome. Mol Cell Biol 25(18):8368–8378

Thiriet C, Hayes JJ (2005) Chromatin in need of a fix: phosphorylation of H2AX connects chromatin to DNA repair. Mol Cell 18(6):617–622

Thomson S, Clayton AL, Hazzalin CA, Rose S, Barratt MJ, Mahadevan LC (1999) The nucleosomal response associated with immediate-early gene induction is mediated via alternative MAP kinase cascades: MSK1 as a potential histone H3/HMG-14 kinase. Embo J 18(17):4779–4793

Trivier E, De Cesare D, Jacquot S, Pannetier S, Zackai E, Young I, Mandel JL, Sassone-Corsi P, Hanauer A (1996) Mutations in the kinase Rsk-2 associated with Coffin-Lowry syndrome. Nature 384(6609):567–570

Unal E, Arbel-Eden A, Sattler U, Shroff R, Lichten M, Haber JE, Koshland D (2004) DNA damage response pathway uses histone modification to assemble a double-strand break-specific cohesin domain. Mol Cell 16(6):991–1002

Utley RT, Lacoste N, Jobin-Robitaille O, Allard S, Cote J (2005) Regulation of NuA4 histone acetyltransferase activity in transcription and DNA repair by phosphorylation of histone H4. Mol Cell Biol 25(18):8179–8190

van Attikum H, Fritsch O, Hohn B, Gasser SM (2004) Recruitment of the INO80 complex by H2A phosphorylation links ATP-dependent chromatin remodeling with DNA double-strand break repair. Cell 119(6):777–788

Van Hooser A, Goodrich DW, Allis CD, Brinkley BR, Mancini MA (1998) Histone H3 phosphorylation is required for the initiation, but not maintenance, of mammalian chromosome condensation. J Cell Sci 111 (Pt 23):3497–3506

Vega FM, Gonzalo P, Gaspar ML, Lazo PA (2003) Expression of the VRK (vaccinia-related kinase) gene family of p53 regulators in murine hematopoietic development. FEBS Lett 544(1–3):176–180

Vega FM, Sevilla A, Lazo PA (2004) p53 Stabilization and accumulation induced by human vaccinia-related kinase 1. Mol Cell Biol 24(23):10366–10380

Ward IM, Chen J (2001) Histone H2AX is phosphorylated in an ATR-dependent manner in response to replicational stress. J Biol Chem 276(51):47759–47762

Wei Y, Mizzen CA, Cook RG, Gorovsky MA, Allis CD (1998) Phosphorylation of histone H3 at serine 10 is correlated with chromosome condensation during mitosis and meiosis in Tetrahymena. Proc Natl Acad Sci USA 95(13):7480–7484

Wei Y, Yu L, Bowen J, Gorovsky MA, Allis CD (1999) Phosphorylation of histone H3 is required for proper chromosome condensation and segregation. Cell 97(1):99–109

Wyatt HR, Liaw H, Green GR, Lustig AJ (2003) Multiple roles for Saccharomyces cerevisiae histone H2A in telomere position effect, Spt phenotypes and double-strand-break repair. Genetics 164(1):47–64

Yamamoto Y, Verma UN, Prajapati S, Kwak YT, Gaynor RB (2003) Histone H3 phosphorylation by IKK-alpha is critical for cytokine-induced gene expression. Nature 423(6940):655–659

Young ID (1988) The Coffin-Lowry syndrome. J Med Genet 25(5):344–348

Zeitlin SG, Shelby RD, Sullivan KF (2001) CENP-A is phosphorylated by Aurora B kinase and plays an unexpected role in completion of cytokinesis. J Cell Biol 155(7):1147–1157

Zhang Y, Griffin K, Mondal N, Parvin JD (2004) Phosphorylation of histone H2A inhibits transcription on chromatin templates. J Biol Chem 279(21):21866–21872

CHAPTER 15

REGULATION AND FUNCTION OF H3K9 METHYLATION

YOICHI SHINKAI*

Experimental Research Center for Infectious Diseases, Institute for Virus Research, Kyoto University, 53 Kawara-cho, Shogoin, Sakyo-ku, Kyoto, Kyoto 606-8507, Japan

Abstract: First histone lysine methyltransferase (HLMTase) was discovered in 2000. Since then, there are reports of dozens of novel HLMTases in different eukaryotes including plant, fungus, insect, nematode and vertebrate. The enzymes and their specific histone-lysine modifications have enormous impacts on the regulation of chromatin structure and function. Furthermore, various histone methyl-lysine demethylases (HLDMases) have been identified recently. In this chapter, histone H3 lysine 9 specific methyltransferases will be discussed as model enzymes involved in the regulation of chromatin function

1. INTRODUCTION

In eukaryotes, DNA is wrapped around core histones to form nucleosome particles and condensed chromatin structures with various nuclear molecules. Thus, regulation of chromatin structure and dynamics is a critical step for genomic functions. Covalent histone modifications play crucial roles in regulating these processes. Since the first report of *h*istone *l*ysine *m*ethyl*t*ransfer*ase* (HLMTase) by Thomas Jenuwein's group in 2000 (Rea *et al.*, 2000), dozens of novel HLMTases have been identified in different eukaryotes. Their roles in genomic functions have also been studied extensively (reviewed by Martin and Zhang, 2005; Lachner *et al.*, 2004; Kouzarides, 2002). The resulting data strongly suggest that histone lysine methylation has significant impacts on various chromatin-associating functions including transcriptional regulation, heterochromatin formation, DNA repair and recombination. Two major roles are proposed regarding the functional mode of histone lysine methylation. The first one suggests that histone lysine methylation recruits/regulates different biological reactions by means of a control of the protein (histone)-protein interaction in a similar fashion as the phosphorylation regulates protein function (reviewed by Martin and Zhang, 2005; Sims and Reinberg, 2006; Lachner *et al.*, 2004; Fischle *et al.*, 2003). It has been shown that each methylated lysine

* E-mail: yshinkai@virus.kyoto-u.ac.jp

T.K. Kundu and D. Dasgupta (eds.), Chromatin and Disease, 337–350.

residue of H3 and H4 is utilized differentially and may recruit different functional molecules involved in different chromatin-associating processes. The second role of histone lysine methylation is to act together with other chemical modifications of histones and thereby, regulates different chromatin functions. For example, histone H3 serine 10 (H3S10) phosphorylation coupled with the activation of early-responsible genes is suppressed by the pre-existing H3K9 methylation, which is mostly associated with transcriptional silencing (Rea *et al.*, 2000). It is obvious that HLMTases and their histone marks are key players in the histone code network. In this chapter the mechanisms of action of H3K9 methyltransferases and their functional consequences to regulate the chromatin dynamics will be discussed.

2. H3K9 HLMTase FAMILY

2.1. Su(var)3-9, Clr4, Suv39h1, 2

Pea ribulose-1,5-bisphosphate carboxylase/oxygenase *l*arge *s*ubunit N-*m*ethyl*t*ransferase Rubisco LSMT) was reported as the first protein lysine methyltransferase in 1995 (Klein and Houtz, 1995). Rubisco LSMT and Su(var)3-9 family molecules (Su(var)3-9, Clr4 and Suv39h) contain the SET-domain. Based on these informations, Su(var)3-9 family molecules have been demonstrated as the first histone lysine methyltransferases (reviewed by Jenuwein, 2006; Rea *et al.*, 2000). Since then, dozens of molecules have been described with histone lysine methyltransferase activity, and all except one molecule named Dot1/DOT1L belong to the members of the SET-domain proteins. *Su(var)3-9* was originally identified by the modifier screening of the position-effect-variegation (PEV) phenotypes in *Drosophila* (reviewed by Reuter and Spierer, 1992; Weiler and Wakimoto, 1995; Wallrath, 1998). *Clr4* was shown to be essential for silencing of centromeres and the mating-type loci in fission yeast (reviewed by Klar, 1998; Allshire, 1995). Human *SUV39H1* and murine *Suv39h1, 2* were identified as the homolog of *Su(var)3-9* (Aagaard *et al.*, 1999; O'Carroll *et al.*, 2000). Suv39h1 is ubiquitously expressed, but Suv39h2 expression is more restricted to the germ-lineage and early embryonic-stage cells. All the Su(var)3-9 family HLMTases mainly regulate H3K9 methylation on "heterochromatin".

2.2. G9a, dG9a, GLP/Eu-HMT1

G9a was originally described as a gene mapped in the human major histocompatibility complex locus (Milner and Campbell, 1993) and then characterized as the *H3K9 HLMTase* gene (Tachibana *et al.*, 2001). GLP/Eu-HMT1 is a single G9a-related molecule in mammal, exists as a heteromeric complex with G9a in a wide variety of cell types (Tachibana *et al.*, 2005). They cooperatively exert H3K9 methyltransferase function *in vivo*. In contrast to Suv39h, G9a/GLP targets H3K9 mostly on "euchromatin" (Tachibana *et al.*, 2002; Rice *et al.*, 2003;

Peters *et al.*, 2003; Tachibana *et al.*, 2005). dG9a was recently reported as Drosophila homolog of G9a (Mis *et al.*, 2006). dG9a exerts HLMTase activity against H3K9 and the *dG9a* mutation is a dominant suppressor of PEV.

2.3. SETDB1/ESET, SETDB2

SETDB1/ESET was originally reported as the gene encoding a bifurcated SET domain protein in mammal (Harte *et al.*, 1999). H3K9 methyltransferase activity of SETDB1/ESET was assessed by the characterization of the KAP-1 transcriptional co-repressor complexes containing SETDB1/ESET (Schultz *et al.*, 2002). Although SETDB1/ESET is essential for mouse early embryogenesis (Dodge *et al.*, 2004), it remains unclear how much SETDB1/ESET contributes to the H3K9 methylation *in vivo*. There is one *SETDB1*-related gene, *SETDB2* in mammal, but function of SETDB2 is not demonstrated yet.

2.4. DIM-5

DIM-5 was identified as the responsible gene for one of DNA hypomethylation mutants in *Neurospora crassa* (Tamaru and Selker, 2001). This is the first report that histone (H3K9) methylation regulates DNA methylation. Furthermore, it is reported that this regulation is mediated through the HP1 recruitment to the tri-methylated H3K9 loci (Freitag *et al.*, 2004), but it remains to be elucidated whether DNA methylation (DNA methyltransferase) is directly controlled (recruited) by HP1.

2.5. KYP/SUVH4

KRYPTONITE (KYP/SUVH4) was also identified as the responsible gene for one of DNA (non-CG) hypomethylation mutants in *Arabidopsis thaliana* (Jackson *et al.*, 2002). This is another clear example for the interaction between histone methylation and DNA methylation. But, in the case of arabidopsis, non-coding double-stranded RNA (dsRNA) also regulates H3K9 methylation (therefore controls DNA methylation) similar with that heterochromatin formation is coordinately regulated and maintained by the dsRNA/histone H3K9 methylation system in fission yeast (reviewed by Mathieu and Bender, 2004; Grewal and Rice, 2004; Verdel and Moazed, 2005). Other arabidopsis SET-domain proteins (SUVH2, 5, 6) were also shown to be involved in some of H3K9 methylation and regulation of DNA methylation (Ebbs and Bender, 2006 2005; Naumann *et al.*, 2005).

2.6. RIZ1/PRDM-2

RIZ1 was originally isolated as a molecule associating with the retinoblastoma protein, Rb (Buyse *et al.*, 1995). RIZ1 is classified as a member of the PR-domain family (PRDM-2) (the PR-domain family is now a sub-family of the SET-domain family proteins). It exerts H3K9 methyltransferase activity (Derunes *et al.*, 2005).

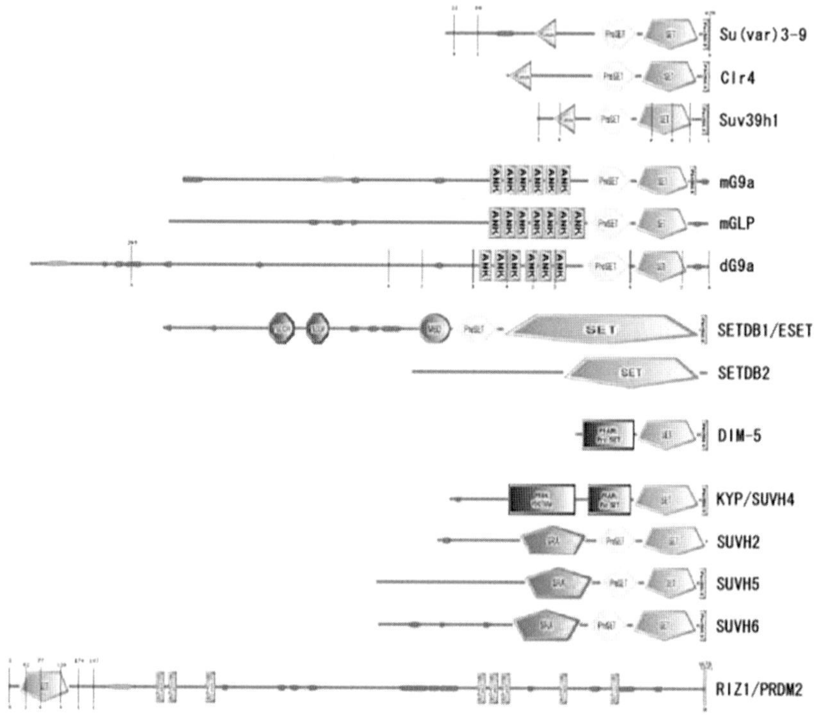

Figure 1. H3K9 HLMTases. (See Colour Plate 19.)

It is reported that various kinds of human tumor cells loose or alter the RIZ1 expression, and inactivation of mouse *Riz1* induces B cell lymphoma suggesting that *RIZ1* acts as a tumor suppressor gene (reviewed by Canote *et al.*, 2002. Structures of the described H3K9 HLMTases are shown in Fig. 1

3. FUNCTIONS OF H3K9 METHYLATION

H3K9 methylation catalyzed by the aforementioned HLMTases is mostly associated with transcriptional silencing and heterochromatin formation. For all characterized species, chromodomain of HP1 or HP1 homologues shows high affinity to the methyl-H3K9 (Nakayama *et al.*, 2001; Lachner *et al.*, 2001; Bannister *et al.*, 2001; Jacobs *et al.*, 2001; Freitag *et al..*, 2004). The recruited HP1 is involved in the heterochromatin formation and DNA methylation. Recent studies indicate that heterochromatin accumulation of mammalian HP1 is positively regulated by both methylated H3K9 and the HP1-Suv39h interaction (Stewart *et al.*, 2005). Molecular details are not clarified yet, but H3K9 methylation suppresses the H3S10 phosphorylation (Rea *et al.*, 2000). On the other hand, H3S10 phosphorylation suppresses the HP1-methylated H3K9 interaction (Fischle *et al.*, 2005; Hirota *et al.*, 2005;

Flanagan *et al.*, 2005). Although H3K9 methylation is mostly associated with transcriptional silencing and the H3K9 HLMTases are frequently seen in the transcriptional silencing complexes containing HDAC or Methyl-CpG binding proteins, some of H3K9 methylation is linked with transcriptional activation (Vakoc *et al.*, 2005; Lee *et al.*, 2006). This implies that H3K9 methylation itself is dispensable for the transcriptional regulation in the certain contexts.

4. RECRUITMENT MECHANISMS OF THE H3K9 HLMTASES

Whereas the roles of H3K9 methylation in heterochromatin formation and epigenetic gene regulation are studied extensively, it remains unclear how H3K9 HLMTases are recruited to their target chromatin loci. Given below is the current molecular mechanism regarding the recruitment of each H3K9 HLMTase.

4.1. CLR4

Elegant genetic screening in fission yeast has identified various genes involved in the heterochromatin formation. Based on these studies, it is proposed that two parallel pathways seem to exist (reviewed by Verdel and Moazed, 2005; Martienssen and Zaratiegui, 2005). One is the RNAi-mediated pathway and the DNA-binding proteins regulate the other one. It is not very clear how each component of these pathways is critical for the recruitment of Clr4 to the target heterochromatic loci. However, recent biochemical studies provide new insights into the Clr4 recruitment processes. Rik1, which is essential for the heterochromatin formation in fission yeast, was shown to associate with Clr4 (Sadaie *et al.*, 2004). Since Rik1 is a protein related to DNA damage binding protein DDB1 and possesses the WD β-propeller domain present in the RNA processing molecules, it is suggested that Rik1 may recruit Clr4 to the target heterochromatic regions though the Rik1-RNA (RNAi) interaction. Furthermore, it was reported that Clr4 also associates with the ubiquitin ligase component, Cul4, and the mutant of Cul4 lacing Nedd8 modification, which is essential for the ubiquitin ligase activity, disrupts H3K9 methylation and heterochromatin formation (Jia *et al.*, 2005; Horn *et al.*, 2005). Cul4-mediated protein ubiquitination might play an important role in the Clr4 recruitment step.

4.2. SU(VAR)3-9

As the RNAi machineries regulate Clr4-mediated H3K9 methylation, it was proposed that H3K9 methylation and heterochromatin formation also involve the RNAi machineries in Drosophila (Pal-Bhadra *et al.*, 2004). Although it remains unclear how Su(var)3-9 is recruited to the heterochromatic regions, the findings of yeast and fruit fly suggest that similar mechanisms exist for the targeting of Su(var)3-9. Another PEV modifier, Su(var)3-7 has been shown to interact with

Su(var)3-9 in the yeast two-hybrid interaction assay and increase the Su(var)3-9-mediated H3K9 methylation on autosomes and the male X chromosome. Su(var)3-7, contains seven widely spaced zinc-finger motifs, is involved in the recruitment of Su(var)3-9 (Schotta *et al.*, 2002; Delattre *et al.*, 2004).

4.3. SUV39H

Suv39h is one of the most characterized H3K9 HLMTases, but the recruitment mechanism is not well known. HP1 is the only known molecule ubiquitously associated with Suv39h, but so far HP1 seems not to be essential for the recruitment of Suv39h into pericentromeric heterochromatin. Su(var)3-9 family proteins are highly conserved among fission yeast, Drosophila and mammals. The involvement of RNAi machineries for the Suv39h recruitment is quite possible even though the phenotypes of the *Dicer*-deficient cells are currently bit controversial for the heterochromatin formation and H3K9 methylation (Murchison *et al.*, 2005; Kanellopoulou *et al.*, 2005; Cobb *et al.*, 2005). Some Suv39h interacting molecules including Rb and HTLV-1 Tax are reported, but they are not major players for the heterochromatin recruitment of Suv39h molecules (Nielsen *et al.*, 2001; Kamoi *et al.*, 2006)

4.4. G9a/GLP

Recent biochemical analysis of the G9a complex revealed that G9a exists as the G9a/GLP/Wiz triple complex in various kinds of human and murine cells (Ueda *et al.*, 2006). Wiz is a molecule containing six Kruppel (C_2H_2)-type zinc finger motifs in a widely-interspaced manner (similar with Su(var)3-7) and seems to bind the G9a/GLP heterodimer complex. Current data suggest that Wiz plays a role in the stabilization of G9a. In addition, it links the G9a/GLP heteromeric complex to the CtBP co-repressor machinery. Furthermore, Wiz shows single- and double-stranded DNA binding activities *in vitro*, therefore suggesting that Wiz may also contribute to the recruitment of the G9a/GLP complex to their target chromatin loci. G9a can bind multiple another zinc-finger molecules including Prdm1/Blimp-1(Gyory *et al.*, 2004), NRSF/REST (Roopra *et al.*, 2004), Gfi1 (Duan *et al.*, 2005), Gif1b (Vassen *et al.*, 2006) and PRIZM/PRDM6 (Davis *et al.*, 2006). Therefore, specific DNA binding activities of these multi zinc-finger molecules might also contribute to the target specificities of G9a(/GLP).

4.5. SETDB1/ESET

SETDB1 was originally described as a binding partner of the transcriptional co-repressor molecule KAP-1 for the KRAB-Zinc finger protein superfamily (Schultz *et al.*, 2002). Therefore, the recruitment of SETDB1/ESET to the target loci is potentially mediated by the specific DNA bindings of these zinc-finger motifs. Furthermore, methyl-CpG binding protein MBD1 form a complex

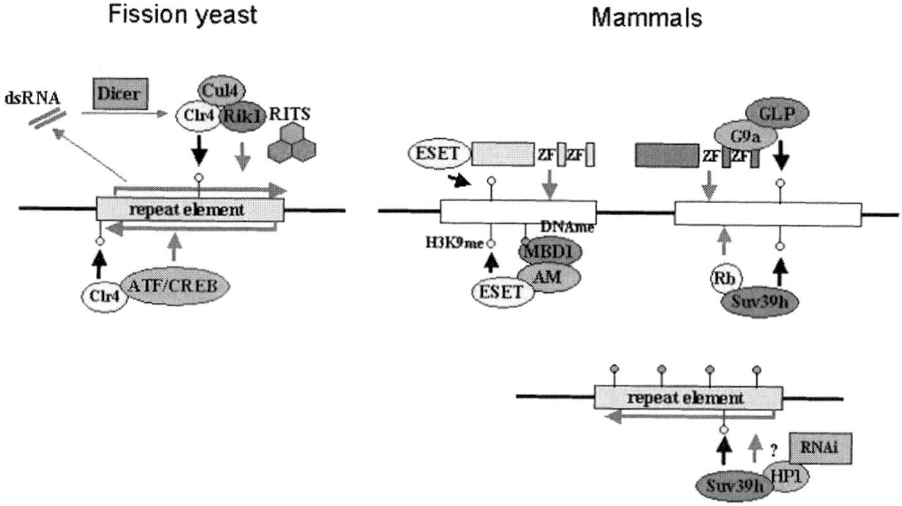

Figure 2. Potential recruitment pathways of yeast and mammalian H3K9 HLMTases. (See Colour Plate 20.)

with SETDB1/ESET (possibly) mediated by the SETDB1 partner AM (ATFa-associated *m*odulator)/MCAF (*M*BD1-containing *c*hromatin-*a*ssociated *f*actor) (Wang *et al.*, 2003; Ichimura *et al.*, 2005). Furthermore, the SETDB1/ESET-AM/MCAF-MBD1 triple complex can transiently interact with the chromatin assembly factor CAF-1 in the S-phase specific manner (Sarraf and Stancheva, 2004). Interaction of these molecules may also contribute to the target specificities or cell-cycle specific recruitment of SETDB1/EST. Potential recruitment mechanism of each of the H3K9 HLMTases in yeast and mammals is described in Fig. 2

4.6. ATSUVH

There are about 15 different Su(var)3-9 homologues (AtSUVH1-10 and AtSUVR1-5) in *Arabidopsis* and some of them were shown to contribute to the H3K9 methylation at heterochromatin. RNAi machineries regulate heterochromatin modifications including H3K9 methylation (reviewed by Gendrel and Colot, 2005; Mathieu and Bender, 2004; Baulcombe, 2004; Matzke *et al.*, 2004). However, details of the connection between RNAi machineries and H3K9 methylation are not elucidated yet.

5. PROPERTIES OF H3K9 HLMTase ACTIVITIES

Based on the biochemical studies, HLMTases can be classified into two categories. One type of the enzymes preferentially methylates histones in a nucleosomal context. This category includes the H3K27 HLMTase EZH2 (EED–EZH2 complex)

(Cao and Zhang, 2004), and the H4K20 HLMTases pr-SET7/SET8 (Fang *et al.*, 2002; Nishioka *et al.*, 2002) and Suv4-20h1,h2 (Schotta *et al.*, 2004). The other type of HLMTases methylates histone octamers, but exhibits a low rate of catalysis for nucleosomal histones. Interestingly, all of the characterized H3K9 HLMTases in mammals (i.e., Suv39h, G9a, GLP/ Eu-HMTase1, and SETDB1/ESET) belong to the latter group (Ogawa *et al.*, 2002; Wang *et al.*, 2003; Schotta *et al.*, 2004; Tachibana *et al.*, 2005). Substrate specificities of pr-SET7 and G9a are shown in Fig. 3

Not only the substrate preferences, some HLMTases also exert the context-dependent enzymatic properties. H3K27 HLMTase EZH2 also shows that enzymatic activity is only functional in the form of complex with other associating molecules including EED (Cao and Zhang, 2004; Pasini *et al.*, 2004). G9a, GLP and SET/DB1 are also context – dependent H3K9 HLMTases. Even G9a and GLP independently express H3K9 HLMTase activities *in vitro*, both are only functional *in vivo* if they can form a heterodimer complex (Tachibana *et al.*, 2005). The stability of G9a in the G9a/GLP complex may partly explain the nature of the G9a/GLP complex *in vivo*. In the case of SETDB1/ESET, this molecule associates with the binding partner AM/MCAF and AM/MCAF enhances the enzymatic activity of SETDB1/ESET (especially conversion of dimethyl to trimethyl H3K9) (Wang *et al.*, 2003).

Recently, it was reported that the SETDB1/ESET-MBD1 complex binds to CAF-1 in an S-phase specific manner (Sarraf and Stancheva, 2004). This finding implies that SETDB1 may exert HLMTase activities when coupled with DNA replication. In this context, it is possible that methyl - transfer of H3K9 is tightly regulated in the cell cycle and mostly targeted to the newly synthesized H3 when deposited into nucleosomes during replication. Methylated histone H3 (at K9) may be replaced by H3.3 which is mostly associated with the induction of H3K4 methylation (Ahmad and Henikoff, 2002; Janicki *et al.*, 2004; McKittrick *et al.*, 2004; Schwartz and Ahmad, 2005; Mito *et al.*, 2005; Henikoff and Ahmad, 2005). In contrast to the H3K9 HLMTases, the nucleosome – preferring HLMTases can continuously transfer

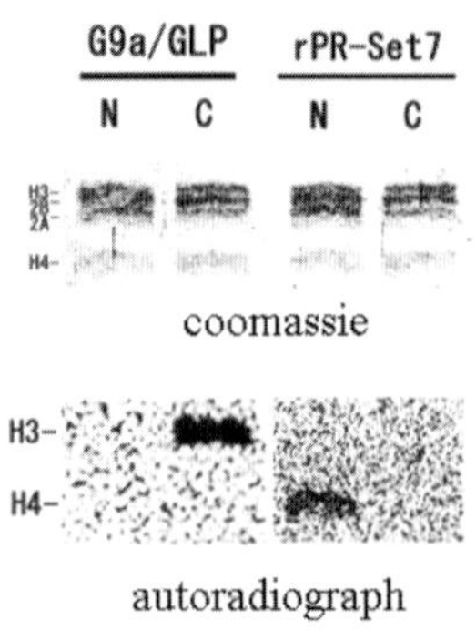

Naked histone preferring HLMTase
Suv39h (H3K9)
ESET/SET-DB1 (H3K9)
G9a/GLP (H3K9)
DIM-5 (H3K9)
SET7/9 (H3K4)

Nucleosome histone preferring HLMTase
rPR-Set7 (H4K20)
Suv4-H20 (H4K20)(controlled by Suv39-mediated H3K9)
Ezh2 (H3K27)
AtSUVH2 (H3K9, H3K27 and H4J20)

Figure 3. H3K9 HLMTases are naked histone preferring enzymes.

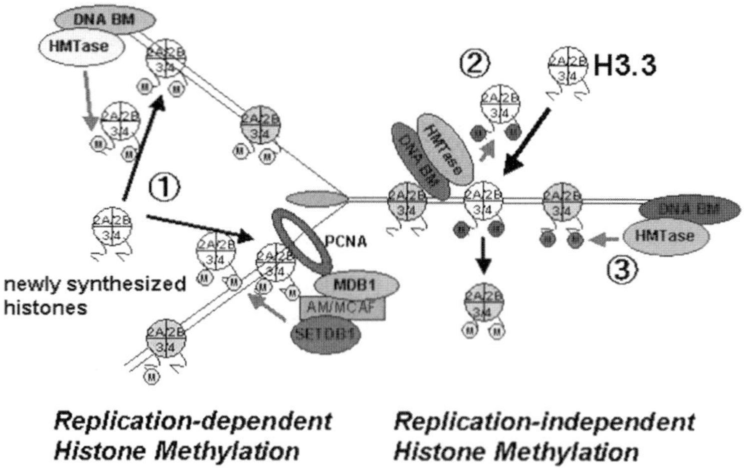

Figure 4. Possible deposition timing of histone lysine methylation

the methyl groups except during the S phase of cell cycle if their *in vitro* enzymatic properties reflect their *in vivo* nature. It is not well characterized yet, but the issues of the methylation timing are quite important for elucidating how cellular memories of histone methyl-lysine codes are maintained though replication. Potential methylation timings of different histone lysine residues are illustrated in Fig. 4.

6. INTERACTION BETWEEN H3K9 METHYLATION AND DEMETHYLATION

The concept of histone methyl-lysine codes as a cellular memory is undergoing changes in the last two years. Especially, emergence of various histone lysine demethylase (HLDMase) activities has made significant impacts on our knowledge for the regulation of histone methylation (Shi *et al.*, 2003; Whetstine *et al.*, 2006; Yamane *et al.*, 2006; Tsukada *et al.*, 2006; Fodor *et al.*, 2006). So far, three enzymatic activities were shown to antagonize the H3K9 methylation. Two of them, JMJD2 group and JHDM2A are members of the jmjC class of hydroxylases and antagonize H3K9 tri- and mono-/di-methylation of H3K9, respectively (Whetstine *et al.*, 2006; Fodor *et al.*, 2006; Yamane *et al.*, 2006). Furthermore, recent data show that methylated H3K4 is recognized by the double tudor domain of JMJD2A (Huang *et al.*, 2006), suggesting that the H3K4 methylation potentially antagonize the H3K9 methylation mediated by the recruitment of JMJD2A, H3K9 HLDMase. In mammals, about 30 different jmjC-domain containing molecules exist (Trewick *et al.*, 2005; Takeuchi *et al.*, 2006). It is quite possible that another substrate specific HLDMase(s) exists in the remaining jmjC - domain molecules.

7. CONCLUSION

Recent new and exciting findings for the genome-wide analysis of the target loci of H3K27 methylation in the embryonic stem cells and other cell types are timely (Lee *et al.*, 2006; Boyer *et al.*, 2006; Bernstein *et al.*, 2006; Bracken *et al.*, 2006,). Not only H3K27 methylation, but other histone methyl-lysine codes including H3K9 methylation are coordinately integrated into the histone code network and regulate various kinds of biological reactions including transcriptional regulation, cell growth, differentiation/pluripotency and DNA repair. In this context, our further studies and understanding of the dynamics and regulation of histone lysine methylation would be crucial for our health and the cure diseases.

ACKNOWLEDGEMENTS

I'd like to thank Jun-ichi Nakayama for his valuable information and Makoto Tachibana for his critical comments.

GLOSSARY

post translational modification, gene expression, histone, chromatin, development, differentiation, methylation, homeostasis, epigenetics

REFERENCES

Aagaard L, Laible G, Selenko P, Schmid M, Dorn R, Schotta G, Kuhfittig S, Wolf A, Lebersorger A, Singh PB *et al* (1999) Functional mammalian homologs of the *Drosophila* PEV-modifier Su(var)3-9 encode centromere-associated proteins which complex with the heterochromatin component M31. EMBO J 18:1923–1938

Ahmad K, Henikoff S (2002) The histone variant H3.3 marks active chromatin by replication-independent nucleosome assembly. Mol. Cell 9:1191–1200

Allshire RC (1995) Elements of chromosome structure and function in fission yeast. Semin Cell Biol 6:55–64

Baulcombe D (2004) RNA silencing in plants. Nature 431:356–363

Bannister AJ, Zegerman P, Partridge JF, Miska EA, Thomas JO, Allshire RC, Kouzarides T (2001) Selective recognition of methylated lysine 9 on histone H3 by the HP1 chromo domain. Nature 410:120–124

Bernstein BE, Mikkelsen TS, Xie X, Kamal M, – Lander ES (2006) A bivalent chromatin structure marks key developmental genes in embryonic stem cells. Cell 125:315–326

Boyer LA, Plath K, Zeitlinger J, Brambrink T, Medeiros LA, Lee TI, Levine SS, Wernig M, Tajonar A, Ray MK *et al* (2006) Polycomb complexes repress developmental regulators in murine embryonic stem cells. Nature 19 April; 441:349–353

Bracken AP, Dietrich N, Pasini D, Hansen KH, Helin K (2006) Genome-wide mapping of Polycomb target genes unravels their roles in cell fate transitions. Genes Dev 20:1123–1136

Buyse IM, Shao G, Huang S (1995) The retinoblastoma protein binds to RIZ, a zinc-finger protein that shares an epitope with the adenovirus E1A protein. Proc Natl Acad Sci USA 92:4467–4471

Canote R, Du Y, Carling T, Tian F, Peng Z, Huang S (2002) The tumor suppressor gene RIZ in cancer gene therapy. Oncol Rep 9:57–60

Cao R, Zhang Y (2004) SUZ12 is required for both the histone methyltransferase activity and the silencing function of the EED–EZH2 complex. Mol Cell 15:57–67

Cobb BS, Nesterova TB, Thompson E, Hertweck A, O'Connor E, Godwin J, Wilson CB, Brockdorff N, Fisher AG, Smale ST, Merkenschlager M (2005) T cell lineage choice and differentiation in the absence of the RNase III enzyme Dicer. J Exp Med 201(9):1367–1373

Davis CA, Haberland M, Arnold MA, Sutherland LB, McDonald OG, Richardson JA, Childs G, Harris S, Owens GK, Olson EN (2006) PRISM/PRDM6, a transcriptional repressor that promotes the proliferative gene program in smooth muscle cells. Mol Cell Biol 26:2626–2636

Delattre M, Spierer A, Jaquet Y, Spierer P (2004) Increased expression of Drosophila Su(var)3-7 triggers Su(var)3-9-dependent heterochromatin formation. J Cell Sci 117:6239–6247

Derunes C, Briknarova K, Geng L, Li S, Gessner CR., Hewitt K, Wu S, Huang S, Woods VI Jr, Ely KR (2005) Characterization of the PR domain of RIZ1 histone methyltransferase. Biochem Biophys Res Commun 333:925–934

Dodge JE, Kang YK, Beppu H, Lei H, Li E (2004) Histone H3-K9 methyltransferase ESET is essential for early development. Mol Cell Biol 24:2478–2486

Duan Z, Zarebski A, Montoya-Durango D, Grimes HL, Horwitz M (2005) Gfi1 coordinates epigenetic repression of p21Cip/WAF1 by recruitment of histone lysine methyltransferase G9a and histone deacetylase 1. Mol Cell Biol 280:13928–13935

Ebbs ML, Bartee L, Bender J (2005) H3 lysine 9 methylation is maintained on a transcribed inverted repeat by combined action of SUVH6 and SUVH4 methyltransferases. Mol Cell Biol 25:10507–10515

Ebbs ML, Bender J (2006) Locus-specific control of DNA methylation by the Arabidopsis SUVH5 histone methyltransferase. Plant Cell 18(5):1166–1176

Fang J, Feng Q, Ketel CS, Wang H, Cao R, Xia L, Erdjument-Bromage H, Tempst P, Simon JA, Zhang Y (2002) Purification and functional characterization of SET8, a nucleosomal histone H4-lysine 20-specific methyltransferase. Curr Biol 12:1086–1099

Fischle W, Wang Y, Allis CD (2003) Histone and chromatin cross-talk. Curr Opin Cell Biol 15:172–183

Fischle W, Tsseng BS, Dormann HL, Ueberheide BM, Garcia BA, Shabanowitz J, hunt DF, Funabini H, Allis CD (2005) Regulation of HP1-chromatin binding by histone H3 methylation and phosphorylation. Nature 438:1116–1122

Flanagan JF, Mi LZ, Chruszcz M, Cymborowski M, Clines KL, Kim Y, Minor W, Rastinejad F, Khorasanizadeh S (2005) Double chromodomains cooperate to recognize the methylated histone H3 tail. Nature 438:1181–1185

Fodor BD, Kubicek S, Yonezawa M, O'Sullivan RJ, Sengupta R, Perez-Burgos L, Opravil S, Mechtler K, Schotta G, Jenuwein T (2006) Jmjd2b antagonizes H3K9 tri-methylation at pericentric heterochromatin in mammalian cells. Genes Dev 20:1557–1562

Freitag M, Hickey PC, Khlafallah TK, Reed ND, Selker EU (2004) HP1 is essential for DNA methylation in neurospora. Mol Cell 13:427–434

Gendrel AV, Colot V (2005) Arabidopsis epigenetics: When RNA meets chromatin. Curr Opin Plant Biol 8:142–147

Grewal SI, Rice JC (2004) Regulation of heterochromatin by histone methylation and small RNAs. Curr Opin Cell Biol 16:230–238

Gyory I, Wu J, Fejer G, Seto E, Wright KL (2004) PRDI-BF1 recruits the histone H3 methyltransferase G9a in transcriptional silencing. Nat Immunol 5:299–308

Harte PJ, Wu W, Carrasquillo MM, Matera AG (1999) Assignment of a novel bifurcated SET domain gene, SETDB1, to human chromosome band 1q21 by in situ hybridization and radiation hybrids. Cytogenet Cell Genet 84:83–86

Henikoff S, Ahmad K (2005) Assembly of variant histones into chromatin. Annu Rev Cell Dev Biol 21:133–153

Hirata T, Lipp JJ, Toh BH, Peters JM (2005) Histone H3 serine 10 phosphorylation by Aurora B causes HP1 dissociation from heterochromatin. Nature 438:1176–1180

Horn PJ, Bastie JN, Peterson CL (2005) A Rik1-associated, cullin-dependent E3 ubiquitin ligase is essential for heterochromatin formation. Genes Dev 19:1705–1714

Huang Y, Fang J, Bedford MT, Zhang Y, Xu RM (2006) Recognition of histone H3 lysine-4 methylation by the double tudor domain of JMJD2A. Science 312:748–751

Ichimura T, Watanabe S, Sakamoto Y, Aoto T, Fujita N, Nakao M (2005) Transcriptional repression and heterochromatin formation by MBD1 and MCAF/AM family proteins. J Biol Chem 280(14):13928–13935

Jackson JP, Lindroth AM, Cao X, Jacobsen SE (2002) Control of CpNpG DNA methylation by the KRYPTONITE histone H3 methyltransferase. Nature416:556–560

Jacobs SA, Taverna SD, Zhang Y, Briggs SD, Li J, Eissenberg JC, Allis CD, Khorasanizadeh S (2001) Specificity of the HP1 chromo domain for the methylated N-terminus of histone H3. EMBO J 20(18):5232–5241

Janicki SM, Tsukamoto T, Salghetti SE, Tansey WP, Sachidanandam R, Prasanth KV, Ried T, Shav-Tal Y, Bertrand E, Singer RH, Spector DL (2004). From silencing to gene expression: Real-time analysis in single cells. Cell 116:683-698

Jenuwein T (2006) The epigenetic magic of histone lysine methylation. FEBS J. 273:3121–3135

Jia S, Kobayashi R, Grewal SI (2005) Ubiquitin ligase component Cul4 associates with Clr4 histone methyltransferase to assemble heterochromatin. Nat Cell Biol 7:1007–1013

Kamoi K, Yamamoto K, Misawa A, Miyake A, Ishida T, Tanaka Y, Mochizuki M, Watanabe T (2006) SUV39H1 interacts with HTLV-1 Tax and abrogates Tax transactivation of HTLV-1 LTR. Retrovirology 3:5

Kanellopoulou C, Muljo SA, Kung AL, Ganesan S, Drapkin R, Jenuwein T, Livingston DM, Rajewsky K (2005) Dicer-deficient mouse embryonic stem cells are defective in differentiation and centromeric silencing. Genes Dev 19:489–501

Klar AJ (1998) Propagating epigenetic states through meiosis: where Mendel's gene is more than a DNA moiety. Trends Genet 14:299–301

Klein RR, Houtz RL (1995) Cloning and developmental expression of pea ribulose-1,5-bisphosphate carboxylase/oxygenase large subunit N-methyltransferase. Plant Mol Biol 27(2):249–261

Kouzarides T (2002) Histone methylation in transcriptional control. Curr Opin Genet Dev 12:198–209

Lachner M, O'Carroll D, Rea S, Mechtler K, Jenuwein T (2001) Methylation of histone H3 lysine 9 creates a binding site for HP1 proteins. Nature 410:116–120

Lachner M, Sengupta R, Schotta G, Jenuwein T (2004) Trilogies of histone lysine methylation as epigenetic landmarks of the eukaryotic genome. Cold Spring Harb Symp Quant Biol 69:209–218

Lee DY, Northrop JP, Kuo MH, Stallcup MR (2006) Histone H3 lysine 9 methyltransferase G9a is a transcriptional coactivator for nuclear receptors. J Biol Chem 281:8476–8485

Lee TI, Jenner, RG, Boyer LA, Guenther MG, Levine SS, – Young RA (2006) Control of developmental regulators by Polycomb in human embryonic stem cells. Cell 125:301–313

Martienssen RA, Zaratiegui M, Goto DB (2005) RNA interference and heterochromatin in the fission yeast Schizosaccharomyces pombe. Trends Genet 21(8):450–456

Martin C, Zhang Y (2005) The diverse functions of histone lysine methylation. Nat Rev Mol Cell Biol 6:838–849

Mathieu O, Bender J (2004) RNA-directed DNA methylation. J Cell Sci 117:4881–4888

Matzke M, Aufsatz W, Kanno T, Daxinger L, Papp I, Mette MF, Matzke AJ (2004) Genetic analysis of RNA-mediated transcriptional gene silencing. Biochem Biophys Acta 1677(1–3):129–141

McKittrick E, Gafken PR, Ahmad K, Henikoff S (2004) Histone H3.3 is enriched in covalent modifications associated with active chromatin. Proc Natl Acad Sci USA 101:1525–1530

Milner CM, Campbell RD (1993) The G9a gene in the human major histocompatibility complex encodes a novel protein containing ankyrin-like repeats. Biochem J 290:811–818

Mis J, Ner SS, Grigliatti TA (2006) Identification of three histone methyltransferases in Drosophila: dG9a is a suppressor of PEV and is required for gene silencing. Mol Genet Genomics 275:513–526

Mito Y, Henikoff JG, Henikoff S (2005) Genome-scale profiling of histone H3.3 replacement patterns. Nat Genet 7:1090–1097

Murchison EP, Partridge JF, Tam OH, Cheloufi S, Hannon GJ (2005)Characterization of Dicer-deficient murine embryonic stem cells. Proc Natl Acad Sci USA 102:12135–12140

Nakayama J, Rice JC, Strahl BD, Allis CD, Grewal SI (2001) Role of histone H3 lysine 9 methylation in epigenetic control of heterochromatin assembly. Science 292:110–113
Naumann K, Fischer A, Hofmann I, Krauss V, Phalke S, Irmler K, Hause G, Aurich AC, Dorn R, Jenuwein T, Reuter G (2005) Pivotal role of AtSUVH2 in heterochr EMBO J 24:1418–1429
Nielsen SJ, Schneider R, Bauer UM, Bannister AJ, Morrison A, O'Carroll D, Firestein R, Cleary M, Jenuwein T, Herrera RE, Kouzasrides T (2001) Rb targets histone H3 methylation and HP1 to promoters. Nature 201(9):1367–1373
Nishioka K, Rice JC, Sarma K, Erdjument-Bromage H, Werner J, Wang Y, Chuikov S, Valenzuela P, Tempst P, Steward R *et al* (2002) PR-Set7 is a nucleosome-specific methyltransferase that modifies lysine 20 of histone H4 and is associated with silent chromatin. Mol Cell 9:1201–1213
O'Carroll D, Scherthan H, Peters AH, Opravil S, Haynes AR, Laible G, Rea S, Schmid M, Lebersorger A, Jerratsch M *et al* (2000) Isolation and characterization of Suv39h2, a second histone H3 methyltransferase gene that displays testis-specific expression. Mol Cell Biol 20:9423–9433
Ogawa H, Ishiguro K, Gaubatz S, Livingston DM, Nakatani Y (2002) A complex with chromatin modifiers that occupies E2F- and Myc-responsive genes in G0 cells. Science 296:1132–1136
Pal-Bhadra M, Leibovitch BA, Gandhi SG, Rao M, Bhadra U, Birchler JA, Elgin SC (2004). Heterochromatic silencing and HP1 localization in Drosophila are dependent on the RNAi machinery. Science 303:669–672
Pasini D, Bracken AP, Jensen MR, Lazzerini DE, Helin K (2004) Suz12 is essential for mouse development and for EZH2 histone methyltransferase activity. EMBO J 23:4061–4071
Peters AH, Kubicek S, Mechtler K, O'Sullivan RJ, Derijck AA, Perez-Burgos L, Kohlmaier A, Opravil S, Tachibana M, Shinkai Y *et al* (2003) Partitioning and plasticity of repressive histone methylation states in mammalian chromatin. Mol Cell 12:1577–1589
Rea S, Eisenhaber F, O'Carroll D, Strahl BD, Dun Z-W, Schmid M, Opravil S, Mechtler K, Ponting CP, Allis CD, Jenuwein T (2000) Regulation of chromatin structure by site-specific histone H3 methyltransferases. Nature 406:593–599
Reuter G, Spierer P (1992) Position effect variegation and chromatin proteins. Bioessays 14:605–612
Rice JC, Briggs SD, Ueberheide B, Barber CM, Shabanowitz J, Hunt DF, Shinkai Y, Allis CD (2003) Histone methyltransferases direct different degrees of methylation to define distinct chromatin domains. Mol Cell 12:1591–1598
Roopra A, Qazi R, Schoenike B, Daley TJ, Morrison JF (2004) Localized domains of g9a-mediated histone methylation are required for silencing of neuronal genes. Mol Cell 14:727–738
Sadaie M, Iida T, Urano T, Nakayama J (2004) A chromodomain protein, Chp1, is required for the establishment of heterochromatin in fission yeast. EMBO J 23:3825–3835
Sarraf SA, Stancheva I (2004) Methyl-CpG binding protein MBD1 couples histone H3 methylation at lysine 9 by SETDB1 to DNA replication and chromatin assembly. Mol Cell 15:595–605
Schotta G, Ebert A, Krauss V, Fischer A, Hoffmann J, Rea S, Jenuwein T, Dorn R, Reuter G (2002) Central role of Drosophila SU(VAR)3-9 in histone H3-K9 methylation and heterochromatic gene silencing. EMBO J 21(5):1121–1131
Schotta G, Lachner M, Sarma K, Ebert A, Sengupta R, Reuter G, Reinberg D, Jenuwein T (2004) A silencing pathway to induce H3-K9 and H4-K20 trimethylation at constitutive heterochromatin. Genes Dev 18:1251–1262
Schultz DC, Ayyanathan K, Negorev D, Maul GG, Rauscher III FJ (2002) SETDB1: A novel KAP-1-associated histone H3, lysine 9-specific methyltransferase that contributes to HP1-mediated silencing of euchromatic genes by KRAB zinc-finger proteins. Genes Dev 16:919–932
Schwartz BE, Ahmad K (2005) Transcriptional activation triggers deposition and removal of the histone variant H3.3. Genes Dev 19:804–814
Shi Y, Lan F, Matson C, Mulligan P, Whetstine JR, Cole PA, Casero RA, Shi Y (2004) Histone demethylation mediated by the nuclear amine oxidase homolog LSD1. Cell 119:941–953
Sims RJ 3rd, Reinberg D (2006) Histone H3 Lys 4 methylation: Caught in a bind? Genes Dev. 20:2779–2786
Stewart MD, Li J, Wong J (2005) Relationship between histone H3 lysine 9 methylation, transcription repression, and heterochromatin protein 1 recruitment. Mol Cell Biol 25:2525–2538

Tachibana M, Sugimoto K, Fukushima T, Shinkai Y (2001) Set domain-containing protein, G9a, is a novel lysine-preferring mammalian histone methyltransferase with hyperactivity and specific selectivity to lysines 9 and 27 of histone H3. J Biol Chem 276:25309–25317

Tachibana M, Sugimoto K, Nozaki M, Ueda J, Ohta T, Ohki M, Fukuda M, Takeda N, Niida H, Kato H *et al* (2002) G9a histone methyltransferase plays a dominant role in euchromatic histone H3 lysine 9 methylation and is essential for early embryogenesis. Genes Dev 16:1779–1791

Tachibana M, Ueda J, Fukuda M, Takeda N, Ohta T, Iwanari H, Sakihama T, Kodama T, Hamakubo T, Shinkai Y (2005) Histone methyltransferases G9a and GLP form heteromeric complexes and are both crucial for methylation of euchromatin at H3-K9. Genes Dev 19:815–826

Takeuchi T, Watanabe Y, Takano-Shimizu T, Kondo S (2006) The roles of *jumonji* and *jumonji* family genes in chromatin. Dev Dynamics 235:2449–2459

Tamaru H, Selker EU (2001) A histone H3 methyltransferase controls DNA methylation in Neurospora crassa. Nature 414:277–283

Trewick SC, Mclaughlin RJ, Allshire RC (2005) Methylation: Lost in hydroxylation? EMBO Rep 6:315–320

Tsukada Y-I, Fang J, Erdjument-Bromage H, Warren ME, Borchers CH, Tempst P, Zhang Y (2006) Histone demethylation by a family of JmjC domain-containing proteins. Nature 439:811–816

Ueda J, Tachibana M, Ikura T, Shinkai Y (2006) Zinc finger protein wiz links G9a/GLP histone methyltransferases to the co-repressor molecule CtBP. J Biol Chem 281:20120–20128

Vakoc CR, Mandat SA, Olenchock BA, Blobel GA (2005) Histone H3 lysine 9 methylation and HP1gamma are associated with transcription elongation through mammalian chromatin. Mol Cell 19:381–391

Vassen L, Fiolka K, Moroy T (2006). Gfi1b alters histone methylation at target gene promoters and sites of gamma-satellite containing heterochromatin. EMBO J 25:2409–2419

Verdel A, Moazed D (2005) RNAi-directed assembly of heterochromatin in fission yeast. FEBS Lett 579(26):5872–5878

Wallrath LL (1998) Unfolding the mysteries of heterochromatin. Curr Opin Genet Dev 8(2):147–153

Wang H, An W, Cao R, Xia L, Erdjument-Bromage H, Chatton B, Tempst P, Roeder RG, Zhang Y (2003) mAM facilitates conversion by ESET of dimethyl to trimethyl lysine 9 of histone H3 to cause transcriptional repression. Mol Cell 12:475–487

Weiler KS, Wakimoto BT (1995) Heterochromatin and gene expression in Drosophila. Annu Rev Genet 29:577–605

Whetstine JR, Nottke A, Lan F, Huarte M, Smolikov S, Chen Z, Spooner E, Li E, Zhang G, Colaiacovo M, Shi Y (2006) Reversal of histone lysine trimethylation by the JMJD2 family of histone demethylases. Cell 125:467–481

Yamane K, Toumazou C, Tsukada Y-I, Erdjument-Bromage H, Tempst P, Wong J, Zhang Y (2006) JHDM2A, a JmjC-containing H3K9 demethylase, facilitates transcription activation by androgen receptor. Cell 125:483–495

CHAPTER 16

HISTONE ACETYLATION AND METHYLATION

Combinatorial players for transcriptional regulation

WOOJIN AN

Department of Biochemistry & Molecular Biology, USC/Norris Comprehensive Cancer Center, 1501 San Pablo Street, ZNI 241, MC 2821, Los Angeles, California 90089-2821

Abstract: Post-synthetic modification of histone proteins in chromatin architecture plays a central role in the epigenetic regulation of transcription. Histone acetylation and methylation are the two major modifications that function as a specific transcription regulator in response to various cellular signals. Albeit the mechanism of action of these modifications in transcription is not well understood, recent discovery of histone acetyltransferase (HAT) and methyltransferase (HMT) activities within transcriptional regulators has an important implication for histone modification to be a key player for the precise regulation of transcription processes. Here, we discuss recent advances made on histone acetylation and methylation as a fundamental process to modulate gene transcription, with a particular emphasis on their combinatorial effects in transcriptional control

1. INTRODUCTION

In eukaryotic cells, the DNA is compacted into the nucleus as a complex structure known as chromatin. The fundamental repeating unit of chromatin is the nucleosome, which consists of 146 bp of DNA wrapped around a core histone octamer containing pairs of each of the histone proteins, H2A, H2B, H3 and H4 (van Holde, 1988; Luger *et al.*, 1997; Kornberg and Lorch, 1999). Structured central domains of core histones mediate histone-histone interactions to stabilize the histone octamer within the nucleosome core particle. Each core histone also contains unstructured N-terminal tail domains ranging from 15 (H2A) to 35 (H3) amino acids, which are extended out of the nucleosome particle (Luger and Richmond, 1998). Although histone tails at the surface of the nucleosome do not contribute to structure and stability of nucleosomes, they are required for the internucleosomal interaction to facilitate the higher-order folding of chromatin fibers (Hansen, 2002).

*E-mail:woojinan@usc.edu

T.K. Kundu and D. Dasgupta (eds.), Chromatin and Disease, 351–369.

A high proportion of the positively charged basic amino acids lysine and arginine within these flexible tails are frequent targets for extensive posttranslational modifications (Berger, 2002). Such modifications include the acetylation of lysine residues, the methylation of lysine and arginine residues, the ubiquitination of lysine residues, the phosphorylation of serine and threonine residues, the sumoylation of lysine residues, and the poly ADP-ribosylation of glutamic acid residues.

Acetylation and methylation of specific lysine or arginine residues in histones H3 and H4 are reversible post-translational processes directly linked to either active or repressive states of gene transcription (Peterson and Laniel, 2004). Although it remains unclear to what extent, it also has been well documented that these modifications mutually affect each other in many cases to regulate specific transcription-based processes (Zhang and Reinberg, 2001; Berger, 2002; Fischle *et al.*, 2003). In the wake of several pioneering techniques, it has been feasible to characterize the contribution of particular histone modifications in transcriptional regulatory networks and to profile the genome-wide status of a specific modification in a massively parallel way. The results of such analyses will be useful in confirming functional consequences of epigenetic signals within histone tails and will ultimately be valuable for diagnostic, therapeutic and drug development purposes. In this article, I will review the individual and combinatorial characters of histone acetylation and methylation in transcription and discuss the implications of epigenetic approaches that target HMT and HAT for the treatment of major pathologies.

2. HISTONE ACETYLATION AND TRANSCRIPTION

Histone acetylation is one of the major histone modifications that takes place at the ε-amino groups on specific lysine residues at the N-terminus of histone proteins (Fig. 1) (Sterner and Berger, 2000; Roth *et al.*, 2001). This modification occurs on all four core histones and reduces the net positive charge on the histone proteins. Earliest studies on this modification suggested that the addition of an acetyl group to a histone N-terminal tail plays some special role in chromatin reorganization for efficient transcription (Allfrey *et al.*, 1964). In support of this idea, histones within active chromatin regions appear to be acetylated to a higher degree than those in inactive chromatin regions (Grunstein, 1997; Kuo and Allis, 1998). Further evidence in favor of a role for histone acetylation in transcription has been obtained by identification of histone acetyltransferases capable of adding acetyl groups to histone tails (Sterner and Berger, 2000; Roth *et al.*, 2001). Observation that many known transcription cofactors possess an intrinsic HAT activity supported the causal relationship between histone acetylation and gene transcription (Table 1). Histone acetylation is a highly dynamic event as a consequence of the presence of histone deacetylases (HDACs) which can induce the removal of acetyl groups from histones (Yang and Seto, 2003); treatment of cells with HDAC inhibitor sodium butyrate has been shown to increase DNase I sensitivity of chromatin and to activate previously repressed genes (Roth *et al.*, 2001). Hence, albeit imperfectly proven, there is a great deal of information to support a positive role for histone acetylation in transcription.

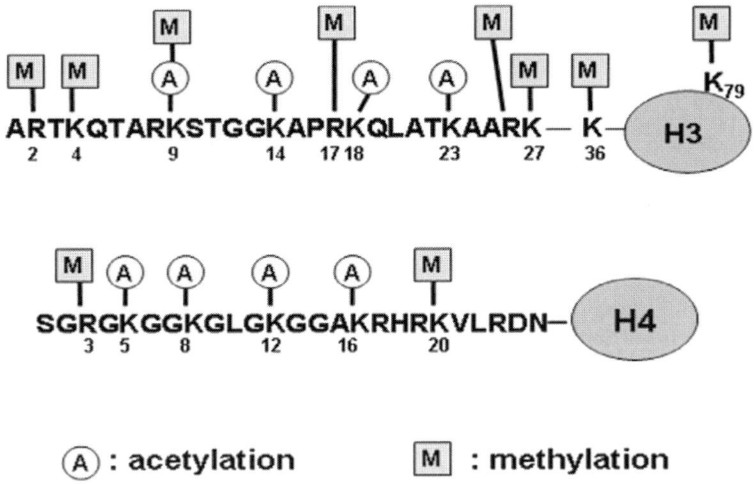

Figure 1. Schematic summary of sequences of H3 and H4 tails. "A" and "M" indicate major *in vivo* acetylation and methylation sites, respectively. (See colour Plate 21.)

HAT proteins are grouped into different family classes through sequence homologies and similarities in their biological functions (Table 1). The GNAT family is the best understood group of nuclear HATs including GCN5 and PCAF and related proteins. In addition to a C-terminal catalytic HAT domain, each

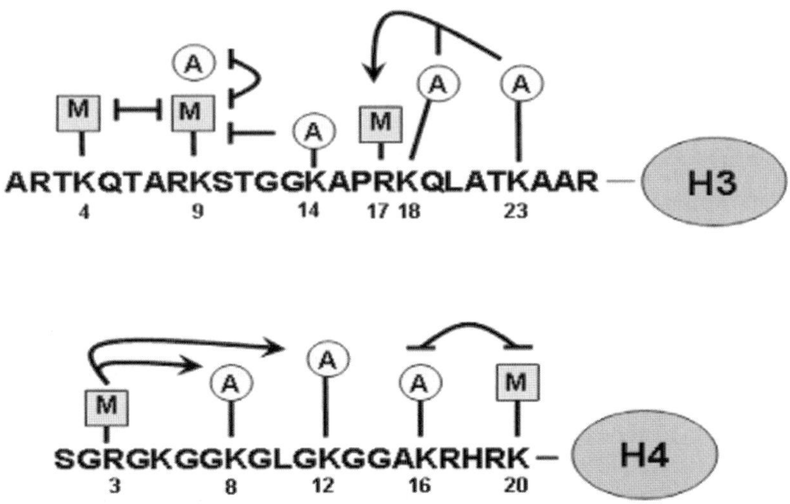

Figure 2. Interplay between histone acetylation and methylation. Acetylated (A) and methylated (M) residues are indicated. Positive effects are represented by normal arrows; negative effects are represented by tee arrow. (See colour Plate 22.)

Table 1. List of known histone acetyltransferases (HATs)

Family	HAT	Complex	Organism	Substrate	Transcriptional functions
GNAT	Gcn5L	STAGA, TFTC	Humans, Mice	H3/H2B	Activation
	PCAF	PCAF	Humans, Mice	H3/H4	Activation
	Gcn5	SAGA, ADA, SLIK, SALSA, HAT-A2	Yeast	H3/H2B	Activation
	Elp3	Elongator	Yeast	H3/H4	Activation
	ATF-2		Humans, Mice	H4/H2B	Activation
	HAT1	HAT-B	Humans, Yeast	H4/H2A	Repression
	Hpa2		Yeast	H3/H4	ND
MYST	TIP60	TIP60	Humans, Mice	H3/H4	Activation
	Sas2	SAS	Yeast	H4	Repression
	Sas3	NuA3	Yeast	H3	Activation
	Esa1	NuA4	Yeast	H4/H2A	Cell cycle control
	Mof	MSL	Drosophila	H4	Dosage compensation
	HBO1	HBO1	Humans	H3/H4	Repression
	MOZ		Humans	H3/H4	Activation
	MORF		Humans	H3/H4	Activation
p160	SRC-1		Humans, Mice	H3/H4	Activation
	ACTR		Humans, Mice	H3/H4	Activation
Others	p300/CBP		Humans, Mice	H3/H4/H2A/H2B	Activation
	TAFII250 (TAF 1)	TFIID	Humans	H3/H4	RNA Pol II transcription
	TFIIIC90/110/220	TFIIIC	Humans	H3/H4/H2A	RNA Pol III transcription
	Nut1	Mediator	Yeast	H3/H4	Activation

"HAT" indicates the catalytic HAT subunit; "complex" indicates the purified HAT complex. The substrates refer to the *in vitro* specificity of the HAT subunit. Known functions of each modification in transcription are listed in the last column. ND: not determined

member possesses a bromodomain of about 110 residues for specific interaction with lysine-acetylated histone tails (Sterner and Berger, 2000; Roth *et al.*, 2001). GCN5L and PCAF are the best characterized human HATs among the GNAT family as transcriptional coactivators. Although PCAF and GCN5L share significant homology, only homozygous GCN5L null mutations exhibit mouse embryonic lethality (Xu *et al.*, 2000), suggesting different physiological roles of different GNAT family members. Members of the MYST family are grouped together on the basis of their possession of a particular highly conserved 370 residue MYST domain, which has a catalytic mechanism different from that shared by other families of HATs. These proteins are involved in a much broader range of biological processes in various organisms. TIP60 was the first identified human MYST HAT and was recently reported as a key regulator for DNA repair and apoptosis (Ikura *et al.*, 2000). Another group of HATs is p160 coactivators which interact with nuclear hormone receptors to upregulate receptor-dependent gene transcription in a ligand dependent manner (Sterner and Berger, 2000; Roth *et al.*, 2001). In addition to these three major groups of HATs, several other proteins such as p300/CBP and TAFII250 have also been shown to possess HAT activity to regulate transcription of a wide range of genes (Bannister and Kouzarides, 1996; Mizzen *et al.*, 1996; Ogryzko *et al.*, 1996). The essential role of p300/CBP for development is also confirmed by embryonic lethality of CBP and p300 mutated mice (Goodman and Smolik, 2000).

HATs are recruited to certain gene promoters through interactions with DNA-binding regulatory factors, which leads to the targeted acetylation and subsequent activation of transcription (Kundu *et al.*, 2000; An *et al.*, 2002). The importance of histone acetylation in gene regulation began to be intensively investigated when several known transcription coactivators such as p300/CBP and GCN5 were found to have intrinsic HAT activities. As listed in Table 1, subsequent studies identified many additional proteins to possess HAT activity. Most HAT enzymes are assembled into multiprotein complexes for their proper recruitment by sequence-specific transcription activators to target genes and for their substrate specificity on nucleosomes. These HAT complexes have recently been purified from human and yeast cells and functionally characterized as positive regulators of transcription (Table 1). Similar to HATs, the HDACs are also part of large multisubunit complexes (Yang and Seto, 2003). Therefore, the level of acetylation in a specific gene is maintained by a dynamic competition between the activities of HAT and HDAC after their recruitments, and aberrant regulation of this competition has been shown to cause inappropriate gene expression.

Although histone acetylation status of promoters and regulatory elements is generally correlated with HAT-mediated transcription (Roth *et al.*, 2001; Liang *et al.*, 2004; Roh *et al.*, 2005), many studies have proven that some of the HATs such as p300 and pCAF/GCN5 can also acetylate non-histone proteins (Sterner and Berger, 2000; Glozak *et al.*, 2005). Thus, caution has been emphasized when interpreting a requirement for a HAT to conclude that a histone acetylation event is responsible for transcriptional activation. In most cases, the evidence for the effect of histone acetylation by HAT proteins in transcription is provided by the inactivation of their HAT activities. This approach may assist in determining the role of histone acetylation in transcription; however, it is likely that a great deal of functional redundancy exists between histone and non-histone modifying activities of HAT proteins. Thus care should be taken when interpreting HAT mutation/inactivation data with regards to its role in transcription. The best approach for proving the contribution of histone acetylation per se in the regulation of gene activation would be to create and test mutations of histone genes within the nucleus of multicellular eukaryotes. However, since several clusters of replication-dependent histone genes are present in all metazoans (Mosammaparast *et al.*, 2001, 2002), it has been impossible to use specific mutations in histone genes to modify cellular histones. To clarify this uncertainty, a reconstituted chromatin transcription system has recently been developed by utilizing recombinant chromatin templates that are assembled from recombinant histone proteins (Loyola *et al.*, 2001; Agalioti *et al.*, 2002; An *et al.*, 2002; Georges *et al.*, 2002; Levenstein and Kadonaga, 2002). The power of the recombinant chromatin template assay is that it allows one to preclude the histone modification of interest by making amino acid substitutions at the physiological modification sites. Therefore a failure to promote transcription upon mutation of modification sites from histones will directly verify that the histone modifications are essential for transcription activation. Using this experimental system, a direct connection between targeted acetylations of H3-H4 and p300-mediated transcription has been established based on the transcription-inhibiting effects of substitution of major lysine substrates within H3-H4 (An *et al.*, 2002). Although a specific combination of HAT and activator has been used in these experiments, the results clearly highlight the requirement of acetylated histone tails to activate transcription.

One major point still to be clarified concerns the molecular mechanism underlying the effects exerted by histone acetylation in regulating gene transcription. There are at least two possible mechanisms by which histone acetylation can facilitate transcription activation; charge neutralization effects versus protein recognition/recruitment effects. The effect of charge neutralization of histone tails by acetylation is consistent with the hypothesis that the chromatin fiber is stabilized by charge dependent interactions between basic histone tails and acidic patches on adjacent nucleosomes (Wade *et al.*, 1997; Hansen *et al.*, 1998). Thus histone acetylation neutralizes the positive charges of lysine side chains to affect the interaction between the lysine residues and the negatively charged DNA backbone. As a consequence, compacted chromatin would be destabilized to allow efficient binding of transcription machinery to the transcription initiation site. In this case, the contribution of acetylation to transcription should be rather cumulative (Ren and Gorovsky, 2001). Indeed, evidence of a direct effect of histone acetylation on the stability of nucleosomal arrays has been provided from *in vitro* studies by several groups (Garcia-Ramirez *et al.*, 1995; Tse *et al.*, 1998). A recent microarray study also showed the interchangeability of all acetylated lysines (K5, K8, K12) except K16 in H4 tails for transcription in yeast, providing support for a rather simple mechanism for the effect of histone acetylation in gene regulation (Dion *et al.*, 2005).

In contrast to simple charge neutralization effects, the effects on protein recognition/recruitment are collectively referred to as the "histone code". This hypothesis predicts that specific patterns of histone tail acetylations and other modifications serve as epigenetic marks for distinct sets of regulatory proteins to differentially modulate chromatin structure and function (Strahl and Allis, 2000; Turner, 2000; Jenuwein and Allis, 2001). Indeed, several recent findings have demonstrated that histone acetylation creates a signal for the binding of a bromodomain which has been found to be present in many chromatin and transcription regulators such as GCN5, PCAF, p300/CBP, TAF250 and others (Zeng and Zhou, 2002). The most prominent example is that acetylation of specific lysines of H3-K9, K14 and H4-K8 functions as a cognate mark for the bromodomain-containing proteins BRG1 and TAFII250 to recruit SWI/SNF and TFIID complexes during *in vitro* activation of the IFN-ß gene (Agalioti *et al.*, 2002). Selective recognition of acetylated histones by bromodomains has also been established in the intact nuclei of living cells by FRET analysis. This *in vivo* study showed that the bromodomain of transcriptional regulator BRD2 specifically recognizes acetylated H4-K12 whereas those of TAFII250 and PCAF recognize acetylated H3 and H4 with broader acetyl-lysine specificity (Kanno *et al.*, 2004). Another bromodomain-containing protein GCN5 has also been found to be required for the yeast SAGA complex to associate with acetylated nucleosomes during chromatin remodeling (Hassan *et al.*, 2002). All these results indicate that the physical interaction of acetylated histones with different bromodomain proteins is rather specific, supporting highly selective recognition/retention of a subpopulation of bromodomains by a certain acetylation mark in chromatin. However it is currently unknown whether all known bromodomains are able to recognize acetylated lysine.

Another important question to be addressed is whether different genes share a similar pattern of acetylation to get expressed or a unique pattern of acetylation is associated with a distinct group of genes. A variety of HAT activities exhibit unique substrate specificity, discriminating between individual lysine residues of histones (Sterner and Berger, 2000; Roth *et al.*, 2001); thus the interaction of a HAT with a gene specific activator that itself binds to a distinct promoter could explain the phenomenon of specificity of HAT and lysine acetylation in certain transcription processes (Deckert and Struhl, 2001). The result that mutations of the acetylated residues of individual histones in yeast influence the expression of particular genes supports the specificity of HAT as a component of the transcription process (Grunstein *et al.*, 1995). The most prominent example of such gene specificity includes the acetylations of H3-K9, K14 and H4-K8 by GCN5/PCAF, shown to be important for activation of the human interferon-β gene upon viral infection (Agalioti *et al.*, 2002). Additionally, it appears that acetylation of histone tails can also be important for the recruitment of proteins without any bromodomain motif as in the case for interaction of acetylated H3 tails with CARM1 (Daujat *et al.*, 2002). Therefore, it will be important to determine the specificity of histone acetylation in differential gene activation and how these different patterns of acetylation impart their selectivity. Considering that most recent studies utilized antibodies against di-acetylated H3 (on lysines 9 and 14) and tetra-acetylated H4 (on lysines 5, 8, 12 and 16) to identify histone acetylation at different genes (e.g., β-globin, PHO8, CATD, INO1 and HO), it is our challenge to investigate genome-wide acetylation profiles of all lysine substrates of the four core histones in human cells, which will contribute to a full understanding of the role of histone acetylation in transcription as well as transcriptional regulation per se.

3. HISTONE METHYLATION AND TRANSCRIPTION

Histone methylation is another posttranslational modification which involves a transfer of a methyl group from the methyl donor S-adenosyl methionine (SAM) to lysine or arginine residues (Fig. 1). In sharp contrast with histone acetylation, this modification occurs particularly in histones H3 and H4 with a remarkable specificity (Kouzarides, 2002; Shilatifard, 2006) (Fig. 1, Table 2). Another feature of histone methylation is that a large fraction of histones in mature chromatin is

Table 2. List of known histone methyltransferases (HMTs)

Family	HMT	Complex	Organism	Substrate	Transcriptional functions
Lysine HMT	MLL1	MLL, MLL1-WDR5, MENIN	Humans	H3-K4	Activation
	MLL2/4	MENIN	Humans	H3-K4	Activation
	ySET1	SET1/COMPASS	Yeast	H3-K4	Activation
	hSET1	SET1/ASH2/CFP1	Humans	H3-K4	Activation
	Trx	TAC1	Drosophila	H3-K4	Activation
	SMYD3		Humans	H3-K4	Activation
	SET9/7		Humans	H3-K4	Activation
	SUV39H1/2	E2F1, E2F4	Humans, Mice	H3-K9	Repression
	Eu-HMTase1	E2F6	Humans	H3-K9	Repression
	SETDB1/ESET	ESET/mAM	Humans, Mice	H3-K9	Repression
	Clr4	CLRC	Yeast	H3-K9	Repression
	G9a		Humans, Mice	H3-K9	Repression/ activation
	EZH2	EED-EZH2	Humans	H3-K27	Repression
	E(z)	ESC-E(Z)	Drosophila	H3-K27	Repression
	SET2	SET2	Yeast	H3-K36	Activation
	DOT1L/Dot1		Humans, Mice, Yeast	H3-K79	Activation
	PR-Set7/SET8		Humans	H4-K20	Repression
	SUV4-20		Humans	H4-K20	Repression
	Riz1		Humans	H4-K9	Repression
Arginine HMT	PRMT1	AR, NUMAC, NCOA2, PCAF	Humans, Mice	H4-R3	Activation
	CARM1/PRMT4	Methylosome	Humans, Mice	H3-K2, K17, K26	Activation
	PRMT5/JBP1		Humans, Mice	H3/H4	Repression
	PRMT7		Humans, Mice	H2A	ND
	HSL7		Yeast	H2A	ND

"HMT" indicates the catalytic HMT subunit; "complex" indicates the purified HMT complex. The substrates refer to the *in vitro* specificity of the HMT subunit. ND: not determined already methylated at some of the sites shown in the Fig. 1. Histone methylation can modulate nucleosomal function and ultimately contributes to different biological processes. However, unlike histone acetylation which in general correlates with transcriptional activation, histone methylation can result in either transcription activation or repression, depending on the modified residue within histone proteins (Table 2). Histone methylation is predicted to be much more stable than other modifications under physiological conditions, and this stability raised the possibility that histone methylation is an irreversible modification. However, recent studies have identified some of enzymes (JHDM1/2A, LSD1, PAD4) that remove lysine methylations at H3-K4/K9, H3-K36 and arginine methylations at H3/H4 for proper regulation of histone methylation (Cuthbert *et al.*, 2004; Shi *et al.*, 2004; Wang *et al.*, 2004; Whetstine *et al.*, 2006; Yamane *et al.*, 2006).

Lysine can be mono-, di-, or tri-methylated by accepting one, two, or three methyl groups on histone H3 at K4, K9, K27, K36 and K79 and on histone H4 at K20. With the exception of Dot1 catalyzing H3-K79 methylation, all the histone lysine methyltransferases identified so far contain a conserved 140 amino acid catalytic domain known as the SET domain. The past decade of biochemical and cellular studies support the idea that HMTs are the key players in transcription regulation. Among the many different sites of methylation, methylations at H3-K4 and H3-K9 are the best studied example of histone methylation-mediated transcriptional regulation. Di- and tri-methylation of H3-K4, often in combination with H3-K14 acetylation, have been primarily linked to transcriptional activation (Santos-Rosa *et al.*, 2002). Conversely, di- and tri-methylation of H3-K9 have been associated with chromatin condensation for transcriptional repression (Noma *et al.*, 2001). Due to their apparent functional contradiction, these two modifications are mutually exclusive in their localization within chromatin. Some of the first identified human HMTs include SET7/9 and SUV39H1 responsible for H3-K4 and H3-K9 modifications, respectively (Rea *et al.*, 2000; Wang *et al.*, 2001a; Nishioka *et al.*, 2002). Similar to K4 methylation, methylations of H3-K36 and H3-K79 have also been implicated as an active mark for transcription. In contrast, methylation of H4-K20 is involved in chromatin condensation and thus linked to transcription repression.

Recent investigations also discovered histone arginine methylation which can be either mono- or di-methylated with the latter in symmetric or asymmetric configurations on histone H3 at R2, R17, R26 and histone H4 at R3 (Lee *et al.*, 2005). At this point, five arginine methyltransferases with a highly conserved catalytic domain have been identified. The best characterized include PRMT1 and CARM1, which respectively methylate H4-R3 and H3-R2, R17, R26 (Chen *et al.*, 1999; Schurter *et al.*, 2001; Wang *et al.*, 2001b). As several laboratories have shown, these arginine methyltransferases correlate with gene activation (Chen *et al.*, 1999; Wang *et al.*, 2001b; Bauer *et al.*, 2002; An *et al.*, 2004). Other less characterized arginine methyltransferases are PRMT5/JBP1, PRMT7, HSL7 but their effects on transcription have not been identified. Compared to lysine methylation, less has been reported regarding the effect of histone arginine methylation in transcription. Recent *in vivo* studies proposed methylations of H4-R3 (by PRMT1) and H3-R17 (by CARM1) acting as the mark of active chromatin. However, considering the intrinsic cofactor activities of PRMT1 and CARM1, it has been difficult to determine if histone modification by these coactivators is required for activation of target gene transcription. As for the effect of p300-mediated histone acetylation, the role of CARM1- and PRMT1-mediated methylation of histone proteins has been investigated by a recent study with recombinant chromatin templates (An *et al.*, 2004). Remarkably, blocking the methylation by mutating histone proteins abolished the ability of CARM1 and PRMT1 to activate p53-dependent transcription, confirming the importance of the coactivator mediated histone methylation in transcriptional activation.

Although all these results allowed us to recognize the effect of histone methylation in transcription, the precise mechanism of action of the modification during transcription still remains unclear. It has been shown that several transcriptional regulators contain a region known as the chromodomain which binds to histone tails with a specific lysine methylation; hence the histone code hypothesis described above can be extended to protein-protein interaction between regulatory proteins and the methylated histones. The best characterized example is the chromodomain of the histone binding protein HP1 (heterochromatin protein 1) which can specifically recognize tri-methylated H3-K9 (Bannister *et al.*, 2001; Lachner *et al.*, 2001; Nakayama *et al.*, 2001; Jacobs *et al.*, 2002). This interaction triggers chromatin reorganization which is responsible for formation of constitutive heterochromatin for gene silencing, possibly by spreading HP1 over wide chromatin regions. In contrast, methylation of H3-K4 creates a binding site for chromodomain-containing protein Chd1 which is a component of the SAGA and the SLIK complexes to activate transcription (Pray-Grant *et al.*, 2005). H3-K4 methylation is also known to be involved in the recruitment of Isw1p ATPase (Santos-Rosa *et al.*, 2003) that catalyzes the chromatin remodeling process during transcription activation.

Another emerging paradigm is that different histone modifications could mutually affect each other, thereby enabling a single modification to exert distinct effects depending on its context (Zhang and Reinberg, 2001; Berger, 2002; Fischle *et al.*, 2003). This so-called "cross-talk" among different histone modifications can take place within the same tail (in cis) or between different tails (in trans). Several studies demonstrated that the context of preexisting phosphorylation on H3-S10 or methylation on H3-K4 promotes histone acetylation, especially at H3-K14, and these ordered activities play an important role in regulating transcription of specific genes (Lo *et al.*, 2000, Wang *et al.*, 2001a). It was also shown that histone H3-K9 trimethylation is required for the induction of H4-K20 trimethylation which functions as a repressive mark in gene-silencing mechanisms (Kourmouli *et al.*, 2004; Schotta *et al.*, 2004). This cross-talk between multiple modifications is best-illustrated by the interdependency of p300-mediated lysine acetylation and PRMT1/CARM1-mediated arginine methylation of histones H3 and H4 (Wang *et al.*, 2001b; Daujat *et al.*, 2002; An *et al.*, 2004). These modifications, each of which exerts positive influences on transcription, are cooperative in p53-mediated transcription from the GADD45 gene (An *et al.*, 2004). However, the issue of their cooperativity is further complicated by the result that a specific order of actions of p300, PRMT1 and CARM1 is required for optimal histone modification and transcription activities. Thus methylation of H4-R3 by PRMT1 facilitates acetylation by p300 on H4-K8, K12 and this enhanced acetylation has a stimulatory effect on methylation of H3-R17 by CARM1; these joint effects of PRMT1, p300 and CARM1 play an important role in transcriptional regulation (An *et al.*, 2004). Given this functional synergy between different histone modifying activities, it would be possible that different transcriptional regulators employ a distinct combination of cofactors to induce a unique pattern of histone modifications in different signal transduction pathways, activating the transcription of specific target genes.

4. ABBERANT HISTONE MODIFICATION AND HUMAN DISEASE

Consistent with the critical role of HAT/HMT in transcription processes, molecular and genetic studies revealed that dysfunction of these epigenetic activities results in many human disorders (Table 3). With respect to histone acetylation, there are some diseases which are caused by inactivation or suppression of HAT activities. Among known HATs, p300/CBP and PCAF have been characterized as a key player in cell differentiation, growth, transformation and apoptosis and somatic mutations of these proteins have been identified in primary human tumors (Table 3) (Goodman and Smolik, 2000; Chan and La Thangue, 2001; Timmermann *et al.*, 2001). Germ line

Table 3. Selected list of HAT and HMT disruptions in human diseases

Gene	Function	Alteration	Disease
CBP	HAT	Mutation	Colon, lung and stomach cancer
		MLL gene fusion	Leukemia
		MOZ/MORF gene fusion	Leukemia
		Deletion/Point mutation	Rubinstein-Taybi syndrom
P300	HAT	Mutation	Colon, breast and stomach cancer
		MLL gene fusion	Leukemia
		MOZ gene fusion	Rubinstein-Taybi syndrom
PCAF	HAT	Mutation	Epithellal cancer
MOZ	HAT	Translocation	Leukemia
MORF	HAT	Translocation	Uterine leiomyomata, Leukemia
MLL1	HMT	Translocation	Leukemia, Gastric cancer
EZH2	HMT	Overexpression	Prostate, breast cancer
		Deletion	Myeloid disorder
ESET	HMT	Deletion	Breast cancer
G9a	HMT	Deletion	Breast cancer
RIZ1	HMT	Mutation, CpG hypermethylation	Lymphoma, Liver and colorectal cancer
SET7/9	HMT	Deletion	Liver and brain cancer
SMYD3	HMT	Suppression	Breast, colon and liver cancer

mutations of the CBP locus were reported to be the genetic basis for Rubinstein–Taybi syndrome, a developmental haploinsufficiency disorder (Petrij *et al.*, 1995). Patients with this syndrome exhibit an increased risk of developing malignant tumors at an early age with symptoms of wide thumbs, broad toes, craniofacial defects, and mental retardation (Murata *et al.*, 2001; Gibbons, 2005). Mutations of the p300 locus have also been detected in primary solid tumors and tumor cell lines of epithelial origin (Muraoka *et al.*, 1996; Gayther *et al.*, 2000) and these specific mutations have been shown to lack HAT activity (Ito *et al.*, 2001). Furthermore, expression of ectopic p300 has a repressive effect on the growth of human carcinoma cells (Suganuma *et al.*, 2002), qualifying p300 as a primary tumor suppressor protein.

With more information on the structure and function of different HATs, the regulation of these enzymatic activities has become an important topic in developing epigenetic drugs. Several inhibitors and enhancers of HAT have been explored to be effective in modulating HAT activities. Analogues of H3 tail and acetyl-CoA have been designed as inhibitors specific for the PCAF and p300, respectively (Lau *et al.*, 2000). A recent study also identified several molecules including anacardic acid and garcinol as strong inhibitors of p300 and PCAF (Balasubramanyam *et al.*, 2004a). However, these inhibitors generally have low cell permeability and high metabolic instability *in vivo*, which prevents their clinical application. Upon search for new molecules, a further study identified a natural compound curcumin as a potent, cell permeable inhibitor for p300 HAT activity and p300-mediated transcription (Balasubramanyam *et al.*, 2004b). Therefore the key challenge for current studies is to facilitate its cellular transfer and to increase its cellular stability for more efficient deregulation of transcription.

Abnormalities of global and targeted histone acetylation have also been associated with inappropriate HDAC activities which will modify chromatin by removing acetyl groups from histone tails; thus synthetic or natural HDAC inhibitors have a great potential utility for alleviating aberrant transcriptional repression (de Ruijter *et al.*, 2003; Marks *et al.*, 2003). Structural characterization of HDAC/HDAC inhibitor complexes showed HDAC inhibitors blocking substrate access of HDAC by binding to the catalytic site (Finnin *et al.*, 1999). Currently, several inhibitors have been identified with *in vivo* capability to inhibit many of the known intracellular HDAC activities (Yoo and Jones, 2006). For example, a study of the transcriptional effects of a HDAC inhibitor SAHA in tumor cells demonstrated a marked change in gene transcription which coincides with the accumulation of acetylated histones (Mitsiades *et al.*, 2004). Although the molecular mechanism for their anticancer selectivity remains obscure, the HDAC inhibitors are believed to alter chromatin structure that leads to expression of tumor suppressor genes. Altered transcriptional status would also include the repression of genes involved in cell cycle and proliferation as well as in antiapoptotic processes. Furthermore, some HDAC inhibitors have the potential to modulate additively or synergistically the activity of other therapeutic agents; thus HDAC inhibitors could be combined with other types of drugs for more efficient regulation of uncontrolled gene transcription. Defining downstream effects of HDAC inhibition should provide a mechanistic rationale for the use of combinational therapies with HDAC inhibitors.

Similar to HATs/HDACs, misregulation of HMTs is also connected to several human diseases. The best recognized case is the early age translocation of HMT mixed lineage leukemia (MLL) gene which is a characteristic of human leukemia (Hess, 2004). Most of these chromosomal rearrangements involve the amino terminal part of MLL fused to diverse fusion partners. This kind of translocation leads to the misregulation of fusion proteins and causes inappropriate changes of gene expression. For example, the N-terminal part of MLL gene is translocated onto the genes for CBP, p300, MORF and MOZ by leukemia-associated chromosomal abnormalities, giving rise to different forms of fusion proteins. Such fusion proteins exhibit properties different from their wild-type counterparts, and these abnormal activities generate aberrant changes of histone methylation and acetylation in leukemia.

Increasing evidence also supports the conclusion that lysine methylations are closely related to various human cancer. Among the HMTs involved in these processes, the best characterized is SUV39H1 which is able to trimethylate H3-K9. This specific modification serves as a binding site for the heterochromatin protein-1 (HP1) to establish a stable heterochromatin configuration and gene silencing. Thus the deregulation of SUV39H1 could lead to destabilized heterochromatin structure with increased risk of cancer. The role of SUV39H1 in oncogenesis is also supported by the study demonstrating that SUV39H-deficient mice have an increased risk of tumor and B-cell lympomas as a result of chromatin instability (Peters *et al.*, 2001). Recent reports also have shown that H3-K4 HMT SMYD3 plays a crucial role in hepatocellular and colorectal carcinoma. Since SMYD3 appears to upregulate oncogene expression, suppression of SMYD3 causes inhibition of cancer cell growth (Hamamoto *et al.*, 2004). Another link between HMT and cancer has been found in HMT EZH2 which specifically methylates H3-K27 as a component of human EED-EZH2 complex. EZH2 is overexpressed in metastatic prostate cancer, lymphomas and breast cancer (Varambally *et al.*, 2002; Kleer *et al.*, 2003). These striking observations rationalize current efforts to regulate HMT activities as a new class of epigenetic therapy. Although some S-adenosylhomocysteine (SAH) analogues have been designed to target some protein methyltransferases (Chie *et al.*, 2003), inhibitors specific for HMTs have not yet been described. The development of specific inhibitors to counteract HMT activity should facilitate our characterization of HMT function *in vivo* and provide more efficient disease protection.

Although the exact mechanism has not been elucidated, histone acetylation and methylation have also been liked to DNA repair. Studies of TIP60 HAT complex in human cells indicated that TIP60-mediated acetylation of H2A and H4 is a key process for efficient repair of double strand DNA break (DBS) (Ikura *et al.*, 2000; Murr *et al.*, 2006). DOT1-mediated methylation of H3-K79 has also been identified as an interaction signal for tandem tudor domain of DSB check point protein 53BP1, and this interaction mediates the 53BP1 targeting to sites of DNA DSBs (Huyen *et al.*, 2004). Another study, conducted by Sanders *et al.,* also found an interesting connection between newly identified HMT Set9 and cellular DNA damage response; methylation of H4-K20 by Set9 results in the recruitment of DNA damage checkpoint protein Crb2 to maintain cell survival after DNA double strand breaks (Sanders *et al.*, 2004). Recent studies also demonstrated that the cooperative function of multiple HATs and HMTs is required for the gene specific regulatory action of tumor suppressor p53 in response to various stress signals (Coutts and La Thangue, 2005). Notably, the CARM1/PRMT1 HMTs and p300 HAT have been shown to physically interact with p53 and cooperatively mediate p53-dependent transcriptional induction of the GADD45 gene during the checkpoint response to DNA damage (An *et al.*, 2004).

Therefore our major challenge on aberrant histone modifications is to understand the biological functions of diverse HATs and HMTs and their involvement in various human diseases. In spite of the fact that most of what we investigate regarding the role of histone modifications in the transcription process has been obtained indirectly from *in vivo* experimental data, we are beginning to unravel the complexities of gene transcription induced by diverse patterns of histone modifications. Since histone acetylation and methylation appear to create signals for many transcription and chromatin regulators, the maintenance of a specific pattern of histone modification will be crucial for proper cell proliferation. Perhaps *in vitro* transcription studies with recombinant chromatin, in which numerous histone modifications are specifically regulated, will be useful to determine the most important HAT and/or HMT in a specific gene transcription pathway. This type of technology should also provide the means for systematic investigations of regulatory mechanisms and actions of HAT/HMT in transcription during pathologic processes, such as cancer. Although this review has narrowly focused on histone acetylation and methylation, the more complicated interplay of multiple modifications as a dominant mechanism of transcription regulation is also an important question to be addressed.

5. CONCLUSIONS

The regulation of gene expression is obviously an extremely complex process involving multiple regulatory factors. Much of the investigation on various histone modifications so far has been centered on the identification of histone-modifying factors and their effect in gene transcription. However, in spite of the continued effort in analyzing the function and regulation of HAT and HMT, more work is required to unarguably determine the outcome of histone modification in transcription. Such functional assays might include the assessment of transcription under conditions employing wild-type and mutant histone proteins. This approach will clearly answer if histones are the primary target of modifications for gene regulation, or if histone modifying cofactors have some other targets for their dynamic role. If modified histones are identified to be crucial for the control of the transcription process, enabling or blocking of histone modification should be an excellent therapeutic strategy. Moreover, the identification of molecular mechanisms by which different histone modifications exert their effects in transcription will yield important insights on the development of more targeted therapy with less side effects.

ACKNOWLEDGEMENTS

Given the limitation in space, I apologize for not being able to cite all the original works in this manuscript. I am grateful to Dr. Michael Stallcup for critical reading of this manuscript. This work was supported in part by grants from the "V" foundation, "Wright" foundation and Margaret E. Early Medical Research Trust to WA.

REFERENCES

Agalioti T, Chen G, Thanos D (2002) Deciphering the transcriptional histone acetylation code for a human gene. Cell 111:381–392

Allfrey VG, Faulkner R, Mirsky AE (1964) Acetylation and methylation of histones and their possible role in the regulation of RNA synthesis. Proc Natl Acad Sci USA 51:786–794

An W, Kim J, Roeder RG (2004) Ordered cooperative functions of PRMY1, p300, and CARM1 in transcriptional activation by p53. Cell 117:1–20

An W, Palhan VB, Karymov MA, Leuba SH, Roeder RG (2002) Selective requirements for histone H3 and H4 N termini in p300-dependent transcriptional activation from chromatin. Mol Cell 9:811–821

Balasubramanyam K, Altaf M, Varier RA, Swaminathan V, Ravindran A, Sadhale PP, Kundu TK (2004a) Polyisoprenylated benzophenone, garcinol, a natural histone acetyltransferase inhibitor, represses chromatin transcription and alters global gene expression. J Biol Chem 279:33716–33726

Balasubramanyam K, Varier RA, Altaf M, Swaminathan, V, Siddappa NB, Ranga U, Kundu TK (2004b) Curcumin, a novel p300/CREB-binding protein-specific inhibitor of acetyltransferase, represses the acetylation of histone/nonhistone proteins and histone acetyltransferase-dependent chromatin transcription. J Biol Chem 279:51163–51171

Bannister AJ, Kouzarides T (1996) The CBP co-activator is a histone acetyltransferase. Nature 384: 641–643

Bannister AJ, Zegerman P, Partridge JF, Miska EA, Thomas JO, Allshire RC, Kouzarides T (2001) Selective recognition of methylated lysine 9 on histone H3 by the HP1 chromo domain. Nature 410:120–124

Bauer UM, Daujat S, Nielsen SJ, Nightingale K, Kouzarides T (2002) Methylation at arginine 17 of histone H3 is linked to gene activation. EMBO Rep 3:39–44

Berger SL (2002) Histone modifications in transcriptional regulation. Curr Opin Genet Dev 12:142–148

Chan HM, La Thangue NB (2001) p300/CBP proteins: HATs for transcriptional bridges and scaffolds. J Cell Sci 114:2363–2373

Chen D, Ma H, Hong H, Koh SS, Huang SM, Schurter BT, Aswad DW, Stallcup MR (1999) Regulation of transcription by a protein methyltransferase. Science 284:2174–2177

Chie L, Cook JR, Chung D, Hoffmann R, Yang Z, Kim Y, Pestka S, Pincus MR (2003) A protein methyl transferase, PRMT5, selectively blocks oncogenic ras-p21 mitogenic signal transduction. Ann Clin Lab Sci 33:200–207

Coutts AS, La Thangue NB (2005) The p53 response: emerging levels of co-factor complexity. Biochem Biophys Res Commun 331:778–785

Cuthbert GL, Daujat S, Snowden AW, Erdjument-Bromage H, Hagiwara T, Yamada M, Schneider R, Gregory PD, Tempst P, Bannister AJ, Kouzarides T (2004) Histone deimination antagonizes arginine methylation. Cell 118:545–553

Daujat S, Bauer UM, Shah V, Turner B, Berger S, Kouzarides T (2002) Crosstalk between CARM1 methylation and CBP acetylation on histone H3. Curr Biol 12:2090–2097

de Ruijter AJ, van Gennip AH, Caron HN, Kemp S, van Kuilenburg AB (2003) Histone deacetylases (HDACs): characterization of the classical HDAC family. Biochem J 370:737–749

Deckert J, Struhl K (2001) Histone acetylation at promoters is differentially affected by specific activators and repressors. Mol Cell Biol 21:2726–2735

Dion MF, Altschuler SJ, Wu LF, Rando OJ (2005) Genomic characterization reveals a simple histone H4 acetylation code. Proc Natl Acad Sci USA 102:5501–5506

Finnin MS, Donigian JR, Cohen A, Richon VM, Rifkind RA, Marks PA, Breslow R, Pavletich NP (1999) Structures of a histone deacetylase homologue bound to the TSA and SAHA inhibitors. Nature 401:188–193

Fischle W, Wang Y, Allis CD (2003) Histone and chromatin cross-talk. Curr Opin Cell Biol 15:172–183

Garcia-Ramirez M, Rocchini C, Ausio J (1995) Modulation of chromatin folding by histone acetylation. J Biol Chem 270:17923–17928

Gayther SA, Batley SJ, Linger L, Bannister A, Thorpe K, Chin SF, Daigo Y, Russell P, Wilson A, Sowter HM, Delhanty JD, Ponder BA, Kouzarides T, Caldas C (2000) Mutations truncating the EP300 acetylase in human cancers. Nat Genet 24:300–303

Georges SA, Kraus WL, Luger K, Nyborg JK, Laybourn PJ (2002) p300-mediated tax transactivation from recombinant chromatin: histone tail deletion mimics coactivator function. Mol Cell Biol 22:127–137

Gibbons RJ (2005) Histone modifying and chromatin remodelling enzymes in cancer and dysplastic syndromes. Hum Mol Genet 14:R85–92

Glozak MA, Sengupta N, Zhang X, Seto E (2005) Acetylation and deacetylation of non-histone proteins. Gene 363:15–23

Goodman RH, Smolik S (2000) CBP/p300 in cell growth, transformation, and development. Genes Dev 14:1553–1577

Grunstein M (1997) Histone acetylation in chromatin structure and transcription. Nature 389:349–352

Grunstein M, Hecht A, Fisher-Adams G, Wan J, Mann RK, Strahl-Bolsinger S, Laroche T, and Gasser S (1995) The regulation of euchromatin and heterochromatin by histones in yeast. J Cell Sci Suppl 19:29–36

Hamamoto R, Furukawa Y, Morita M, Iimura Y, Silva FP, Li M, Yagyu R, Nakamura Y (2004) SMYD3 encodes a histone methyltransferase involved in the proliferation of cancer cells. Nat Cell Biol 8:731–740

Hansen JC (2002) Conformational dynamics of the chromatin fiber in solution: determinants, mechanisms, and functions. Annu Rev Biophys Biomol Struct 31:361–392

Hansen JC, Tse C, Wolffe AP (1998) Structure and function of the core histone N-termini: more than meets the eye. Biochemistry 37:17637–17641

Hassan AH, Prochasson P, Neely KE, Galasinski SC, Chandy M, Carrozza MJ, Workman JL (2002) Function and selectivity of bromodomains in anchoring chromatin-modifying complexes to promoter nucleosomes. Cell 111:369–379

Hess JL (2004) MLL: a histone methyltransferase disrupted in leukemia. Trends Mol Med 10:500–507

Huyen Y, Zgheib O, Ditullio RA Jr, Gorgoulis VG, Zacharatos P, Petty TJ, Sheston EA, Mellert HS, Stavridi ES, Halazonetis TD (2004) Methylated lysine 79 of histone H3 targets 53BP1 to DNA double-strand breaks. Nature 432:406–411

Ikura T, Ogryzko VV, Grigoriev M, Groisman R, Wang J, Horikoshi M, Scully R, Qin J, Nakatani Y (2000) Involvement of the TIP60 histone acetylase complex in DNA repair and apoptosis. Cell 102:463–473

Ito, A, Lai CH, Zhao X, Saito S, Hamilton MH, Appella E, Yao TP (2001) p300/CBP-mediated p53 acetylation is commonly induced by p53-activating agents and inhibited by MDM2. EMBO J 20:1331–1340

Jacobs SA, Khorasanizadeh S (2002) Structure of the HP1 chromodomain bound to a lysine 9-methylated histone H3 tail. Science 295:2080–2083

Jenuwein T, Allis CD (2001) Translating the histone code. Science 293:1074–1079

Kanno T, Kanno Y, Siegel RM, Jang MK, Lenardo MJ, Ozato K (2004) Selective recognition of acetylated histones by bromodomain proteins visualized in living cells. Mol Cell 13:33–43

Kleer CG, Cao Q, Varambally S, Shen R, Ota I, Tomlins SA, Ghosh D, Sewalt RG, Otte AP, Hayes DF, Sabel MS, Livant D, Weiss SJ, Rubin MA, Chinnaiyan AM (2003) EZH2 is a marker of aggressive breast cancer and promotes neoplastic transformation of breast epithelial cells. Proc Natl Acad Sci USA 100:11606–11611

Kornberg RD, Lorch Y (1999) Twenty-five years of the nucleosome, fundamental particle of the eukaryote chromosome. Cell 98:285–294

Kourmouli N, Jeppesen P, Mahadevhaiah S, Burgoyne P, Wu R, Gilbert DM, Bongiorni S, Prantera G, Fanti L, Pimpinelli S, Shi W, Fundele R, Singh PB (2004) Heterochromatin and tri-methylated lysine 20 of histone H4 in animals. J Cell Sci 117:2491–2501

Kouzarides T (2002) Histone methylation in transcriptional control. Curr Opin Genet Dev 12:198–209

Kundu TK, Palhan VB, Wang Z, An W, Cole PA, Roeder RG (2000) Activator-dependent transcription from chromatin *in vitro* involving targeted histone acetylation by p300. Mol Cell 6:551–561

Kuo MH, Allis CD (1998) Roles of histone acetyltransferases and deacetylases in gene regulation. Bioessays 8:615–626

Lachner M, O'Carroll D, Rea S, Mechtler K Jenuwein T (2001) Methylation of histone H3 lysine 9 creates a binding site for HP1 proteins. Nature 410:116–120

Lau OD, Kundu TK, Soccio RE, Ait-Si-Ali S, Khalil EM, Vassilev A, Wolffe AP, Nakatani Y, Roeder RG, Cole PA (2000) HATs off: selective synthetic inhibitors of the histone acetyltransferases p300 and PCAF. Mol Cell 5:589–595

Lee DY, Teyssier C, Strahl BD, Stallcup MR (2005) Role of protein methylation in regulation of transcription. Endocr Rev 26:147–170

Levenstein ME, Kadonaga JT (2002) Biochemical analysis of chromatin containing recombinant Drosophila core histones. J Biol Chem 277:8749–8754

Liang G, Lin JC, Wei V, Yoo C, Cheng JC, Nguyen CT, Weisenberger DJ, Egger G, Takai D, Gonzales FA, Jones PA (2004) Distinct localization of histone H3 acetylation and H3-K4 methylation to the transcription start sites in the human genome. Proc Natl Acad Sci USA 101:7357–7362

Lo WS, Trievel RC, Rojas JR, Duggan L, Hsu JY, Allis CD, Marmorstein R, Berger SL (2000) Phosphorylation of serine 10 in histone H3 is functionally linked *in vitro* and *in vivo* to Gcn5-mediated acetylation at lysine 14. Mol Cell 5:917–926

Loyola A, LeRoy G, Wang YH, Reinberg D (2001) Reconstitution of recombinant chromatin establishes a requirement for histone-tail modifications during chromatin assembly and transcription. Genes Dev 15:2837–2851

Luger K, Richmond TJ (1998) The histone tails of the nucleosome. Curr Opin Genet Dev. 8:140–146

Luger K, Maeder AW, Richmond RK, Sargent DF, Richmond TJ (1997) X-ray structure of the nucleosome core particle at 2.8 A resolution. Nature 389:251–259

Marks PA, Miller T, Richon VM (2003) Histone deacetylases. Curr Opin Pharmacol 3:344–351

Mitsiades CS, Mitsiades NS, McMullan CJ, Poulaki V, Shringarpure R, Hideshima T, Akiyama M, Chauhan D, Munshi N, Gu X, Bailey C, Joseph M, Libermann TA, Richon VM, Marks PA, Anderson KC (2004) Transcriptional signature of histone deacetylase inhibition in multiple myeloma: biological and clinical implications. Proc Natl Acad Sci USA 101:540–545

Mizzen CA, Yang XJ, Kokubo T, Brownell JE, Bannister AJ, Owen-Hughes T, Workman J, Wang L, Berger SL, Kouzarides T, Nakatani Y, Allis CD (1996) The $TAF_{II}250$ subunit of TFIID has histone acetyltransferase activity. Cell 87:1261–1270

Mosammaparast N, Guo Y, Shabanowitz J, Hunt DF, Pemberton LF (2002) Pathways mediating the nuclear import of histones H3 and H4 in yeast. J Biol Chem 277:862–868

Mosammaparast N, Jackson KR, Guo Y, Brame CJ, Shabanowitz J, Hunt DF, Pemberton LF (2001) Nuclear import of histone H2A and H2B is mediated by a network of karyopherins. J Cell Biol 153:251–262

Muraoka M, Konishi M, Kikuchi-Yanoshita R, Tanaka K, Shitara N, Chong JM, Iwama T, Miyaki M (1996) p300 gene alterations in colorectalgastric carcinomas. Oncogene 12:1565–1569

Murata T, Kurokawa R, Krones A, Tatsumi K, Ishii M, Taki T, Masuno M, Ohashi H, Yanagisawa M, Rosenfeld MG, Glass CK, Hayashi Y (2001) Defect of histone acetyltransferase activity of the nuclear transcriptional coactivator CBP in Rubinstein-Taybi syndrome. Hum Mol Genet 10:1071–1076

Murr R, Loizou JI, Yang YG, Cuenin C, Li H, Wang ZQ, Herceg Z (2006) Histone acetylation by Trrap-Tip60 modulates loading of repair proteinsrepair of DNA double-strand breaks. Nat Cell Biol 8:91–99

Nakayama J, Rice JC, Strahl BD, Allis CD, Grewal SI (2001) Role of histone H3 lysine 9 methylation in epigenetic control of heterochromatin assembly. Science 292:110–113

Nishioka K, Chuikov S, Sarma K, Erdjument-Bromage H, Allis CD, Tempst P, Reinberg D (2002) Set9, a novel histone H3 methyltransferase that facilitates transcription by precluding histone tail modifications required for heterochromatin formation. Genes Dev 16:479–489

Noma K, Allis CD, Grewal SI (2001) Transitions in distinct histone H3 methylation patterns at the heterochromatin domain boundaries. Science 293:1150–1155

Ogryzko VV, Schiltz RL, Russanova V, Howard BH, Nakatani Y (1996) The transcriptional coactivators p300 and CBP are histone acetylases. Cell 87:953–959

Peters AH, O'Carroll D, Scherthan H, Mechtler K, Sauer S, Schofer C, Weipoltshammer K, Pagani M, Lachner M, Kohlmaier A, Opravil S, Doyle M, Sibilia M, Jenuwein T (2001) Loss of the Suv39h histone methyltransferases impairs mammalian heterochromatin and genome stability. Cell 107:323–337

Peterson CL, Laniel MA (2004) Histones and histone modifications. Curr Biol 14:R546–551

Petrij F, Giles RH, Dauwerse HG, Saris JJ, Hennekam RC, Masuno M, Tommerup N, van Ommen GJ, Goodman RH, Peters DJ, Breuning MH (1995) Rubinstein-Taybi syndrome caused by mutations in the transcriptional co-activator CBP. Nature 376:348–351

Pray-Grant MG, Daniel JA, Schieltz D, Yates JR 3rd, Grant PA (2005) Chd1 chromodomain links histone H3 methylation with SAGA- and SLIK-dependent acetylation. Nature 433:434–438

Rea S, Eisenhaber F, O'Carroll D, Strahl BD, Sun ZW, Schmid M, Opravil S, Mechtler K, Ponting CP, Allis CD, Jenuwein T (2000) Regulation of chromatin structure by site-specific histone H3 methyltransferases. Nature 406:593–599

Ren Q, Gorovsky MA (2001) Histone H2A.Z acetylation modulates an essential charge patch. Mol Cell 7:1329–1335

Roh TY, Cuddapah S, Zhao K (2005) Active chromatin domains are defined by acetylation islands revealed by genome-wide mapping. Genes Dev 19:542–552

Roth SY, Denu JM, Allis CD (2001) Histone acetyltransferases. Annu Rev Biochem 70:81–120

Sanders SL, Portoso M, Mata J, Bahler J, Allshire RC, Kouzarides T (2004) Methylation of histone H4 lysine 20 controls recruitment of Crb2 to sites of DNA damage. Cell 119:603–614

Santos-Rosa H, Schneider R, Bannister AJ, Sherriff J, Bernstein BE, Emre NC, Schreiber SL, Mellor J, Kouzarides T (2002) Active genes are tri-methylated at K4 of histone H3. Nature 419:407–411

Santos-Rosa H, Schneider R, Bernstein BE, Karabetsou N, Morillon A, Weise C, Schreiber SL, Mellor J, Kouzarides T (2003) Methylation of histone H3 K4 mediates association of the Isw1p ATPase with chromatin. Mol Cell 12:1325–1332

Schotta G, Lachner M, Sarma K, Ebert A, Sengupta R, Reuter G, Reinberg D, Jenuwein T (2004) A silencing pathway to induce H3-K9 and H4-K20 trimethylation at constitutive heterochromatin. Genes Dev 18:1251–1262

Schurter BT, Koh SS, Chen D, Bunick GJ, Harp JM, Hanson BL, Henschen-Edman A, Mackay DR, Stallcup MR, Aswad DW (2001) Methylation of histone H3 by coactivator-associated arginine methyltransferase 1. Biochemistry 40:5747–5756

Shi Y, Lan F, Matson C, Mulligan P, Whetstine JR, Cole PA, Casero RA, Shi Y (2004) Histone demethylation mediated by the nuclear amine oxidase homolog LSD1. Cell 119:941–953

Shilatifard A (2006) Chromatin Modifications by Methylation and Ubiquitination: Implications in the Regulation of Gene Expression. Annu Rev Biochem 75:243–269

Sterner DE, Berger SL (2000) Acetylation of histones and transcription-related factors. Microbiol. Mol Biol Rev 64:435–459

Strahl BD, Allis CD (2000) The language of covalent histone modifications. Nature 403:41–45

Suganuma T, Kawabata M, Ohshima T, Ikeda MA (2002) Growth suppression of human carcinoma cells by reintroduction of the p300 coactivator. Proc Natl Acad Sci USA 99:13073–13078

Timmermann S, Lehrmann H, Polesskaya A, Harel-Bellan A (2001) Histone acetylation and disease. Cell Mol Life Sci 58:728–736

Tse C, Sera T, Wolffe AP, Hansen JC (1998) Disruption of higher-order folding by core histone acetylation dramatically enhances transcription of nucleosomal arrays by RNA polymerase III. Mol Cell Biol 18:4629–4638

Turner BM (2000) Histone acetylation and an epigenetic code. Bioessays 22:836–845

van Holde KE (1988) Chromatin. Springer-Verlag, New York
Varambally S, Dhanasekaran SM, Zhou M, Barrette TR, Kumar-Sinha C, Sanda MG, Ghosh D, Pienta KJ, Sewalt RG, Otte AP, Rubin MA, Chinnaiyan AM (2002) The polycomb group protein EZH2 is involved in progression of prostate cancer. Nature 419:624–629
Wade PA, Pruss D, Wolffe AP (1997) Trends Biochem Sci 22:128–132
Wang H, Cao R, Xia L, Erdjument-Bromage H, Borchers C, Tempst P, Zhang Y (2001a) Purification and functional characterization of a histone H3-lysine 4-specific methyltransferase. Mol Cell 8:1207–1217
Wang H, Huang ZQ, Xia L, Feng Q, Erdjument-Bromage H, Strahl BD, Briggs SD, Allis CD, Wong J, Tempst P, Zhang Y (2001b) Methylation of histone H4 at arginine 3 facilitating transcriptional activation by nuclear hormone receptor. Science 293:853–857
Wang Y, Wysocka J, Sayegh J, Lee YH, Perlin JR, Leonelli L, Sonbuchner LS, McDonald CH, Cook RG, Dou Y, Roeder RG, Clarke S, Stallcup MR, Allis CD, Coonrod SA (2004) Human PAD4 regulates histone arginine methylation levels via demethylimination. Science 306:279–283
Whetstine JR, Nottke A, Lan F, Huarte M, Smolikov S, Chen Z, Spooner E, Li E, Zhang G, Colaiacovo M, Shi Y (2006) Reversal of histone lysine trimethylation by the JMJD2 family of histone demethylases. Cell 125:467–481
Xu W, Edmondson DG, Evrard YA, Wakamiya M, Behringer RR, Roth SY (2000) Loss of Gcn5l2 leads to increased apoptosis and mesodermal defects during mouse development. Nat Genet 26:229–232
Yamane K, Toumazou C, Tsukada YI, Erdjument-Bromage H, Tempst P, Wong J, Zhang Y (2006) JHDM2A, a JmjC-Containing H3K9 Demethylase, Facilitates Transcription Activation by Androgen Receptor. Cell 125:483–495
Yang XJ, Seto E (2003) Collaborative spirit of histone deacetylases in regulating chromatin structure and gene expression. Curr Opin Genet Dev 13:143–153
Yoo CB, Jones PA (2006) Epigenetic therapy of cancer: past, present and future. Nat Rev Drug Discov 5:37–50
Zeng L, Zhou MM (2002) Bromodomain: an acetyl-lysine binding domain. FEBS letters 513:124–128
Zhang Y, Reinberg D (2001) Transcription regulation by histone methylation: interplay between different covalent modifications of the core histone tails. Genes Dev 15:2343–2360

CHAPTER 17

CHROMATIN-ASSOCIATED REGULATION OF HIV-1 TRANSCRIPTION

Implications for the development of therapeutic strategies

VINCENT QUIVY* STEPHANE DE WALQUE*, AND CARINE VAN LINT[†]

Université Libre de Bruxelles (ULB), Institut de Biologie et de Médecine Moléculaires (IBMM), Laboratoire de Virologie Moléculaire, 12, rue des Professeurs Jeener et Brachet 6041 Gosselies, Belgium

Abstract: Human Immunodeficiency Virus type 1 (HIV-1) infection can now be treated effectively in many patients in the developed world, using combinations of antiretroviral therapeutics, called Highly Active Anti-Retroviral Therapy (HAART). However, despite prolonged treatment with HAART, the persistence of latently HIV-1-infected cellular reservoirs harboring transcriptionally silent but replication-competent proviruses represents the major hurdle to virus eradication. These latently infected cells are a permanent source for virus reactivation and lead to a rebound of the viral load after interruption of HAART. Therefore, a greater understanding of the molecular mechanisms regulating proviral latency and reactivation should lead to rational strategies aimed at purging these cellular reservoirs of HIV-1. This review summarizes our current knowledge and understanding of the elements involved in HIV-1 transcriptional reactivation: (1) the site of integration; (2) the transcription factor NF-κB, which is induced by proinflammatory cytokines (such as TNFα) and binds to two κB sites in the HIV-1 promoter region; (3) the specific remodeling of a single nucleosome (called nuc-1 and located immediately downstream of the HIV-1 transcription start site under latency conditions) upon activation of the HIV-1 promoter; (4) post-translational acetylation of histones and of non-histone proteins (following treatment with deacetylases inhibitors, which induce viral transcription and nuc-1 remodeling); and (5) the viral *trans*-activator Tat, which promotes transcription by mediating the recruitment to the HIV-1 promoter of histone-modifying enzymes and ATP-dependent chromatin remodeling complexes required for nucleosome disruption and transcriptional processivity. Finally, this review highlights experimental therapies aimed at administrating HIV-1 gene expression activators (such as HDAC inhibitors) combined with an effective HAART in order to reactivate and decrease/eliminate the pool of latently HIV-1-infected cellular reservoirs

---

*Both authors contributed equally to this work

[†]E-mail: cvlint@ulb.ac.be

T.K. Kundu and D. Dasgupta (eds.), Chromatin and Disease, 371–396.

1. INTRODUCTION

HIV-1 is the etiologic agent responsible for AIDS, a syndrome characterized by depletion of $CD4^+$ T-lymphocytes and collapse of the immune system. People with AIDS are prone to opportunistic infections easily defended against by a normal immune system. Generally, it takes several years, post-infection, to progress to AIDS.

Since the 1990s, the development of Highly Active Anti-Retroviral Therapies (HAART) has dramatically improved the survival and life quality of HIV-1-infected individuals. Unfortunately, whereas these treatments significantly reduce the levels of viral RNA in plasma and lymphoid tissues, cessation of even prolonged highly suppressive HAART regimens results in viral load rebound to pre-therapy levels, indicating that anti-retroviral therapy of this type is unable to completely eliminate HIV-1 (Pierson *et al.*, 2000; Persaud *et al.*, 2003; Blankson *et al.*, 2002; Marcello, 2006). This failure has been attributed in part to the presence of a long-lived, stable population of latently infected resting memory $CD4^+$ T cells. Since these latently infected cells express no viral proteins, they are immunologically indistinguishable from uninfected cells and are insensitive to HAART (Chun *et al.*, 1997a; Finzi *et al.*, 1997; Wong *et al.*, 1997). While many HIV-1-susceptible cells are fast-turnover cells, this small part of memory T cells are long-lived cells (Michie *et al.*, 1992; Mclean and Michie, 1995). These infected cells can go dormant and stay in tissues for years despite effective HAART, thereby serving as the HIV-1 reservoirs *in vivo* (Wong *et al.*, 1997). These reservoirs have such a slow rate of decay during HAART that their eradication during a human lifespan is unlikely (Finzi *et al.*, 1999; Siliciano *et al.*, 2003).

As with all retroviruses, HIV-1 integrates into the genome of the host cell. As a consequence, HIV-1 is confronted with a unique problem in terms of transcriptional regulation and packaging into chromatin. HIV-1 proviruses can integrate in many different sites within the host cell genome, each site with its own properties susceptible of influencing the degree of viral expression. Most cells infected by HIV-1 are productively infected, that is, they go on, within days, to complete the viral replication cycle, release progeny virus and die. However, a small fraction of incoming HIV-1 enters a latent mode of infection and constitutes a reservoir of infected cells that can produce infectious virus given appropriate stimulation. HIV-1 gene expression in these latently-infected cells can be reactivated by a wide variety of signals including cytokines such as Interleukin-2, TNFα, macrophage colony-stimulating factor (MCSF), antigens and other T-cell mitogens, glucocorticoid and thyroid hormones, bacterial infections, lipopolysaccharides, ... Consequently, if HAART is ceased, viremia rapidly re-emerges, regardless of the duration of drug therapy (reviewed in Blankson *et al.*, 2002).

At the cellular level, two major forms of HIV-1 latency have been described: pre-integration latency and post-integration latency (Bisgrove *et al.*, 2005; Marcello, 2006). The first one which can not be taken into account for the formation of the long-term viral reservoirs, occur when virions fail to undergo integration and remain in the cytoplasm of the infected cell for days as a labile pre-integration complex. This latency form will not be further discussed in this review. Post-integration latency occurs when a provirus fails to effectively express its genome and is reversibly silenced after successful integration. This latent state is exceptionally stable and limited only by the lifespan of the infected cell and its progeny.

It has been proposed that one possible solution to the problem of HIV-1 latency is to purge the latent reservoirs by deliberately forcing HIV-1 gene expression in these latently infected cells in presence of HAART to prevent spreading of the infection by the newly synthesized viruses (Chun *et al.*, 1998). Such type of treatment could reduce the number of latently-infected cells by causing them to be directly killed by the cytopathic action of the virus or to be destroyed by the immune system. The definition of such strategies is clearly dependent on the knowledge of the molecular mechanisms regulating HIV-1 latency and reactivation from latency.

Much progress has recently been made to elucidate the molecular mechanisms underlying HIV-1 post-integration latency, which is intimately tied to HIV-1 transcription level. Among the possible molecular mechanisms behind HIV-1 post-integration latency are: (1) the chromatin status at the integration site; (2) the presence of the repressive nucleosome nuc-1; (3) epigenetic modifications such as acetylation (reviewed in Quivy and Van Lint, 2002); (4) the lack of activation-dependent host transcription factors such as NF-κB in resting cells; and (5) the viral *trans*-activator Tat, which promotes transcription via the recruitment to the HIV-1 promoter of chromatin-modifying complexes.

1.1. Integration Site

The eukaryotic genome is compacted with histones and other proteins to form chromatin, which allows for efficient storage of genetic information. However, this packaging also prevents the transcription machinery from gaining access to the DNA template (reviewed in Workman and Kingston, 1998; Felsenfeld and Groudine, 2003). The repeating unit of chromatin is the nucleosome core, composed of about 146 bp of DNA tightly wrapped, in 1.65 turns, in a left-handed superhelix around a central histone octamer which contains two molecules of each of the four core histones : H2A, H2B, H3 and H4. Two adjacent nucleosome cores are separated by a region of linker DNA (10–60 bp) that is associated with a single molecule of histone H1. Histones have a common structure, the "histone fold", consisting of two short α-helices and a long central helix separated by β-bridges, which are required for histone-DNA and histone-histone interactions (reviewed in Hansen, 2002).

The packaging of genes into chromatin is increasingly recognized as an important component in the regulation of transcription initiation and elongation (Wolffe, 1999). Chromatin is heterogeneous in the nucleus: transcriptionally active genes are characterized by a more diffuse chromatin structure (active chromatin or euchromatin), whereas inactive genes are packaged in a highly condensed chromatin configuration that impairs access to the underlying DNA (inactive chromatin or heterochromatin) (Craig, 2005).

The histone N-terminal tails are highly basic functional domains, which are subject to multiple post-translational modifications, including acetylation, phosphorylation, methylation, poly-ADP-ribosylation, ubiquitinylation and sumoylation (reviewed in Peterson and Laniel, 2004; Margueron *et al.*, 2005). The post-translational modifications of the histone tails can potentially alter the interaction between DNA and the histone octamers, leading to decondensation of the chromatin fiber. The "histone code" hypothesis postulates that the modifications of the histone tails are interdependent and that various combination serve to categorize regions of chromatin as transcribed, heterochromatic or centromeric (Fischle *et al.*, 2003). For example, methylation of lysine 4 of histone H3 has been associated with active gene expression, whereas methylation of lysine 9 of histone H3 has been associated with transcriptional silencing.

The molecular mechanism underlying the specific integration of a provirus into the host cell chromatin is still poorly understood. But many studies have shown that the integration site and its corresponding chromatin environment affect HIV-1 gene expression (Nahreini and Mathews, 1997; Jordan *et al.*, 2003). Jordan *et al.* developed a system to select for latently infected cellular clones (Jordan *et al.*, 2003). The integration site of 8 latently-infected HIV cell lines, which have no basal HIV transcription, was sequenced. Half of the clones were integrated within or near alphoid repeat elements in heterochromatin. In this experiment, the latently-infected clones represented less than 0.06% of the original population. This study suggests integration into heterochromatin as a mechanism leading to HIV latency.

It has been demonstrated, however, that HIV integrate preferentially within actively transcribed genes (Schroder *et al.*, 2002; Wu *et al.*, 2003). An analysis of integration sites in purified resting $CD4^+$ T cells from patients on HAART found the majority (93%) of silent provirus located within the coding region of host genes (Han *et al.*, 2004). In this case, transcriptional interference provides another potential explanation for HIV-1 latency. Transcriptional interference can occur through several different mechanisms including enhancer trapping, promoter occlusion, or steric hindrance (reviewed in Lassen *et al.*, 2004; Bisgrove *et al.*, 2005). However, the replication competence of the proviruses was not analyzed in this study. It has been estimated that there are 100 defective proviruses for each latent provirus, making it likely that many of the viral integration sites represent replication-incompetent defective provirus (Chun *et al.*, 1997a,b).

Lewinsky *et al.* have compared the integration sites of stably expressed proviruses with those of latent but TNFα-inducible proviruses (Lewinsky *et al.*, 2005). Three chromosomal features corresponded with inducible expression: centromeric heterochromatin, gene deserts, and highly active host transcription units. This study is consistent with both heterochromatic silencing and transcriptional interference contributing to latency. It must be noted, however, that these two mechanisms appear to account for only a portion of latent proviruses. Other chromosomal environments unfavorable for expression may yet be found (Lewinsky *et al.*, 2005).

It has been suggested that viral latency might also be a consequence of natural antiviral defenses in mammalian cells (Williams and Greene, 2005). The genome of HIV-1 contains numerous dsRNA regions which might silence proviral transcription via RNA interference (RNAi). HIV-1 does encode a virus specific small interfering RNA precursor which elicits antiviral restriction in human cells (Bennasser *et al.*, 2005). It was found, however, that HIV-1 evades elicited RNAi through a suppressor of RNA silencing (SRS) function encoded in its Tat protein.

1.2. Nucleosomal Organisation of the 5' LTR

Given that HIV-1 retroviral DNA is integrated into the human genome and that cellular gene expression is controlled by local chromatin structure, viral transcription is likewise modulated by local chromatin structure.

DNase I digestion of chromatin shows that different regions of chromatin are differently susceptible to endonuclease cleavage. Small regions of the genome are exquisitely sensitive to digestion by nucleases and are called nuclease hypersensitive sites. Such sites are thought to represent nucleosome-free or -disrupted regions of chromatin which are bound by *trans*-acting factors, and they are generally found associated with regions of the genome that are important for the regulation of gene expression. Our laboratory has studied the chromatin organization of HIV-1 proviruses integrated in five different latently-infected cell lines by a nuclease digestion method (Verdin *et al.*, 1993). Independently of the site of integration, two major hypersensitive sites (HS) are present in the promoter region (or Long Terminal Repeat (5' LTR)): HS2+3 and HS4 (Fig. 1b); and one major hypersensitive site in the *pol* gene: HS7 (Fig. 1a) (see Sections 1.4 and 1.7). Identification of these hypersensitive sites allowed our laboratory to establish the nucleosomal organisation of the 5' LTR. Two nucleosomes (called nuc-0 and nuc-1) are positioned at the viral promoter DNA at precise locations with respect to regulatory elements (Verdin *et al.*, 1993). Nuc-0 is positioned immediately upstream of the modulatory region and nuc-1 immediately downstream of the viral transcription start site (Fig. 1b). These nucleosomes define two open regions of chromatin corresponding, respectively, to the modulatory region plus the enhancer/core promoter region (nt 200–465) (HS2+3) and to a regulatory domain in the leader region downstream of the transcription start site (nt 610–720) (HS4), where transcription factors have been found to bind *in vitro* and *in vivo* (Fig. 1b and 1c) (reviewed in al Harthi and Roebuck, 1998; Rohr *et al.*, 2003). The location of the nucleosomes in the promoter region was also verified *in vitro* using the HIV-1 promoter reconstituted into chromatin and analyzed by DNase I footprinting analysis (reviewed in Van Lint , 2000). Importantly, activation of the integrated HIV-1 promoter by cytokine Tumor Necrosis Factor α (TNFα) or phorbol ester TPA is accompanied by the loss or rearrangement of the nucleosome nuc-1 near the transcription start site (Verdin *et al.*, 1993). Therefore, chromatin modifications might result in HIV-1 promoter activation.

Chromatin-modifying complexes are classified into two major groups: (1) enzymes that control covalent modifications of the amino-terminal tails of histones (acetylation, methylation, phosphorylation, ubiquitinylation) (see Sections 1.3 and

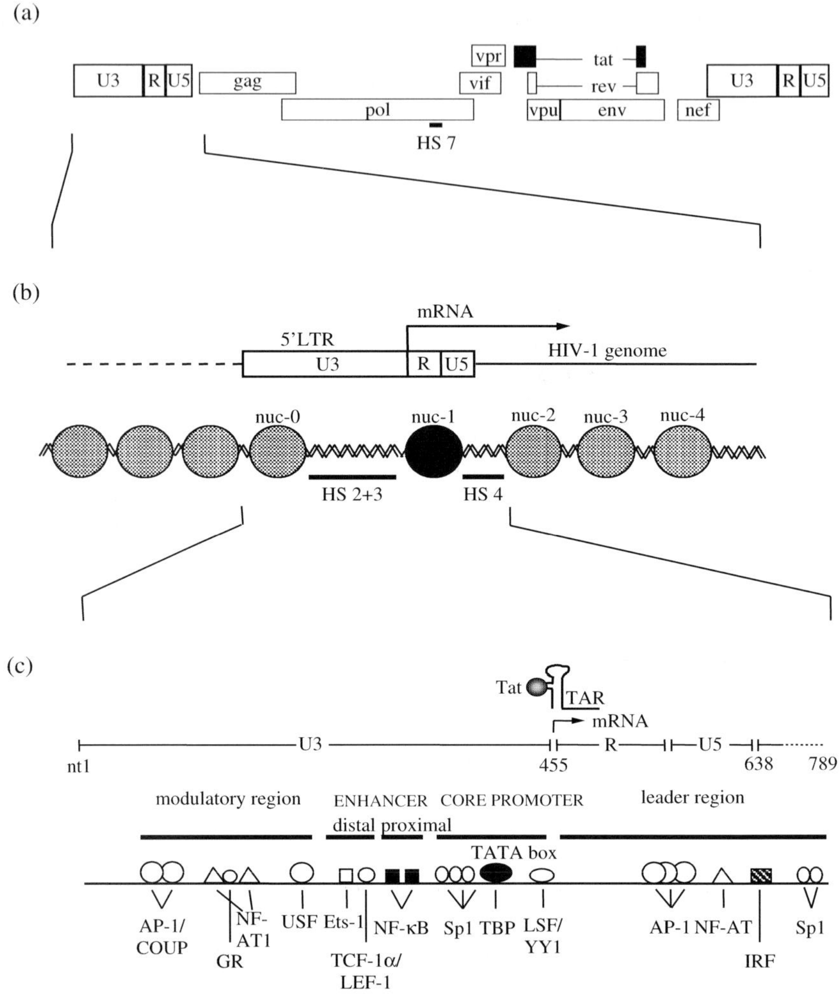

Figure 1. (a) Representation of the HIV-1 genome. Hypersensitive site 7 (HS7) in the *pol* gene is indicated. (b) Mapping of nucleosome positioning in the 5' LTR and the leader region of HIV-1. The location of the transcription initiation site at the U3-R junction is indicated by an arrow. The nucleosomes are represented by spheres. Hypersensitive sites 2 + 3 (HS 2 + 3) and 4 (HS4) are indicated. During transcriptional activation, a single nucleosome (nuc-1 in black) located immediately downstream of the transcription start, is specifically remodelled. (c) Organization of the 5' LTR region of HIV-1. The LTR is composed of the U3, R and U5 regions. The complete LTR and the leader region (nt 1 to 789) (where nt +1 is the start of U3 in the 5' LTR) are shown. The location of the transcription initiation site at the U3-R junction is indicated by an arrow. The viral *trans*-activator Tat binds to a RNA hairpin (TAR) present at the 5' end of all nascent viral transcripts. The LTR region is divided in a modulatory region (possessing binding sites for AP1/COUP, GR, NF-AT1, USF . . .), an enhancer region (possessing binding sites for Ets-1, TCF/LEF, NFκB . . .), the core promoter region (possessing a TATA box and binding sites for Sp1, YY1 . . .) and a leader region (possessing binding sites for AP-1, NF-AT, IRF, Sp1 . . .)

1.4); and (2) ATP-dependent remodeling complexes, which use the energy of ATP hydrolysis to alter interactions between histones and DNA (see Section 1.5). These two classes of chromatin-modifying complexes can work together to generate a chromatin structure that is accessible for the transcription apparatus (reviewed in Narlikar *et al.*, 2002).

1.3. Acetylation, nuc-1 and HIV-1 Activation

Acetylation of internal lysine residues of core histone N-terminal domains has been found correlatively associated with gene transcription in eukaryotes for more than four decades. Histone acetylation levels are the result of a competition between two families of enzymes: histone acetyltransferases (HATs) and histone deacetylases (HDACs).

We and others have demonstrated the transcriptional activation of the HIV-1 promoter in response to deacetylase inhibitors (HDAC inhibitors) [such as trichostatin A (TSA), trapoxin (TPX), valproic acid (VPA) and sodium butyrate (NaBut)]. This occurs in cells transiently or stably transfected with HIV-1 LTR promoter reporter constructs (el Kharroubi *et al.*, 1998; Kiernan *et al.*, 1999; Jordan *et al.*, 2001), in latently HIV-1-infected cell lines (Van Lint *et al.*, 1996a), on *in vitro* chromatin-reconstituted HIV-1 templates (Sheridan *et al.*, 1997; Steger *et al.*, 1998), as well as in the context of a *de novo* infection (Quivy *et al.*, 2002). HDAC inhibitors provoke a global histone hyperacetylation. But, remarkably, the only detectable modification at the level of the HIV-1 chromatin is the disruption of nuc-1 in the 5' LTR (Van Lint *et al.*, 1996a,b), whereas other nucleosomes remain unaffected. This disruption is accompanied by an activation of virus production, which occurs at the transcriptional level (Van Lint *et al.*, 1996a). Using a Chromatin ImmunoPrecipitation assay (ChIP), histone acetylation surrounding nuc-1 was observed to be significantly increased following HDAC inhibitors treatment (He and Margolis, 2002; Ylisastigui *et al.*, 2004; Kiefer *et al.*, 2004). Interestingly, a similar increase in histone hyperacetylation was observed in the nuc-1 region after activation by Tat, phorbol esters or following cell cycle arrest in G2 (He and Margolis, 2002; Lusic *et al.*, 2003; Thierry *et al.*, 2004).

These observations suggest that under latency conditions, the viral promoter is poised for transcription but repressed by the presence of a single nucleosome, nuc-1. The fact that inhibition of deacetylase activity is sufficient for transcriptional activation suggests that nuc-1 is constitutively deacetylated by targeted HDACs. This model is supported by data showing that HDACs are recruited to the HIV-1 promoter region in the absence of activation (Fig. 2a) (Coull *et al.*, 2000, 2002; Imai and Okamoto, 2006; Williams *et al.*, 2006). In accordance, following activation, cellular HATs (including CBP, P/CAF and GCN5) are recruited to the HIV-1 promoter region (Fig. 2b) (Lusic *et al.*, 2003; Thierry *et al.*, 2004). Accordingly, several transcription factors binding to the HIV-1 LTR were demonstrated to recruit HDAC and/or HAT (see Section 1.4).

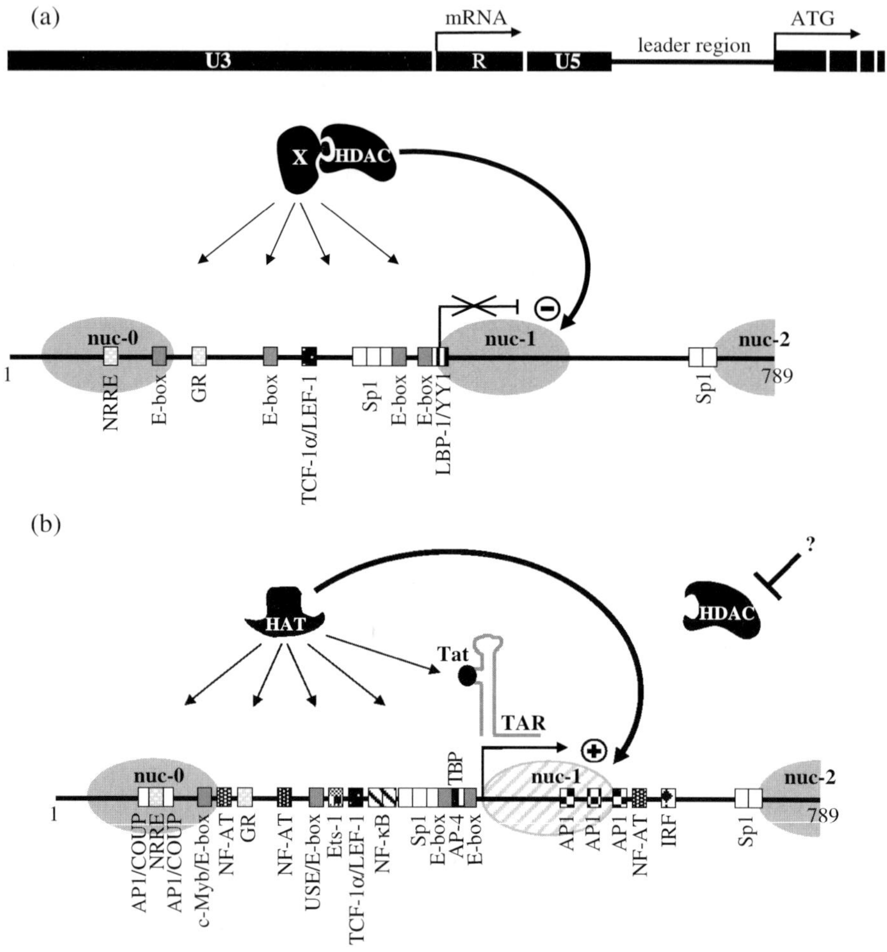

Figure 2. (a) Model for transcriptional repression of the HIV-1 provirus during latency. Under latency conditions, nuc-1 blocks transcriptional initiation and/or elongation because this nucleosome is maintained hypoacetylated by nearby deacetylases (HDAC). The targeting of these HDACs to nuc-1 is mediated by interactions with transcription factors binding to DNA in the HIV-1 promoter region. These interactions take place directly or indirectly through another protein X. (b) Model for transcriptional activation of the HIV-1 provirus. Nuc-1 is a major obstacle to transcription. To activate transciption, nuc-1 must be remodeled. Nuc-1 remodeling could happen by inhibiting the deacetylases. How this is accomplished *in vivo* is unknown. The disruption could also happen by local recruitment of acetyl-transferases (HATs) by DNA-binding factors, and/or by the viral protein Tat, which binds to the neosynthesized RNA hairpin TAR. This would result in nuc-1 hyperacetylation and remodeling, thereby eliminating the block to transcription

1.4. Deacetylase(s) and Acetyltransferase(s) Recruitment to the HIV-1 Promoter

The 5' LTR has been extensively characterized *in vitro*, and binding sites for several cellular transcription factors have been identified using DNase I *in vitro* footprinting and gel retardation assays (Fig. 1c) (see for reviews Roebuck and Saifuddin, 1999; Pereira *et al.*, 2000; Rohr *et al.*, 2003).

On the one hand, several of these factors, including AP-1, c-Myb, glucocorticoid receptor (GR), CCAAT/enhancer binding proteins (C/EBP), NF-AT, E-box binding proteins, Ets-1, TCF-1α/LEF-1, NF-κB, Sp1, IRF and the HIV-1 *trans*-activator Tat (see Section 1.6) have been shown to interact with acetyltransferases (Fig. 2b) (reviewed in Quivy and Van Lint, 2002). On the other hand, several transcription factors that bind to the LTR, including unliganded nuclear hormone receptors, GR, E-box binding proteins, YY1, Sp1, TCF-1α/LEF-1, AP-4, NF-κB have been shown to interact with deacetylases (Fig. 2a) (reviewed in Quivy and Van Lint, 2002). Among these factors, some of them were demonstrated to recruit HAT and/or HDAC specifically to the HIV-1 5' LTR, thereby regulating the acetylation level of histones (in particular the histones of nuc-1).

1.4.1. YY1

The host transcription factor YY1 is able to mediate transcription activation or repression, dependent on promoter context (Gordon *et al.*, 2006). YY1 and LSF cooperate in repression of HIV-1 expression from both integrated reporter and infected cell lines (Romerio *et al.*, 1997). Studies from David Margolis group have demonstrated that the host factors YY1 and LSF cooperatively recruit histone deacetylase 1 (HDAC1) to the HIV-1 LTR and inhibit transcription by maintaining nuc-1 in a hypoacetylated state (Coull *et al.*, 2002; He and Margolis, 2002). In agreement with their data, pyrole-imidazole polyamides, which block the binding of the LSF-YY1 complex to the LTR and consequently the recruitment of HDAC-1 close to nuc-1, have been shown to induce reactivation of HIV-1 expression (Coull *et al.*, 2002; He and Margolis, 2002). These experiments highlight the repressive role played by the nucleosome nuc-1 in the generation of post-integration latency.

1.4.2. Thyroid hormone receptor

Thyroid Hormone Receptor (TR) can bind to the LTR *in vivo* independently of its ligand and regulates promoter activity. ChIP assays with anti-acetylated-histone antibodies revealed that unliganded TR reduce the local histone acetylation levels at the HIV-1 LTR, while thyroid hormone treatment reverses this induction (Hsia and Shi, 2002). Accordingly, unliganded TR recruits co repressors and at least one HDAC (Hsia and Shi, 2002).

1.4.3. C/EBP

C/EBP proteins have been shown to be required for HIV-1 transcription in macrophages (Henderson and Calame, 1997). C/EBPβ recruits co activators to the

HIV-1 promoter and physically interacts with HAT complexes (Lee *et al.*, 2002). Overexpression of a C/EBP dominant-negative inhibits disruption of the nucleosome nuc-1 (Lee *et al.*, 2002).

1.4.4. NF-κB

An important candidate for the specific targeting of HDACs and HATs to the HIV-1 5' LTR is the transcription factor NF-κB. NF-κB plays a central role in the activation pathway of the HIV-1 provirus (reviewed in Rabson and Lin, 2000). Various studies have reported that the two NF-κB-binding sites in the HIV-1 enhancer as well as the NF-κB proteins are critical for LTR promoter activity and important for optimal HIV-1 replication (reviewed in Rabson and Lin, 2000). The p65 subunit of NF-κB stimulates transcriptional elongation from the HIV-1 promoter (West *et al.*, 2001). Moreover, on the one hand NF-κB interact with HDACs (principally HDAC1 and 3) and on the other hand NF-κB-dependent gene expression requires the function of transcriptional co-activator proteins, including p300/CBP, P/CAF, SRC-1, which possess acetyltransferase activity (reviewed in Quivy and Van Lint, 2004).

The absence of a transactivation-competent NF-κB heterodimer in the nucleus of latently infected resting memory $CD4^+$ T cells could contribute to latency. Activation of the NF-κB pathway leading to migration of a transactivating heterodimer such as p50/p65 could allow viral reactivation. In the absence of induction, NF-κB p50-HDAC1 complexes constitutively bind the latent HIV-1 LTR (Williams *et al.*, 2006). NF-κB p50 does not possess a transactivation domain. These p50-HDAC1 complexes induce histone deacetylation and repressive changes in chromatin structure of the HIV-1 LTR (Williams *et al.*, 2006). Knockdown of p50 expression reduces HDAC1 binding to the latent HIV-1 LTR and induces RNA polymerase II recruitment (Williams *et al.*, 2006). Concomitantly with HIV-1 transcriptional activation, the p65 subunit and different HATs are recruited to the viral promoter (Lusic *et al.*, 2003; Thierry *et al.*, 2004).

1.4.5. AP-4

AP-4 represses HIV-1 gene expression by recruiting HDAC1 as well as by masking TATA-binding protein to TATA box (Imai and Okamoto, 2006). AP-4 interacts both *in vivo* and *in vitro* with HDAC1, and ChIP assays revealed that AP-4 and HDAC1 are present in the HIV-1 5'LTR promoter in latently infected cell lines and are dissociated from the promoter upon TNFα stimulation (Imai and Okamoto, 2006).

1.4.6. Sp1

The ubiquitous transcription factor Sp1 is critical for both basal and Tat-induced transcription of the HIV-1 LTR (Kamine and Chinnadurai, 1992; Sune and Garcia-Blanco, 1995; Yedavalli *et al.*, 2003). Sp1 could also play an important role in nuc-1-dependent HIV-1 regulation. Indeed, Sp1 has been demonstrated to interact with the acetyltransferase p300 and to act as a co-activator for Sp1-mediated transcriptional activation (Billon *et al.*, 1999; Suzuki *et al.*, 2000; Xiao *et al.*, 2000). Sp1 has also been shown to directly interact with HDAC1 (Doetzlhofer *et al.*, 1999). Transcriptional activation by HDAC inhibitor can also be mediated through Sp1 interactions (Sowa *et al.*, 1997; Butler *et al.*, 2002). Whether these interactions have functional implications in the context of the HIV-1 promoter activity remains however to be established.

1.5. ATP-Dependent Chromatin Remodeling Complexes

Packaging of DNA into chromatin modifies its accessibility and geometry in a manner that interferes with transcription initiation and elongation (reviewed in Workman and Kingston, 1998; Wolffe, 2001). In order for the transcription machinery to gain access to DNA, the compacted chromatin structure needs to be altered. In addition to posttranslational covalent modifications of the histone amino-terminal tails, ATP-dependent chromatin remodeling by protein complexes is a highly conserved mechanism used by eukaryotic cells to alter chromatin structure (reviewed in Narlikar *et al.*, 2002). Chromatin-remodeling complexes carry out key enzymatic activities, changing chromatin structure by altering DNA-histone contacts within a nucleosome in an ATP-dependent manner (reviewed in Martens and Winston, 2003; de la Serma *et al.*, 2006). Several ATP-dependent nucleosome remodeling complexes have been purified from different organisms, including the mammalian SWI/SNF complex from human. SWI/SNF complexes are recruited to promoters by DNA bound activators or repressors (Peterson and Workman, 2000). The SWI2/SNF2 ATPase component conserved in all SWI/SNF complexes contains a bromodomain that can bind acetylated lysine residues in histone N-terminal tails *in vitro* (Hassan *et al.*, 2002). How SWI/SNF disrupts histone-DNA contacts to facilitate transcription factor binding is not clear, but several recent studies support a translocation model for remodeling. In this model, a wave of accessible DNA caused by translocation-induced topological stress is transmitted around the nucleosome (Narlikar *et al.*, 2001; Aoyagi and Hayes, 2002; Saha *et al.*, 2002).

Chromatin is an important regulatory component of HIV-1 transcription. Indeed, we have shown that activation of the integrated HIV-1 promoter is accompanied by a disruption of the nucleosome nuc-1 near the transcription start site (see Sections 1.1 and 1.2) (Van Lint *et al.*, 1996a). Following activation, there is a correlation between the recruitment of cellular HATs (including p300/CBP, P/CAF and GCN5) to the HIV-1 promoter region and acetylation of both histones H3 and H4 (Lusic *et al.*, 2003; Thierry *et al.*, 2004). Acetylated histones may favor the recruitment of SWI/SNF to the LTR promoter. Indeed, the ATPase subunit of SWI/SNF, Brg1, is recruited to the 3' boundary of nuc-1 following phorbol ester stimulation (Henderson *et al.*, 2004). This latter study has shown that ATF-3, which binds to the AP-1(III) site located downstream from the transcription start site in the HS4 region (see Section 1.7), recruits Brg1 *in vitro* to the HIV-1 LTR.

Transcriptional activation of the HIV-1 promoter by Tat is accompanied by the remodeling of nuc-1 (El Kharroubi *et al.*, 1998; Jordan *et al.*, 2003). Recently, it has been shown that, via its arginine-rich motif, Tat is able to interact with

Brm, another ATPase subunit of the SWI/SNF complex (Tréand *et al.*, 2006). This interaction is regulated by the acetylation of Tat at lysine 50. Acetylated Tat recruits the SWI/SNF complex to the LTR *in vivo*, leading to the activation of the integrated HIV-1 promoter (Tréand *et al.*, 2006). This Brm recruitment could facilitate Tat-mediated processive elongation, by opening the chromatin structure to facilitate RNAPII progression.

1.6. Tat

The HIV-1 Tat protein plays an important role in regulating viral gene expression (reviewed in Brigati *et al.*, 2003; Barboric and Peterlin, 2005; Brady and Kashanchi, 2005). Full-length Tat is encoded by two exons on a spliced transcript and is 86–101 aa in length, depending on the viral isolate. Transcription of the HIV-1 provirus is characterized by an early, Tat-independent phase and a late, Tat-dependent phase. During the Tat-independent phase, the HIV-1 promoter is strictly under the control of the local chromatin environment and cellular transcription factors binding to *cis*-acting elements in the viral promoter region. In the absence of Tat, transcription is initiated normally but elongation is inefficient and results in short abortive transcripts (~60 nucleotides) that cannot support viral replication. The negative transcription elongation factor (N-TEF) complex, composed of the negative elongation factor (NELF) and the DRB sensitivity inducing factor (DSIF), has been proposed to be responsible for premature stopping and termination of the initiated polymerase (Yamaguchi *et al.*, 1999; Ping and Rana, 2001). Despite this elongation defect, occasional full-length genomic transcripts are generated, leading to the synthesis of a few Tat molecules sufficient to stimulate HIV transcription elongation.

The viral *trans*-activator protein Tat is an atypical transcriptional activator that functions through binding, not to DNA, but to a short leader RNA, called TAR (*trans*-activating response element), present at the 5' end of all nascent viral transcripts. Tat enhances the processivity of transcribing RNAPII complex by recruiting cellular proteins, the positive elongation factor complex b, P-TEFb (also called tat-associated kinase, TAK) (reviewed in Price, 2000; Sims *et al.*, 2004). Cyclin T1, a component of P-TEFb, interacts cooperatively with both the N-terminal transactivation domain of Tat (amino acid 1-48) and the loop sequence at the top of TAR stem-loop structure. As a consequence, P-TEFb and Tat bind TAR with higher affinity and specificity than Tat alone. The cyclin dependent kinase 9 (CDK9) is then recruited to the HIV-1 promoter where it phosphorylates the C-terminal domain (CTD) of previously bound RNA polymerase II complexes (RNAPII). This process is known to occur as RNAPII converts from an intiating (RNAPIIa) to an elongating enzyme (RNAPo). Additionally, P-TEFb phosphorylates components of the N-TEF, relieving its block on transcriptional elongation (Bourgeois *et al.*, 2002; Fujinaga *et al.*, 2004). Furthermore, the splicing-associated c-Ski-interacting protein, SKIP, associates with Tat/P-TEFb/TAR complexes and stimulates transcription elongation (Bres *et al.*, 2005). In this way, Tat increases production of viral mRNA by ~ 100-fold. These observations suggest that, unlike typical activators, Tat stimulates transcriptional elongation rather than initiation. However, a recent report has shown that Tat and P-TEFb stimulate transcription complex (TC) assembly (Raha *et al.*, 2005). The TC formed on the HIV-1 LTR is atypical and contains TATA-box binding protein (TBP) but not TBP-associated factors (TAFs) (Raha *et al.*, 2005). Tat may thus promote transcription initiation as well as elongation.

The optimal activity of Tat is furthermore dictated by its association with different acetyltransferases, including GCN5, p300/CBP and P/CAF (reviewed in Nakatani, 2002; Quivy and Van Lint, 2002). p300/CBP, P/CAF and GCN5 activate the HIV-1 5'LTR transcriptional activity. This activation could be mediated through acetylation of histones (Lussic *et al.*, 2003). This acetylation could derepress the chromatinized viral promoter and thus explain, at least in part, how Tat participates in initiating access and formation of the RNAPII complex.

Furthermore, it has been shown that Tat itself is modified by direct acetylation (Fig. 3). P/CAF acetylates lysine 28 in the activation domain of Tat (Kiernan *et al.*, 1999). Mechanistically, acetylation at Lys28 by P/CAF enhances the recruitment of the cyclinT1/CDK9 complex by Tat (Kiernan *et al.*, 1999; Bres *et al.*, 2002). This would result in enhanced CTD phosphorylation by this TAK complex and in enhanced transcriptional elongation. p300/CBP and GCN5 acetylate lysine 50 and weakly lysine 51 in the TAR RNA binding domain of Tat (Kiernan *et al.*, 1999; Ott *et al.*, 1999: Deng *et al.*, 2000; Col *et al.*, 2001). Acetylation at lysine 50 of Tat promotes the dissociation of Tat from TAR RNA. Once dissociated from TAR, the Lys50-acetylated Tat binds to the bromodomain of P/CAF (Dorr *et al.*, 2002; Mujtaba *et al.*, 2002). This would recruit P/CAF to the elongating RNAPIIo and likely facilitates chromatin remodeling. Interestingly, Tat acetylation can be reverted by the protein deacetylase SIRT1, allowing Tat to reassemble with P-TEFb on TAR (Pagans *et al.*, 2005). This would recycle functional Tat protein and enhance its apparent concentration. This might be especially advantageous in the early stage of HIV infection, when the amount of Tat is still limited. Thus, cycles of Tat acetylation and deacetylation regulate critical steps in HIV-1 transcription: binding to the RNAPII CTD-kinase, release from the TAR RNA, and, after elongation, recycling to the TAR RNA (Fig. 3). Finally, it has been shown that Tat ubiquitinylation also stimulates its transcriptional properties (Bres *et al.*, 2003).

1.7. Two Regulatory Regions Located Downstream of the HIV-1 Transcription Initiation Site are Associated with DNase I-Hypersensitive Sites

Structural studies of the chromatin organization of integrated HIV-1 have identified two DNase I-Hypersensitive Sites (HS4 and HS7) outside of the promoter/enhancer region in U3, that were likely to play a role in controlling viral transcription by virtue of their accessibility in chromatin (Fig. 1a and 1b) (Verdin, 1991).

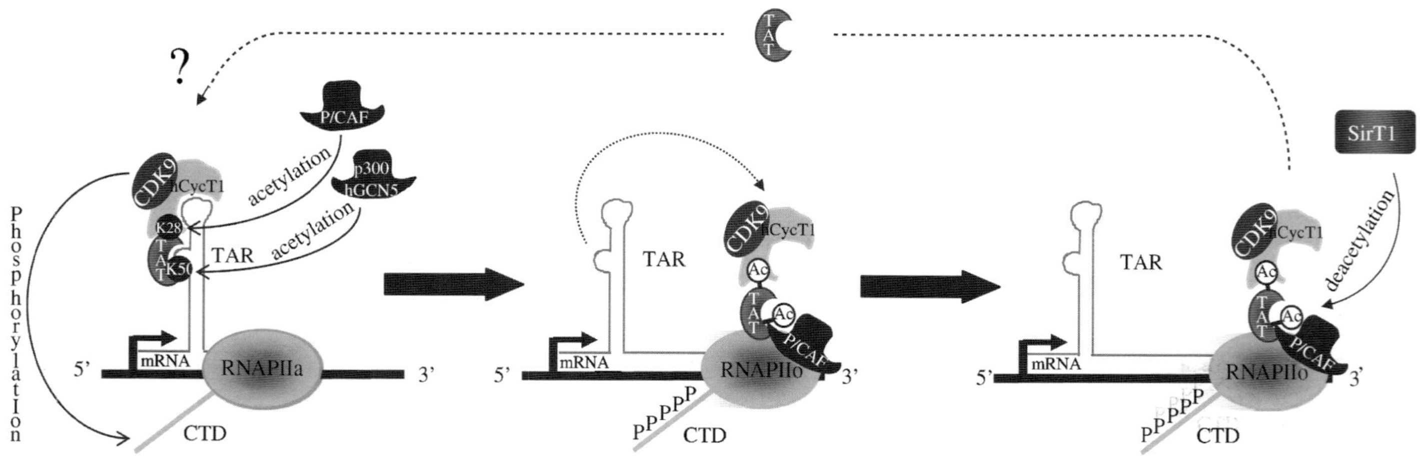

Figure 3. Model for the acetylation-dependent regulation of Tat. P/CAF acetylates Lys28 in the *trans*-activation domain of Tat and this enhances the recruitment of CycT1/CDK9. This leads to CTD phosphorylation and transcriptional elongation. p300 acetylates Lys50 in the RNA-binding domain of Tat and this promotes dissociation of Tat from TAR RNA. Acetylated Tat binds P/CAF via its bromodomain and recruits it to elongating RNAPIIo, possibly facilitating chromatin remodeling. Finally, Lys-50 acetylated Tat is deacetylated by SIRT1, allowing Tat to reassemble on TAR

1.7.1. Characterization of the DNase I-hypersensitive site HS4 associated with the 5' untranslated region of HIV-1

The transcriptionally active HIV-1 promoter is characterized by a large open chromatin region encompassing both the promoter/enhancer region (HS2 + HS3) and a 255-nt region downstream of the transcription start site (nt 465–720) (nuc-1 region and HS4) (Fig. 1b). *In vivo* and *in vitro* footprint analysis of the latter region has identified recognition sites for several constitutive and inducible transcription factors (reviewed in al-Harti and Roebuck, 1998): an NF-κB binding site and three AP-1 binding sites (I, II, III) which lie in the region protected by nuc-1 in basal conditions; a NF-AT binding site; an interferon-stimulated response element (ISRE); and juxtaposed Sp1 binding sites (Fig. 1c).

It has been shown that this cluster of transcription factors binding sites located downstream of the transcription start site play a crucial role in HIV-1 infectivity and transcriptional regulation (Van lint *et al.*, 1997). The AP-1(III) and NF-AT sites and the ISRE lie at the 3' boundary of nuc-1 while the Sp1 sites lie at the 5' boundary of nuc-2. These sites may thus play a role in the positioning of nucleosomes in the 5' LTR. This could occur either indirectly through a boundary effect or directly through an interaction between a nucleosomal component and one of these factors. In good agreement with this hypothesis, a LTR containing mutations in most of the HS4 binding sites (AP-1(III), NF-AT, IRF, and Sp-1) and stably integrated into HeLa cells shows reduced promoter activity accompanied by the disappearance of nuclease-hypersensitive site HS4 (El Kharroubi and Martin, 1996). These experiments demonstrate that these sites collectively contribute to the establishment of a nucleosome-free region corresponding to HS4.

In addition to c-fos and junD, the three AP-1 binding sites (I, II, III) in the 5' untranslated region also bind transcription factors of the CREB/ATF family, permitting induction of the HIV-1 5'LTR via activation of the cAMP-dependent protein kinase A (PKA) signaling pathway (Rabbi *et al.*, 1997). The efficiency of AP-1 binding to the AP-1 (III) site is also selectively modulated by the High-Mobility-Group protein I (HMG I) (Henderson *et al.*, 2000). The ISRE located downstream of the transcription start site has been shown to bind interferon regulatory factor 1 (IRF-1) (Sgarbanti *et al.*, 2002). Furthermore, IRF-1 activates LTR-driven transcription in the absence of the viral transactivator Tat. IRF-1 may thus effectively activate transcription of Tat, leading to efficient HIV-1 transcription and virus replication. Interestingly, IRF-1 is stimulated early after virus infection and before expression of Tat in both cell lines and primary $CD4^+$ T-lymphocytes. These results define IRF-1 as a positive regulator of HIV-1 transcription and replication that may play a role during viral reactivation from latency (Sgarbanti *et al.*, 2002).

1.7.2. Characterization of the DNase I-hypersensitive site HS7 associated with the HIV-1 pol gene

Analysis of the chromatin organization of integrated HIV-1 identified a single major nuclease-hypersensitive site in the 8 kb region located between the two LTRs (Verdin, 1991; Van Lint *et al.*, 1994). This hypersensitive site, centered around nt

4490–4766, is located in the part of the *pol* gene encoding the integrase protein (Fig. 1a). This constitutive hypersensitive site is present only in a cell line from monocytic origin (U1) and not in two cell lines from lymphoid origin (8E5 and ACH2), suggesting a cellular specificity associated with this intragenic element (Verdin, 1991). A 500 bp fragment (nt 4481–4982) encompassing the *pol* gene hypersensitive site region (called HS7 region) positively regulates transcription from the HIV-1 5'-LTR in transient transfection experiments (Van Lint *et al.*, 1994). Our laboratory has identified by *in vitro* binding studies several transcription factors binding sites in the HS7 region including two Oct-1/-2 sites, a PU.1 site and a Sp1/Sp3 sites (Van Lint *et al.*, 1994; Goffin *et al.*, 2005). Infection studies with infectious clones of HIV-1 mutated individually or in combination in the identified transcription factors binding sites have demonstrated that they are important for HIV-1 replication in human $CD4^+$ cell lines, indicating a positive regulatory function for the *pol* intragenic region (Goffin *et al.*, 2005).

DNA regulatory regions in eukaryotic genomes frequently adopt a nuclease-hypersensitive configuration (Wolffe, 1999). We have shown that PU.1 occupies *in vivo* the HS7 region in U1 cells and could therefore be involved in the nuclease hypersensitivity of the intragenic region (Goffin *et al.*, 2005). The PU.1 protein is an ETS family transcription factor that is highly expressed in B-lymphocytes, myeloid cells and immature erythrocytes (Scott *et al.*, 1994; McKercher *et al.*, 1996). Indeed, studies on the B-cell-specific μ and κ3' enhancers have shown that PU.1 is involved in chromatin remodeling events by functioning in some contexts as an accessibility factor (i.e. a protein capable of binding its site in the repressive context of chromatin, thereby increasing accessibility of the region) (Pongubala *et al.*, 1992; Nelsen *et al.*, 1993; Marecki *et al.*, 2004). Therefore, the presence in the *po*l gene of HIV-1 of a regulatory region containing a binding site for the macrophage and B-cell-specific factor PU.1 and associated with a monocyte-specific nuclease hypersensitive site suggests that PU.1 could act as an accessibility factor triggering the open chromatin configuration in the *pol* region. Moreover, the positive regulatory element associated with HS7 could play an important role in the macrophage tropism of HIV-1. Although previous studies have demonstrated that the HIV-1 macrophage tropism is primarily determined at the level of the entry (O'Brien *et al.*, 1990; Hwang *et al.*, 1991; Shioda *et al.*, 1991), several reports have highlighted additional determinants of macrophage tropism present at the level of transcription. Indeed, two motifs located in the HIV-1 LTR bind members of the C/EBP family of transcription factors and are required for virus replication in macrophages, but not T-cells (Henderson *et al.*, 1997). The tropism of other viruses for their target cells is determined in part at the transcriptional level (Erselius *et al.*, 1990; Hines *et al.*, 2004).

The HS7 region is composed of multiple factor binding sites; some of these factors are ubiquitously expressed, whereas others are cell-specific factors, supporting the notion that the HS7 transcriptional activity is determined by a combinatorial control: the combined action of tissue-restricted and ubiquitously expressed proteins. The HS7 regulatory region described in this report could either bring additional cellular specificity, or increase the strength of the promoter/enhancer unit located in the HIV-1 LTR, or allow viral responses to a broader variety of exogenous stimuli. The positive *cis*-regulatory element located in the transcribed region of the HIV-1 genome brings an additional factor in an already complex network of regulators affecting the level of HIV-1 replication. Such a complexity could allow a finer-tuned regulation than a simple 'On/Off' switching mechanism; this fine-tuning might find its purpose when HIV-1 transcription needs to be moderately and/or transiently modified.

1.8. HDAC Inhibitors: A New Hope for HIV-1 Eradication?

HAART is unable to completely clear the pool of latently infected cells. The residual viremia in patients on HAART would be due to low-level reactivation of latently-infected cells. Therefore, elimination of the latent pool might be achieved by inducing HIV replication in latently-infected cells, while maintaining the patient on HAART to prevent a spreading infection. Following viral reactivation, latently-infected cells would then be eliminated by the immune system or virus-mediated cell lysis. Several clinical trials to test this strategy have been performed to date. Studies with agents activating nonspecifically T-cells, such as IL-2 alone, or in combination with TNFα and/or OKT3 (a TCR activating monoclonal antibody), have shown little, if any, reduction in viral rebound after cessation of HAART (for review, see Blankson *et al.*, 2002). An ideal compound for viral purging would induce expression of HIV-1 in a wide variety of infected cells without inducing global T-cell activation to prevent the generation of new target cells for the neo-synthesized virus. A number of potential drugs under investigation meet these criteria. IL-7 effectively induces HIV-1 replication in peripheral blood mononuclear cells (PBMCs) of infected patients, while having minimal effects on T-cell phenotype (Smithgall *et al.*, 1996; Scripture-Adams *et al.*, 2002). Other interesting HIV-1 inducing agents are prostratin, a nontumor-promoting phorbol ester, and the related 12-deoxyphorbol 13-phenylacetate (DPP). Further study revealed that prostratin upregulated HIV-1 transcription in both lymphoid (Korin *et al.*, 2002) and myeloid (Kulkosky *et al.*, 2001) cell lineages, while simultaneously preventing subsequent infection by downregulating $CD4^+$, the cell-surface receptor used for HIV attachment (Biancotto *et al.*, 2004). Unfortunately, attempts to develop prostratin as a drug have encountered severe dose-limiting toxicities in primates.

Based on our results, we have proposed HDAC inhibitors as an adjuvant therapy with HAART (Quivy *et al.*, 2002; Demonté *et al.*, 2004). HDAC inhibitors present several advantages in the context of an anti-HIV-1 adjuvant therapy. First, HDAC inhibitors do not induce proliferation or activation of T cells (Byrd *et al.*, 1999; Marks *et al.*, 2001; Piekarz *et al.*, 2001). Recent data even suggest that HDAC inhibitors inhibit $CD4^+$ T-cell proliferation in a dose-dependent manner (Skov *et al.*, 2003). Second, although infected resting $CD4^+$ T cells represent the major long-term HIV-1 reservoirs, other cell types (macrophages, dendritic cells and other non-T cells) clearly contribute to the persistence of HIV-1 during HAART (Schrager and D'Souza, 1998; Collman *et al.*, 2003). As viral suppression in T cells becomes today increasingly effective, these alternative reservoirs may take on even greater relative importance as sites for viral persistence and as targets for purging. HDAC inhibitors have been shown to act in a broad spectrum of cell lines and therefore, in contrast to agents that specifically induce T cells, they could target the different cellular HIV-1 reservoirs. Third, this class of agents is safely administered since several years for other diseases, including beta-chain hemoglobinopathies (such as beta-thalassemia and sickle cell anemia) (Dover *et al.*, 1994; Collins *et al.*, 1995), epilepsy and bipolar disorders (Tunnicliff, 1999; Johannessen, 2000; Phiel *et al.*, 2001). Moreover, HDAC inhibitors are potent inducers of apoptosis and growth inhibition in transformed cells originating from lymphoid cells (Byrd *et al.*, 1999; Marks *et al.*, 2001; Piekarz *et al.*, 2001). As HDAC inhibitors are relatively non-toxic to normal cells, they are now considered as good candidates for novel cancer therapy (reviewed in Yoo and Jones, 2006). Among the different agents that have been identified as having HDAC inhibitors activity (including TSA, TPX, oxamflatin, apicidin, phenylbutyrate, valproic acid, suberoylanilide hydroxamic acid (SAHA), depsipeptide, MS-275), at least the latter five are currently used in clinical trials for evaluation of their anticancer efficacy (reviewed in Minucci and Pelicci, 2006). The fourth element in favor of the use of HDAC inhibitors in anti-HIV treatment is the ability of these drugs to induce transcriptional activation of several HIV-1-subtypes LTRs (Quivy *et al.*, 2002). Indeed, an increasing number of non-B HIV-1 subtype infections is currently diagnosed. We have shown that, in addition to the prototypical subtype B LTR, the LTRs from subtypes A through G of the HIV-1 major group M were also activated by HDAC inhibitors (Quivy *et al.*, 2002). These data suggest that HDAC inhibitors could be used to induce HIV-1 expression in a subtype-non specific manner. A fifth element is that HDAC inhibitors potently repress CXCR4 chemokine receptor expression and function (Crazzolara *et al.*, 2002). This could impede viral expansion. A sixth important element is the fact that HDAC inhibitors are already safely administrated to HIV-1 patients for years. Indeed, it is well documented that patients with HIV-1 are at an increased risk for the development of multiple neurological manifestations including seizures (Wong *et al.*, 1990). Therefore, HIV-positive patients typically receive long-term anticonvulsant therapy following an initial episode of seizure activity (Romanelli and Pomeroy, 2003). Historically, these patients have been treated with many anticonvulsant agents including valproic acid which possess, in addition to its neurological properties, HDAC inhibitors activities. Supporting our hypothesis based on the ability of HDAC inhibitors to induce HIV-1 replication *in vivo*, an increase in viral load of some HIV-positive patients receiving both HAART and valproic acid (VPA) has been observed (Jennings and Romanelli, 1999). However, this phenomenon has not been observed in all patients receiving this combinatory therapy (Romanelli and Pomeroy, 2003), suggesting that, in some patients, increase in viral load can be controlled by the anti-retroviral treatment.

Margolis and colleagues have demonstrated that VPA, in presence of IL2, induced outgrowth of HIV-1 from purified resting $CD4^+$ cells obtained from aviremic patients (Ylisastigui *et al.*, 2004). In addition, our preliminary experiments performed on CD8-depleted PBMCs isolated from aviremic patients suggest that HDAC inhibitors are able to reactivate HIV-1 expression in latently infected cells even in the absence of IL2 (Van Lint, unpublished results). Recently, the Margolis group has performed a proof-of-concept study indicating that VPA is a promising agent to accelerate clearance of HIV-1 from resting $CD4^+$ T cells (Lehrman *et al.*, 2005). In three of four patients, they have observed a significant decline in the frequency of resting cell infection after addition of enfuvirtide (an inhibitor of HIV-1 entry) and VPA (Lehrman *et al.*, 2005).

Interestingly, we have demonstrated a synergistic effect of TNFα and HDAC inhibitor on reactivation of HIV-1 expression in the latently infected U1 cell line (Quivy *et al.*, 2002). Mechanistically, we have demonstrated that HDAC inhibitor prolongs TNFα-induced NF-κB binding to DNA (Quivy *et al.*, 2002; Adam *et al.*, 2003). It is important to note that an array of cytokines, including the proinflammatory cytokines TNFα and interleukin-1 (inducers of NF-κB), are already copiously expressed in the microenvironment of the lymphoid tissues, which harbor latent viral reservoirs (Navikas *et al.*, 1995) and could therefore, amplify the reactivation potential of HDAC inhibitors used in patients.

In the future, studies testing co-administration of HDAC inhibitor and an activator such as IL7 or prostratin-like agent, together with intensified HAART, could offer a new hope for HIV-1-infected patients. This type of clinical protocol could provoke a drastic HIV-1 activation leading to a strong decline of HIV-1 reservoir level, sufficiently to allow HIV-1 remission.

ACKNOWLEDGEMENTS

The authors apologize to those whose work could not be cited directly because of space limitations. This work was supported by grants to C.V.L. from the "Fonds National de la Recherche Scientifique" (FNRS, Belgium), the Télévie-Program, the "Université Libre de Bruxelles" (ULB, ARC program n° 04/09-309), the International Brachet Stiftung, the CGRI-INSERM cooperation, the Région Wallonne-Commission Européenne FEDER (Intergene Project, program Interreg III), the Agence Nationale de Recherches sur le SIDA (ANRS, France), the Theyskens-Mineur Foundation, the "Fortis Banque Assurance" and the "Fédération Belge contre le Cancer". C.V.L. is "Directeur de Recherches" of the FNRS. S.dW. is supported by a post-doctoral fellowship from the "Région Wallonne" (Program WALEO 616295).

REFERENCES

Adam E, Quivy V, Bex F, Chariot A, Collette Y, Vanhulle C, Schoonbroodt S, Goffin V, Nguyên TL-A, Carrard G, Friguet B, de Launoit Y, Burny A, Bours V, Piette J, Van Lint C (2003) Potentiation of TNFα-induced NF-kappaB activation by deacetylase inhibitors is associated with a delayed cytoplasmic reappearence of IκBalpha. Mol Cell Biol 23:6200–6209

al Harthi L, Roebuck KA (1998) Human immunodeficiency virus type-1 transcription: role of the 5'-untranslated leader region. Int J Mol Med 1:875–881

Aoyagi S, Hayes JJ (2002) hSWI/SNF-catalyzed nucleosome sliding does not occur solely via a twist-diffusion mechanism. Mol Cell Biol 22:7484–7490

Barboric M, Peterlin BM (2005) A new paradigm in eukaryotic biology: HIV Tat and the control of transcriptional elongation. PLoS Biol 3:200–203

Bennasser Y, Le SY, Benkirane M, Jeang, KT (2005) Evidence that HIV-1 encodes an siRNA and a suppressor of RNA silencing. Immunity 22:607–619

Biancotto A, Grivel JC, Gondois-Rey F, Bettendroffer L, Vigne R, Brown S, Margolis LB, Hirsch I (2004) Dual role of prostratin in inhibition of infection and reactivation of human immunodeficiency virus from latency in primary blood lymphocytes and lymphoid tissue. J Virol 78:10507–10515

Billon N, Carlisi D, Datto MB, van Grunsven LA, Watt A, Wang XF, Rudkin BB (1999) Cooperation of Sp1 and p300 in the induction of the CDK inhibitor p21WAF1/CIP1 during NGF-mediated neuronal differentiation. Oncogene 18:2872–2882

Bisgrove D, Lewinski M, Bushman F, Verdin E (2005) Molecular mechanisms of HIV-1 proviral latency. Expert Rev Anti Infect Ther 3:805–814

Blankson JN, Persaud D, Siliciano RF (2002) The challenge of viral reservoirs in HIV-1 infection. Annu Rev Med 53:557–593

Bourgeois CF, Kim YK, Churcher MJ, West MJ, Karn J (2002) Spt5 cooperates with human immunodeficiency virus type 1 Tat by preventing premature RNA release at terminator sequences. Mol Cell Biol 22:1079–1093

Brady J, Kashanchi F (2005) Tat gets the "green" light on transcription initiation. Retrovirology 2

Bres V, Tagami H, Peloponese JM, Loret E, Jeang KT, Nakatani Y, Emiliani S, Benkirane M, Kiernan RE (2002) Differential acetylation of Tat coordinates its interaction with the co-activators cyclin T1 and PCAF. EMBO J 21:6811–6819

Bres V, Kiernan RE, Linares LK, Chable-Bessia C, Plechakova O, Treand C, Emiliani S, Peloponese JM, Jeang KT, Coux O, Scheffner M, Benkirane M (2003) A non-proteolytic role for ubiquitin in Tat-mediated transactivation of the HIV-1 promoter. Nat Cell Biol 5:754–761

Bres V, Gomes N, Pickle L, Jones KA (2005) A human splicing factor, SKIP, associates with P-TEFb and enhances transcription elongation by HIV-1 Tat. Genes Dev 19:1211–1226

Brigati C, Giacca M, Noonan DM, Albini A (2003) HIV Tat, its TARgets and the control of viral gene expression. FEMS Microbiol Lett 220:57–65

Butler LM, Zhou X, Xu WS, Scher HI, Rifkind RA, Marks PA, Richon VM (2002) The histone deacetylase inhibitor SAHA arrests cancer cell growth, up-regulates thioredoxin-binding protein-2, and down-regulates thioredoxin. Proc Natl Acad Sci USA 99:11700–11705

Byrd JC, Shinn C, Ravi R, Willis CR, Waselenko JK, Flinn IW, Dawson NA, Grever MR (1999) Depsipeptide (FR901228): a novel therapeutic agent with selective, in vitro activity against human B-cell chronic lymphocytic leukemia cells. Blood 94:1401–1408

Chun TW, Stuyver L, Mizell SB, Ehler LA, Mican JA, Baseler M, Lloyd AL, Nowak MA, Fauci AS (1997a) Presence of an inducible HIV-1 latent reservoir during highly active antiretroviral therapy. Proc Natl Acad Sci USA 94:13193–13197

Chun TW, Carruth L, Finzi D, Shen X, DiGiuseppe JA, Taylor H, Hermankova M, Chadwick K, Margolick J, Quinn TC, Kuo YH, Brookmeyer R, Zeiger MA, Barditch-Crovo P, Siliciano RF (1997b) Quantification of latent tissue reservoirs and total body viral load in HIV-1 infection. Nature 387:183–188

Chun TW, Engel D, Mizell SB, Ehler LA, Fauci AS (1998) Induction of HIV-1 replication in latently infected CD4+ T cells using a combination of cytokines. J Exp Med 188:83–91

Col E, Caron C, Seigneurin-Berny D, Gracia J, Favier A, Khochbin S, (2001) The histone acetyltransferase, hGCN5, interacts with and acetylates the HIV transactivator, Tat. J Biol Chem 276:28179–28184

Collins AF, Pearson HA, Giardina P, McDonagh KT, Brusilow SW, Dover GJ (1995) Oral sodium phenylbutyrate therapy in homozygous beta thalassemia: a clinical trial. Blood 85:43–49

Collman RG, Perno CF, Crowe SM, Stevenson M, Montaner LJ (2003) HIV and cells of macrophage/dendritic lineage and other non-T cell reservoirs: new answers yield new questions. J Leukoc Biol 74:631–634

Coull JJ, Romerio F, Sun JM, Volker JL, Galvin KM, Davie JR, Shi Y, Hansen U, Margolis DM (2000) The human factors YY1 and LSF repress the human immunodeficiency virus type 1 long terminal repeat via recruitment of histone deacetylase 1. J Virol 74:6790–6799

Coull JJ, He G, Melander C, Rucker VC, Dervan PB, Margolis DM (2002) Targeted derepression of the human immunodeficiency virus type 1 long terminal repeat by pyrrole-imidazole polyamides. J Virol 76:12349–12354

Craig JM (2005) Heterochromatin-many flavours, common themes. Bioessays 27:17–28

Crazzolara R, Johrer K, Johnstone RW, Greil R, Kofler R, Meister B, Bernhard D (2002) Histone deacetylase inhibitors potently repress CXCR4 chemokine receptor expression and function in acute lymphoblastic leukaemia. Br J Haematol 119:965–969

de la Serma I, OhkawaY, Imbalzano AN (2006) Chromatin remodelling in mammalian differentiation: lessons from ATP-dependent remodellers. Nat Rev Genet 7:461–473

Demonté D, Quivy V, Colette Y, Van Lint C (2004) Administration of HDAC inhibitors to reactivate HIV-1 expression in latent cellular reservoirs: implications for the development of therapeutic strategies. Biochem Pharmacol 68:1231–1238

Deng L, de la Fuente C, Fu P, Wang L, Donnelly R, Wade JD, Lambert P, Li H, Lee CG, Kashanchi F (2000) Acetylation of HIV-1 Tat by CBP/P300 increases transcription of integrated HIV-1 genome and enhances binding to core histones. Virology 277:278–295

Doetzlhofer A, Rotheneder H, Lagger G, Koranda M, Kurtev V, Brosch G, Wintersberger E, Seiser C (1999) Histone deacetylase 1 can repress transcription by binding to Sp1. Mol Cell Biol 19:5504–5511

Dorr A, Kiermer V, Pedal A, Rackwitz HR, Henklein P, Schubert U, Zhou MM, Verdin E, Ott M (2002) Transcriptional synergy between Tat and PCAF is dependent on the binding of acetylated Tat to the PCAF bromodomain. EMBO J 21:2715–2723

Dover GJ, Brusilow S, Charache S (1994) Induction of fetal hemoglobin production in subjects with sickle cell anemia by oral sodium phenylbutyrate. Blood 84:339–343

El Kharroubi A, Martin MA (1996) cis-acting sequences located downstream of the human immunodeficiency virus type 1 promoter affect its chromatin structure and transcriptional activity. Mol Cell Biol 16:2958–2966

El Kharroubi A, Piras G, Zensen R, Martin MA (1998) Transcriptional activation of the integrated chromatin-associated human immunodeficiency virus type 1 promoter. Mol Cell Biol 18:2535–2544

Erselius JR, Jostes B, Hatzopoulos AK, Mosthaf L, Gruss P (1990) Cell-type-specific control elements of the lymphotropic papovavirus enhancer. J Virol 64:1657–1666

Felsenfeld G, Groudine M (2003) Controlling the double helix. Nature 421:448–453

Finzi D, Hermankova M, Pierson T, Carruth LM, Buck C, Chaisson RE, Quinn TC, Chadwick K, Margolick J, Brookmeyer R, Gallant J, Markowitz M, Ho DD, Richman DD, Siliciano RF (1997) Identification of a reservoir for HIV-1 in patients on highly active antiretroviral therapy. Science 278:1295–1300

Finzi D, Blankson J, Siliciano JD, Margolick JB, Chadwick K, Pierson T, Smith K, Lisziewicz J, Lori F, Flexner C, Quinn TC, Chaisson RE, Rosenberg E, Walker B, Gange S, Gallant J, Siliciano RF (1999) Latent infection of CD4+ T cells provides a mechanism for lifelong persistence of HIV-1, even in patients on effective combination therapy. Nat Med 5:512–517

Fischle W, Wang Y, Allis CD (2003) Histone and chromatin cross-talk. Curr Opin Cell Biol 15:172–183

Fujinaga K, Irwin D, Huang Y, Taube R, Kurosu T, Peterlin BM (2004) Dynamics of human immunodeficiency virus transcription: P-TEFb phosphorylates RD and dissociates negative effectors from the transactivation response element. Mol Cell Biol 24:787–795

Goffin V, Demonte D, Vanhulle C, de Walque S, de Launoit Y, Burny A, Collette Y, Van Lint C (2005) Transcription factor binding sites in the pol gene intragenic regulatory region of HIV-1 are important for virus infectivity. Nucleic Acids Res 33:4285–4310

Gordon S, Akopyan G, Garban H, Bonavida B (2006) Transcription factor YY1: structure, function and therapeutic implications in cancer biology. Oncogene 25:1125–1142

Han Y, Lassen K, Monie D, Sedaghat AR, Shimoji S, Liu X, Pierson TC, Margolick JB, Siliciano RF, Siliciano JD (2004) Resting CD4+ T cells from human immunodeficiency virus type 1 (HIV-1)-infected individuals carry integrated HIV-1 genomes within actively transcribed host genes. J Virol 78:6122–6133

Hansen JC (2002) Conformational dynamics of the chromatin fiber in solution: determinants, mechanisms, and functions. Annu Rev Biophys Biomol Struct 31:361–392

Hassan AH, Prochasson P, Neely KE, Galasinski SC, Chandy M, Carrozza MJ, Workman JL (2002) Function and selectivity of bromodomains in anchoring chromatin-modifying complexes to promoter nucleosomes. Cell 111:369–379

He G, Margolis DM (2002) Counterregulation of chromatin deacetylation and histone deacetylase occupancy at the integrated promoter of human immunodeficiency virus type 1 (HIV-1) by the HIV-1 repressor YY1 and HIV-1 activator Tat. Mol Cell Biol 22:2965–2973

Henderson AJ, Calame KL (1997) CCAAT/enhancer binding protein (C/EBP) sites are required for HIV-1 replication in primary macrophages but not CD4+ T cells. Proc Natl Acad Sci USA 94:8714–8719

Henderson A, Bunce M, Siddon N, Reeves R, Tremethick DJ (2000) High-mobility-group protein I can modulate binding of transcription factors to the U5 region of the human immunodeficiency virus type 1 proviral promoter. J Virol 74:10523–10534

Henderson A, Holloway A, Reeves R, Tremethick DJ (2004) Recruitment of SWI/SNF to the human immunodeficiency virus type 1 promoter. Mol Cell Biol 24:389–397

Hines R, Sorensen BR, Shea MA, Maury W (2004) PU.1 binding to ets motifs within the equine infectious anemia virus long terminal repeat (LTR) enhancer: regulation of LTR activity and virus replication in macrophages. J Virol 78:3407–3418

Hsia SC, Shi YB (2002) Chromatin disruption and histone acetylation in regulation of the human immunodeficiency virus type 1 long terminal repeat by thyroid hormone receptor. Mol Cell Biol 22:4043–4052

Hwang SS, Boyle TJ, Lyerly HK, Cullen BR (1991) Identification of the envelope V3 loop as the primary determinant of cell tropism in HIV-1. Science 253:71–74

Imai K, Okamoto T (2006) Transcriptional repression of Human Immunodeficiency Virus Type 1 by AP-4. J Biol Chem 281:12495–12505

Jennings HR, Romanelli F (1999) The use of valproic acid in HIV-positive patients. Ann Pharmacother 33:1113–1116

Johannessen CU (2000) Mechanisms of action of valproate: a commentatory. Neurochem Int 37:103–110

Jordan A, Defechereux P, Verdin E (2001) The site of HIV-1 integration in the human genome determines basal transcriptional activity and response to Tat transactivation. EMBO J 20:1726–1738

Jordan A, Bisgrove D, Verdin E (2003) HIV reproducibly establishes a latent infection after acute infection of T cells in vitro. EMBO J 22:1868–1877

Kamine J, Chinnadurai G (1992) Synergistic activation of the human immunodeficiency virus type 1 promoter by the viral Tat protein and cellular transcription factor Sp1. J Virol 66:3932–3936

Kiefer HL, Hanley TM, Marcello JE, Karthik AG, Viglianti GA (2004) Retinoic acid inhibition of chromatin remodeling at the human immunodeficiency virus type 1 promoter. Uncoupling of histone acetylation and chromatin remodeling. J Biol Chem 279:43604–43613

Kiernan RE, Vanhulle C, Schiltz L, Adam E, Xiao H, Maudoux F, Calomme C, Burny A, Nakatani Y, Jeang KT, Benkirane M, Van Lint C (1999) HIV-1 tat transcriptional activity is regulated by acetylation. EMBO J 18:6106–6118

Korin YD, Brooks DG, Brown S, Korotzer A, Zack JA (2002) Effects of prostratin on T-cell activation and human immunodeficiency virus latency. J Virol 76:8118–8123

Kulkosky J, Culnan DM, Roman J, Dornadula G, Schnell M, Boyd MR, Pomerantz RJ (2001) Prostratin: activation of latent HIV-1 expression suggests a potential inductive adjuvant therapy for HAART. Blood 98:3006–3015

Lassen K, Han Y, Zhou Y, Siliciano J, Siliciano RF (2004) The multifactorial nature of HIV-1 latency. Trends Mol Med 10:525–531

Lee ES, Sarma D, Zhou H, Henderson AJ (2002) CCAAT/enhancer binding proteins are not required for HIV-1 entry but regulate proviral transcription by recruiting coactivators to the long-terminal repeat in monocytic cells. Virology 299:20–31

Lehrman G, Hogue IB, Palmer S, Jennings C, Spina CA, Wiegand A, Landay AL, Coombs RW, Richman DD, Mellors JW, Coffin JM, Bosch RJ, Margolis DM (2005) Depletion of latent HIV-1 infection in vivo: a proof of concept study. Lancet 366:523–524

Lewinski MK, Bisgrove D, Shinn P, Chen H, Hoffmann C, Hannenhalli S, Verdin E, Berry CC, Ecker JR, Bushman FD (2005) Genome-wide analysis of chromosomal features repressing human immunodeficiency virus transcription. J Virol 79:6610–6619

Lusic M, Marcello A, Cereseto A, Giacca M (2003) Regulation of HIV-1 gene expression by histone acetylation and factor recruitment at the LTR promoter. EMBO J 22:6550–6561

Marcello A (2006) Latency: the hidden HIV-1 challenge. Retrovirology 3:7

Marecki S, McCarthy KM, Nikolajczyk BS (2004) PU.1 as a chromatin accessibility factor for immunoglobulin genes. Mol Immunol 40:723–731

Margueron R, Trojer P, Reinberg D (2005) The key to development: interpreting the histone code ? Curr Opin Genet Dev 15:163–176

Marks PA, Richon VM, Breslow R, Rifkind RA (2001) Histone deacetylase inhibitors as new cancer drugs. Curr Opin Oncol 13:477–483

Martens JA, Winston F (2003) Recent advances in understanding chromatin remodeling by Swi/Snf complexes. Curr Opin Genet Dev 13:136–142

McKercher SR, Torbett BE, Anderson KL, Henkel GW, Vestal DJ, Baribault H, Klemsz M, Feeney AJ, Wu GE, Paige CJ, Maki RA (1996) Targeted disruption of the PU.1 gene results in multiple hematopoietic abnormalities. EMBO J 15:5647–5658

Mclean AR, Michie CA (1995) In vivo estimates of division and death rates of human T lymphocytes. Proc Natl Acad Sci USA 92:3707–3711

Michie CA, McLean A, Alcock C, Beverley PC (1992) Lifespan of human lymphocyte subsets defined by CD45 isoforms. Nature 360:264–265

Minucci S, Pelicci PG (2006) Histone deacetylase inhibitors and the promise of epigenetic (and more) treatments for cancer. Nat Rev Cancer 6:38–51

Mujtaba S, He Y, Zeng L, Farooq A, Carlson JE, Ott M, Verdin E, Zhou MM (2002) Structural basis of lysine-acetylated HIV-1 Tat recognition by PCAF bromodomain. Mol Cell 9:575–586

Nahreini P, Mathews MB (1997) Transduction of the human immunodeficiency virus type 1 promoter into human chromosomal DNA by adeno-associated virus: effects on promoter activity. Virology 234:42–50

Nakatani Y (2002) HIV-1 transcription: activation mediated by acetylation of Tat. Structure 10:443–444

Narlikar GJ, Phelan ML, Kingston RE (2001) Generation and interconversion of multiple distinct nucleosomal states as a mechanism for catalyzing chromatin fluidity. Mol Cell 8:1219–1230

Narlikar GJ, Fan HY, Kingston RE (2002) Cooperation between complexes that regulate chromatin structure and transcription. Cell 108:475–487

Navikas V, Link J, Persson C, Olsson T, Hojeberg B, Ljungdahl A, Link H, Wahren B (1995) Increased mRNA expression of IL-6, IL-10, TNF-alpha, and perforin in blood mononuclear cells in human HIV infection. J Acquir Immune Defic Syndr Hum Retrovirol 9:484–489

Nelsen B, Tian G, Erman B, Gregoire J, Maki R, Graves B, Sen R (1993) Regulation of lymphoid-specific immunoglobulin mu heavy chain gene enhancer by ETS-domain proteins. Science 261:82–86

O'Brien WA, Koyanagi Y, Namazie A, Zhao JQ, Diagne A, Idler K, Zack JA, Chen IS (1990) HIV-1 tropism for mononuclear phagocytes can be determined by regions of gp120 outside the CD4-binding domain. Nature 348:69–73

Ott M, Schnolzer M, Garnica J, Fischle W, Emiliani S, Rackwitz HR, Verdin E (1999) Acetylation of the HIV-1 Tat protein by p300 is important for its transcriptional activity. Curr Biol 9:1489–1492

Pagans S, Pedal A, North BJ, Kaehlcke K, Marshall BL, Dorr A, Hetzer-Egger C, Henklein P, Frye R, McBurney MW, Hruby, H., Jung, M., Verdin, E., and Ott, M. 2005. SIRT1 regulates HIV transcription via Tat deacetylation. PLoS. Biol. 3: e41

Pereira LA, Bentley K, Peeters A, Churchill MJ, Deacon NJ (2000) A compilation of cellular transcription factor interactions with the HIV-1 LTR promoter. Nucleic Acids Res 28:663–668

Persaud D, Zhou Y, Siliciano JM, Siliciano RF (2003) Latency in human immunodeficiency virus type 1 infection: no easy answers. J Virol 77:1659–1665

Peterson CL, Workman JL (2000) Promoter targeting and chromatin remodeling by the SWI/SNF complex. Curr Opin Genet Dev 10:187–192

Peterson CL, Laniel MA (2004) Histones and histone modifications. Curr Biol 14:R546–R551

Phiel CJ, Zhang F, Huang EY, Guenther MG, Lazar MA, Klein PS (2001) Histone deacetylase is a direct target of valproic acid, a potent anticonvulsant, mood stabilizer, and teratogen. J Biol Chem 276:36734–36741

Piekarz RL, Robey R, Sandor V, Bakke S, Wilson WH, Dahmoush L, Kingma DM, Turner ML, Altemus R, Bates SE (2001) Inhibitor of histone deacetylation, depsipeptide (FR901228), in the treatment of peripheral and cutaneous T-cell lymphoma: a case report. Blood 98:2865–2868

Pierson T, McArthur J, Siliciano RF (2000) Reservoirs for HIV-1: mechanisms for viral persistence in the presence of antiviral immune responses and antiretroviral therapy. Annu Rev Immunol 18:665–708

Ping YH, Rana, TM (2001) DSIF and NELF interact with RNA polymerase II elongation complex and HIV-1 Tat stimulates P-TEFb-mediated phosphorylation of RNA polymerase II and DSIF during transcription elongation. J Biol Chem 276:12951–12958

Pongubala JM, Nagulapalli S, Klemsz MJ, McKercher SR, Maki RA, Atchison ML (1992) PU.1 recruits a second nuclear factor to a site important for immunoglobulin kappa 3' enhancer activity. Mol Cell Biol 12:368–378

Price DH (2000) P-TEFb, a cyclin-dependent kinase controlling elongation by RNA polymerase II. Mol Cell Biol 20:2629–2634

Quivy V, Adam E, Collette Y, Demonte D, Chariot A, Vanhulle C, Berkhout B, Castellano R, de Launoit Y, Burny A, Piette J, Bours V, Van Lint C (2002) Synergistic activation of HIV-1 promoter activity by NF-kappaB and inhibitors of deacetylases: potential perspectives for the development of therapeutic strategies. J Virol 76:11091–11103

Quivy V, Van Lint C (2002) Diversity of acetylation targets and roles in transcriptional regulation: the human immunodeficiency virus type 1 promoter as a model system. Biochem Pharmacol 64:925–934

Quivy V, Van Lint C (2004) Regulation at multiple levels of NF-kappaB-mediated transactivation by protein acetylation. Biochem Pharmacol 68:1221–1229

Rabbi MF, Saifuddin M, Gu DS, Kagnoff MF, Roebuck KA (1997) U5 region of the human immunodeficiency virus type 1 long terminal repeat contains TRE-like cAMP-responsive elements that bind both AP-1 and CREB/ATF proteins. Virology 233:235–245

Rabson AB, Lin HC (2000) NF-kappa B and HIV: linking viral and immune activation. Adv Pharmacol 48:161–207

Raha T, Cheng SW, Green MR (2005) HIV-1 Tat stimulates transcription complex assembly through recruitment of TBP in the absence of TAFs. PLoS Biol 3:e44

Roebuck KA, Saifuddin M (1999) Regulation of HIV-1 transcription. Gene Expr 8:67–84

Rohr O, Marban C, Aunis D, Schaeffer E (2003) Regulation of HIV-1 gene transcription: from lymphocytes to microglial cells. J Leukoc Biol 74:736–749

Romanelli F, Pomeroy C (2003) Concurrent use of antiretrovirals and anticonvulsants in human immunodeficiency virus (HIV) seropositive patients. Curr Pharm Des 9:1433–1439

Romerio F, Gabriel MN, Margolis DM (1997) Repression of human immunodeficiency virus type 1 through the novel cooperation of human factors YY1 and LSF. J Virol 71:9375–9382

Saha A, Wittmeyer J, Cairns BR (2002) Chromatin remodeling by RSC involves ATP-dependent DNA translocation. Genes Dev 16:2120–2134

Schrager LK, D'Souza MP (1998) Cellular and anatomical reservoirs of HIV-1 in patients receiving potent antiretroviral combination therapy. JAMA 280:67–71

Schroder AR, Shinn P, Chen H, Berry C, Ecker JR, Bushman F (2002) HIV-1 integration in the human genome favors active genes and local hotspots. Cell 110:521–529

Scott EW, Simon MC, Anastasi J, Singh H (1994) Requirement of transcription factor PU.1 in the development of multiple hematopoietic lineages. Science 265:1573–1577

Scripture-Adams DD, Brooks DG, Korin YD, Zack JA (2002) Interleukin-7 induces expression of latent human immunodeficiency virus type 1 with minimal effects on T-cell phenotype. J Virol 76:13077–13082

Sgarbanti M, Borsetti A, Moscufo N, Bellocchi MC, Ridolfi B, Nappi F, Marsili G, Marziali G, Coccia EM, Ensoli B, Battistini A (2002) Modulation of human immunodeficiency virus 1 replication by interferon regulatory factors. J Exp Med 195:1359–1370

Sheridan PL, Mayall TP, Verdin E, Jones KA (1997) Histone acetyltransferases regulate HIV-1 enhancer activity in vitro. Genes Dev 11:3327–3340

Shioda T, Levy JA, Cheng-Mayer C (1991) Macrophage and T cell-line tropisms of HIV-1 are determined by specific regions of the envelope gp120 gene. Nature 349:167–169

Siliciano JD, Kajdas J, Finzi D, Quinn TC, Chadwick K, Margolick JB, Kovacs C, Gange SJ, Siliciano RF (2003) Long-term follow-up studies confirm the stability of the latent reservoir for HIV-1 in resting CD4+ T cells. Nat Med 9:727–728

Sims RJ, Belotserkovskaya R, Reinberg D (2004) Elongation by RNA polymerase II: the short and long of it. Genes Dev 18:2437–2468

Skov S, Rieneck K, Bovin LF, Skak K, Tomra S, Michelsen BK, Odum N (2003) Histone deacetylase inhibitors: a new class of immunosuppressors targeting a novel signal pathway essential for CD154 expression. Blood 101:1430–1438

Smithgall MD, Wong JG, Critchett KE, Haffar OK (1996) IL-7 up-regulates HIV-1 replication in naturally infected peripheral blood mononuclear cells. J Immunol 156:2324–2330

Sowa Y, Orita T, Minamikawa S, Nakano K, Mizuno T, Nomura H, Sakai T (1997) Histone deacetylase inhibitor activates the WAF1/Cip1 gene promoter through the Sp1 sites. Biochem Biophys Res Commun 241:142–150

Steger DJ, Eberharter A, John S, Grant PA, Workman JL (1998) Purified histone acetyltransferase complexes stimulate HIV-1 transcription from preassembled nucleosomal arrays. Proc Natl Acad Sci USA 95:12924–12929

Sune C, Garcia-Blanco MA (1995) Sp1 transcription factor is required for in vitro basal and Tat-activated transcription from the human immunodeficiency virus type 1 long terminal repeat. J Virol 69:6572–6576

Suzuki T, Kimura A, Nagai R, Horikoshi M (2000) Regulation of interaction of the acetyltransferase region of p300 and the DNA-binding domain of Sp1 on and through DNA binding. Genes Cells 5:29–41

Thierry S, Marechal V, Rosenzwajg M, Sabbah M, Redeuilh G, Nicolas JC, Gozlan J (2004) Cell cycle arrest in G2 induces human immunodeficiency virus type 1 transcriptional activation through histone acetylation and recruitment of CBP, NF-kappaB, and c-Jun to the long terminal repeat promoter. J Virol 78:12198–12206

Tréand C, du Chêne I, Bres V, Kiernan R, Benarous R, Benkirane M, Emiliani S (2006) Requirement for SWI/SNF chromatin-remodeling complex in Tat-mediated activation of the HIV-1 promoter. EMBO J 25:1690–1699

Tunnicliff G (1999) Actions of sodium valproate on the central nervous system. J Physiol Pharmacol 50:347–365

Van Lint C, Ghysdael J, Paras P Jr, Burny A, Verdin E (1994) A transcriptional regulatory element is associated with a nuclease-hypersensitive site in the pol gene of human immunodeficiency virus type 1. J Virol 68:2632–2648

Van Lint C, Emiliani S, Ott M, Verdin E (1996a) Transcriptional activation and chromatin remodeling of the HIV-1 promoter in response to histone acetylation. EMBO J 15:1112–1120

Van Lint C, Emiliani S, Verdin E (1996b) The expression of a small fraction of cellular genes is changed in response to histone hyperacetylation. Gene Expr 5:245–53

Van Lint C, Amella CA, Emiliani S, John M, Jie T, Verdin E (1997) Transcription factor binding sites downstream of the human immunodeficiency virus type 1 transcription start site are important for virus infectivity. J Virol 71:6113–6127

Van Lint C, (2000) Role of chromatin in HIV-1 transcriptional regulation. Adv Pharmacol 48:121–160

Verdin E (1991) DNase I-hypersensitive sites are associated with both long terminal repeats and with the intragenic enhancer of integrated human immunodeficiency virus type 1. J Virol 65:6790–6799

Verdin E, Paras P Jr, Van Lint C (1993) Chromatin disruption in the promoter of human immunodeficiency virus type 1 during transcriptional activation. EMBO J 12:3249–3259

West MJ, Lowe AD, Karn J (2001) Activation of human immunodeficiency virus transcription in T cells revisited: NF-kappaB p65 stimulates transcriptional elongation. J Virol 75:8524–8537

Williams SA, Greene WC (2005) Host factors regulating post-integration latency of HIV. Trends Microbiol 13:137–139

Williams SA, Chen LF, Kwon H, Ruiz-Jarabo CM, Verdin E, Greene WC (2006) NF-kappaB p50 promotes HIV latency through HDAC recruitment and repression of transcriptional initiation. EMBO J 25:139–149

Wolffe A (1999) Chromatin: Structure and Function. Academic Press, San Diego, CA

Wolffe AP (2001) Transcriptional regulation in the context of chromatin structure. Essays Biochem 37:45–57

Wong MC, Suite ND, Labar DR (1990) Seizures in human immunodeficiency virus infection. Arch Neurol 47:640–642

Wong JK, Hezareh M, Gunthard HF, Havlir DV, Ignacio CC, Spina CA, Richman DD (1997) Recovery of replication-competent HIV despite prolonged suppression of plasma viremia. Science 278:1291–1295

Workman JL, Kingston RE (1998) Alteration of nucleosome structure as a mechanism of transcriptional regulation. Annu Rev Biochem 67:545–579

Wu X, Li Y, Crise B, Burgess SM (2003) Transcription start regions in the human genome are favored targets for MLV integration. Science 300:1749–1751

Yamaguchi Y, Takagi T, Wada T, Yano K, Furuya A, Sugimoto S, Hasegawa J, Handa H (1999) NELF, a multisubunit complex containing RD, cooperates with DSIF to repress RNA polymerase II elongation. Cell 97:41–51

Yedavalli VS, Benkirane M, Jeang KT (2003) Tat and trans-activation-responsive (TAR) RNA-independent induction of HIV-1 long terminal repeat by human and murine cyclin T1 requires Sp1. J Biol Chem 278:6404–6410

Ylisastigui L, Archin NM, Lehrman G, Bosch RJ, Margolis DM (2004) Coaxing HIV-1 from resting CD4 T cells: histone deacetylase inhibition allows latent viral expression. AIDS 18:1101–1108

Yoo CB, Jones PA (2006) Epigenetic therapy of cancer: past, present and future. Nat Rev Drug Discovery 5:37–50

Xiao H, Hasegawa T, Isobe K (2000) p300 collaborates with Sp1 and Sp3 in p21(waf1/cip1) promoter activation induced by histone deacetylase inhibitor. J Biol Chem 275:1371–1376

CHAPTER 18

SMALL MOLECULE MODULATORS IN EPIGENETICS

Implications in gene expression and therapeutics

V. SWAMINATHAN*, B.A. ASHOK REDDY*, RUTHROTHA SELVI B*, SUKANYA M.S. AND TAPAS K. KUNDU†

Transcription and Disease Laboratory, Molecular Biology and Genetics Unit Jawaharlal Nehru Centre for Advanced Scientific Research, Jakkur, P O, Bangalore-560064, INDIA

Abstract: Altered gene expression resulting from changes in the post-translational modification patterns of the histones and DNA is collectively termed epigenetics. Such changes are inherited albeit there are no alterations in the DNA sequence. Epigenetic regulation of gene expression is implemented by a wide repertoire of histone and DNA modifying enzymes including the acetyltransferases and deacetylases, the methyltransferases and kinases among others. Therefore, a regulation of these enzyme activities affords a tighter regulation of gene expression. Conversely, aberrant enzymatic activities lead to unregulated gene expression, resulting in several diseases such as RTS (loss of CBP HAT activity) and Spinal and Bulbar muscular atrophy (HATs and HMTases), apart from several forms of cancers, particularly myeloid leukemia (RAR-PML or RAR-PLZF fusion proteins resulting in the mistargeting of HDACs). Thus these enzymes have emerged as novel targets for the design of therapeutics. In this direction, several small molecule modulators (activators and inhibitors) of HATs, HDACs and HMTases are being reported in literature. This chapter introduces the different histone modifying enzymes involved in gene regulation, their connection to disease manifestation and focuses on the role of small molecule modulators in understanding enzyme function and also the design and the evolution of chromatin therapeutics

1. INTRODUCTION

Heritable changes in gene expression, brought about by DNA and histone modifications without altering the coding sequences is collectively termed **'epigenetics'**. These changes, in most cases affect the chromatin structure, thereby regulating

* Equal Contribution

† E-mail: tapas@jncasr.ac.in

T.K. Kundu and D. Dasgupta (eds.), Chromatin and Disease, 397–428.

the gene expression profile. Research into the components that establish epigenetic marks have revealed interrelationships, which have led to the evolution of newer concepts in the field of gene regulation. Perturbation of the 'codes' set by these epigenetic marks results in the inappropriate expression or silencing of genes leading to 'epigenetic diseases'. Deregulation of the epigenetic balance results in the development of carcinogenic malignancies, neurodegenerative diseases, diabetes and several airway diseases amongst many more. Epigenetic codes are set up by modifications on the DNA (methylation) or on the histones (acetylation, methylation, phosphorylation, etc.), by different classes of enzymes in a precise and targeted manner. Effector proteins that recognize these marks bring about localized alterations in the chromatin structure, shaping the outcome of these codes. DNA methylation marks are recognized by methyl cytosine binding proteins MBDs or MeCP2, initiating a series of events that eventually shut off a gene. Acetylation marks are recognized by bromodomain containing proteins (found on transcriptional coactivators) whereas methylation marks, particularly H3K9 and other repressive marks are recognized by chromodomain containing proteins (found on silencers and repressors). These effector proteins are as important as the epigenetic marks in regulating the transcriptional status. This review focuses to introduce the enzymes that establish the epigenetic pattern, summarize the different epigenetic diseases and discuss the possible therapeutic potential of epigenetic therapy with examples.

2. EPIGENETIC REGULATION OF GENE EXPRESSION

2.1. Histone Acetyltransferases

Histone acetyltransferases (HATs) transfer an acetyl group from acetyl coenzyme A onto the ε-amino group of lysine residues in the amino terminal tails of histones. Histone acetylation is associated with transcriptionally active chromatin, a fact that has been established over 40 years ago (Littau *et al.*, 1964). The different HATs identified have been found as a part of transcriptional coactivator complexes, strengthening the active transcription-HAT connection (Eberharter *et al.*, 2002). In addition to acetylation of histones, a wide variety of proteins including transcription factors and non-histone proteins are modified by HATs, which earns them the new title of Factor Acetyltransferases (FATs). The acetylation of the histone tails was surmised to result in a decreased affinity of the histone for the DNA, on account of the decreasing positive charge, establishing an 'open' chromatin state. The transcriptionally active state may be mediated via the transient formation of $(H3\text{-}H4)_2$ tetrameric particle that could adopt an open structure only when H3 and H4 tails are acetylated (Morales *et al.*, 2000). Recent data suggests that acetylation could regulate diverse functions like protein–protein interaction and protein stability (Kouzarides, 2000), thereby affecting the recruitment of the transcriptional regulatory complexes. Various HATs can be classified in to broad families: (a) the GNAT superfamily that includes Hat1, Gcn5, PCAF, Elp3, Hpa2; (b) MYST family that consists of Sas2, Sas3, Esa1, MOF, Tip60, MOZ, MORF, HBO1; (c) p300/CBP family; (d) Nuclear receptor coactivators like SRC-1, ACTR, TIF2; (e) TAFII250 family; and (f) TFIIIC family (Berger *et al.*, 2000). HAT complexes are targeted to promoter regions through their interaction with transcriptional activators (Lau *et al.*, 2000).

2.2. Histone Deacetylases

Acetylation is a reversible process. Different classes of deacetylases are involved in the deacetylation of histones and nonhistone proteins. Histone deacetylation leads to transcriptionally inactive state of the chromatin. This balance between acetylation, (transcriptional activation) and deacetylation (transcriptional repression) is critical in regulating gene expression. Mammalian HDACs are classified into three classes based on their homology to yeast HDACs. Class I HDACs (HDAC1, 2, 3, 8, and 11) are homologues of *Sacharomyces cerevisiae* histone deacetylase Rpd 3 (reduced potassium dependency 3) and those with greater similarity to yeast Hda1, are class II HDACs. (Gray and Ekstrom, 2001; Gao *et al.*, 2002; Kao *et al.*, 2002) All these enzymes possess a highly conserved catalytic domain of approximately 390 amino acids and appear to deacetylate their substrates by the same mechanism. Class II is further divided into two sub classes namely II a (HDAC-4, 5, -7, and 9) and II b (HDAC6, containing as unique feature of two homologous deacetylase domains, and HDAC10, more similar to HDAC6 than to HDAC4, 5, -7, and 9) (Grozinger *et al.*, 1999; Kao *et al.*, 2002). Class I HDACs, were found as corepressor complexes with other partners, such as Sin3-HDAC and NuRD-Mi2-NRD complexes. Whereas class II HDACs interact with one or more transcription factors like BCL6, PLZF, MEF-2, and TR2; with transcriptional corepressors such as N-COR, SMRT, B-CoR, CtBP, and with heterochromatin binding protein, HP1. (Bertos *et al.*, 2001; Yang *et al.*, 2003). In addition, class II HDACs differs from class I HDAC proteins depending on their protein expression, subcellular localization, and consequently biological roles. Class I HDACs are ubiquitously expressed, whereas class II enzymes display tissue-specific expression in humans and mice. For example, HDAC4 is more abundant in skeletal muscles, moderately express in brain, heart and ovary, but not detectable in liver, lung, spleen, and placenta whereas HDAC5 is expressed in mouse heart, brain, liver, and skeletal muscles but not spleen (Bertos *et al.*, 2001, Fischle *et al.*, 2001). In contrast to class I HDACs, which are mainly nuclear enzymes (except HDAC3), (Yoon *et al.*, 2003), under different cellular conditions class II HDACs shuttle between the nucleus and the cytoplasm. HDAC4 actively shuttles between the nucleus and the cytoplasm *in vivo*, and phosphorylation and/or overexpression of 14-3-3 chaperone promotes its cytoplasmic accumulation. (Fischle *et al.*, 2001).

Class III HDACs are called Sirtuins, which are homologoues of yeast sir2 (*s*ilence *i*nformation *r*egulator). To date, seven sirtuins (1–7) have been identified in humans. The sirtuins family can be divided into five classes based on their primary structure (Frye, 2000). *S. cerevisiae* has five sirtuins, all the five belongs to class I proteins, as well as the human SIRT 1–3. SIRT4 is a class II sirtuins, SIRT5 is a class III, SIRT6 and SIRT7 are class IV proteins. The sirtuin deacetylase contains a conserved 275 amino acid catalytic domain, which is structurally and mechanistically different from that of HDACs (class I and II). Most significantly the class III deacetylases require NAD+ as a cosubstrate to deacetylate several histone and nonhistone substrate, including tumor suppressor p53 (Chang *et al.*, 2002; for more recent review see Blander and Guarente, 2004).

2.3. Histone Methyltransferases

Histone methylation is brought about by histone methyltransferases (HMTases), which catalyze the transfer of methyl group from the donor SAM (S-adenosyl methionine) predominantly to the lysine or arginine residues on the N-terminal histone tails. Based on the basis of amino acids that get modified, they are classified into the lysine methyltransferases (Martin *et al.*, 2005) and arginine methyltransferases (Bedford *et al.*, 2005). The residues can be mono, di or trimethylated, which further increase the scope and range of methylation-mediated regulation. Arginine methyltransferase have an additional level of regulation in catalyzing the formation of asymmetricdimethylarginine (aDMA) or symmetricdimethylarginine (sDMA).

HMTases are versatile enzymes with their modifications being involved in both activation and repression. This dual nature is possible because of these multiple levels of regulation (Zhang *et al.*, 2001). The exact residue on the histone tails that gets modified determines transcriptional activation or repression. The lysine methyltransferases are involved in transcriptional activation (Methylation on H3K4, H3K36 and H3K79) as well as transcription repression (Methylation of H3K9, H3K27 and H4K20), while the arginine methyltransferases so far have been shown to be involved in transcriptional activation. Several lysine methyltransferases have been found to be involved in transcriptional silencing in association with the polycomb group of proteins (e.g. EZH2 that methylates H3 K9 and K27) and the heterochromatin protein 1 (HP1) (e.g. SUV39H that trimethylates H3 K9 prominently) leading to the heterochromatinization of a particular locus. These HMTases contain a common SET (*S*uVar39, *E*nhancer of Zeste and *T*rithorax) domain that has been demonstrated to possess the HMTase activity (Martin *et al.*, 2005). Arginine methyltransferase PRMT4/CARM1 has been shown to bring out transcriptional activation of p53 and NF-κB responsive genes. (Covic *et al.*, 2005).

2.4. Histone Demethylases

The covalent modifications of histone tails such as acetylation, phosphorylation, and ubiquitination have been shown to be reversible. This reversibility help the cells to respond to these regulatory modifications and thereby, influence the gene expression. Methylation of histones however, has been considered to be a relatively stable and irreversible mark on histones. Nevertheless active turnover of methyl groups on histones do exist. One of the possible mechanism of removal of methyl group from histone tail could be replacement of methylated histone with an unmodified version on chromatin by (presumably) histone chaperone as found in case of deposition of histone H3.3 variant at rDNA (Ahmad and Henikoff, 2002) or *HSP70* promoter in Drosophila (Schwartz and Ahmad, 2005). In addition to histone replacement, two classes of enzymes were recently reported which are required either to inhibit the histone methylation or demethylation of histone tails. The first report was the identification of PADI4 (Bannister *et al.*, 2002; Cuthbert *et al.*, 2004; Wang *et al.*, 2004), which is a deiminase and antagonizes arginine methylation of histone. However, deimination does not assure the demethylation of arginine suggesting that histone replacement could be the major pathway. A significant break through was made in this field by the discovery of histone H3 lysine demethylases (Shi *et al.*, 2004; Kubicek and Jenuwein, 2004). It was found that KIAA0601 (BHC110) protein which is a riboflavin binding protein possess this activity. The recombinant BHC110/LSD1 (lysine-specific demethylase 1) catalyze the amine oxidation of methylated histone H3 (K4) to generate unmodified lysine and formaldehyde. It is note worthy that LSD1 can only demethylate mono or dimethylated lysines, not trimethylated lysine or other methylated lysines. Nevertheless, this pioneering discovery suggests that all the histone modifications are dynamic in nature, which is essential to regulate the gene expression and probably also DNA repair.

2.5. Histone Kinases

Phosphorylation has a major influence on chromatin dynamics either by altering the charge or by facilitating the binding of various chromatin remodeling factors or transcription factors that regulate gene expression (Strahl and Allis, 2000). Among the core histones, histone H3 and H2A predominantly get phosphorylated with functional consequences. The linker histone H1 is also subjected to phosphorylation in a cell cycle dependent manner. Inducible phosphorylation of histone H3 at serine 10 residue has been studied much more extensively as compared to serine 28. There are several kinases which may phosphorylate serine 10 upon mitogenic stimulation or stress. These include mitogen stimulated protein (MAP) kinase, p90 ribosomal S6 kinase 2 (RSK2) and a src family tyrosine kinase (Fyn). Phosphorylation of H3 serine 28 is also inducible (such as by UV) and mediated by MAP kinase, MSK1/2 and/or mixed lineage triple kinase alpha (MLTK-a) (Dong and Bode, 2006). Phosphorylation of histone H3 serine 10 is positively correlated to transcription activation from chromatin whereas, serine 28 phosphorylation is coupled with mitotic chromosome condensation in diverse mammalian cell lines. Recently it has been shown that tissue transglutaminase 2 (TG2) possess an intrinsic serine/threonine kinase activity and can phosphorylate serine 10 residue of histone H3 (Mishra *et al.*, 2006). Interestingly, it is also a H1 kinase. However, the *in vivo* significance of TG2 histone kinase activity is yet to be shown. Phosphorylation of histone H2AX occurs rapidly in response to DNA damage at S129 (for yeast) and

S139 (for mammals). The responsible kinases are members from phosphatidylinositol 3-kinase (PI3K)-like family kinases, which include ataxia telangiectasia mutated (ATM), AT-related (ATR), and DNA dependent protein kinase (DNA-PK) (Burma *et al.*, 2001; Stiff *et al.*, 2004; van Attikum and Gasser, 2005). Histone H2AX phosphorylation is directly related to repair of damaged chromatin (for details see chapter on 'Role of histone phosphorylation in chromatin dynamics and its implication in diseases').

2.6. DNA Methyltransferases

DNA methyltransferases or DNMTs catalyze the methylation at the C5 position of cytosine residues, which are located 5' to the guanine residue constituting CpG dinucleotide stretches. Interestingly, these dinucleotides are underrepresented in the human genome, and wherever present occur in clusters that are termed CpG islands. These CpG islands have been found associated with promoter regions. Methylation of these regions is associated with long term gene silencing, and is thought to be a mechanism to limit the DNA instability by transposons and repetitive DNA elements (Yoder *et al.*, 1997). The exact mechanism of targeting of DNMTs to specific regions is not very clearly understood. As with most methylation reactions, DNMTs also use S-adenosyl methionine as the methyl group donor.

DNA methylation is established during early embryogenesis and continues through different generations of cell cycle and development. Different DNMTs are involved in this process. DNMT3 family is necessary for the *de novo* methylation of the DNA in the early embryonic stages, and thereby is indispensable for the development of the embryo. These enzymes not only establish new methylation patterns but also methylate specific regions such as pericentric repetitive sequences and CpG islands on the inactive X-chromosomes, thereby initiating the imprinting process. DNMT1 on the other hand, is necessary for the maintenance of methylation. This methyltransferase has a preference for the 'hemimethylated DNA' created during the semi-conservative mode of replication, thereby maintaining the imprinting patterns across generations. DNMT1 is associated with the DNA replication machinery, which helps it to be targeted to regions of active replication (Vertino *et al.*, 2002). These two classes of DNMTs show interdependence, as the functions of DNMT1 is taken over by DNMT3 in DNMT1 deficient cell lines.

2.7. Epigenetic Interplay in the Regulation of Gene Expression

Individual histone modifications lead to specific transcriptional outcomes as has been discussed in the preceding sections. Recent work demonstrates that these marks do not act individually, rather do so in combination to achieve a particular transcriptional state. Such a combinatorial approach to gene regulation augments the requirement of the cell to regulate the expression of genes at varying levels. Several examples exist of histone modifications modulating functions of each other either synergistically or antagonistically. One of the earliest examples in gene activation is the phosphorylation of H3S10 that facilitates acetylation of H3K14 and methylation of H3K4 (Cheung *et al.*, 2000). It also results in the acetylation of H3K9, preventing its methylation and thereby preventing repression (Rea *et al.*, 2000). H3K4 trimethylation is generally associated with the 5' coding regions of the active genes, facilitating the hyper acetylation of H3K9, 14, 17, 23 and H4K8. This has been elucidated from studies on the chicken and mammalian β globin locus. (Kim and Dean, 2004; Schneider, 2004). On the other hand methylation of H3K9, prevents the hyperacetylation of H3 and H4. Promoter DNA methylation is generally accompanied by the concomitant deacetylation resulting in silencing of the gene. This is accomplished by the recruitment of the methyl CpG binding proteins (MDB or MeCP2), which in turn recruit HDACs setting up a repressive environment.

Yet another example of combinatorial histone codes come to fore in yeast, wherein H2B K123 monoubiquitination was shown to be a necessary prerequisite for H3K24 and H3K79 methylation resulting in opening of the chromatin (Sun and Allis, 2002). Combination of histone marks in the regulation of physiologically relevant processes such as embryonic development and lymphocyte maturation have also been described.

An ordered sequence of histone acetylation and arginine methylation has been observed in estrogen responsive genes. Upon estrogen stimulation of the pS2 gene, CBP mediated acetylation of H3K18 and K23 is followed by the methylation of H3R17 by CARM1. CARM1 is directly recruited to the gene by the acetylation marks, as an acetylase deficient CBP is unable to recruit CARM1 (Daujat, 2002). In case of p53 responsive genes, this scheme is extended to include PRMT1, which

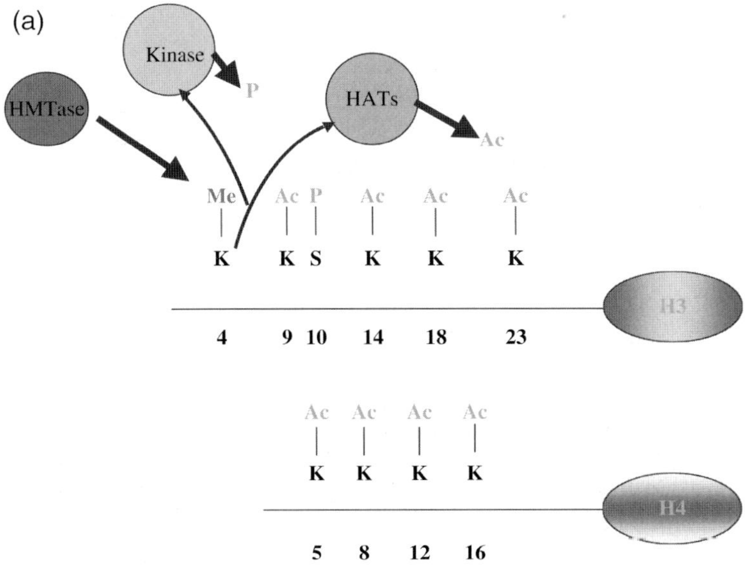

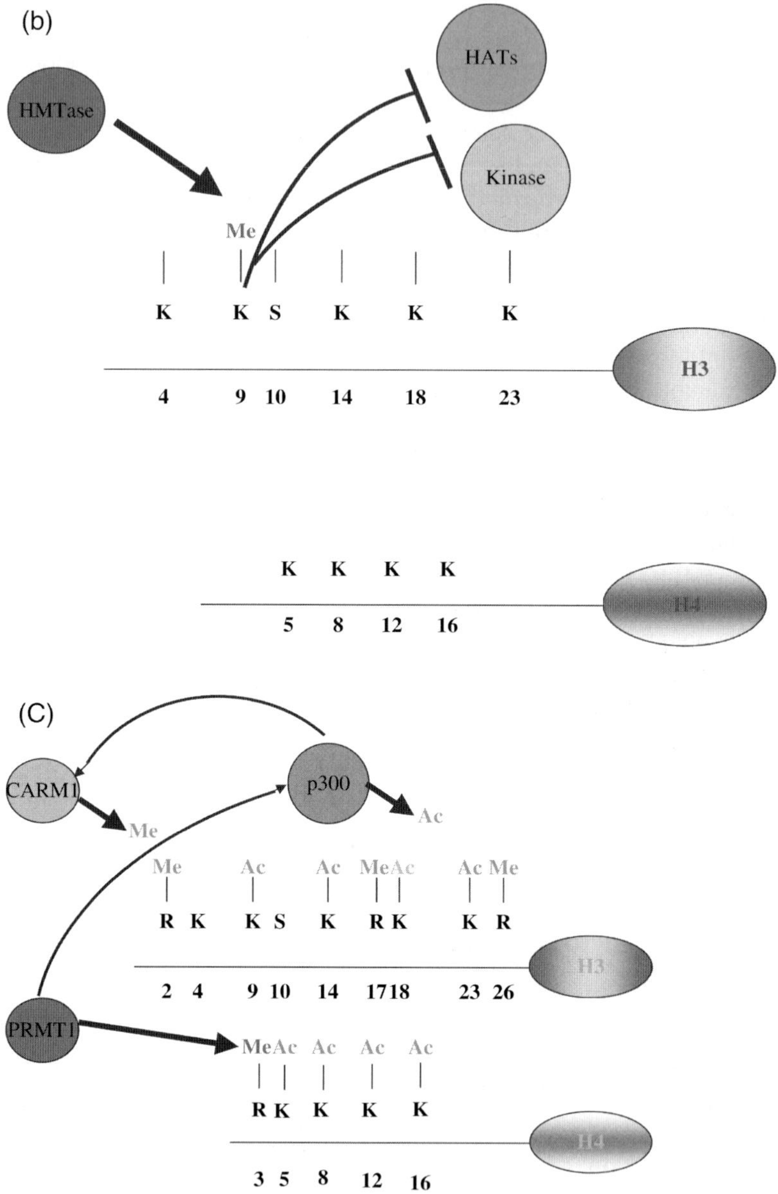

Figure 1. Epigenetic interplay in the regulation of transcription. (A) Methylation of H3K4 and phosphorylation of H3S10 facilitates the acetylation of H3 and H4 leading to transcriptional activation. (B) Methylation of H3K9 inhibits the phosphorylation of H3S10 as well as acetylation leading to transcriptional repression(C) Methylation of H4R3 by PRMT1 facilitates the acetylation by p300 which in turn acts as a favorable substrate for CARM1 mediated methylation of H3 R2, R17 and R26. (See Colour Plate 23.)

methylates H4R3 resulting in the recruitment of p300 followed by CARM1 (An *et al.*, 2004). Interestingly, methylation of Arg 3 of histone H4 by PRMT1 facilitates histone H4 acetylation leading to gene activation (Wang *et al.*, 2001). (Fig. 1). Recent insights in histone demethylases have revealed the existence of interplay between different histone marks. LSD1, a flavin dependent amine oxidase that acts on trimethylated H3K4 returning it to the ground unmethylated state is negatively regulated by H3K9 acetylation and H3S10 phosphorylation. Thus deacetylation and dephosphorylation are prerequisites for the demethylation by LSD1. (Forneris *et al.*, 2005).

3. EPIGENETIC BASIS OF DISEASE

The previous section lists and enumerates the importance of the different groups of proteins that serve to establish the epigenetic code of gene transcription. As discussed before, the cross talk between the different epigenetic marks forms the basis for a tight regulation of gene expression and thereby sets a stage for the controlled maintenance of cellular functions. Thus the integrity and stability of epigenetic gene regulation is necessary for cellular homeostasis. Perturbations to this delicate balance leads to altered gene expression, ultimately resulting in the manifestation of diseased conditions. A clear understanding of the molecular basis of epigenetic diseases could help in the design of novel therapeutics, opening vistas for the medical treatment of these diseases.

3.1. Diseases Associated with Histone Acetyltransferases

Mistargeting and mutations in HATs and HDACs are major factors leading to diseases and disorders. A classic example of one such disorder is the Rubinstein-Taybi syndrome (RSTS), which is a consequence of a single mutation in the gene encoding the HAT CREB binding protein (CBP), located on chromosome 16p13.3. The mutation occurs in the conserved aminoacid residue E, changing it to K at position 1278. This mutation alters the structure of the PHD type zinc finger in the HAT domain of *CBP*. Interestingly, this mutation in the zinc finger also abolishes its HAT activity. Yet another point mutation in CBP that converts an arginine into a proline abolishes the HAT activity and transcriptional activation of the CREB protein. Several deletions and intragenic duplication are also reported in patients with RSTS. In addition to the most common cause of RSTS, i.e., mutation in the CBP gene, mutations in p300 gene has also been reported to cause this disorder. RSTS is congenital and characterized by mental retardation and a wide range of typical dysmorphic features. Clues into the possible mode of action of CBP have come from studies in Drosophila where CBP regulates the decapentaplegic (dpp) gene expression downstream of the hedgehog signaling pathway (Akimaru *et al.*, 1997). Dpp is the homolog of the mammalian bone morphogenetic protein BMP, indicating that skeletal deformities in RSTS could be due to the improper expression of the

BMPs. Consistent with this surmise is the finding that *CBP* knockout mice show reduced Bmp7 expression and skeletal deformities (Tanaka *et al.*, 1997).

Disorders result even as a consequence of translocation events in genes encoding HATs. This results in its mistargeting and misregulation, which leads to altered chromatin acetylation and hence improper regulation of gene expression. Interestingly, most translocations involving HATs commonly cause hematological malignancies. A classic example of one such translocation is the MOZ-CBP fusion proteins. This fusion protein is expressed due to a translocation t(8;16)(p11;p13), associated with a subtype of acute monocytic leukemia (AML M5). Another example is the Mixed Lineage Leukemia (MLL) gene fusions. In therapy related AML, myelodysplastic syndrome or chronic myelomonocytic leukemia, MLL-CBP fusion proteins are expressed in translocation t(11;16)(q23;p13). The fusion protein has A/T hooks and cysteine-rich DNA recognition domain of MLL fused to intact CBP and fails to recruit SWI/SNF to its target as it lacks the SET domain, which is important for interacting with hSNF5 (Taki *et al.*, 1997). MORF gene fusions are expressed in t(10;16)(q22;p13) in childhood AML. Benign uterine leiomyomata are the most common tumors in women of the reproductive age. This is caused by t(10;17)(q22;q21)leading to the disruption of the histone acetyltransferase MORF (Moore *et al.*, 2000).

Different HATs, particularly CBP and p300 act as transcriptional coactivators for a whole host of cellular oncoproteins such as c-jun, c-fos, c-myb, BRCA1, p53 and pRb. These proteins are critical regulators of gene expression in the cell. A wide variety of viral oncoproteins, such as E1A, E6 and SV40 large T antigen, also associate with the HATs generally resulting in the suppression of the HAT activity. CBP and p300 have been suggested to act as tumor suppressors, since both genes are mutated or lost during carcinogenesis. Mutations in CBP, p300 and PCAF genes have been associated with colon, lung and stomach cancer.

HATs have also been implicated in cardiac hypertrophy, which is characterized by an increased HAT activity (Gusterson *et al.*, 2002). Consistent with this observation, p300 knockout mice show reduced expression of muscle proteins and cardiac structural defects (Yao *et al.*, 1998). Apart from the acetylation of histone tails, p300 serves as a transcriptional coactivator for hypertrophy responsive transcriptional factors such as GATA-4, SRF and MEF2. Besides, cardiac hypertrophy, several airway diseases such as asthma and chronic obstructive pulmonary disease (COPD), have an indirect dependence on HATs through the pro-inflammatory transcription factor, NF-κB. NF-κB is known to associate with different HATs such as CBP/p300, PCAF, GRIP-1 and SRC1.

3.2. Diseases Associated with Histone Deacetylases

The epigenetic codes set by HATs are countered by the activity of the histone deacetylases. This delicate balance has been documented to maintain the cellular homeostasis. HDACs seem to play dual opposing roles in disease manifestation. Dysfunctioning of HDACs are involved in several cancers. In some cases HDACs suppress the expression of some tumor suppressor genes, while in other cases it aids the function of tumor suppressor proteins. The retinoblastoma tumor suppressor protein recruits HDAC1, (Magnaghi-Jaulin L *et al.*, 1998, Chan *et al.*, 2001), p53 protein complexes with Sin3-HDAC (Murphy *et al.*, 1999) and BRCA1, the breast cancer susceptibility protein, associates with HDAC1 and 2. (Yarden *et al.*, 1999). Although mutations in HDACs have not been reported in carcinogenesis, mistargeting of the repressive complexes containing HDACs have been reported in leukemias and lymphomas (Fenrick *et al.*, 1998; He *et al.*, 2001). The example of mistargeting of HDACs is best demonstrated in case of acute pro-myelolytic leukemia (APL) where the retinoic acid receptor RARa undergoes translocation with oncoproteins such as PML and PLZF. The retinoic acid (RA) receptor (RAR) is a transcriptional regulator that is important for myeloid differentiation. Under normal conditions, unliganded RARα associates with retinoic acid response elements (RARE) and represses genes with the concomitant recruitment of the corepressor complexes N-CoR and SMRT. In the presence of retinoic acid, RAR complexes with HATs and results in the initiation of myeloid differentiation. In the event of a translocation the fusion protein binds to RARE and recruits HDACs thereby repressing promyelolytic differentiation (see review Mai *et al.*, 2005).

Apart from this HDACs are also associated with a number of other epigenetic repression mechanisms including histone methylation, polycomb group of proteins and DNA methylation (discussed later). The class II HDACs have been found to be involved in muscle development, particularly HDAC 5 and 9 knockouts or mutants show evidence of cardiac hypertrophy in a age or stress dependent manner (Zhang *et al.*, 2002; Chang *et al.*, 2004).

3.3. Diseases Associated with Histone Methyltransferases

In cancer, there is genome-wide DNA hypomethylation as well as gene specific DNA hypermethylation (Esteller *et al.*, 2001). Aberrant gene activation and repression both result as an interplay of different histone modifying enzymes and DNA methylation enzymes. Deregulation of this interplay or the mistargeting leads to neoplasia. Cancer and epigenetics are very closely linked, the involvement of HMTases, more so of DNMTs (DNA methyltransferase) is observed in cancer (Zhang *et al.*, 2005). Enlisted below are the different HMTases and their implication in cancer both direct and indirect.

Retinoblastoma (Rb) protein is a tumor suppressor with multiple interacting protein partners. Deregulation of the tumor suppressor Retinoblastoma protein is a major contributor to carcinogenesis. This is mutated in nearly 90% of human solid tumors. Rb has been shown to interact with HMTases Suv39H1 and RIZ both enzymes associated with bringing about general repression, mediated by H3K9 methylation (Nielsen *et al.*, 2001). Homeotic gene expression is maintained by two groups of proteins, the Polycomb group and the Trithorax group of proteins (Mahmoudi *et al.*, 2001).

Ezh2 is a polycomb group protein contains EED and *Enhancer of Zeste* homologs 1 and 2 which recruit HDACs and thus bring about repression by the cooperative methylation and deacetylation. Ezh2 is over expressed in prostate cancer. Prostate cancer is a leading cause of death in males second only to lung cancer. Ezh2 has been observed to be over expressed in hormone refractory metastasis prostate cancer (Varambally *et al.*, 2002). Dysregulated expression of Ezh2 is involved in the progression of prostate cancer as well as acts as a marker that distinguishes indolent prostate cancer from those at risk of lethal progression. Breast cancer is a leading cause of cancer-related death in women; EZH2 protein levels are strongly associated with breast cancer aggressiveness. Overexpression of EZH2 in immortalized human mammary epithelial cell lines promotes anchorage-independent growth and cell invasion (Kleer *et al.*, 2003).

There is a school of thought, which believes that HMTases are tumor suppressors especially the lysine methyltransferases because of the loss of SET domain proteins in tumor conditions, exceptions do exist like Ezh2. The well-known example of the above is RIZ1, which interacts with Rb protein (again the same tumor suppressor). RIZ-1 is in chromosome 1p36 region, which is commonly deleted, in more than a dozen different types of human cancers. Riz-1 expression is commonly silenced in many tumors including breast cancer, liver cancer, colon cancer, neuroblastoma, melanoma, lung cancer and osteosarcoma (Kim *et al.*, 2003).

MLL is a H3K4 methyltransferase and therefore associated with activation but it is also known to recruit PcG proteins, thus the final activity of MLL is basically a balance between activation and repression. The most widely known target of this enzyme is the Homeobox (Hox) genes, which play important roles in axial morphogenesis, patterning and hematopoeitic differentiation. Expression of the Hox genes is upregulated in human leukemias with MLL rearrangements; transformation by the MLL fusion proteins requires the above activation. The MLL gene is associated with many chromosomal translocations associated with ALL and AML; the balance of its activity gets disturbed in such translocations. In leukemia the SET domain of MLL is deleted and the PHD finger domain of MLL that is responsible for recruiting the repressive factors are lost, thereby resulting in loss of enzymatic activity. But, the fusion as with CBP, targets the genes of MLL and brings about transcriptional activation mediated by acetylation. Other such theories are also proposed but the underlined fact being that MLL targets the proteins and the other partner of the fusion brings about the activation (Hake *et al.*, 2004).

Yet another SET domain protein, which is directly related to cancers, is *SMYD3* that gets upregulated in colorectal cancer and hepatocellular carcinoma, a leading cause of death in developed countries (Hamamoto *et al.*, 2004).

The human homolog of yeast Dot1, hDOT1L, is also aH3 K79 specific methyltransferase, and recent studies relate this protein to cancer hDOT1L physically interacts with AF10, which is an MLL fusion partner in certain leukemias (Okada *et al.*, 2005).

Lysine methyltransferases have a well-established role in cancer whereas the arginine methyltransferases role is relatively less explored. Inspite of that there is enough evidence to suggest their role both directly and indirectly with cancer (Bedford *et al.*, 2005). CARM1/PRMT4 and PRMT1 exhibit cooperativity with the histone acetyltransferase p300 in the transcriptional activation of p53 responsive genes (An *et al.*, 2004) and p53 is again a well-known tumor suppressor that is mutated in many of the cancers. Arginine methyltransferases are coactivators for nuclear receptors, CARM1 over expression correlates with androgen independence in human prostate carcinoma (Hong *et al.*, 2004). Over expression of PRMT5 promotes anchorage independent cell growth and therefore is a probable target to be deregulated in transformed cellular states (Pal *et al.*, 2004).

3.4. Diseases Associated with DNA Methyltransferases

A growing number of diseases caused by the disturbance of DNA methylation patterns underscore the importance of it as a pharmacological target. DNA methylation has long been known to be associated with cancer, wherein the cancer cell genome was shown to be hypomethylated in comparison to its normal counterparts. More recently it has been demonstrated that while global hypomethylation leads to oncogene expression and genomic instability, cancer cells also show local hypermethylation on the promoters of tumor suppressor genes. Several genes are over expressed in cancers, as for example, melanoma antigens of testis cancer (DeSmet *et al.*, 1999) and the serine protease inhibitor gene SERPIN B5 in gastric cancer, are frequently found to be demethylated in the cancerous cells. It is presumed that the global hypomethylation early in tumorogenesis aids in the evolution of the tumor cells by making the genome amiable for further genetic changes. DNA hypermethylation of promoter regions accompanied by the associated histone hypoacetylation leads to the silencing of the tumor suppressor genes (Song *et al.*, 2005). [Interestingly the disordered DNA methylation can be utilized as a unique marker in the early detection of cancer.] The p16INK4A tumor suppressor gene has to be widely studied in this context.

Apart from cancers, DNA methylation defects lead to diseases with defective imprinting patterns. While loss of imprinting on some loci (e.g. IGF2/H19 locus) leads to several cancers including lung, colon, liver, ovarian and Wilms' tumor, it could also result in several different syndromes. Prader Willi syndrome characterized by a failure to thrive during infancy, obesity, mental retardation and behavioral problems caused by a loss of imprinting in the region. Angelman syndrome, arising out of a loss of imprinting in the maternally expressed ubiquitin protein ligase E3A (UBE3A) gene, leads to speech impairment and mental retardation.

As has been mentioned previously, DNA methylation plays an important role in stabilizing the genome particularly at the repeat sequences. In case of the Fragile X syndrome, an increase in the trinucleotide repeat length leads to silencing of the FMR1 gene locus. In contrast to this, decrease in repeat lengths of the D4Z4 repeat leads to hypomethylation at 4q3S locus leading to Facioscapulohumeral muscular dystrophy (FSHD). Systemic lupus erythmatosus (SLE) is caused by a loss of DNA methylation in T cells due to reduced activity of DNMT1 while

ICF syndrome (Immunodeficiencies centromeric instability and facial anomalies) is caused by a mutation in the DNMT3B gene. An interesting example highlighting the interplay between different chromatin remodeling factors is seen in the ATRX (alpha thalasemia/mental retardation syndrome, X-linked) wherein the affected gene ATRX encodes ATP dependent chromatin remodeling factor. Interestingly, it contains a PHD domain similar to that of DNMT3A, and patients with this syndrome show DNA methylation defects.

4.1. Therapeutic Measures Targeting the Epigenetic Regulation

The cause of most epigenetic diseases can be traced to the enzymes that establish them. A great deal of research has gone into the discovery of the modulators of these enzymes, which not only leads to a better understanding of the mechanism, but also to therapeutic possibilities.

From Rubinstein-Taybi syndrome to MOZ gene fusions to Cancers, mutations in HATs and their mistargeting explain their significance, which leads to several disorders including neurodegenerative ones resulting in improper regulation of gene expression. Their general role as important intermediates involved in regulating gene expression makes them attractive as most potent targets for therapeutics. Activators or inhibitors of such HATs will be the focus in designing drugs for therapeutic purposes. Several small molecule activators and inhibitors of HATs, which can find their lead in such therapy have been discovered. Inactivation of tumor suppressor genes is central to the development of cancer. Silencing of these genes occurs by epigenetic means and inhibition of these factors leads to reversal of tumor suppressor gene silencing and inhibition of tumorigenesis (Gibbons *et al.*, 2005). Among the enzymes involved in the regulation of gene expression, the CREB binding protein CBP displays important functions during central nervous system development. CBP loss of function has been reported in several disorders. New drugs targeted at counteracting CBP loss of function could stand as a valid therapeutic approach in neurodegenerative disorders (Rouaux *et al.*, 2004, see also Chapter 'Chromatin acetylation status in the manifestation of neurodegenerative diseases: HDAC inhibitors as therapeutic tools?'). Among the synthetic HAT modulators, the first to be designed were the peptide CoA conjugates. Of these, Lys-CoA was specific for p300 and H3-CoA was specific for PCAF. These compounds are now used as biological tools to study HATs in transcription (Lau *et al.*, 2000). Recently isothiazolone based modulators have also been designed that inhibit p300 and PCAF (Stimons *et al.*, 2005). The search for natural small molecule modulators of HATs succeeded when anacardic acid from cashew nut shell liquid was proved as a potent inhibitor of p300 and PCAF. Furthermore, CTPB (N-(4-chloro-3-trifluoromethyl-phenyl)-2-ethoxy-6-pentadecyl-benzamide), an amide was derived using anacardic acid as synthon, which remarkably activates p300 HAT activity.

These compounds may find its use as biological switching molecules to study the role of p300 on transcription and for the development of therapeutics against cancer (Balasubramanyam *et al.*, 2003). Another lead to the development of therapeutic drugs using small molecular modulators of HATs was garcinol, a polyisoprenylated benzophenone derivative from *Garcinia indica* fruit rind. Garcinol was a potent inhibitor of p300 both *in vitro* and *in vivo* and a potent inducer of apoptosis in HeLa cells. Garcinol was also found to alter global gene expression, mainly by down regulating genes. As it is known that alterations in histone acetylation patterns leads to several diseases like asthma, AIDS etc., garcinol or its derivatives may serve as lead compounds for designing therapeutics for diseases in addition to cancers (Balasubramanyam *et al.*, 2004). HIV is most dreadful to the mankind. The conventional therapeutic approach, Highly Active Anti Retroviral Therapy (HAART), leads to latency of the virus. This latency could be reverted using inhibitors of histone deacetylases. The new therapeutic approach to treat HIV infection aims at better understanding of targeting HATs and HDACs, which may lead to a better and economical anti-HIV combinatorial therapeutics (Varier and Kundu, 2006). Curcumin (diferuloylmethane), a major curcuminoid in turmeric is a specific inhibitor of p300/CBP HAT activity. Curcumin could also inhibit the p300 mediated acetylation of HIV-Tat protein in vitro and also the proliferation of the virus. Hence, Curcumin may serve as a lead compound in combinatorial HIV therapeutics (Balasubramanyam *et al.*, 2004). There are few well characterized HAT inhibitors enlisted in the table 1.

Structurally different classes of HDAC inhibitors (HDACi) have been reported, which selectively inhibits cancer cell growth invitro. These compounds have shown to inhibit cell proliferation by induction of cell cycle arrest in G1 and/or G2 phase, terminal differentiation, and/or apoptosis of transformed cells in culture for a large range of transformed cell types including solid tumor and hematological neoplastic cell lines. Moreover, they inhibit the growth of several types of cancers in tumor bearing animal models, being some of them actually in clinical trials. The treatment of normal and tumor cells with HDAC inhibitors causes a similar accumulation of acetylated histones H3, H4, H2A and H2B (Marks *et al.*, 2000, 2001; Kramer OH *et a.l*, 2001; Vigushin *et al.*, 2002). Nevertheless, tumor cells appear to be much more sensitive to growth inhibition and apoptotic effects of these agents than normal cells (Parson *et al.*, 1997; Qui *et al.*, 1999; Butler *et al.*, 2000, 2001). To date number of HDAC inhibitors have been reported as useful tools for the study of

Table 1. HAT modulators

S.No.	HAT modulator	Target	Reference
1	Lysyl-CoA	p300	Lau *et al.*, 2000
2	H3-CoA-20	PCAF	Lau *et al.*, 2000
3	Anacardic acid	p300/PCAF/CBP	Balasubramanyam *et al.*, 2003
4	CTPB	p300	Balasubramanyam *et al.*, 2003
5	Garcinol	p300/PCAF/CBP	Balasubramanyam *et al.*, 2004
6	Curcumin	p300/CBP	Balasubramanyam *et al.*, 2004
7	γ-butyrolactones	CBP/Gcn5	Biel *et al.*, 2004
8	Isothiazolones	p300/PCAF	Stimons *et al.*, 2005

function of chromatin acetylation and deacetylation, and gene expression. Based on their structure, HDACi can be divided into four major classes as enlisted in table 2.

The exact molecular mechanism of antitumor activity of HDAC inhibitors has not yet been unraveled. One model suggests that hyperacetylation of histones activates tumor suppressor genes and repress oncogenes. For example, HDAC inhibitors have been shown to induce cyclin dependent kinases (CDK) inhibitors such as p21, which are responsible for cell cycle arrest and subsequent cell differentiation (Rocchi *et al.*, 2005). Another model suggests that HDACi can also activates the death receptor mediated and intrinsic apoptotic pathways (Nebbioso *et al.*, 2005; Peart *et al.*, 2005). Furthermore they can also alter the expression of genes involved in angiogenesis (Michaelis *et al.*, 2004) and Metastasis (Joseph *et al.*, 2004). Trichostatin A (TSA) can inhibit hypoxia-induced expression of vascular endothelial growth factor (VEGF) and suppress angiogenesis, both *in vitro* and *in vivo*. So, augmentation of the host immune response and inhibition of tumor angiogenesis might markedly suppress the growth of primary tumors and impede metastasis. Another possible mechanism is hyperacetylation of histones leads to genomic instability, which ultimately leads to cell cycle arrest. One probable more mechanism of action of HDACi is by altering the expression of thioredoxin-binding protein 2 (TBP2). Its expression is low in a number of human cancers. Suberoylanilide hydroxamic acid (SAHA) selectively increases the expression of TBP2, which in turn binds reduced thioredoxin and inactivates this

(a)

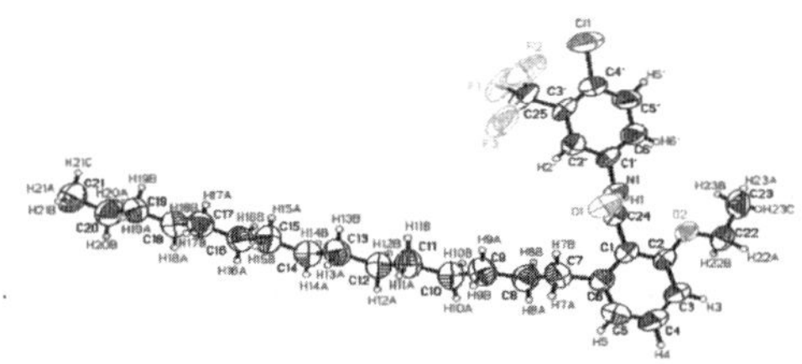

X-ray crystal structure CTPB showing the ORTEP view of the compound

Histones	+	+	+	+	+	+	+
p300	-	+	+	+	-	-	-
PCAF	-	-	-	-	+	+	+
DMSO	-	-	+	+	-	+	-
CTPB (μM)	-	-	-	100	-	-	100

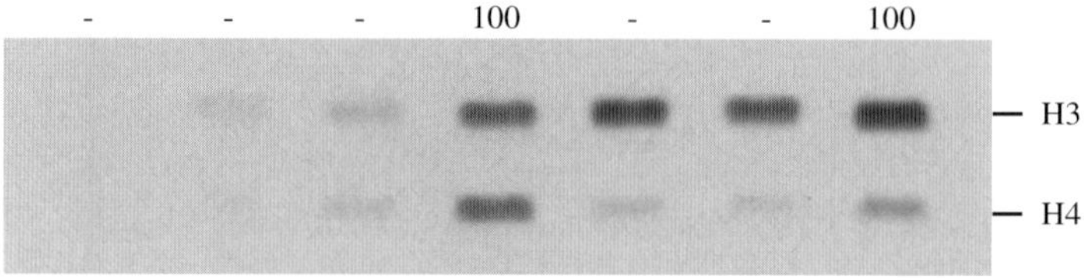

1 2 3 4 5 6 7

(b)

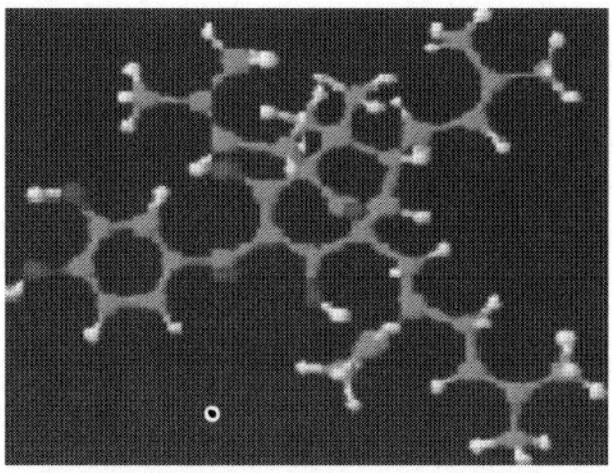

Ball and stick model of Garcinol

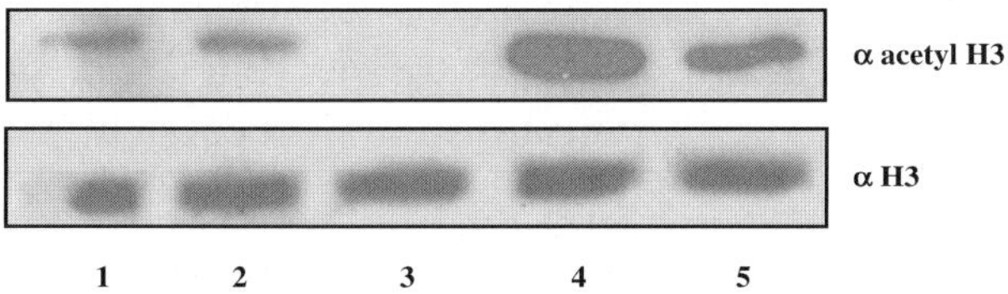

(c)

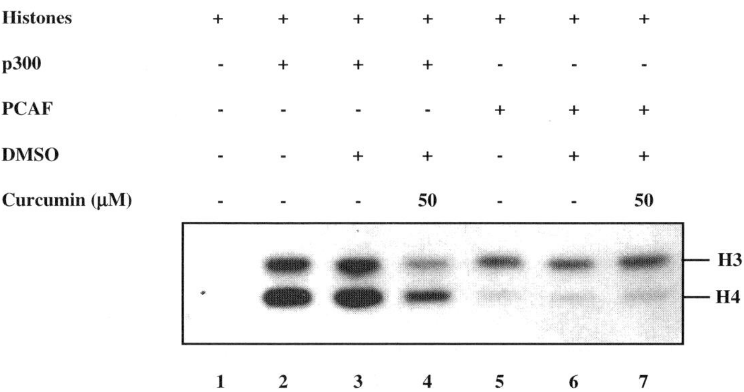

Figure 2. Newly discovered small molecule HAT activator and inhibitors alter histone acetylation *in vitro* and *in vivo*: (a) CTPB activates the p300 histone acetyltransferase activity; Histone acetyltransferase assays were performed in the presence and absence of CTPB using highly purified HeLa core histones (1.8 μg) and either with p300 (5 ng) or PCAF (15 ng) and processed for fluorography. Fluorographic analysis of acetylated histones by p300 and PCAF in the presence of CTPB; Lane 1, core histones without any HAT; lanes 2 and 5 histones with p300 and PCAF respectively; lane 3 and 6 with DMSO as a control, lane 4 and 7 with 100μM concentration of CTPB. (b) Garcinol inhibits the histone acetylation *in vivo* in HeLa cells: HeLa cells were treated for 24 h; *Lane 1*, histones extracted from untreated cells; *lane 2*, Me2SO (solvent control) treated cells; *lane 3*, garcinol 100μM treated cells; *lane 4*, NaBu (1mM), TSA (2 μM); *lane* 5, cells treated with NaBu (1mM) and TSA (2 μM) and garcinol (100 μM). The acid-extracted histones were resolved over 12% SDS-PAGE and analyzed by Western blot using antibodies against acetylated histone H3 (Calbiochem). Loading and transfer of equal amounts of protein were confirmed by immunodetection of histone H3. (c) Curcumin inhibits the HAT activity of p300 but not PCAF. Fluorographic analysis of acetylated histones by p300 and PCAF in the presence and absence of curcumin; Lane 1, core histones without any HAT; lane 2 and histones with p300 and PCAF respectively.; lanes 3 and 6 with DMSO as a solvent control, lane 4 and 7 with 50 μM concentrations of curcumin. (See Colour Plate 24.)

Table 2. HDAC inhibitors

HDACi	Target	References
Short Chain Fatty Acids		
Butyrate	Class I/II	Vidali *et al.*, 1978
Valproic acid	Class I/II	Phiel *et al.*, 2001
Hydroxamic acids		
m-Carboxy cinnamic acid bishydroxamic acid (CBHA)	Class I/II	Richon *et al.*, 1998
Oxamflatin	Class I/II	Kim *et al.*, 1999
Pyroxamide	Class I/II	Butler *et al.*, 2001
Scripataid	Class I/II	Su *et al.*, 2000
Suberoylanilide hydroxamic acid (SAHA)	Class I/II	Richon *et al.*, 1998
Trichostatin A (TSA)	Class I/II	Yoshida *et al.*, 1990
LBH 589	Class I/II	George *et al.*, 2005
NVP-LAQ824	Class I/II	Catley *et al.*, 2003
Cyclic tetrapetides		
Apicidin	Class I/II	Darkin-Rattray *et al.*, (1996)
Depsipetide (FK-228, FR901228)	Class I/II	Nakajima *et al.*, 1998
TPX-HA analogue (CHAP)	Class I/II	Furumai *et al.*, 2001
Trapoxin	Class I/II	Kijima *et al.*, 1993
Benzamides		
CI-994 (N-acetyl dinaline)	Class I/II	Kraker *et al.*, 2003
MS-275	Class I/II	Saito *et al.*, 1999
Sirtuin inhibitors		
Sirtionol	Class III	Grozinger *et al.*, 2001
Splitomycin	Class III	Bedalov *et al.*, 2001
Nicotinamide	Class III	Jackson *et al.*, 2003
Cambinol	SIRT1 and SIRT2	Heltweg *et al.*, 2006

important cellular redox regulatory protein, increasing sensitivity to oxidative stress and apoptosis. The activation of TBP2 and probably other genes repressed in tumor cells by HDAC inhibitors could contribute to cell growth arrest, terminal differentiation, and/or apoptosis caused by these agents. These agents can cause accumulation of acetylated proteins that are important regulators of cell-cycle progression. Furthermore, HDAC inhibitors have been proved to cause acetylation of heat shock protein (Hsp90) and degradation of its cargo oncoproteins in tumor cells (Atadja *et al.*, 2004).

HDAC inhibitors have been recently proposed as drugs for the treatment of blood parasites like leishmania's and trypanosomes (Belli 2000) (see next section). Epigenetic phenomena are also involved in many psychiatric disorders. For example, HDAC inhibitors have recently been proposed as therapeutic agents in mitigating vulnerability to schizophrenia among high-risk individuals (Impagnatiello *et al.*, 1998). Although many of the genes affected by HDAC inhibition are those encoding proteins known to mediate tumorigenesis or tumor suppression, increasing evidence suggests that modifications of the epigenetic histone code may not represent the primary mechanism for HDAC inhibitor-mediated growth inhibition and apoptosis in cancer cells (Brinkmann *et al.*, 2001; Johnstone *et al.*, 2003). Recent evidences indicate that HDAC inhibitors affect the transcription of a relatively limited set of genes through chromatin remodeling, with lesser than 10% of expressed genes being either activated or repressed (Drummond *et al.*, 2005). Therefore, it is clear indication that some other non epigenetic mechanisms are operating in the cells because of inhibition of HDACs. A number of tumor-associated proteins that mediate proliferation and cell-cycle progression, including p53, Ku70, Hsp90, RelA, and Stats, have been identified as substrates for various HDAC isoforms. Targeting the acetylation status of these signal mediators might underlie the antiproliferative activities of HDAC inhibitors in cancer cells. For example, a recent report indicates that HDAC inhibitors induced cell-cycle arrest/apoptosis in prostate cancer cells through the stabilization of acetylated p53 (Roy *et al.*, 2005). Mechanistically, increased p53 acetylation diminishes Mdm2-mediated ubiquitination and the subsequent proteasome-facilitated degradation (Luo *et al.*, 2000). Similarly, a link between increased Ku70 acetylation and TSA-triggered apoptosis in neuroblastoma cells has also been described (Subramanian *et al.*, 2005). Acetylation of Ku70 at Lys539 and Lys542 would disrupt its association with Bax, a proapoptotic Bcl-2 member. The resulting mitochondrial translocation of Bax facilitates cytochrome-c release and caspase-dependent apoptosis. Moreover, as Hsp90 represents a target of HDAC6, inhibition of HDAC6 resulted in the hyperacetylation and loss of the chaperone activity of Hsp90, suggesting reversible acetylation as a unique mechanism to regulate Hsp90 activity (Bali *et al.*, 2005). Various HDACs have been shown to form complexes with a series of cellular proteins including 14-3-3 proteins, α-tubulin, ubiquitin, and PP1, illustrating the complexity of their biological functions (Grozinger *et al.*, 2000; Hook *et al.*, 2002; Canettieri *et al.*, 2003; Kawaguchi *et al.*, 2003; Brush *et al.*, 2004; Yang *et al.*, 2005). For example, recent evidence shows that HDACs 1 and 6 formed complexes with PP1, of which the combined deacetylase/phosphatase activities underlie the ability of HDAC1 to modulate transcriptional activity of the cAMP responsive element binding protein (CREB) and that of HDAC6 to regulate microtubule dynamics. HDAC–PP1 complexes play a key role in regulating the activation status of Akt and other signaling kinases (Chen *et al.*, 2005). TSA, (S)-HDAC-42, and to a lesser extent, SAHA facilitated the dephosphorylation of Akt and other signaling kinases by altering the dynamics of HDAC–PP1 complexes in U87MG glioblastoma cells (Mai *et al.*, 2005).

4.1.1. Histone methyltransferase inhibitors (HMTasei)

The field of HMTasei is relatively unexplored with just a few examples of which majority are substrate analogues. The only specific inhibitor is Chaetocin, a SU(VAR)3–9 inhibitor (Greiner *et al.*, 2005) and the documented analogue inhibitors are AMI-1, analogue inhibitor of PRMT (Cheng *et al.*, 2004) Sinefungin, is another analogue inhibitor of Arginine methyltransferase (Amur *et al.*, 1986). Recently we identified an Arginine methyltransferase specific inhibitor from natural sources (Selvi, Mantelingu and Kundu, unpublished). Since the role of HMTases in cancer manifestations is well established, these inhibitors will be of great use for cancer treatment.

4.2. Clinical Status of HDACi as a Antineoplastic Drugs

Drugs belonging to several classes of HDACi are in clinical trials. Phenylbutyrate is a fatty acid with HDACi activity that has been studied extensively in patients with solid tumors, leukemia, and myelodysplastic syndromes (MDS). Depsipetide (FK-228) is a cyclic tetrapeptide with potent HDACi activity especially of Class I HDACs. It is a prodrug that is activated by reduction upon cellular uptake. Depsipeptide also has been studied in several clinical trials. SAHA is a potent HDACi belonging to the hydroxamate class. This agent has been studied extensively both *in vitro* and in ongoing Phase I/II studies in patients with solid tumors and hematological malignancies using different schedules and formulations (and oral formulation exists). SAHA inhibits both Class I and II HDACs. SAHA has been shown to have clinical activity in a transgenic animal model of therapy resistant acute promyelocytic leukemia, restoring sensitivity to retinoic acid, and to induce differentiation of human breast cancer cells. Valproic Acid is a short chain fatty acid that is clinically used as an anticonvulsant. It has excellent bioavailability and can be given orally; its elimination half-life is 6–17 hours; and overall, it has a good toxicity profile (see review Garcia-Manero G and Issa JP, 2005).

4.3. Modulators of Histone Modifying Enzymes and Protozoan Diseases

Histones of *Plasmodium falciparum* represent a new target for antimalarials. Apicidin [Cyclo (N-O-methyl–L-tryptophanyl–L–isoleucinyl–D–pipecolinyl–L-2-amino-oxodecanoyl] is a potent, broad spectrum antiprotozoal agent. The target of apicidin is Histone deacetylases HDA (Darkin Rattray *et al.*, 1996), evidenced as hyperacetylation of histones in treated parasites.

Apicomplexan parasites cause life threatening diseases like malaria, cryptosporidiosis, toxoplasmosis, coccidiosis. Suberic acid bisdimethylamide which also inhibits HDA selectively arrests tumor cells as opposed to normal mammalian cells, has an *in vivo* cytostatic effect against the acute murine malaria *Plasmodium berghei* (Andrews *et al.*, 2000).

Neospora caninum is an apicomplexan parasite that causes abortion in cattle. Depudecin is an inhibitor of these parasites without exerting any cytotoxicity on the host cells. This inhibitor targets the protozoal histone deacetylases (Kwon *et al.*, 2003).

Trypanasoma brucei causes sleeping sickness in humans. These flagellated parasites are directly exposed to immune defences as they circulate in the mammalian host bloodstream but maintain persistent infections by undergoing antigenic variation. These parasites have been reported to have certain unusual histone modifications where the N-terminal alanine of H2A, H2B, H4 get monomethylated which is a modification so far unknown for the histones, and therefore the identification of the enzyme bringing about this modification and targeting it to be a very useful venture.

4.4. Combinatorial Therapy

Epigenetic silencing of tumor suppressors is a major cause for malignancies. HDACi have emerged as a tool to reverse the transcriptionally silenced tumor suppressors, which in turn results in the induction of cell differentiation and apoptosis *in vitro* in diverse cellular processes and animal models. This has resulted in the development of several HDACi and subsequent clinical studies. Most of these studies have indicated that the majority of HDACi so far studied have limited clinical activity. Therefore, clinical future of HDACi is in the development of combination therapy with other agents. This is being pursued following different strategies. One includes the combination of HDACi with hypomethylating agents. Aberrant DNA methylation is very common phenomenon in human cancers. Aberrant methylation of promoter associated CpG islands has been also associated with gene silencing. This process can be reverted *in vitro* by inhibitors of DNA methyltransferases resulting in DNA hypomethylation, gene reactivation, and potentially induction of apoptosis. Two hypomethylating agents have been studied extensively. One of them is 5-azacytidine, has been approved by FDA for use in-patients with MDS (Kornblith *et al.*, 2002; Silverman *et al.*, 2002). The other agent, 5-aza 2'deoxy-cytidine currently in clinical trials (Hennessy *et al.*, 2003). Additive to synergetic induction of apoptosis has been reported as a result of sequential incubation of neuroblastoma cell line, BL1521 with a HDACi and gemcitabine (dFdC), a deoxy-cytidine analogue with proven antitumor activity (de Ruijter *et al.*, 2006).

Another combination strategy consists in the combination of HDACi with more classic chemotherapeutic agents, in particular those that have a direct effect on topology of DNA. For instance, a logical combination would be the use of a HDACi with DNA topoisomerase I or II inhibitors. Several reports indicate that this combination has significant *in vitro* antitumor activity (Kim *et al.*, 2003; Marchion *et al.*, 2004). The exposure of tumor bearing mice to valproic acid potentiated the antitumor effects of topoisomerse II inhibitors without enhancing toxicity. The HDACi induced histone acetylation and modulation of heterochromatin correlated with potentiation of epirubilin-mediated DNA damage (Marchion *et al.*, 2005). Synergetic interaction between HDACi and topoisomerase II inhibitors is mediated through cleavable complex formation. The HDACi induced sensitization has also observed in cells with target specific resistance to topoisomerase II poisons (Marchion *et al.*, 2005). HDACi were also coadministered with Cyclin Dependent Kinases (CDKs) inhibitors in human leukemia cells (U937 and HL-60) ectopically expressing Bcl-2/Bcl-X (L) and in primary AML cells. Coadministration of flavopiridol with HDACi synergistically potentiated mitochondrial damage, caspase activation, and cell death (Dasmahapatra *et al.*, 2006) (Fig. 3).

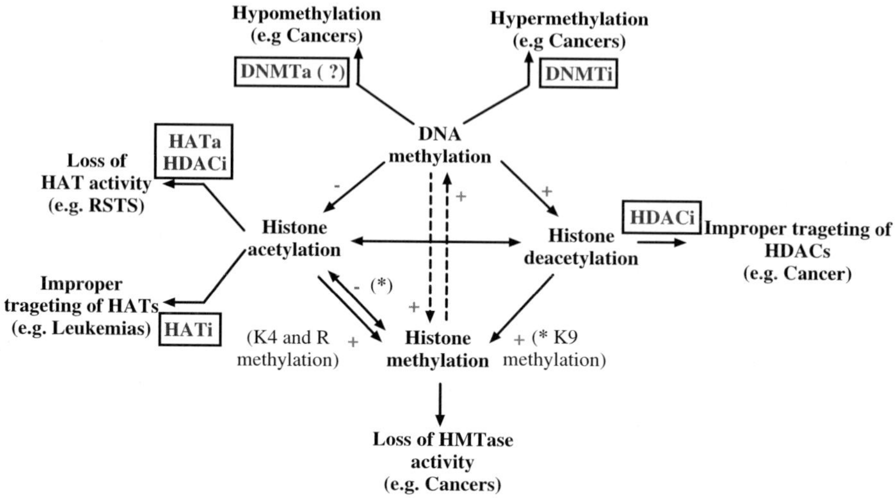

Figure 3. Schematic representation of the interplay of the various epigenetic marks and its therapeutic potential: DNA methylation causes the concomitant deacetylation of the histones, whereby it negatively (−) correlates with histone acetylation and positively (+) with histone methylation, particularly the repressive marks. The active methylation marks correlate positively with histone acetylation. The loss of activity or the loss or mistargeting of these activities are the most common cause of epigenetic diseases. Shown in the boxes are the small molecular modulators (a, activators or i, inhibitors) of the various enzymes that have potential to develop epigenetic therapeutics

Finally, HDACi also have been found to have a paradoxical effect in down regulating the expression of potent oncogenic proteins such as BCR/ABL and Flt-3. Several studies have shown that the combination of an HDAC inhibitor with imatinib mesylate, a potent BCR/ABL inhibitor used as first line therapy in chronic myelogenous leukemia, or a Flt-3 inhibitor have significant antileukemia *in vitro* activity. Of interest, this effect is thought to be mediated by the acetylation of the heat shock protein 90 and not by histone acetylation. Therefore, subsequent studies combining a HDACi with either imatinib or a Flt-3 inhibitor are being developed (Bali *et al.*, 2004). The synergistic combination of HDAC inhibitors with many molecular targets are being tested as potential therapeutic agents, which include HDACi along with the Her-2 antibody trastuzumab (Fuino *et al.*, 2003), TNF-related apoptosis inducing ligand (TRAIL) (Nimmanapalli *et al.*, 2003), the proteasome inhibitors Bortezomib (Yu *et al.*, 2003; Mitsiades *et al.*, 2004), purine analog fludaribine (Maggio *et al.*, 2004), and the Hsp90 antagonist 17-AGG (George *et al.*, 2005). In addition, HDAC inhibitors have also been shown to augment the sensitivity of prostate cancer cells to radiation therapy (Zhang *et al.*, 2004). Mechanistically, this chemo-/radio-sensitizing effect may be mediated through both histone acetylation-dependent and -independent effects of HDAC inhibitors.

5. TARGETED DRUG DELIVERY

We have speculated the use of small molecule modulators of histone modifying enzymes in regulating gene expression and thereby also as effective drugs for various diseases. This is in the wake of the increasing amount of data pointing towards the histone modifying enzymes and their roles in various disease manifestations. This is where chromatin therapy has a huge role to play. And along with this comes the huge responsibility of targeted drug delivery. A quick introspection in this context would quickly reveal how useful as well as how dreadful this therapy can turn into based on how specific it is. It is indeed a double-edged sword if used in a generalized manner, but with the targeted usage it could be a boon in disguise. Histone modifying enzymes are not active only in the diseased condition at that site, rather these are the enzymes required for the normal DNA templated phenomenon of the cell, and therefore the targeted delivery is absolutely essential. A quick review of what is available and what will be available in the coming few years has been provided herewith. Most of the disease conditions are localized at particular regions and therefore, these modulators can be easily targeted to the specific sites by the correct delivery techniques.

Although liposomes have been in use for quite some time now, nanoparticles are fast catching up and almost all the recent drug delivery systems that are being studied are nanoparticle based. Overcoming the blood brain barrier was an immense task and therefore probably for a long time there was no alternative for this target. Few reports on PEGylated nanoparticles encapsulating water-soluble drugs were used for this, a better result has been obtained by the use of modified PEGylated nanoparticles, wherein the surface is anchored with transferrin as a ligand for brain targeting (Mishra *et al.*, 2006).

When targeting to lung is considered, once again nanoparticles have an upper hand. Particulate nanocarriers are extremely advantageous because of avoidance of macrophage clearance mechanisms and long residence times (Dailey *et al.*, 2006).

Histone acetyltransferases and Histone deacetylases should have an important role in hepatocellular carcinoma. To date many modulators of these enzymes are well known. Liposome mediated targeting to the liver cells thus proves to be highly beneficial. This has been made more effective by incorporating an amino acid sequence, which recognizes glycosaminoglycans from *Plasmodium berghei.* This modified approach is based on the fact that several species of Plasmodium effectively target mammalian liver during the initial phase of host invasion (Longmuir *et al.*, 2006). Tat peptide mediated targeting is also a promising technique. Dendrimers are other modes of delivery, which are being extensively researched. These are branched synthetic polymers and by manipulating their synthesis, they can be made effective carriers (Lee *et al.*, 2005). Neuro degenerative diseases are yet another forte quite untouched because of the lack of much success in targeting drugs. Biodegradable polymers proving to be better alternatives as biomaterials directed to intracerebral regions are well tolerated by the host tissue. What adds on to their advantage is the fact that behavioral improvement and normalization of brain morphology has been observed following treatment with such biomaterials in animal models of Parkinson's, Alzheimer's and Huntington's disease (Popovic *et al.*, 2006). Aptamers are also widely studied modes of drug delivery. Aptamers are oligonucleotide ligands that are selected for high affinity binding to molecular targets.

Immunoliposomes are being studied wherein antibodies are used to provide cell specificity to the liposome mediated delivery (Kontermann *et al.*, 2006). Harnessing covalent chemistry for targeted drug delivery and its reversibility has proved to be a boon. Disulfide bonds and hydrazone bonds are the commonly used linkages between the carriers and the drugs. These linkages can be used to selectively release the drugs, disulfides can be cleaved in reducing environment found in intracellular fluids and for Hydrazones, acidic conditions (as in endosomes) release the drug (West *et al.*, 2005).

T-lymphocytes seem to be excellent carriers, inclusion of therapeutical nanoparticles into immune cells as a new strategy for localized therapy (Steinfeld *et al.*, 2006). Nanoparticles are colloidal structures with diameter smaller than 1000nm and can therefore penetrate through fine capillaries into the cells. Several of these are under investigation (Yih *et al.*, 2006).

6. CONCLUSIONS

The epigenetic code is set and maintained efficiently by the intricate interplay of different classes of enzymes that have been reviewed briefly. The various modifications brought about by these enzymes have distinct functional outcomes, which forms the basis of epigenetics. This regulation is disturbed when the enzymatic modifications are intervened either due to deletion or fusion of the enzyme, mistargeting of the enzyme or aberrant expression of the enzyme. All these finally result in an imbalance, which manifests in disease conditions. The significance of small molecule modulators in this context, is immense because of their potential therapeutic value. These modulators which are specific to the enzymes can thus modulate the enzyme activity. The clinical trials of HDACi is probably the best examples of this kind, and the discovery of modulators of HATs and HMTases which are highly specific may bring a new era of EPIGENETICS based drugs. Thus the key players of epigenetics apart from helping in diagnostics are also potential therapeutic targets and the advance in the field of targeted drug discovery is an impetus to proposed specificity.

ACKNOWLEDGEMENT

Research work in authors laboratory was supported by JNCASR, Department of Science and Technology (DST) and Department of Biotechnology (DBT) Government of India and Dabur Research foundation. RS is a CSIR junior research fellow (JRF).

REFERENCES

Ahmad K, Henikoff S (2002) The histone variant H3.3 marks active chromatin by replication-independent nucleosome assembly. Mol Cell 9:1191–1200

Akimaru H, Chen Y, Dai P, Hou DX, Nonaka M, Smolik SM, Armstrong S, Goodman RH, Ishii S (1997) Drosophila CBP is a co-activator of cubitus interruptus in hedgehog signaling. Nature 386:735–738

Amur SG, Shanker G, Cochran JM, Ved HS, Pieringer RA (1986) Correlation between inhibition of myelin basic protein (arginine) methyltransferase by sinefungin and lack of compact myelin formation in cultures of cerebral cells from embryonic mice. J Neurosci Res 16:367–376

An W, Kim J, Roeder RG (2004) Ordered cooperative functions of PRMT1, p300, and CARM1 in transcriptional activation by p53. Cell117:735–748

Andrews KT, Walduck A, Kelso MJ, Fairlie DP, Saul A, Parsons PG (2000) Antimalarial effect of histone deacetylation inhibitors and mammalian tumor cytodifferentiating agents. Int J Paarasitol 30:761–768

Atadja P, Hsu M, Kwon P, Trogani N, Bhalla K, Remiszewski S (2004) Molecular and cellular basis for the anti-proliferative effects of the HDAC inhibitor LAQ824. Novartis Found Symp 259:249–266

Balasubramanyam K, Swamninathan V, RAnganathan A, Kundu TK (2003) Small molecule modulators of histone acetyltransferase p300. J Biol Chem 278:19134–19140

Balasubramanyam K, Altaf M, Varier RA, Swaminathan V, Ravindran A, Sadhale PP, Kundu TK (2004) Polyisoprenylated benzophenone, garcinol, a natural histone acetyltransferase inhibitor, represses chromatin transcription and alters global gene expression. J Biol Chem 279:33716–33726

Balasubramanyam K, Varier RA, Altaf M, Swaminathan V, Siddappa NB, Ranga UK, Kundu TK (2004) Curcumin, a novel p300/CREB-binding protein-specific inhibitor of acetyltransferase, represses the acetylation of histone/nonhistone proteins and histone avetyltransferase-dependent chromatin transcription. J Biol Chem 279:51163–51171

Bali P, George P, Cohen P, Tao J, Guo F, Sigua C, Vishvanath A, Scuto A, Annavarapu S, Fiskus W, Moscinski L, Atadja P, Bhalla K (2004) Superior activity of the combination of histone deacetylase inhibitor LAQ824 and the FLT-3 kinase inhibitor PKC412 against human acute myelogenous leukemia cells with mutant FLT-3. Clin Cancer Res 10:4991–4997

Bali P, Pranpat M, Bradner J, Balasis M, Fiskus W, Guo F, Rocha K, Kumarawsamy S, Boyapalle S, Atadja P, Seto E, Bhalla K (2005) Inhibition of histone deacetylase 6 acetylates and disrupts the chaperone function of heat shock protein 90: a novel basis for antileukemia activity of histone deacetylase inhibitors.: J Biol Chem 280(29):26729–26734

Bannister AJ, Schneider R, Kouzarides T (2002) Histone methylation: Dynamic or static? Cell 109:801–806

Bedalov A, Gatbonton T, Irvine WP, Gottschling DE, Simon JA (2001) Identification of a small molecule inhibitor of Sir2p. Proc Natl Acad Sci USA 98(26):15113–15118

Bedford MT, Richard S (2005) Arginine methylation an emerging regulator of protein function. Mol Cell 18:263–272

Belli SI (2000) Chromatin remodelling during life cycle of trypanosomatids. Int J Parasitol 30:679–687

Berger SL, Sterner DE (2000) Acetylation of histones and transcription-related factors, microbiology and molecular biology reviews, June 435–459

Bertos NR, Wang AH, Yang XY (2001) Class II histone deacetylases: Structure, function, and regulation. Biochem Cell Biol 79:243–252

Biel M, Kretsovali A, Karatzali E, Papamatheakis J, Giannis A (2004) Design, synthesis, and biological evaluation of a small-molecule inhibitor of the histone acetyltransferase Gcn5. Angew Chem Int Ed Engl 43(30):3974–3976

Blander G, Guarente L (2004) The Sir2 family of protein deacetylases. Annu Rev of Biochem 73:417–435

Brinkmann H, Dahler AL, Popa C, Serewko MM, Parsons PG, Gabrielli BG, Burgess AJ, Saunders NA (2001) Histone hyperacetylation induced by histone deacetylase inhibitors is not sufficient to cause growth inhibition in human dermal fibroblasts. J Biol Chem 276:22491–22499

Brush MH, Guardiola A, Connor JH, Yao TP, Shenolikar S (2004) Deacetylase inhibitors disrupt cellular complexes containing protein phosphatases and deacetylases. J Biol Chem 279:7685–7691

Burma S, Chen BP, Murphy M, Kurimasa A, Chen DJ (2001) ATM phosphorylates histone H2AX in response to DNA double-strand breaks. J Biol Chem 276:42462–42467

Butler LM, Agus DB, Scher HI, Higgins B, Rose A, Cordon-Cardo C, Thaler HT, Rifkind RA, Marks PA, Richon VM (2000) Suberoylanilide hydroxamic acid: An inhibitor of histone deacetylase, suppresses the growth of prostate cancer cells in vitro and in vivo. Cancer Res 60: 5165–5170

Butler LM, Webb Y, Agus DB, Higgins B, Tolentino TR, Kutko MC, LaQuaglia MP, Drobnjak M, Cordon-Cardo C, Scher HI, Breslow R, Richon VM, Rifkind RA, Marks PA (2001) Inhibition of transformed cell growth and induction of cellular differentiation by pyroxamide, an inhibitor of histone deacetylase. Clin Cancer Res 7(4): 962–970

Canettieri G, Morantte I, Guzman E, Asahara H, Herzig S, Anderson SD, Yates JR 3rd, Montminy M (2003) Attenuation of a phosphorylation-dependent activator by an HDAC-PP1 complex. Nat Struct Biol 10:175–181

Catley L, Weisberg E, Tai YT, Atadja P, Remiszewski S, Hideshima T, Mitsiades N, Shringarpure R, LeBlanc R, Chauhan D, Munshi NC, Schlossman R, Richardson P, Griffin J, Anderson KC (2003) NVP-LAQ824 is a potent novel histone deacetylase inhibitor with significant activity against multiple myeloma. Blood 102(7):2615–2622

Chan HM, Kastic-Demonacos M, Smith L, Demonacos C, La Thangue NB (2001) Acetylation control of the retinoblastoma tumour- suppressor protein. Nat Cell Biol 3:667–674

Chang JH, Kim HC, Hwang KY, Lee JW, Jackson SP, Bell SD, Cho Y (2002) Structural basis for the NAD dependent deacetylase mechanism of Sir 2. J Biol Chem 277:34489–34498

Chang S, McKinsey TA, Zhang CL, Richardson JA, Hill JA, Olson EN (2004) Histone deacetylases 5 and 9 govern responsiveness of the heart to a subset of stress signals and play redundant roles in heart development. Mol Cell Biol 24:8467–8476

Chen CS, Weng SC, Tseng PH, Lin HP, Chen CS (2005) Histone acetylation-independent effect of histone deacetylase inhibitors on Akt through the reshuffling of protein phosphatase 1 complexes. J Biol Chem 280:38879–38887

Cheng D, Yadav N, King RW, Swanson MS, Weinstein EJ, Bedford MT (2004) Small molecule regulators of protein arginine methyltransferases. J Biol Chem 279:23892–23899

Cheung P, Tanner KG, Cheung WL, Sassone-Corsi P, Denu JM, Allis CD (2000) Synergistic coupling of histone H3 phosphorylation and acetylation in response to epidermal growth factor stimulation. Mol Cell 5:905–915

Covic M, Hassa PO, Saccani S, Buerki C, Meier NI, Lombardi C, Imhof R, Bedford MT, Natoli G, Hottiger MO (2005) Arginine methyltransferase CARM1 is a promoter-specific regulator of NF-kappaB-dependent gene expression. EMBO J 24:85–96

Cuthbert GL, Daujat S, Snowden AW, Erdjument-Bromage H, Hagiwara T, Yamada M, Schneider R, Gregory PD, Tempst P, Bannister AJ, Kouzarides T (2004) Histone deimination antagonizes arginine methylation. Cell 118:545–553

Dailey LA, Jekel N, Fink L, Gessler T, Schmehl T, Wittmar M, Kissel T, Seeger W (2006) Investigation of the proinflammatory potential of biodegradable nanoparticle drug delivery systems in the lung. Toxicol Appl Pharmacol (in press)

Darkin-Rattray SJ, Gurnett AM, Myers RW, Dulski PM, Crumley TM, Allocco JJ, Cannova C, Meinke PT, Colletti SL, Bednarek MA, Singh SB, Goetz MA, Dombrowski AW, Polishook JD, Schmatz DM (1996) Apicidin: A novel antiprotozoal agent that inhibits parasite histone deacetylase. Proc Natl Acad Sci USA 93(23):13143–13147

Dasmahapatra G, Almenara JA, Grant S (2006) Flavopiridol and histone deacetylase inhibitors promote mitochondrial injury and cell death in human leukemia cells that overexpress Bcl-2. Mol Pharmocol 69:288–298

Daujat S, Bauer UM, Shah V, Turner B, Berger S, Kouzarides T (2002) Crosstalk between CARM1 methylation and CBP acetylation on histone H3. Curr Biol 12:2090–2097 de Ruijter AJ, Leen R, Hoebink J, Caron HN, van Kuilenburg AB (2006) Antagonistic effects of sequential administration of BL1521, a histone deacetylase inhibitor, and gemcitabine to neuroblastoma cells. Cancer Lett 233:240–246

De Smet C, Lurquin C, Lethé B, Martelange V, Boon T (1999) DNA methylation is the primary silencing mechanism for a set of germ line- and tumor-specific genes with a CpG-rich promoter. Mol Cell Biol 19:7327–7335

Dong Z, Bode AM (2006) The role of histone H3 phosphorylation (Ser10 and Ser28) in cell growth and cell transformation. Mol Carcinog 45:416–421

Drummond DC, Noble CO, Kirpotin DB, Guo Z, Scott GK, Benz CC (2005) Clinical development of histone deacetylase inhibitors as anticancer agents. Annu Rev Pharmacol Toxicol 45:495–528

Eberharter A, Becker PB (2002) Histone acetylation: A switch between repressive and permissive chromatin. EMBO reports 3:224–229

Esteller M, Corn PG, Baylin SB, Herman JG (2001) A gene hypermethylation profile of human cancer. Cancer Res 61:3225–3229

Fenrick R, Hiebert SW (1998) Role of histone deacetylase in acute leukemia. J Cell Biochem 31:194–202

Fischle W, Kiermer V, Dequiedt F, Verdin E (2001) The emerging role of class II histone deacetylases. Biochem Cell Biol 79:337–348

Forneris F, Binda C, Vanoni MA, Battaglioli E, Mattevi A (2005) Human histone demethylase LSD1 reads the histone code. J Biol Chem 280:41360–41365

Frye RA (2000) Phylogenetic classification of prokaryotic and eukaryotic Sir2-like proteins. Biochem Biophys Res Commun 273:793–798

Fuino L, Bali P, Wittmann S, Donapaty S, Guo F, Yamaguchi H, Wang HG, Atadja P, Bhalla K (2003) Histone deacetylase inhibitor LAQ824 down-regulates Her-2 and sensitizes human breast cancer cells to trastuzumab, taxotere, gemcitabine, and epothilone B. Mol Cancer Ther 2:971–984

Furumai R, Komatsu Y, Nishino N, Khochbin S, Yoshida M, Horinouchi S (2001) Potent histone deacetylase inhibitors built from trichostatin A and cyclic tetrapeptide antibiotics including trapoxin. Proc Natl Acad Sci USA 98(1):87–92

Gao L, Cueto MA, Asselbergs F, Ataoja P (2002) Cloning and functional characterization of HDAC11: A novel member of the human histone deacetylase family. J Biol Chem 277:25748–25755

Garcia-Manero G, Issa JP (2005) Histone deacetylase inhibitors: A review of their clinical status as antineoplastic agents. Cancer Invest 23:635–642

George P, Bali P, Annavarapu S, Scuto A, Fiskus W, Guo F, Sigua C, Sondarva G, Moscinski L, Atadja P, Bhalla K (2005) Combination of the histone deacetylase inhibitor LBH589 and the hsp90 inhibitor 17-AAG is highly active against human CML-BC cells and AML cells with activating mutation of FLT-3. Blood 105(4):1768–1776

Gibbons RJ (2005) Histone modifying and chromatin remodeling enzymes in cancer and dysplastic syndromes. Hum Mol Genet 14:R85–R92

Gray SG, Ekstrom TJ (2001) The human histone deacetylase family. Exp Cell Res 262:75–83

Greiner D, Bonaldi T, Eskeland R, Roemer E, Imhof A (2005) Identification of a specific inhibitor of the histone methyltransferase SU(VAR)3–9. Nat Chem Biol1:143–145

Grozinger CM, Chao ED, Blackwell HE, Moazed D, Schreiber SL (2001) Identification of a class of small molecule inhibitors of the sirtuin family of NAD-dependent deacetylases by phenotypic screening. J Biol Chem 276(42):38837–38843

Grozinger CM, Hassig CA, Schreiber SL (1999) Three proteins define a class of human histone deacetylases related to yeast Hda1p. Proc Natl Acad Sci USA 96:4868–4873

Grozinger CM, Schreiber SL (2000) Regulation of histone deacetylase 4 and 5 and transcriptional activity by 14-3-3-dependent cellular localization. Proc Natl Acad Sci USA 97:7835–7840

Gusterson R, Brar B, Faulkes D, Giordano A, Chrivia J, Latchman D (2002) The transcriptional co-activators CBP and p300 are activated via phenylephrine through the p42/p44 MAPK cascade. J Biol Chem 277:2517–2524

Hake SB, Xiao A, Allis CD (2004) Linking the epigenetic 'language' of covalent histone modifications to cancer. Br J Cancer 90:761–769

Hamamoto R, Furukawa Y, Morita M, Iimura Y, Silva FP, Li M, Yagyu R, Nakamura Y (2004) SMYD3 encodes a histone methyltransferase involved in the proliferation of cancer cells. Nat Cell Biol 6:731–740

He L-Z, Tolentino T, Grayson P. Zhong S, Warrell RP Jr, Rifkind RA, Marks PA, Richon VM, Pandolfi PP (2001) Histone deacetylase inhibitors induce remission in transgenic models of therapy-resistant acute promyelocytic leukemia. J Clin Invest 108:321–330

Heltweg B, Gatbonton T, Schuler AD, Posakony J, Li H, Goehle S, Kollipara R, Depinho RA, Gu Y, Simon JA, Bedalov A (2006) Antitumor activity of a small-molecule inhibitor of human silent information regulator 2 enzymes. Cancer Res 66(8):4368–4377

Hennessy BT, Garcia-Manero G, Kantarjian HM, Giles FJ (2003) DNA methylation in haematological malignancies: the role of decitabine. Expert Opin Investig Drugs 12:1985–1993

Hong H, Kao C, Jeng MH, Eble JN, Koch MO, Gardner TA, Zhang S, Li L, Pan CX, Hu Z, MacLennan GT, Cheng L (2004) Aberrant expression of CARM1, a transcriptional coactivator of androgen receptor, in the development of prostate carcinoma and androgen-independent status. Cancer 101:83–89

Hook SS, Orian A, Cowley SM, Eisenman RN (2002) Histone deacetylase 6 binds polyubiquitin through its zinc finger (PAZdomain) and co purifies with deubiquitinating enzymes. Proc Natl Acad Sci USA 99:13425–13430

Impagnatiello F, Guidotti AR, Pesold C, Dwivedi Y, Caruncho H, Pisu MG, Uzunov DP, Smalheiser NR, Davis JM, Pandey GN, Pappas GD, Tueting P, Sharma RP, Costa E (1998) A decrease of reelin expression as a putative vulnerability factor in schizophrenia. Proc Natl Acad Sci USA 95:15718–15723

Jackson MD, Schmidt MT, Oppenheimer NJ, Denu JM (2003) Mechanism of nicotinamide inhibition and transglycosidation by Sir2 histone/protein deacetylases. J Biol Chem 278(51):50985–50998

Johnstone RW, Licht JD (2003) Histone deacetylase inhibitors in cancer therapy: Is transcription the primary target? Cancer Cell 4:13–18

Joseph J, Mudduluru G, Antony S, Vashistha S, Ajitkumar P, Somasundaram K (2004) Expression profiling of sodium butyrate (NaB)-treated cells: Identification of regulation of genes related to cytokine signaling and cancer metastasis by NaB. Oncogene 23:6304–6315

Kao H-Y, Lee C-H, Komarov A (2002) Isolation and characterization of mammalian HDAC10: A novel histone deacetylase. J Biol Chem 277:187–193

Kawaguchi Y, Kovacs JJ, McLaurin A, Vance JM, Ito A, Yao TP (2003) The deacetylase HDAC6 regulates aggresome formation and cell viability in response to misfolded protein stress. Cell 115:727–738

Kijima M, Yoshida M, Sugita K, Horinouchi S, Beppu T (1993) Trapoxin, an antitumor cyclic tetrapeptide, is an irreversible inhibitor of mammalian histone deacetylase. J Biol Chem 268(30):22429–22435

Kim A, Dean A (2004) Developmental stage differences in chromatin subdomains of the β-globin locus. Proc Natl Acad Sci USA 101:7028–7033

Kim YB, Lee KH, Sugita K, Yoshida M, Horinouchi S (1999) Oxamflatin is a novel antitumor compound that inhibits mammalian histone deacetylase. Oncogene 18(15): 2461–2470

Kim MS, Blake M, Baek JH, Kohlhagen G, Pommier Y, Carrier F (2003) Inhibition of histone deacetylase increases cytotoxicity to anticancer drugs targeting DNA. Cancer Res 63:7291–7300

Kleer CG, Cao Q, Varambally S, Shen R, Ota I, Tomlins SA, Ghosh D, Sewalt RG, Otte AP, Hayes DF, Sabel MS, Livant D, Weiss SJ, Rubin MA, Chinnaiyan AM (2003) EZH2 is a marker of aggressive breast cancer and promotes neoplastic transformation of breast epithelial cells. Proc Natl Acad Sci USA 100:11606–11611

Kontermann RE (2006) Immunoliposomes for cancer therapy. Curr Opin Mol Ther 8:39–45

Kornblith AB, Herndon JE II, Silverman LR, Demakos EP, Odchimar-Reissig R, Holland JF, Powell BL, DeCastro C, Ellerton J, Larson RA, Schiffer CA, Holland JC (2002) Impact of azacytidine on the quality of life of patients with myelodysplastic syndrome treated in a randomized phase III trial: A cancer and leukemia group B study. J Clin Oncol 20:2441– 2452

Kouzarides T (2000) Acetylation: A regulatory modification to rival phosphorylation? EMBO J 19:1176–1179

Kraker AJ, Mizzen CA, Hartl BG, Miin J, Allis CD, Merriman RL (2003) Modulation of histone acetylation by [4-(acetylamino)-N-(2-amino-phenyl) benzamide] in HCT-8 colon carcinoma. Mol Cancer Ther 2(4):401–408

Kramer OH, Gottlicher M, Heinzel T (2001) Histone deacetylase as a therapeutic target. Trends Endocrinol Metab 12:294–300

Kubicek S, Jenuwein T (2004) A crack in histone lysine methylation. Cell 119:903–906

Kwon HJ, Kim JH, Kim M, Lee JK, Hwang WS, Kim DY (2003) Anti-parasitic activity of depudecin on Neospora caninum via the inhibition of histone deacetylases. Vet Parasitol 112:269–276

Lau OD, Kundu TK, Soccio RE, Ait-Si-Sli S, Khail EM, Vassilev A, Wolffe AP, Nakatani Y, Roeder RG, Cole PA (2000) HATs off: Selective synthetic inhibitors of the histone acetyltransferases p300 and PCAF. Mol Cell 5:589–595

Lee CC, MacKay JA, Frechet JM, Szoka FC (2005) Designing dendrimers for biological applications. Nat Biotechnol 23:1517–1526

Littau VC, Allfrey VG, Frenster JH, Mirsky AE (1964) Active and inactive regions of nuclear chromatin as revealed by electron microscope autoradiography. Proc Natl Acad Sci USA 52:93–100

Longmuir KJ, Robertson RT, Haynes SM, Baratta JL, Waring AJ (2006) Effective targeting of liposomes to liver and hepatocytes in vivo by incorporation of a Plasmodium amino acid sequence. Pharm Res 23:759–769

Luo J, Su F, Chen D, Shiloh A, Gu W (2000) Deacetylation of p53 modulates its effect on cell growth and apoptosis. Nature 408:377–381

Maggio SC, Rosato RR, Kramer LB, DaiY, Rahmani M, Paik DS, CzarnikAC, Payne SG, Spiegel S, Grant S (2004) The histone deacetylase inhibitor MS-275 interacts synergistically with fludarabine to induce apoptosis in human leukemia cells. Cancer Res 64:2590–2600

Magnaghi-Jaulin L, Groisman R, Naguibneva I, Robin P, Lorain S, Le Villain JP, Troalen F, Trouche D, Harel-Bellan A (1998) Retinoblastoma protein represses transcription by recruiting a histone deacetylase. Nature 391:601–605

Mahmoudi T, Verrijzer CP (2001) Chromatin silencing and activation by Polycomb and trithorax group proteins. Oncogene 20:3055–3066

Mai A, Massa S, Rotili D, Cerbara I, Valente S, Pezzi R, Simeoni S, Ragno R (2005) Histone deacetylation in epigenetics: an attractive target for anticancer therapy. Med Res Rev 25:261–309

Marchion DC, Bicaku E, Daud AI, Sullivan DM, Munster PN (2005) In vivo synergy between topoisomerase II and histone deacetylase inhibitors: Predictive correlates. Mol Cancer Ther 4:1993–2000

Marchion DC, Bicaku E, Turner JG, Daud AI, Sullivan DM, Munster PN (2005) Synergistic interaction between histone deacetylase and topoisomerase II inhibitors is mediated through topoisomerase IIbeta. Clin Cancer Res 11:8467–8475

Marchion DC, Bicaku E, Daud AI, Richon V, Sullivan DM, Munster PN (2004) Sequence-specific potentiation of topoisomerase II inhibitors by the histone deacetylase inhibitor suberoylanilide hydroxamic acid. J Cell Biochem 92:223–237

Marks PA, Rifkind RA, Richon VM (2000) Deacetylase inhibitors: Inducers of differentiation or apoptosis of transformed cells. J Natl Cancer I 92:1210–1216

Marks PA, Richon VM, Breslow R (2001) Histone deacetylase inhibitors as new cancer drugs. Curr Opin Oncol 13:477–485

Martin C, Zhang Y (2005) The diverse functions of histone lysine methylation. Nat Rev Mol Cell Biol 6:838–849

Michaelis M, Michaelis UR, Fleming I, Suhan T, Cinatl J, Blaheta RA, Hoffmann K, Kotchetkov R, Busse R, Nau H, Cinatl J Jr (2004) Valproic acid inhibits angiogenesis in vitro and in vivo. Mol Pharmacol 65:520–527

Mishra S, Saleh A, Espino PS, Davie JR, Murphy LJ (2006) Phosphorylation of histones by tissue transglutaminase. J Biol Chem 281:5532–5538

Mishra V, Mahor S, Rawat A, Gupta PN, Dubey P, Khatri K, Vyas SP (2006) Targeted brain delivery of AZT via transferrin anchored pegylated albumin nanoparticles. J Drug Target 14:45–53

Mitsiades CS, Mitsiades NS, McMullan CJ, Poulaki V, Shringarpure R, Hideshima T, Akiyama M, Chauhan D, Munshi N, Gu X, Bailey C, Joseph M, Libermann TA, Richon VM, Marks PA,

Anderson KC (2004) Transcriptional signature of histone deacetylase inhibition in multiple myeloma: Biological and clinical implications. Proc Natl Acad Sci USA 101:540–545

Moore SDP, Herrick SR, Ince TA, Klienman MS, Cin PD, Morton C, Quade BJ, 2000; Uterine Leiomyomata with t(10;17) disrupt the histone acetyltransferase MORF. Cancer Res 61:5570–5577

Morales V, Richard-Foy H (2000) Role of Histone N-Terminal Tails and Their Acetylation in Nucleosome Dynamics. Mol Cell Biol 20:7230–7237

Murphy M, Ahn J, Walker KK (1999) Transcriptional repression by wild-type p53 utilizes histone deacetylases, mediated by interaction with mSin3a. Genes Dev 13:2490–2501

Nakajima H, Kim YB, Terano H, Yoshida M, Horinouchi S (1998) FR901228, a potent antitumor antibiotic, is a novel histone deacetylase inhibitor. Exp Cell Res 241(1):126–133

Nebbioso A, Clarke N, Voltz E, Germain E, Ambrosino C, Bontempo P, Alvarez R, Schiavone EM, Ferrara F, Bresciani F, Weisz A, de Lera AR, Gronemeyer H, Altucci L (2005) Tumor-selective action of HDAC inhibitors involves TRAIL induction in acute myeloid leukemia cells. Nature Med 11:77–84

Nielsen SJ, Schneider R, Bauer UM, Bannister AJ, Morrison A, O'Carroll D, Firestein R, Cleary M, Jenuwein T, Herrera RE, Kouzarides T (2001) Rb targets histone H3 methylation and HP1 to promoters. Nature 412:561–565

Nimmanapalli R, Fuino L, Stobaugh C, Richon V, Bhalla K (2003) Cotreatment with the histone deacetylase inhibitor suberoylanilide hydroxamic acid (SAHA) enhances imatinib-induced apoptosis of Bcr-Ablpositive human acute leukemia cells. Blood 101:3236–3239

Okada Y, Feng Q, Lin Y, Jiang Q, Li Y, Coffield VM, Su L, Xu G, Zhang Y (2005) hDOT1L links histone methylation to leukemogenesis. Cell. 121:167–178

Pal S, Vishwanath SN, Erdjument-Bromage H, Tempst P, Sif S (2004) Human SWI/SNF-associated PRMT5 methylates histone H3 arginine 8 and negatively regulates expression of ST7 and NM23 tumor suppressor genes. Mol Cell Biol 24:9630–9645

Parson PG, Hansen C, Fairlie DP, West ML, Danoy PA, Sturm RA, Dunn IS, Pedley J, Ablett EM (1997) Tumor selectivity and transcriptional activation by azelaic bishydroxamic acid in human melanocytic cells. Biochem Pharmacol 53:1719–1724

Peart MJ, Smyth GK, van Laar RK, Bowtell DD, Richon VM, Marks PA, Holloway AJ, Johnstone RW (2005) Identification and functional significance of genes regulated by structurally different histone deacetylase inhibitors. Proc Natl Acad Sci USA 102:3697–3702

Phiel CJ, Zhang F, Huang EY, Guenther MG, Lazar MA, Klein PS (2001) Histone deacetylase is a direct target of valproic acid, a potent anticonvulsant, mood stabilizer, and teratogen. J Biol Chem 276(39):36734–36741

Popovic N, Brundin P (2006) Therapeutic potential of controlled drug delivery systems in neurodegenerative diseases. Int J Pharm 314:120–126

Qui L, Kelso MJ, Hansen C, West ML, Fairlie DP, Parson PG (1999) Antitumor activity in vitro and in vivo of selective differentiating agents containing hydroxamate. Br J Cancer 80:1252–1258

Rea S, Eisenhaber F, O'Carroll D, Strahl BD, Sun ZW, Schmid M, Opravil S, Mechtler K, Ponting CP, Allis CD, Jenuwein T (2000) Regulation of chromatin structure by site-specific histone H3 methyltransferases. Nature 406:593–599

Richon VM, Emiliani S, Verdin E, Webb Y, Breslow R, Rifkind RA, Marks PA (1998) A class of hybrid polar inducers of transformed cell differentiation inhibits histone deacetylases. Proc Natl Acad Sci USA 95(6):3003–3007

Rocchi P, Tonelli R, Camerin C, Purgato S, Fronza R, Bianucci F, Guerra F, Pession A, Ferreri AM (2005) p21Waf1/Cip1 is a common target induced by short-chain fatty acid HDAC inhibitors (valproic acid, tributyrin and sodium butyrate) in neuroblastoma cells. Oncol Rep 13:1139–1144

Rouaux C, Loeffler JP, Boutillier AL (2004) Targeting CREB-binding protein (CBP) loss of function as a therapeutic strategy in neurological disorders. Biochem Pharmacol 68:1157–1164

Roy S, Packman K, Jeffrey R, Tenniswood M (2005) Histone deacetylase inhibitors differentially stabilize acetylated p53 and induce cell cycle arrest or apoptosis in prostate cancer cells. Cell Death Differ 12:482–491

Saito A, Yamashita T, Mariko Y, Nosaka Y, Tsuchiya K, Ando T, Suzuki T, Tsuruo T, Nakanishi O (1999) A synthetic inhibitor of histone deacetylase, MS-27–275, with marked in vivo antitumor activity against human tumors. Proc Natl Acad Sci USA 96(8):4592–4597

Schneider R, Bannister AJ, Myers FA, Thorne AW, Crane-Robinson C, Kouzarides T (2004) Histone H3 lysine 4 methylation patterns in higher eukaryotic genes. Nature 6:73–77

Schwartz BE, Ahmad K (2005) Transcriptional activation triggers deposition and removal of the histone variant H3.3. Genes Dev 19:804–814

Shi Y, Lan F, Matson C, Mulligan P, Whetstine JR, Cole PA, Casero RA, Shi Y (2004) Histone demethylation mediated by the nuclear amine oxidase homolog LSD1. Cell 119:941–953

Silverman LR, Demakos EP, Peterson BL, Kornblith AB, Holland JC, Odchimar-Reissig R, Stone RM, Nelson D, Powell BL, DeCastro CM, Ellerton J, Larson RA, Schiffer CA, Holland JF (2002) Randomized controlled trial of azacitidine in patients with the myelodysplastic syndrome: a study of the cancer and leukemia group. B J Clin Oncol 20:2429–2440

Song F, Smith JF, Kimura MT, Morrow AD, Matsuyama T, Nagase H, Held WA (2005) Association of tissue-specific differentially methylated regions (TDMs) with differential gene expression, Proc Natl Acad Sci USA 102:3336–3341

Steinfeld U, Pauli C, Kaltz N, Bergemann C, Lee HH (2006) T lymphocytes as potential therapeutic drug carrier for cancer treatment. Int J Pharm 311:229–236

Stiff T, O'Driscoll M, Rief N, Iwabuchi K, Lobrich M, Jeggo PA (2004) ATM and DNA-PK function redundantly to phosphorylate H2AX after exposure to ionizing radiation. Cancer Res 64:2390–2396

Stimson L, Rowlands MG, Newbatt YM, Smith NF, RAynaud FI, Rogers P, BAvetsias V, Gorsuch S, Jarman M, Bannister A, Kouzarides T, McDOnald E, Workman P, Aherne GW (2005) Isothiazolones as inhibitors of PCAF and p300 histone acetyltransferase activity. Mol Cancer Ther 4:1521–1532

Strahl BD, Allis CD (2000) The language of covalent histone modifications. Nature 403:41–45

Su GH, Sohn TA, Ryu B, Kern SE (2000) A novel histone deacetylase inhibitor identified by high-throughput transcriptional screening of a compound library. Cancer Res 60(12):3137–3142

Subramanian C, Opipari AW Jr, Bian X, Castle VP, Kwok RP (2005) Ku70 acetylation mediates neuroblastoma cell death induced by histone deacetylase inhibitors. Proc Natl Acad Sci USA 102:4842–4847

Sun ZW, Allis CD (2002) Ubiquitination of histone H2B regulates H3 methylation and gene silencing in yeast. Nature 418:104–108

Taki T, Sako M, Tsuchida M, Hayashi Y (1997) The t(11;16)(q23;p13) translocation in Myelodysplastic syndrome fuses the MLL gene to the CBP gene. Blood 89:3945–3950

Tanaka Y, Naruse I, Maekawa T, Masuya H, Shiroishi T, Ishii S (1997) Abnormal skeletal patterning in embryos lacking a single Cbp allele: A partial similarity with Rubinstein-Taybi syndrome. Proc Natl Acad Sci USA 94:10215–10220

van Attikum H, Gasser SM (2005) The histone code at DNA breaks: A guide to repair? Nat Rev Mol Cell Biol 6:757–765

Varambally S, Dhanasekaran SM, Zhou M, Barrette TR, Kumar-Sinha C, Sanda MG, Ghosh D, Pienta KJ, Sewalt RG, Otte AP, Rubin MA, Chinnaiyan AM (2002) The polycomb group protein EZH2 is involved in progression of prostate cancer. Nature 419:624–629

Varier RA, Kundu TK (2006) Chromatin modifications (acetylation/ deacetylation/methylation) as new targets for HIV therapy. Curr Pharm Des 12:1975–1993

Vertino PM, Sekowski JA, Coll JM, Applegren N, Han S, Hickey RJ, Malkas LH (2002) DNMT1 is a component of a multiprotein DNA replication complex. Cell Cycle 1:416–423

Vidali G, Boffa LC, Bradbury EM, Allfrey VG (1978) Butyrate suppression of histone deacetylation leads to accumulation of multiacetylated forms of histones H3 and H4 and increased DNase I sensitivity of the associated DNA sequences. Proc Natl Acad Sci USA 75(5):2239–2243

Vigushin DM, Coombes RC (2002) Histone deacetylase inhibitors in cancer treatment. Anticancer Drugs 13:1–13

Wang H, Huang ZQ, Xia L, Feng Q, Erdjument-Bromage H, Strahl BD, Briggs SD, Allis CD, Wong J, Tempst P, Zhang Y (2001) Methylation of histone H4 at arginine 3 facilitating transcriptional activation by nuclear hormone receptor. Science 293:853–857

Wang Y, Wysocka J, Sayegh J, Lee YH, Perlin JR, Leonelli L, Sonbuchner LS, McDonald CH, Cook RG, Dou Y, Roeder RG, Clarke S, Stallcup MR, Allis CD, Coonrod SA (2004) Human PAD4 regulates histone arginine methylation levels via demethylimination. Science 306(5694):279–283

West KR, Otto S (2005) Reversible covalent chemistry in drug delivery. Curr Drug Discov Technol 2:123–160

Yang XJ, Gregoire S (2005) Class II histone deacetylases: From sequence to function, regulation, and clinical implication. Mol Cell Biol 25:2873–2884

Yang XY, Seto E (2003) Collaborative spirit of histone deacetylases in regulating chromatin structure and gene expression. Curr Opin Genet Dev 13:143–153

Yao TP, Suk P Oh, Fuchs M, Nai-Dong Zhou, Lian-Ee Ch'ng, Newsome D, Bronson RT, En Li, Livingston DM, Eckner R (1998) Gene dosage–dependent embryonic development and proliferation defects in mice lacking the transcriptional integrator p300. Cell 93:361–372

Yarden RI, Brody LC (1999) BRCA1 interacts with components of the histone deacetylase complex. Proc Natl Acad Sci USA 96:4983–4988

Yih TC, Al-Fandi M (2006) Engineered nanoparticles as precise drug delivery systems. J Cell Biochem 97:1184–1190

Yoder JA, Walsh CP, Bestor TH (1997) Cytosine methylation and the ecology of intragenomic parasites. Trends Genet 13:335–340

Yoon HG, Chan DW, Huang ZQ, Li J, Fondell JD, Quin J, Wong J (2003) Purification and functional characterization of the human N-CoR complex: The roles of HDAC3, TBL1, and TBLR1. EMBO J 22:1336–1346

Yoshida M, Kijima M, Akita M, Beppu T (1990) Potent and specific inhibition of mammalian histone deacetylase both in vivo and in vitro by trichostatin A. J Biol Chem 265(28):17174–17179

Yu C, Rahmani M, Conrad D, Subler M, Dent P, Grant S (2003) The proteasome inhibitor bortezomib interacts synergistically with histone deacetylase inhibitors to induce apoptosis in Bcr/Ablþ cells sensitive and resistant to STI571. Blood 102:3765–3774

Zhang CL, McKinsey TA, Chang S, Antos CL, Hill JA, Olson EN (2002) Class II histone deacetylases act as signal-responsive repressors of cardiac hypertrophy. Cell 110:479–488

Zhang K, Dent SY (2005) Histone modifying enzymes and cancer: Going beyond histones. J Cell Biochem 96:1137–1148

Zhang Y, Adachi M, Zhao X, Kawamura R, Imai K (2004) Histone deacetylase inhibitors FK228, N-(2-aminophenyl)-4-[N-(pyridin-3-yl-methoxycarbonyl)amino-methyl]benzamide and m-carboxycinnamic acid bis-hydroxamide augment radiation-induced cell death in gastrointestinal adenocarcinoma cells. Int J Cancer 110:301–308

Zhang Y, Reinberg D (2001) Transcription regulation by histone methylation: Interplay between different covalent modifications of the core histone tails. Genes Dev 15:2343–2360

INDEX

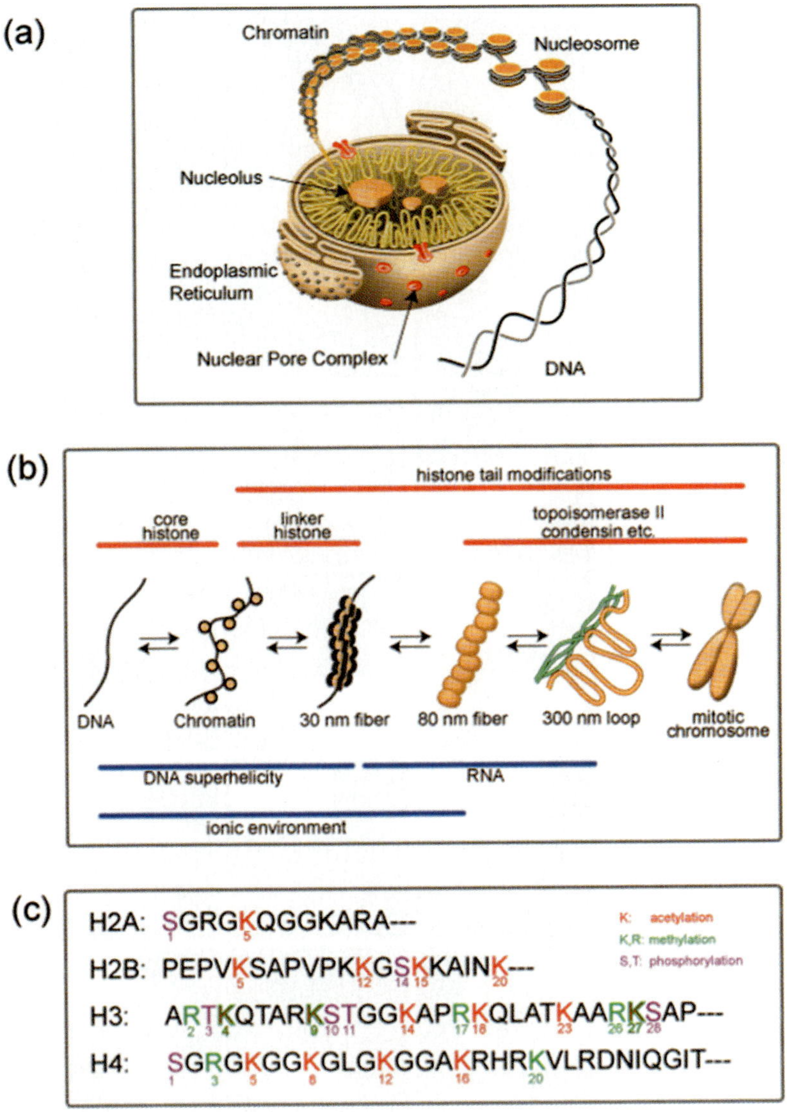

Plate 1. Hierarchical model of chromosome structure. (a) In interphase cells, DNA is packed in a nucleus as forming nucleosome and chromatin. (b) DNA forms nucleosome structure together with core histone octamer, which is then folded up into 30 nm fiber with a help of linker histone H1. This 30 nm fiber is further folded into 80 nm fiber and 300 nm loop structures in a nucleus. In mitosis, chromosome is highly condensed. Proteins which are involved in each folding step are indicated above and non-protein factors are indicated below. (c) The amino acid sequences of histone tails (H2A, H2B, H3 and H4) are shown to indicate acetylation, methylation and phosphorylation sites. (See Chapter 1 Figure 1, p. 4.)

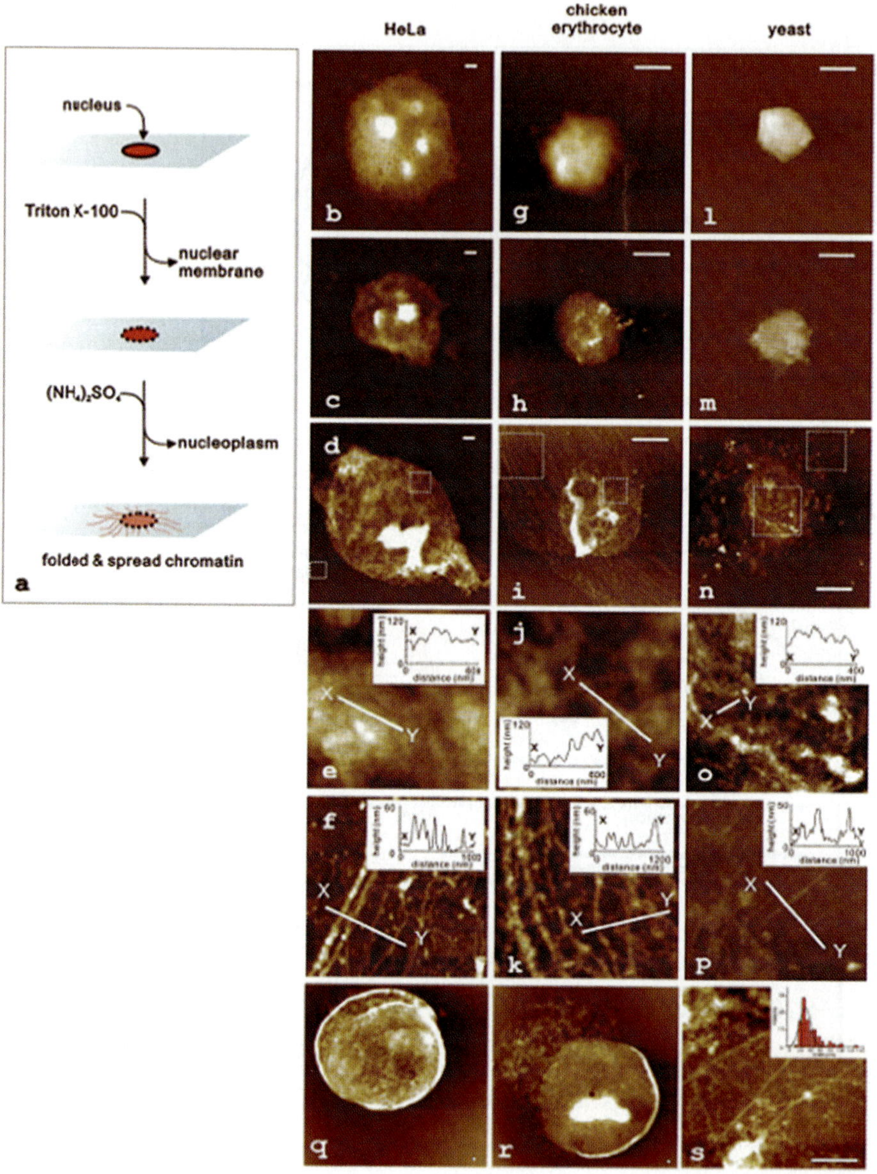

Plate 2. 80 nm fibers and granules. The 'on-substrate' lysis makes it possible to observe a chromatin structure by removing nuclear membrane and nucleoplasm from nuclei on a cover slide (a). HeLa, chicken erythrocyte, and yeast nuclei were subject to the 'on-substrate' lysis (panels b–f for HeLa nucleus, g–k for chicken erythrocyte, l–p for yeast). AFM visualized the isolated nucleus without detergent treatment (b, g, l), after the detergent treatment (c, h, m) and after the high-salt treatment (d, i, n). Scale bars indicate 2 μm. The high-salt treated nuclei (d, i, n) were rescanned for magnification, and shown in different panels (e and f for HeLa, j and k for chicken erythrocyte, and o and p for yeast). (See Chapter 1 Figure 5, p. 17.) *(Plate 2 Caption cont.)*

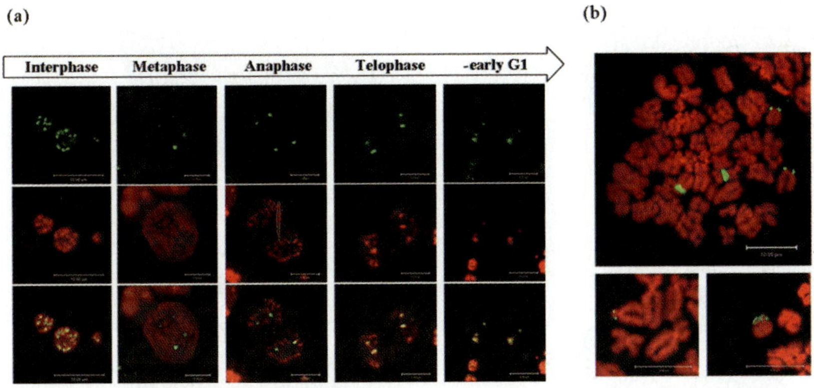

Plate 3. Localization of a nucleolar protein in FC regin. (a) Localization in mitotic cells. Green: antigen, Red: fibrillarin (DFC marker). (b) Localization on the metaphase chromosomes. Green: antigen, Red: DNA (PI). (Scale bars: 10 μm). (See Chapter 1 Figure 8, p. 21.)

Plate 2. (Continued) A section profile obtained along X-Y line shows a typical granular structure in the nucleus (e, j, o), and the peak-to-peak distance between the granular structure was distributed from 60 nm to 120 nm (e). The thickness of the chromatin fibers released out of the nucleus varied possibly due to the assembly of thinner fibers (f, k, p). A section profile for the spread fibers was obtained along X-Y line (f, k, p). Isolated HeLa cell nucleus was treated with (r, s) or without (q) RNase. The treatment releases ~30 nmfiber from the nucleus. The histogram of the fiber width is shown in an inset of (s). Bars, 250 nm.

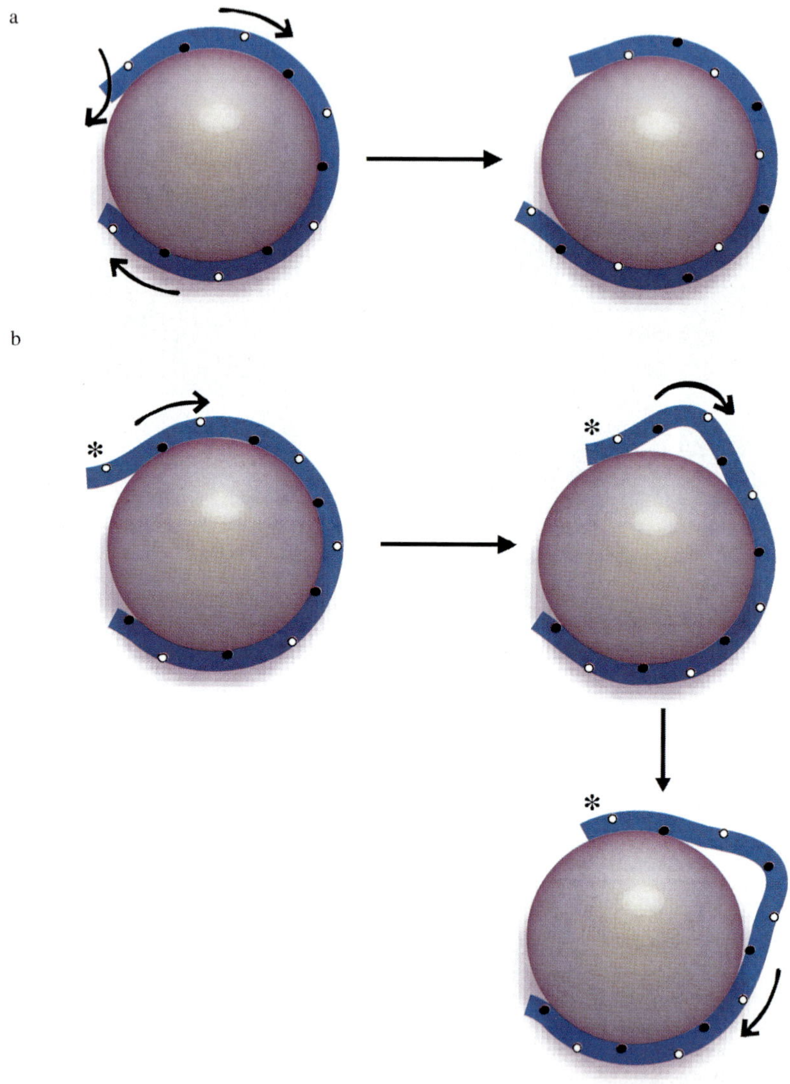

Plate 4. Schematic representation of remodelling mechanisms. (Adapted form Langst and Becker, 2004.) The schemes show nucleosomes from the top. (a) The twist diffusion model – Twisting of DNA moves it over the histone surface in a one base pair increments. This changes the position of the DNA with respect to the histone, as shown by the open and closed circles. (b) The Loop recapture model – Extranucleosomal DNA is pulled into the nucleosomes to replace a DNA segment which consequently loops out. This loop is then propragated over the histone surface like ripples of a wave. The star, *, indicates how this leads to a change in the position of DNA relative to the nucleosome. (See Chapter 2 Figure 1, p. 36.)

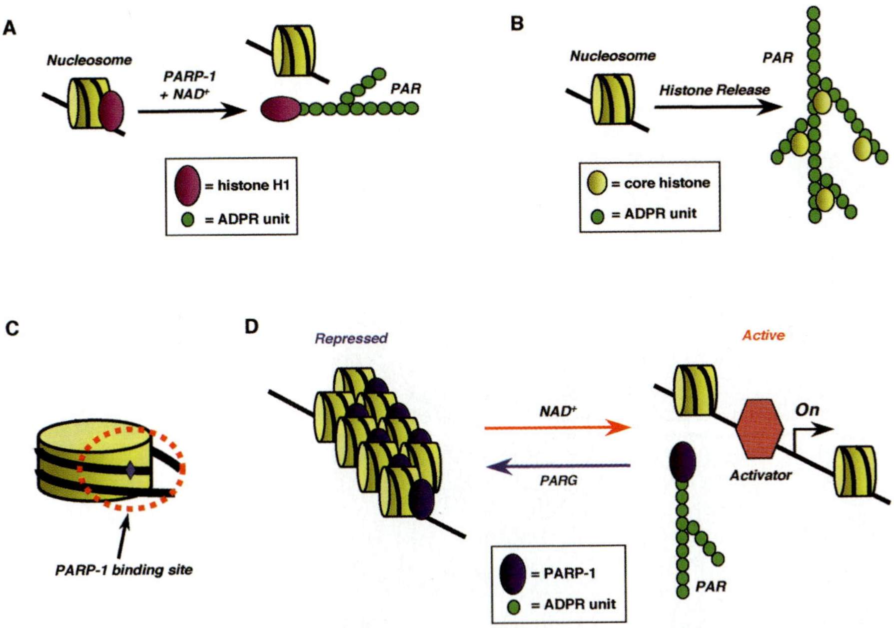

Plate 5. Possible mechanisms for the modulation of chromatin structure by PARP-1 and PAR. Multiple models for the modulation of chromatin structure by PARP-1 and PAR are shown. (a) A number of chromatin proteins are targets for PARylation by PARP-1, including the linker histone H1. PARylation may alter the interaction of these proteins with chromatin (as shown for H1 in this panel) and, as a result, alter chromatin structure. (b) PAR chains may form an attractive anionic scaffold that can bind displaced histones and other chromatin proteins. Removing these proteins from the DNA to a PAR scaffold would allow access for transcription, replication, and repair enzymes. (c) PARP-1 is a nucleosome-binding protein that binds at or near the dyad axis of the nucleosome (blue diamond) and contacts the linker DNA where it exits the nucleosome. (d) PARP-1, through its nucleosome-binding activity, functions as a structural component of chromatin and promotes the compaction of chromatin into transcriptionally repressed structures. Upon autoPARylation in the presence of NAD+, PARP-1 is released from chromatin and the compaction is reversed, promoting the formation of transcriptionally active chromatin structures and allowing activator binding. PARG cleaves the PAR chains from PARP-1, allowing PARP-1 to re-bind to the nucleosomes. (See Chapter 3 Figure 3, p. 51.)

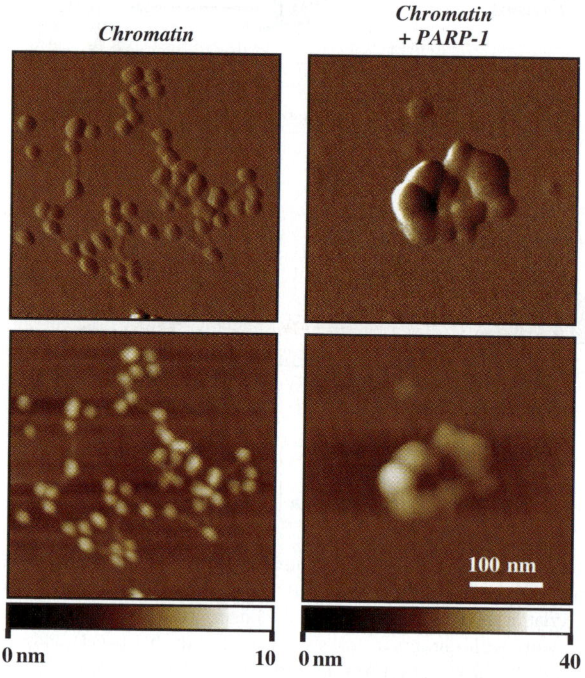

Plate 6. Visualization of PARP-1-mediated chromatin compaction by atomic force microscopy. Chromatin assembled *in vitro* on a circular ∼10.5 kb plasmid DNA was purified, incubated with or without recombinant human PARP-1, and imaged by atomic force microscopy. Two types of images are shown: scan probe oscillation amplitude *(top)* and topography (*bottom*; height scale is indicated). The length scale is indicated. (See Chapter 3 Figure 4, p. 53.)

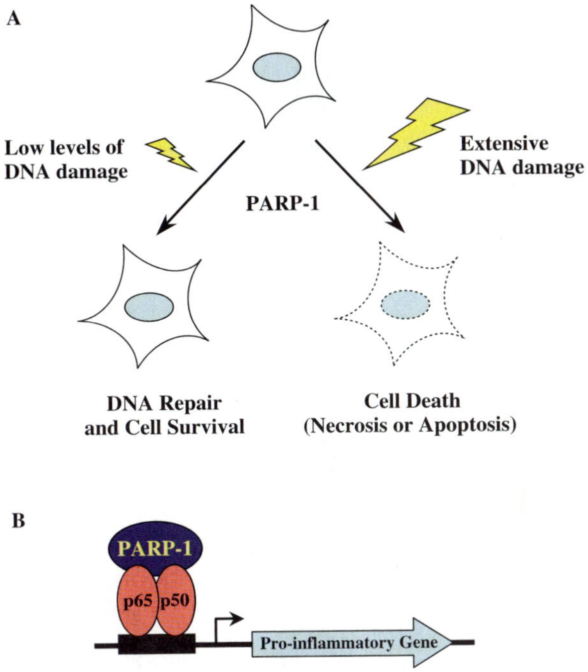

Plate 7. PARP-1 plays important roles in genome maintenance, cell death, and inflammatory responses. PARP-1's roles in DNA repair, cell death pathways, and pro-inflammatory gene expression underlie many of the contributions of PARP-1 to various disease states. (a) PARP-1 has well-characterized roles in DNA repair and cell death. With low levels of DNA damage, PARP-1 acts as a survival factor that can promote the repair of damaged DNA. Under conditions of extensive DNA damage, PARP-1 acts to promote cell death through necrotic and apoptotic pathways. (b) PARP-1 has been implicated in pathophysiological inflammatory responses through its role as a coactivator of transcription factors that regulate immune and inflammatory response genes (*e.g.*, NF-κ B and AP-1). (See Chapter 3 Figure 6, p. 60.)

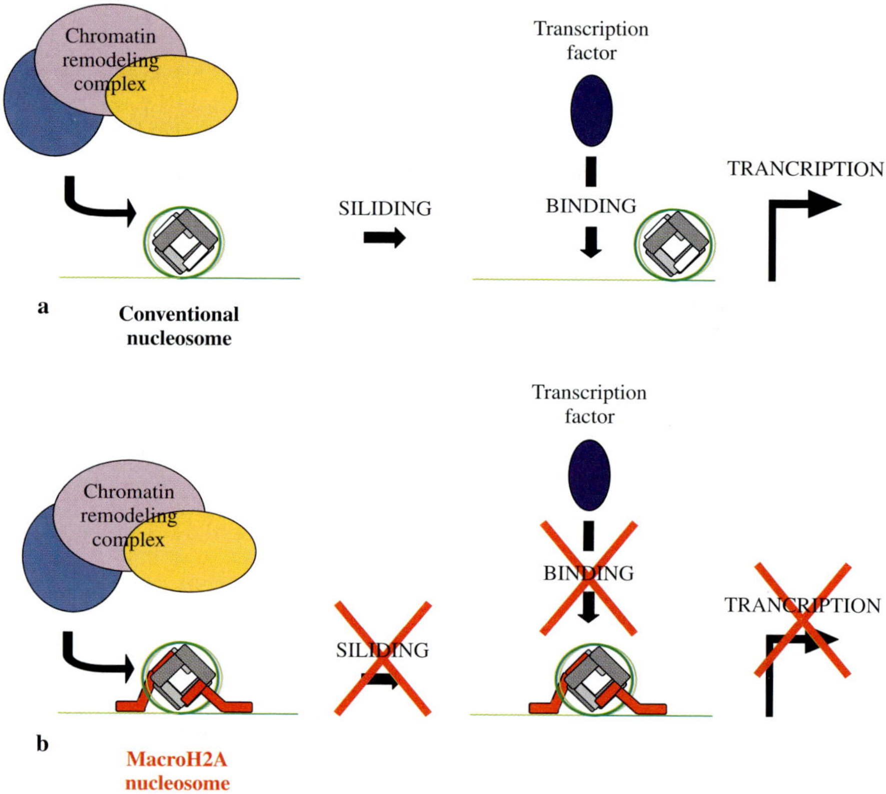

Plate 8. Schematics of the effect of mH2A on transcription. (a) The mobilization of conventional promoter nucleosome by the chromatin remodeling complex generated a nucleosome-free promoter, which allowed the binding of the transcription factors and transcription to proceed. (b) MacroH2A nucleosome cannot be mobilized by the chromatin remodeling complex and the transcriptions factors are unable to bind macroH2A nucleosome containing promoter, which results in inhibition of the initiation of transcription. (See Chapter 4 Figure 1, p. 79.)

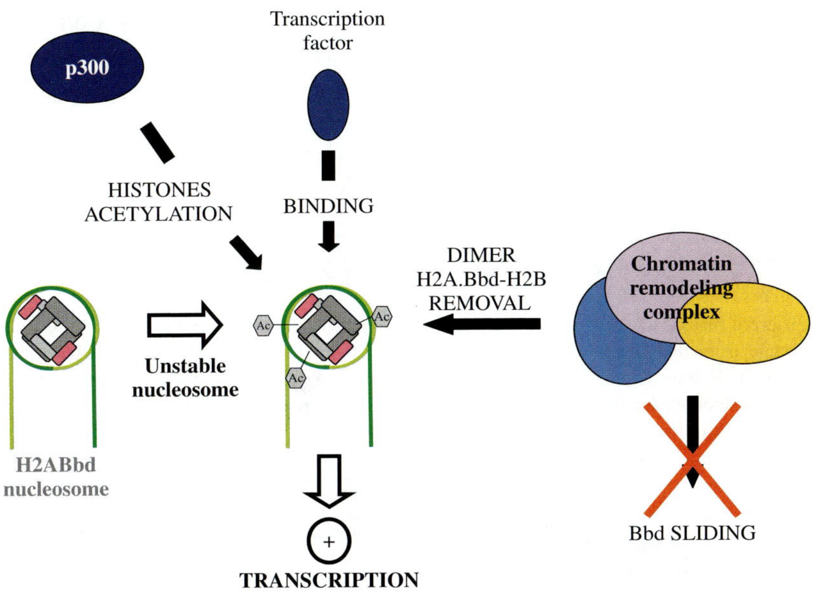

Plate 9. Effect of H2A.Bbd on transcription. The presence of H2A.Bbd confers lower stability and more loose structure to the nucleosomes, which allows the transcription factors binding to this variant nucleosome and thereby recruitment of p300 and acetylation of the promoter proximal histones. The remodeling complex can not mobilize the variant nucleosome, but instead helps in the removal of H2A.Bbd-H2B dimer. All these events facilitate transcription. (See Chapter 4 Figure 2, p. 80.)

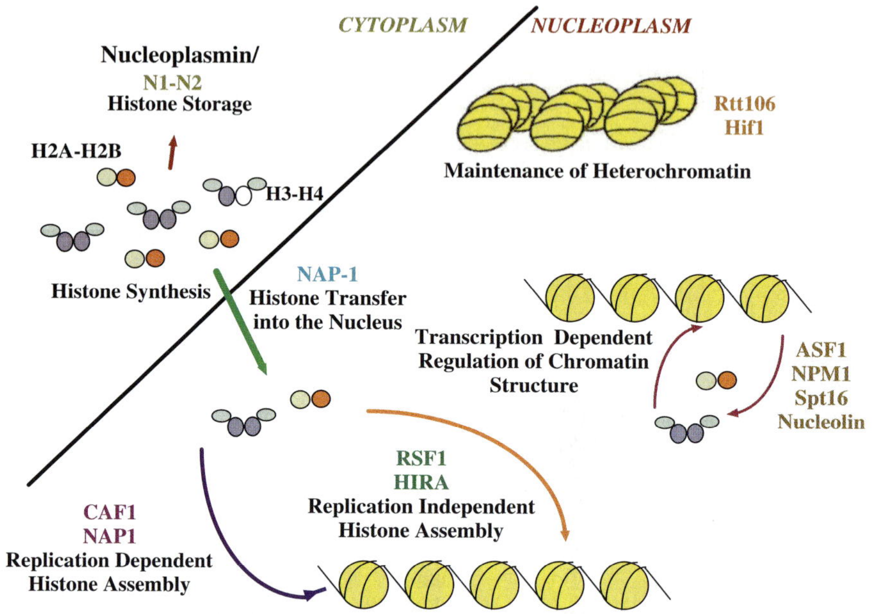

Plate 10. Different histone chaperones in the key histone metabolic pathways: Functions of histone chaperones range from the storage of newly synthesized histones in the cytoplasm, its transfer into the nucleus and in histone assembly into nucleosomes. Apart from this the histone chaperones are also involved in histone exchange, maintenance of heterochromatin and in the regulation of chromatin structure during transcription. (See Chapter 6 Figure 1, p. 114.)

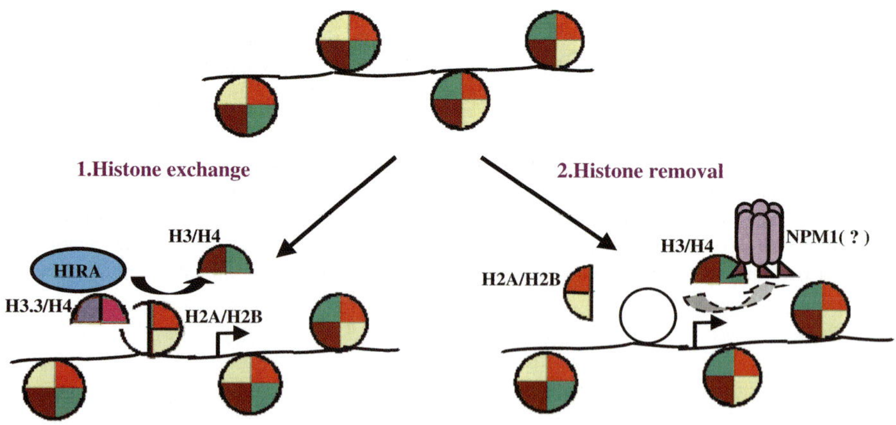

Plate 11. Histone chaperones facilitate favorable chromatin dynamics during transcriptional activation: Transcriptional competence of chromatin template is achieved by the replacement of histone variants and finally removal of histones. Histone chaperone may help in both the process in a replication independent manner. Acetylation of histone and also the chaperone may assist in this process. Recent evidence suggest that NPM1 may participate in these events globally or gene specific manner. (See Chapter 6 Figure 2, p. 115.)

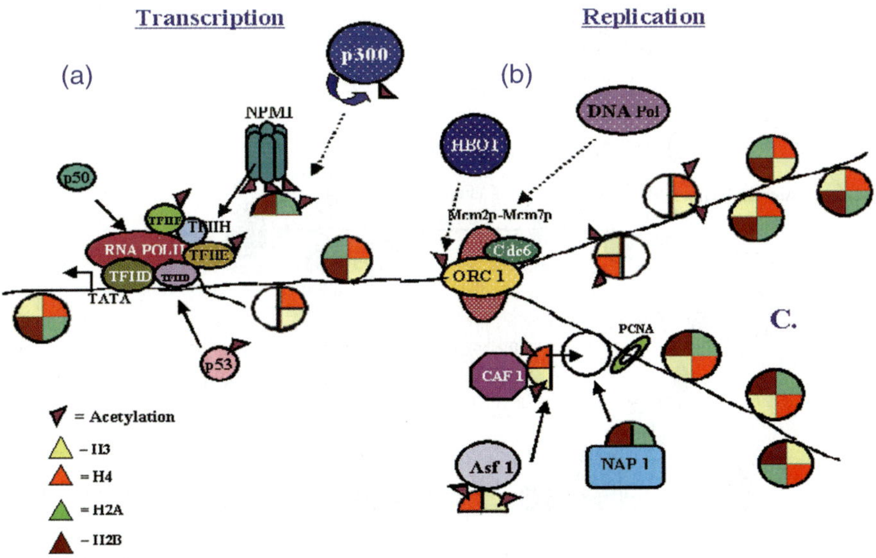

Plate 12. Mechanistic effect of acetylation/deacetylation of histones and nonhistones on chromatin structure.(a) Acetylation of non-histone proteins results in transcriptional activation (b) Acetylation of ORC1 by HBO1 is important for replication. (c) Acetylation of newly synthesized histones necessary for chromatin assembly. (See Chapter 9 Figure 1, p. 195.)

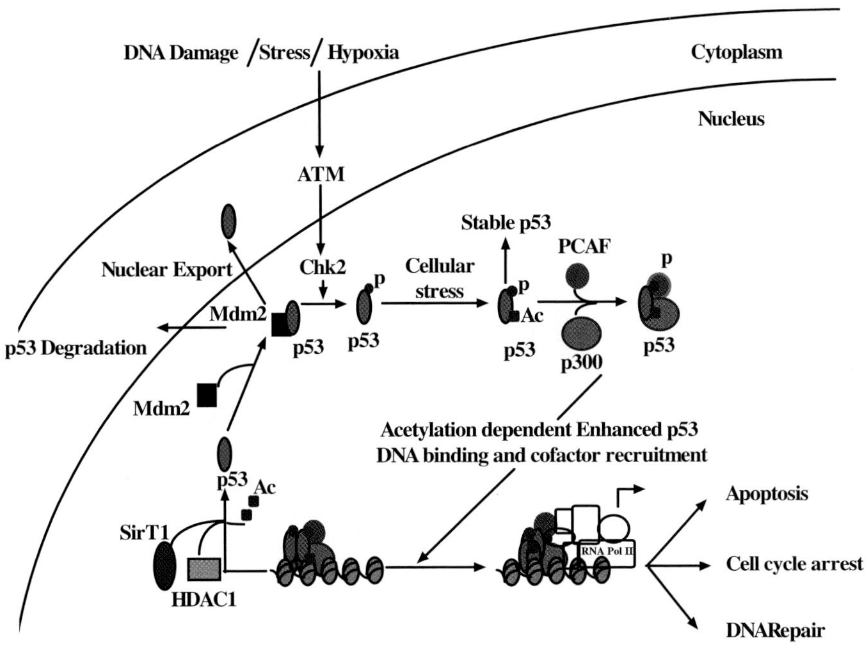

Plate 13. Regulation of p53 function by acetylation / deacetylation: Under stresses conditions p53 gets phosphorylated, acetylated and consequently gets stabilized. Acetylated p53 has enhanced transcriptional ability leading to the activation several p53 responsive genes, which plays important roles in diverse cellular processes. Decateylation of p53 by SirT1 and HDAC1 down regulates p53 activity by enabling interaction with MDM2 followed by nuclear export and p53 degradation. (See Chapter 9 Figure 2, p. 198.)

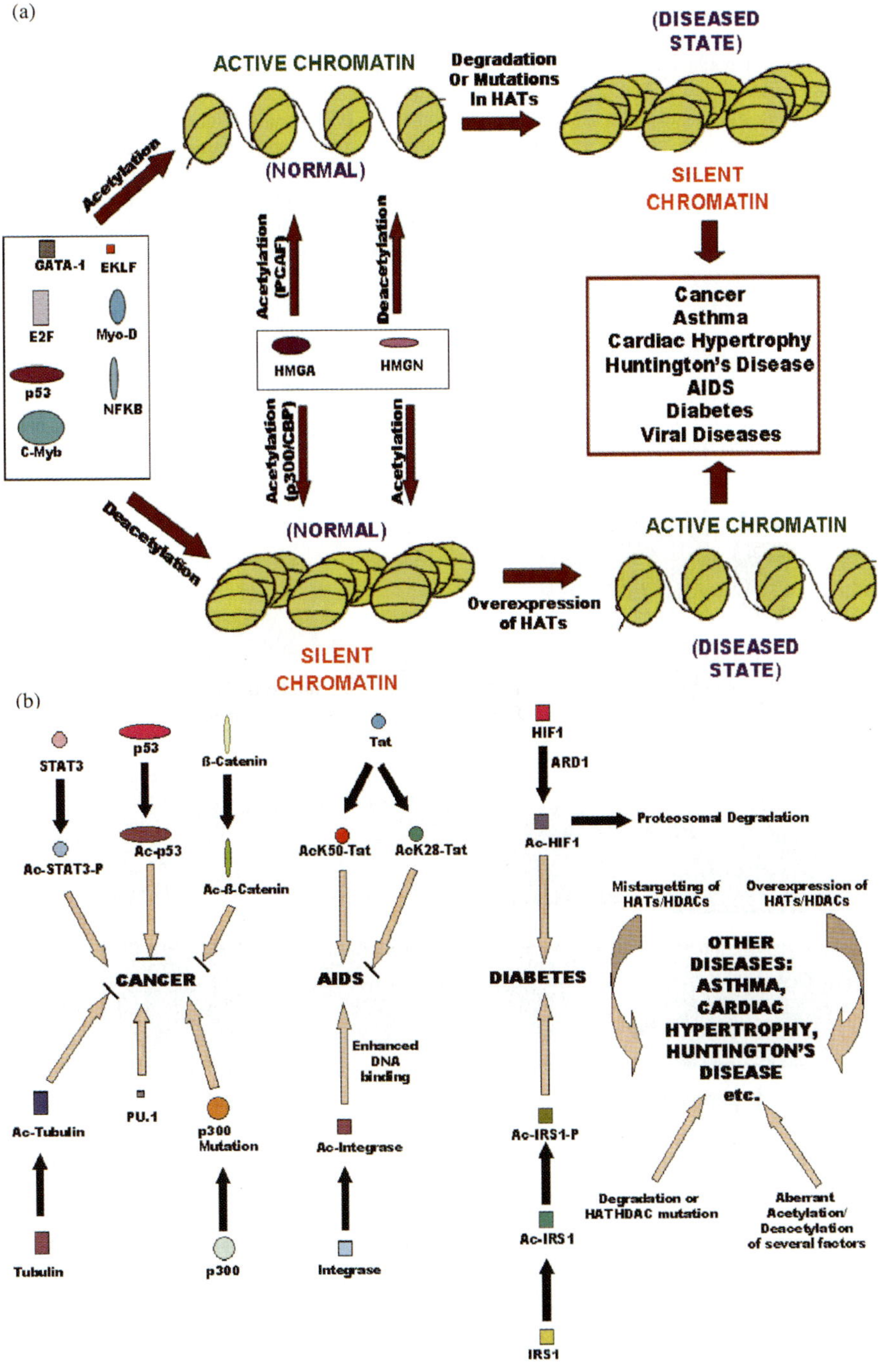
(a)
(DISEASED STATE)
ACTIVE CHROMATIN
Degradation Or Mutations In HATs
Acetylation
(NORMAL)
SILENT CHROMATIN
GATA-1
EKLF
E2F
Myo-D
p53
NFKB
C-Myb
Acetylation (PCAF)
Deacetylation
HMGA
HMGN
Cancer
Asthma
Cardiac Hypertrophy
Huntington's Disease
AIDS
Diabetes
Viral Diseases
Acetylation (p300/CBP)
Acetylation
Deacetylation
(NORMAL)
ACTIVE CHROMATIN
Overexpression of HATs
SILENT CHROMATIN
(DISEASED STATE)
(b)
STAT3
p53
ß-Catenin
Tat
HIF1
ARD1
Ac-STAT3-P
Ac-p53
Ac-ß-Catenin
AcK50-Tat
AcK28-Tat
Ac-HIF1
Proteosomal Degradation
Mistargetting of HATs/HDACs
Overexpression of HATs/HDACs
CANCER
AIDS
DIABETES
OTHER DISEASES: ASTHMA, CARDIAC HYPERTROPHY, HUNTINGTON'S DISEASE etc.
Enhanced DNA binding
Ac-Tubulin
PU.1
p300 Mutation
Ac-Integrase
Ac-IRS1-P
Degradation or HAT HDAC mutation
Aberrant Acetylation/ Deacetylation of several factors
Tubulin
p300
Integrase
Ac-IRS1
IRS1

Plate 14. Role of nonhistone protein acetylation in maintaining cellular homeostasis- mis-regulation and disease connection: (a) Acetylation of nonhistone proteins are associated with active or repressed chromatin architecture as guided by suitable cellular signals for maintenance of gene expression. Misregulation of HAT function leads to diseased state, where chromatin architecture is altered than under normal condition. In a parallel way the posttranslational modification status of these proteins may act as versatile tool to diagnose the various stages of disease manifestation e.g. probable involvement of acetylated NPM1 modulating its stress response function can lead us to use it as a marker for various disease states. (b) Acetylation of nonhistone proteins in connection to diseases like Cancer, AIDS, Diabetes and others. (See Chapter 9 Figure 3, p. 206.)

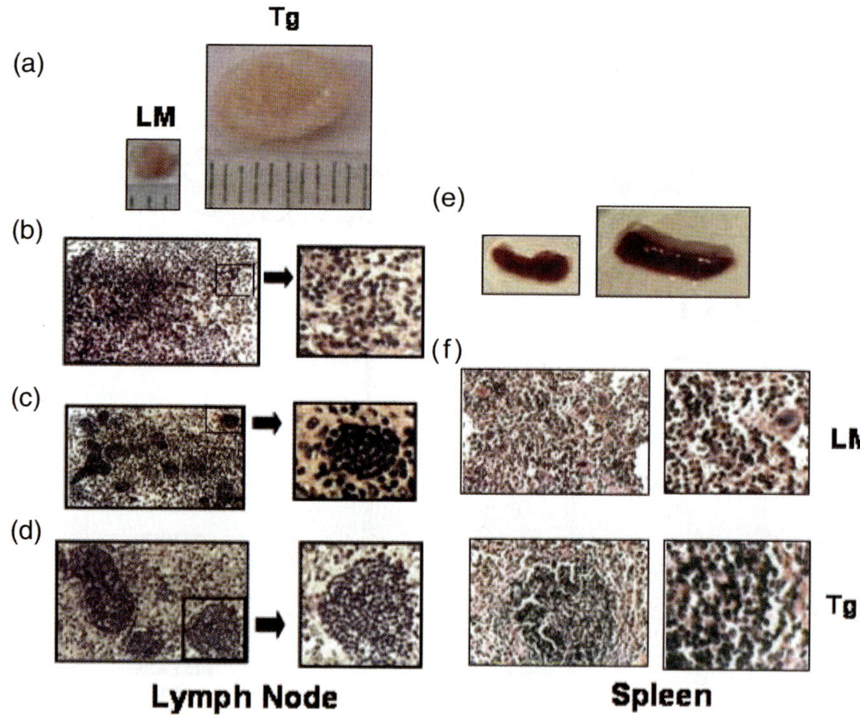

Plate 15. Comparison of lymph node and spleen architecture in SMAR1 transgenic and control littermate mice. (a) Lymph node size of non-transgenic (N) and transgenic (T) mice are shown at the same scale. (b) Histological analysis of lymph node from control at 10X and 40X magnification are displayed. (c and d) Histological sections of Lymph node at 10X and 40X. (e) Enlargement of spleen size shown in transgenic mice compared to control mice. (f) Histological sections of spleen from Littermate normal (LM) and SMAR1 transgenic mice showing strong infiltration of T cells into the lymph node. (See Chapter 10 Figure 4, p. 224.)

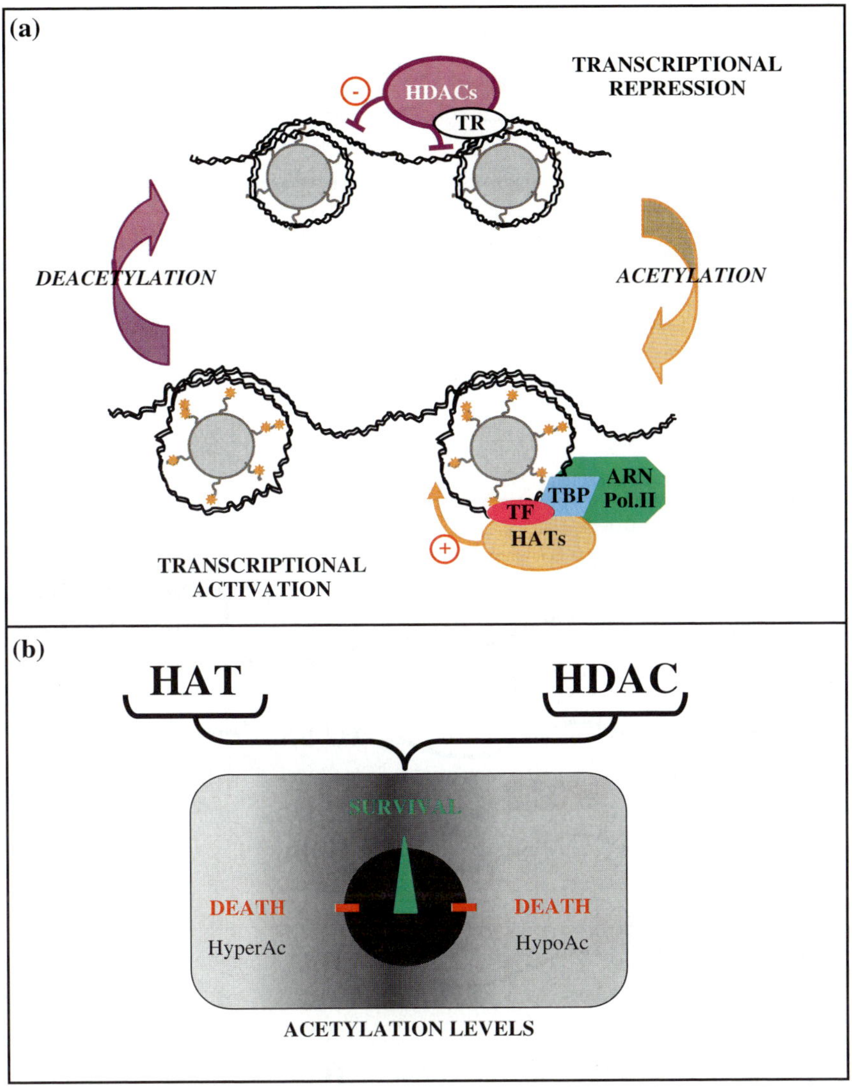

Plate 16. Chromatin acetylation status, transcription and survival: a balance between HAT and HDAC activities. (a) Transcriptional activationlrepression relies on the chromatin acetylation status of histones. TBP: TATA-Binding Protein, TF: Transcription Factor, TR: Transcriptional Repressor. (b) A fine-tuning of HAT/HDAC activities orchestrates neuronal death and survival. On one hand, acetylation levels can be decreased (HypoAc) because of CBP loss of function, as observed during apoptosis and neurodegeneration. On the other hand, when the threshold of acetylation is exceeded (HyperAc), this ultimately leads to neuronal death. (See Chapter 12 Figure 1, p. 265.)

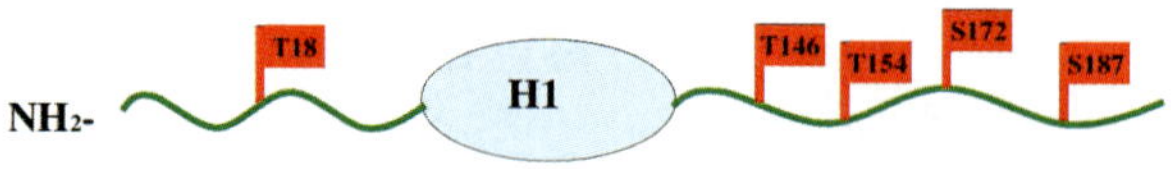

Plate 17. Linker Histone H1 CDK phosphorylation site in mammalian cells. (See Chapter 14 Figure 1, p. 320.)

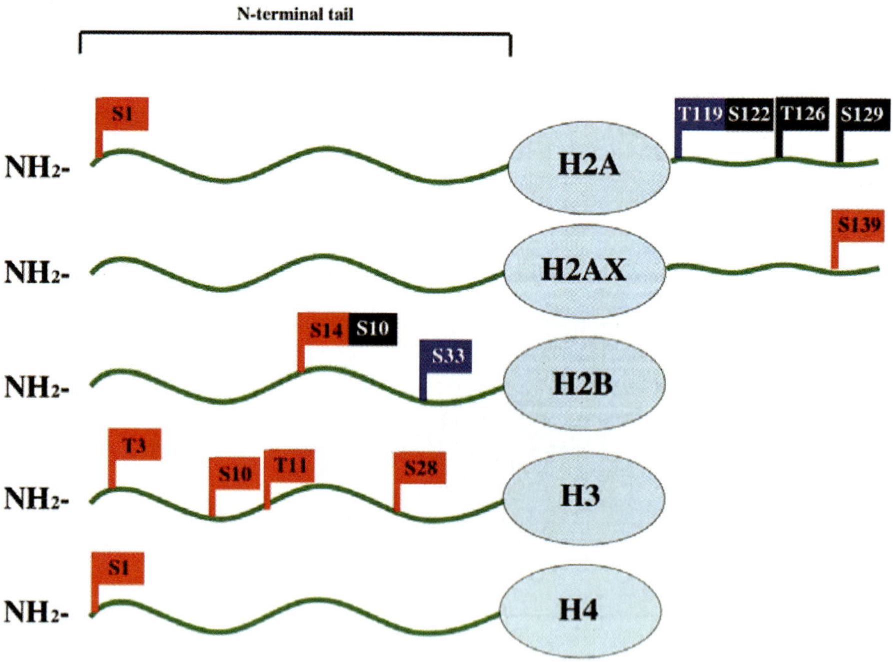

Plate 18. Current view of post-translational histone phosphorylation. Red flag : mammalian specific or common, Blue flag : *Drosophila melanogaster* specific, Black flag : *Saccharomyces cerevisiae* specific. (See Chapter 14 Figure 2, p. 321.)

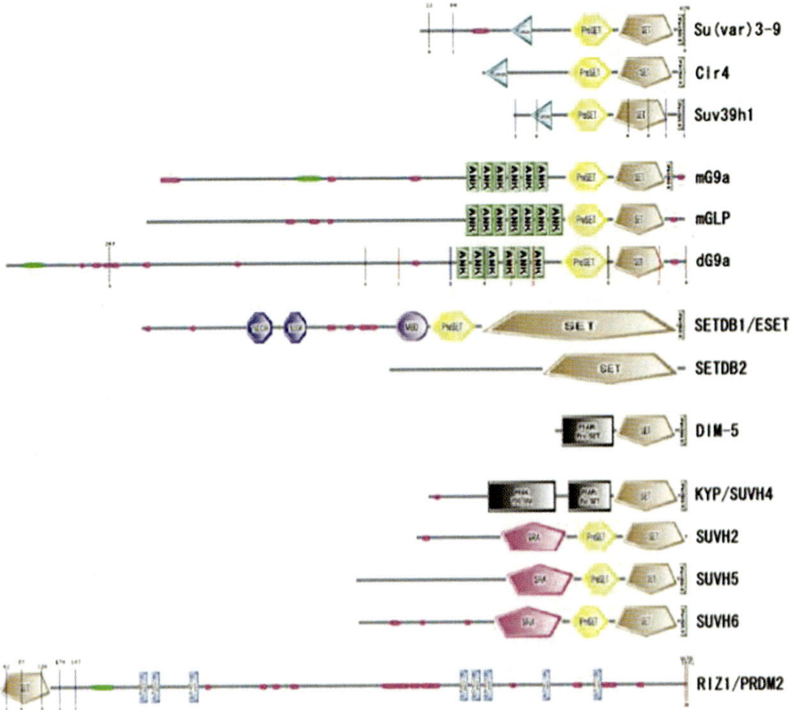

Plate 19. H3K9 HLMTases. (See Chapter 15 Figure 1, p. 340.)

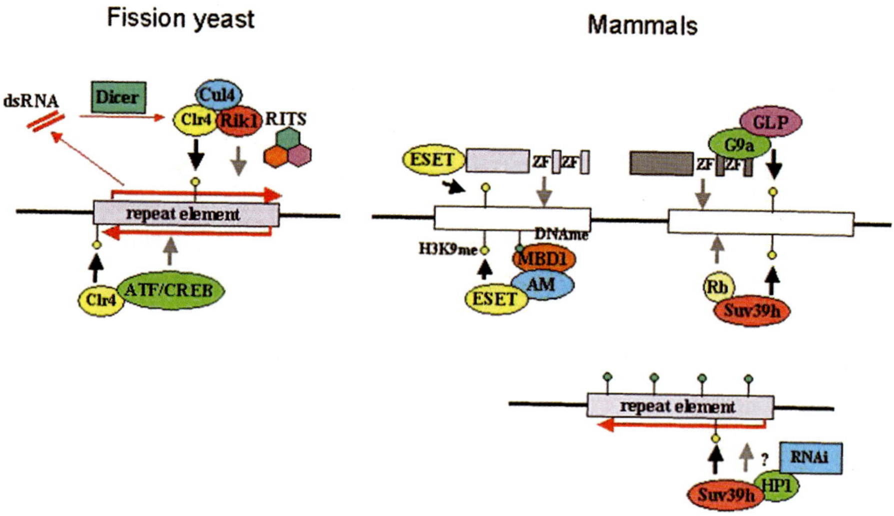

Plate 20. Potential recruitment pathways of yeast and mammalian H3K9 HLMTases. (See Chapter 15 Figure 2, p. 343.)

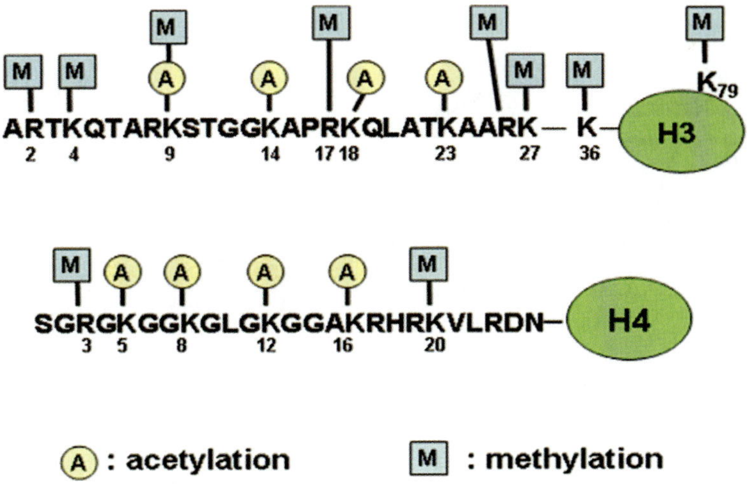

Plate 21. Schematic summary of sequences of H3 and H4 tails. "A" and "M" indicate major *in vivo* acetylation and methylation sites, respectively. (See Chapter 16 Figure 1, p. 353.)

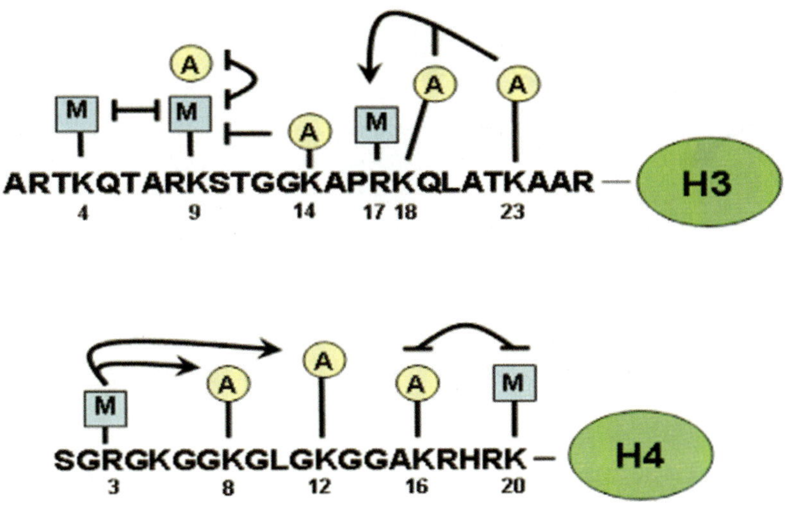

Plate 22. Interplay between histone acetylation and methylation. Acetylated (A) and methylated (M) residues are indicated. Positive effects are represented by normal arrows; negative effects are represented by tee arrow. (See Chapter 16 Figure 2, p. 353.)

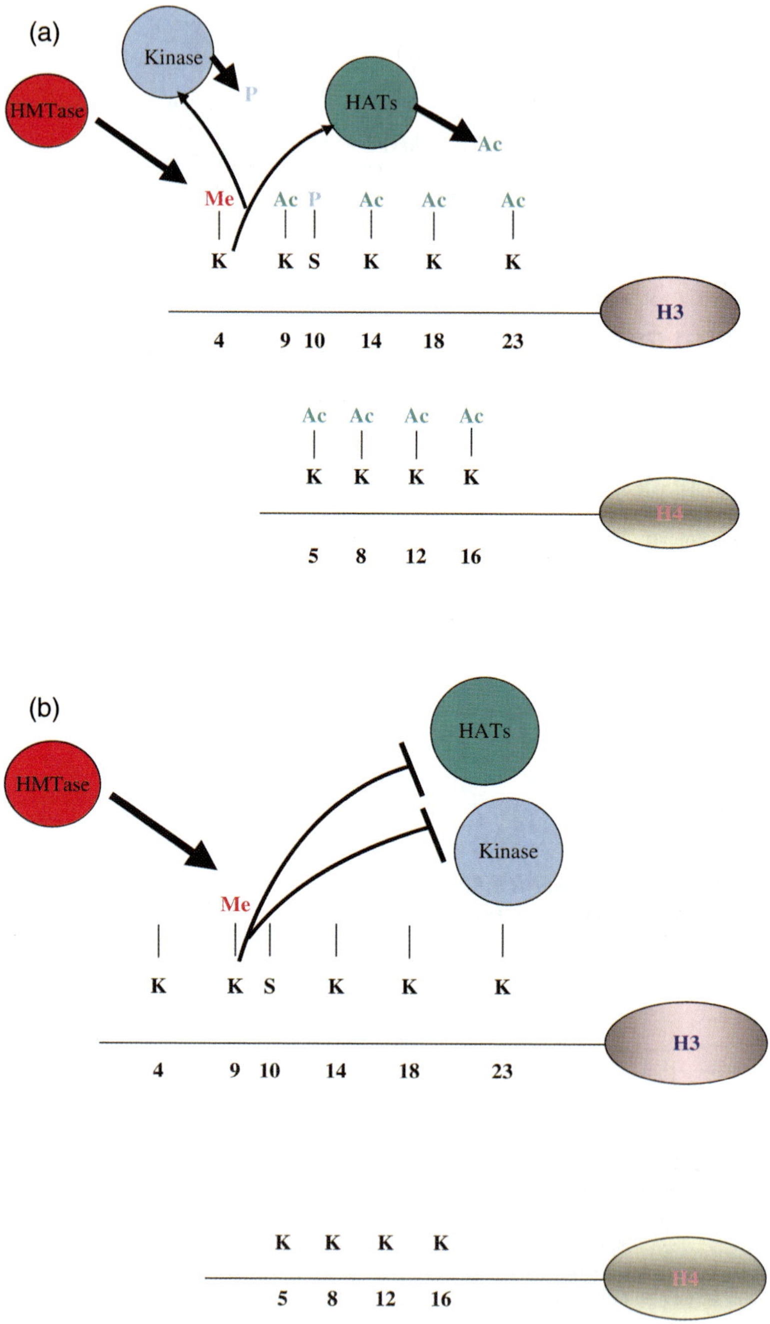
(a)
Kinase
P
HMTase
HATs
Ac
Me
Ac
P
Ac
Ac
Ac
K
K
S
K
K
K
H3
4
9
10
14
18
23
Ac
Ac
Ac
Ac
K
K
K
K
H4
5
8
12
16
(b)
HATs
HMTase
Kinase
Me
K
K
S
K
K
K
H3
4
9
10
14
18
23
K
K
K
K
H4
5
8
12
16

(c)

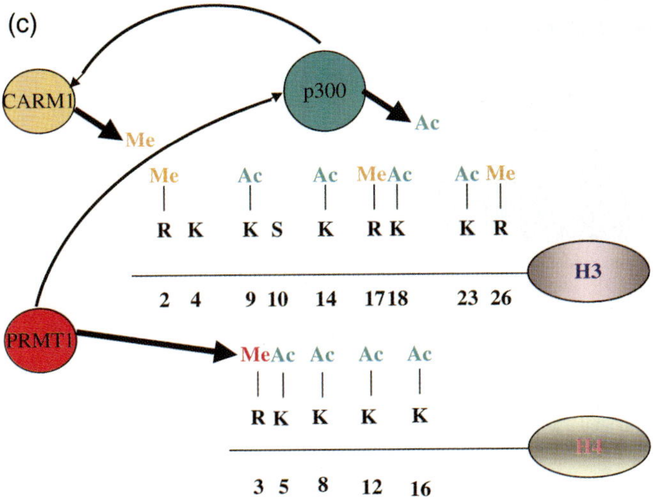

Plate 23. Epigenetic interplay in the regulation of transcription. (A) Methylation of H3K4 and phosphorylation of H3S10 facilitates the acetylation of H3 and H4 leading to transcriptional activation. (B) Methylation of H3K9 inhibits the phosphorylation of H3S10 as well as acetylation leading to transcriptional repression(C) Methylation of H4R3 by PRMT1 facilitates the acetylation by p300 which in turn acts as a favorable substrate for CARM1 mediated methylation of H3 R2, R17 and R26. (See Chapter 18 Figure 1, p. 404.)

(a)

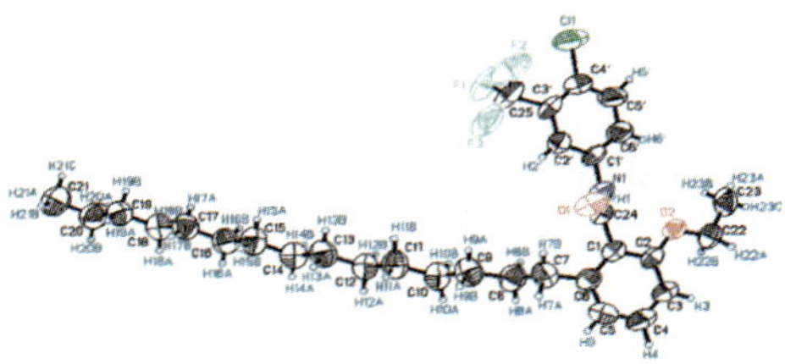

X-ray crystal structure CTPB showing the ORTEP view of the compound

Histones	+	+	+	+	+	+	+
p300	-	+	+	+	-	-	-
PCAF	-	-	-	-	+	+	+
DMSO	-	-	+	+	-	+	-
CTPB (μM)	-	-	-	100	-	-	100

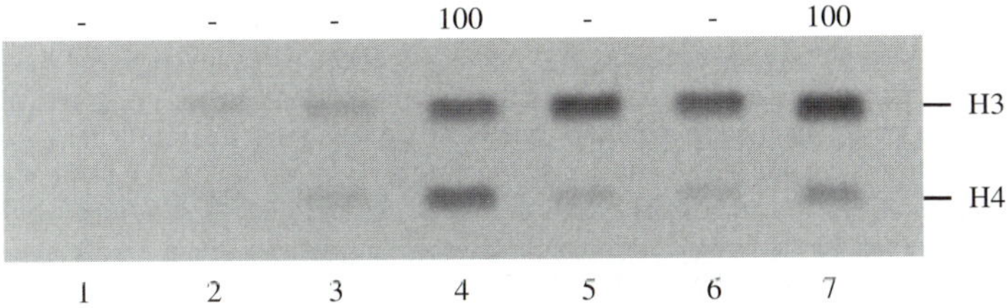

(b)

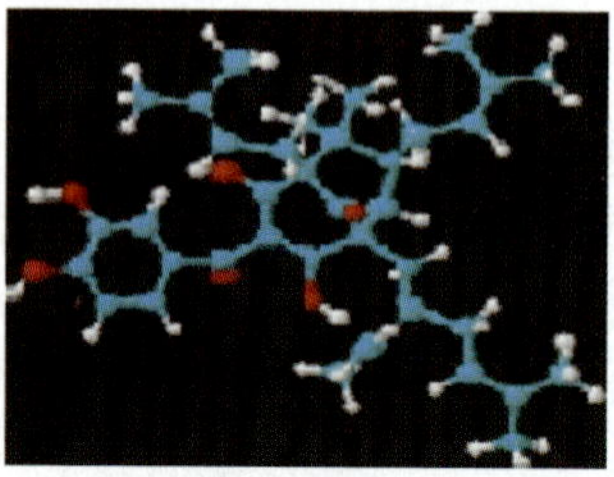

Ball and stick model of Garcinol

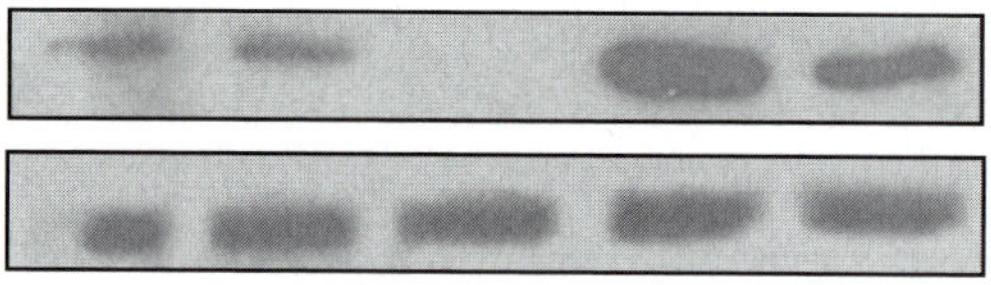

(c)

Histones	+	+	+	+	+	+	+
p300	-	+	+	+	-	-	-
PCAF	-	-	-	-	+	+	+
DMSO	-	-	+	+	-	+	+
Curcumin (μM)	-	-	-	50	-	-	50

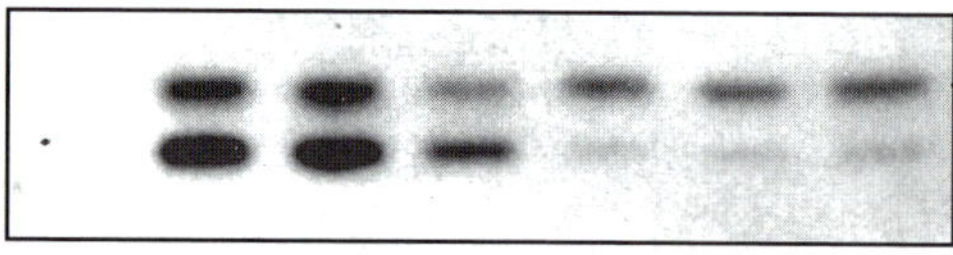

Plate 24. Newly discovered small molecule HAT activator and inhibitors alter histone acetylation in vitro and in vivo: (a) CTPB activates the p300 histone acetyltransferase activity; Histone acetyltransferase assays were performed in the presence and absence of CTPB using highly purified HeLa core histones (1.8 g) and either with p300 (5 ng) or PCAF (15 ng) and processed for fluorography. Fluorographic analysis of acetylated histones by p300 and PCAF in the presence of CTPB; Lane 1, core histones without any HAT; lanes 2 and 5 histones with p300 and PCAF respectively; lane 3 and 6 with DMSO as a control, lane 4 and 7 with100 M concentration of CTPB. (b). Garcinol inhibits the histone acetylation in vivo in HeLa cells: HeLa cells were treated for 24 h; Lane 1, histones extracted from untreated cells; lane 2, Me2SO (solvent control) treated cells; lane 3, garcinol 100 M treated cells; lane 4, NaBu (1mM), TSA (2 M); lane 5, cells treated with NaBu (1mM) and TSA (2 M) and garcinol (100 M). The acid-extracted histones were resolved over 12% SDS-PAGE and analyzed by Western blot using antibodies against acetylated histone H3 (Calbiochem). Loading and transfer of equal amounts of protein were confirmed by immunodetection of histone H3. (c) Curcumin inhibits the HAT activity of p300 but not PCAF. Fluorographic analysis of acetylated histones by p300 and PCAF in the presence and absence of curcumin; Lane 1, core histones without any HAT; lanes 2 and 5 histones with p300 and PCAF respectively; lanes 3 and 6 with DMSO as a solvent control; lanes 4 and 7 with 50 μM concentration of curcumin. (See Chapter 18 Figure 2, p. 413.)